Indian Higher Education
Research Landscape in the Global Context

W0230648

Usha Mujoo Munshi
Angad Munshi

CBS

CBS Publishers & Distributors Pvt Ltd

New Delhi • Bengaluru • Chennai • Kochi • Kolkata • Mumbai
Hyderabad • Nagpur • Patna • Pune • Vijayawada

Disclaimer

Science and technology are constantly changing fields. New research and experience broaden the scope of information and knowledge. The author has tried her best in giving information available to her while preparing the material for this book. Although all efforts have been made to ensure optimum accuracy of the material, yet it is quite possible some errors might have been left uncorrected. The publisher, the printer, and the author will not be held responsible for any inadvertent errors, omissions or inaccuracies.

ISBN: 978-81-239-2947-7

Copyright © Author and Publisher

First Edition: 2016

All rights reserved. No part of this book may be reproduced or transmitted in any form or by any means, electronic or mechanical, including photocopying, recording, or any information storage and retrieval system without permission, in writing, from the author and the publisher.

Published by Satish Kumar Jain and Produced by Varun Jain for

CBS Publishers & Distributors Pvt Ltd

4819/XI Prahlad Street, 24 Ansari Road, Daryaganj, New Delhi 110 002, India.
Ph: 23289259, 23266861, 23266867   Fax: 011-23243014   Website: www.cbspd.com
e-mail: delhi@cbspd.com; cbspubs@airtelmail.in.

Corporate Office: 204 FIE, Industrial Area, Patparganj, Delhi 110 092
Ph: 4934 4934   Fax: 4934 4935   e-mail: publishing@cbspd.com; publicity@cbspd.com

Branches

- **Bengaluru:** Seema House 2975, 17th Cross, K.R. Road, Banasankari 2nd Stage, Bengaluru 560 070, Karnataka
  Ph: +91-80-26771678/79   Fax: +91-80-26771680   e-mail: bangalore@cbspd.com
- **Chennai:** No. 7, Subbaraya Street, Shenoy Nagar, Chennai 600 030, Tamil Nadu
  Ph: +91-44-26680620, 26681266   Fax: +91-44-42032115   e-mail: chennai@cbspd.com
- **Kochi:** Ashana House, 39/1904, AM Thomas Road, Valanjambalam, Ernakulam 682 018, Kochi, Kerala
  Ph: +91-484-4059061-62-64-65   Fax: +91-484-4059065   e-mail: kochi@cbspd.com
- **Kolkata:** No. 6/B, Ground Floor, Rameswar Shaw Road, Kolkata-700 014 (West Bengal), India
  Ph: +91-33-2289-1126, 2289-1127, 2289-1128   e-mail: kolkata@cbspd.com
- **Mumbai:** 83-C, Dr E Moses Road, Worli, Mumbai-400018, Maharashtra
  Ph: +91-22-24902340/41   Fax: +91-22-24902342   e-mail: mumbai@cbspd.com

Representatives

- **Hyderabad**   0-9885175004
- **Pune**   0-9623451994
- **Nagpur**   0-9021734563
- **Vijayawada**   0-9000660880
- **Patna**   0-9334159340

Printed at: Rashtriya Printer, Delhi

FOREWORD

Assessment of quality and quantity of research out put of individuals, institutions, and nations are required for a variety of purposes. These include career progression of individuals and in deciding Honours and Awards including fellowship of learned bodies like National Science Academies. Funding agencies also need this information for assessing research proposals. These are an important consideration in judging the performance of major organizations. Even countries are being ranked on the basis of the quality of their research outs.

Bibliometrics has been playing a major role in assessing the quality and the quantity of research out puts. This is also being utilized by non-academic stakeholders in decision making. However, there is considerable debate about merits and shortcomings of these research evaluation tools. The research analysis is evidence based, with explicitly expressed limitations regarding availability of comprehensive data sets. The significant desire to create awareness and education regarding research assessment in order to ensure that the emergent systems are universal in their acceptability and adoption throughout the academic community is becoming evident now. The assessment has to be relevant universally at international level. Therefore, the objective of this study rests on a comprehensive undertaking of research and research systems on the state of higher education and research in India that can be fed to all stakeholders to enable decision making and guide synergies, investments and the like. To project the research landscape, hardcore data has been analyzed to draw inferences and also status of the global ranking systems of higher education institutions has been kept in view. The evidences are used to draw arguments in favour of both objective and subjective systems of research evaluation.

It is well established that Bibliometric data and analyses work best with large data samples. The study data has been sourced from Thomson Reuters databases through the two time frames of long term and short time respectively that portrays time-series *analysis* in order to extract meaningful statistics and other characteristics of the data. Bibliometric tools are sensitive to the context and interpretation and therefore appropriate normalization wherever necessary, had been worked out in this study to reduce anomalies as far as possible. This procedure specifically focuses on the decadal growth trends across disciplines, key institutions contributing to this growth and pivotal collaborating institutions and India's most favoured countries across disciplines and decades.

Throughout the analyses, data deficit aspects have been specified and it has been observed that in respect of national research evaluation systems, utilizing the right combination of data management and tools suitable to their research domains and objectives should be sought. This is essential because, a performance measurement model for research perspectives (qualitative and quantitative) demands a thorough understanding of the culture, context, advantages and limitations within which systems may work. This in turn, can significantly improve and impact the systematic improvement in higher education and research.

In the fast changing global research scenario, competitiveness among the nations in the emerging knowledge economy of the world is assessed on the basis of research outputs originating from a country. Furthermore, health of research enterprise of various countries is monitored to track key trends in research productivity in order to assess the impact in various subject domains at national and global levels. These assessments require mapping of research landscape with the aid of scientometric tools. Current discourses around evaluation of research performance and debates over meaningful indicators and methods will inevitably continue towards placing rigorous use of tools and models to rational standards of assessment. Clearly, the development mirrors the opportunities and anxieties of the global research communities and policy makers. This research study is a valuable contribution to that emerging environment of deliberations.

National science academies in many countries are seriously concerned about the evaluation procedures for research outputs. Some inter-academy initiatives are underway in this respect.

I congratulate Dr. Usha Mujoo Munshi for her hard work in preparing this important document. It has considerable contemporary value. I hope it will be widely used by organizations/institutions as well as individuals.

Krishan Lal
President, The Association of Academies and Societies of Sciences in Asia AASSA
Immediate Past President, Indian National Science Academy, New Delhi
Past President, ICSU-CODATA
Visiting Professor, University of Delhi, Delhi
Former Director, National Physical Laboratory, New Delhi

PREFACE

Whether or not to measure is not the issue now, how to measure is the key concern. The ability of metrics to represent complex information about research in an accessible format has previously been overlooked in preference to debate about their shortcomings as research evaluation tools. Bibliometrics have the potential to widen scientific participation by allowing non-academic stakeholders to access scientific decision making, thereby increasing the democratization of science[1]. For the use of quantitative performance measures to assess the university/higher education institution's research, a balanced approach is required to be maintained. This is achieved through unpacking of the apparent dichotomies of metrics vs peer review and quality vs impact, while considering the hazards of adopting research evaluation policies in isolation from wider developments in bibliometrics/scientometrics (the science of research evaluation). Unarguably, 'metrics' have their place, and can make the process more efficient and cost-effective, however, peer review must be retained as a central element in any research assessment exercise. The role of metrics acts as 'a trigger to the recognition of anomalies', rather than as a straight replacement for peer review. The Government policies in various sectors, (such as Research and Development, Science, Technology & Medicine, Education) and their commitments in promoting and democratisation in one set of policies have practically ignored developing parallel research policies. Future research policies should take advantage of bibliometrics to foster greater democratization of research to create more socially-reflexive evaluation systems. Another point of deliberation in the course of discourse is fulfillment of societal aspect. Keeping in view, the thrust given to

---

[1] Derrick, Gemma E., and Pavone, Vincenzo, (2013). Democratizing research evaluation: Achieving greater public engagement with bibliometrics-informed peer review. Science and Public Policy 40(5): 563-575. doi:10.1093/scipol/sct007

responsible research and innovation (of late) wherein the moving wall from science to society, to science for society is becoming a buzz, even a norm, if it is not overstating. This has gained more relevance during the past three years due to EU policy relevance, in particular within the European Commission's Monitoring Research and Policy Activities of Science in Society (MASIS) initiative - a major undertaking of the Science in Society (SIS) Programme in the context of the Horizon 2020 Strategy.[2] Monitoring the evolution and benefits of responsible research and innovation with an aim of steering and monitoring the SIS landscape in the European Research Area (ERA), so that EU citizens and society can benefit the most from SIS efforts. Thus, the distinct features that are emerging from associated discourse is an emphasis on the democratic governance of the purposes of research and innovation and their orientation towards the 'right impacts'.

Government investments led to a steady rise in global publication rankings, scientific collaborations and the number of institutions involved. This growth is mainly rooted in fundamental research and public research institutes. Industry involvement and patenting activity are at a nascent stage and developing slowly. Issues that are pertinent to be raised in the Indian context relate to funding, capacity, commercialization, regulation of risks, and the distribution of benefits. The institutional research environment is one of the core constituents of the overall institutional rankings. Of late, the rankings are clearly influential for many Higher Education Institutions (HEI). Although rankings are designed largely for stakeholders outside of higher education, their strongest influence is on those within the higher education field. Thus, creating systems for establishing the foundation for strategic and institutional accounts is *de rigueur*.

International research assessment has emerged as a key component of strategy, planning, management and implementation and as such, a systematized route to maximizing research output and guiding excellence in higher education. However a comprehensive, effective and efficient research assessment model is a complex and context sensitive system, which assists and enables all levels of stakeholders (government, institutions, individuals and society).

For India, while the higher education landscape reveals encouraging advancements, there is yet immense potential for systemic improvements that can create a transformational impact and place India at a competitive position on the international landscape as well as create a quality balance within the nation.

---

[2] European Commission's – Research and Innovation, Science in Society. Available at: *http://ec.europa.eu/ research/science-society/* [Accessed on 20-12-2013].

A major theme underlying the current higher education and research context is the internationalization of higher education and research, whether it is represented by increased competition or collaboration. Systems and initiatives need to work well with global perspective, and promote a culture and spirit of excellence towards personal, national, and international goals. There are myriad international experiences to take a cue from, but it is important to innovate and encourage internally generated solutions that cater specifically to the unique Indian context.

It is also significantly desirable to create awareness and education regarding research assessment and ensure that the emergent systems are universal in their acceptability, access and adoption throughout the academic community. While research assessment is an important tool towards realizing strategic vision, it by no means must be confused with the end goal of research excellence itself and as such should always be valued only in terms of the purpose served within a given context.

There are various key components that need detailed attention including but not limited to data management; assessment methodologies, metrics and parameters; resource and fund allocation and performance assessment. This requires a comprehensive exercise taken forward in conjunction with various stakeholders. Based on the experiences and analysis contained in this document, some indicative suggestions and cautions to mark the way forward have also been incorporated as broadly reflected below.

Data Management: Collection, Accuracy, Accessibility

The analysis presented within this document is robust and evidence based, with explicitly expressed limitations regarding availability of fairly comprehensive data sets. However, it is clear that there remains some data inadequacy that needs to be addressed. There is a need to increase the focus on data and actively engage with it through systems and structures to assess and analyze from within. Currently data available is primarily though commercial organizations that collect and collate data and provide systems within the framework of their respective organizational objectives. While this has certainly proved useful and governments and institutions world over accept and involve these services, they remain external to the system and as such have their limitations when it comes to serving the Indian higher education context holistically.

Data and data management systems should evolve into a 'dynamic' system within which data and analysis is a continuous process, intrinsic to institutions, policy makers and individuals, rather than a periodic exercise conducted when required. With the progress towards online systems, having a web-based interface is both essential and desirable as it can successfully ensure accessibility and transparency.

Metrics and Methodologies

The introduction and increasing popularity of quantitative metrics for performance assessment in higher education has fueled a persistent debate between peer-review based methodologies and quantitative data driven treatments. International trends, such as the transition from RAE (Research Assessment Exercise) to REF (Research Excellence Framework) in UK, are testament to this global shift in research assessment. One senses, that the optimal balance requires the appropriate inclusion of both these approaches. While one is perceptual, the other is observational and neither of them can continue to exist in denial of the advantages of the other. Especially, considering the current research space and international trends, it is highly desirable to work towards an inclusive system that constructs a synergy leveraging the strengths of both these approaches.

Usability: Strategic Planning, Funding etc.

Robust data and analytic systems can prove a powerful tool to drive decision-making. Beyond funding, data-driven decisions can enhance strategic planning at policy, institutional and individual levels. Decisions such as focusing on specific disciplines, setting up of new institutions or fuelling excellence can receive a much-desired boost, if aided by dependable data-driven analysis. Not only does it promote efficiency, it also improves transparency and accountability within the system - both of which are well recognized as the need of the hour in India.

Internationalization: Competition and Collaboration

There is no doubt that the new face of higher education is international and global in character. Be that in terms of increased competition or growing collaboration across borders. The Indian Higher Education landscape also must take cognizance of the fact that it is part of a larger international system and its survival, and success both depend on developing global synergies. Currently high-end research can be seen to have a largely collaborative character and while much collaboration and exchange is still at an individual level, there is a need to elevate collaboration to an institutional level. Creating partnerships, collaborations and clusters again require sophisticated assessment systems for planning and execution. Further, it can propel cross-disciplinary research by enabling a macro view of research, especially as directed towards problem solving. Research is being increasingly understood now as the domain of multiple disciplines coming together towards common goals. As the world is recognizing shared problems – be that climate change, sustainable energy or terrorism, there is an international calling towards borderless research. It is important for India to build systems to enable this kind of big level thinking to take the Indian Higher Education and Research to the next frontier.

Third Mission

In the current scenario, Higher Education Institutions (HEIs) are expected to serve three missions, viz. teaching, research and service. While mission of teaching (or education) and research (or knowledge) have dominance, the third or the service mission seems to be independent though emerging as equally (if not more) important. Community engagement may often actually contribute to improvements in HEIs, specially to their teaching and research functions. Research and Knowledge mobilization have two broad streams of research to focus on - innovations in community based research methodologies and synthesizing local experiences with professional expertise. Eventually the synergy of such interactions leads to the co-creation of new knowledge on a variety of issues facing the people and communities (variously emphasized under the rubric of MDGs, Human Development, Sustainable Livelihoods, etc.), that has practical and theoretical relevance and resonance. The approaches in social responsibility lies in the demand for public accountability and local relevance of higher, post-secondary education that is growing rapidly in many societies. This demand is being responded in many different ways by different institutions through service learning and student internships, by co-production of knowledge (with partnering local communities). Moreover, by bringing in the experiences of communities and practitioners in designing curricula and teaching new problem and issue-centred courses. This social responsibility is expressed both inside and outside the institutions.

Centre for Excellence in Research Performance

With such a direction in mind, a holistic effort needs to be followed to build and manage the systems, processes and cultures around research assessment. While India has capable organizations such as the UGC and AICTE with their respective mandates, there seems to be good cause for an a new centre or organization dedicated to promoting, managing and propelling various aspects of research output and higher education in India with a view to maximizing data, assessment and decision making. For accelerating the research excellence, building institutional profiles for evolving measures for ranking and evidence based metrics ought to be in place.

There is a well-known adage that tells us "if you can't beat them, join them". So if rankings are to be a permanent part of the academic landscape, we should encourage the ongoing debate and establish further guidelines and criteria on how best to conduct and interpret higher education quality assessments.

Thus, the present work is unique and empirical in its content and has endeavoured to factor objectivity to eclipse anomalies inherent in subjective approaches.

Globally, emphasis is increasingly being given to research in Higher Education (HE) as a key motor to advance in the knowledge society and economy. India, with a vast set up of higher education institutions should aspire to have at least a few of its universities in the global top-league and make a mark in the global research landscape.

The objectives of the present study solely rests on a comprehensive undertaking of research and research systems on the state of Higher Education (HE) and research in India that can be fed in to all stakeholders to enable decision making and guide strategies, investment, and the like.

Thus the prime focus herein is to focus on - (i) accelerating research performance and ranking; (ii) to encourage peaks of excellence in Higher Education Research (HER), in an endeavour to make India competitive on the global map; (iii) to increase collaborations and harmonize institutional excellence across levels, while facilitating constructive competition; and finally, (iv) to provide policy makers, institutions and individuals with access to data, mapping and ranking of the higher education system.

The scope of the document attempts projecting research landscape of Indian higher education system through multipronged approach parameters so as to portray meaningful performance metrics and procedure in Indian HE system.

The study consists of major three parts besides, Introduction and Conclusion, recommendations and limitations. The discussions in the document are corroborated with the schematic representations and the hard core data depicted in tabulations appended at the end. The document is also complimented with the additional bibliography for further reading.

Thus, the document has eloquently deliberated these issues while taking on board evidence based metrics in three parts viz. Part 1: Research Performance Management, Part 2, Reading the Research and Part 3, Health of Research Landscape in India. The document is substantially supported with the data in tables and depicting trends schematically.

The present document is thus a comprehensive undertaking of research and research systems on the state of higher education and research in India.

The volume thus, covers interest of a wide range of readers including students, library science experts, professionals, researchers, faculty and policy makers. Basically the book is a reference book and will find way in most of the institutions in general and by and large all institutions of higher education in the country and all types of libraries. Efforts have been made to minimize typo and other types of error. Hope this volume finds space in the minds of the higher education stakeholders and sketches new landscape in future.

It is my duty to acknowledge with gratitude, the encouragement and support given from time to time by Shri T.N. Chaturvedi, Former Governor, Karnataka and Chairman, IIPA and Dr. Tishyarakshit Chatterjee, Former Secretary, Ministry of Environment and Forests and Director, IIPA for undertaking the academic and research work. Research is a painstaking endeavour and one cannot overemphasize the importance of carefully and holistically taking on board suggestions and advice of the experts for validating expected and observed outcomes, for which I am grateful to Prof. S.K. Tandon, former Pro-Vice Chancellor, Delhi University. I also express my sincere appreciation to M/s Thomson Reuters for permitting me to use their database for data collection; Mr. Bibhuti Sahoo (IIT, Delhi) and Ms. Kruti Trivedi (INFLIBNET, Ahmedabad) for their assistance in data collection; Ms. Aditi Roy, IAMR for going through the text; Dr. Roma Mitra Debnath (IIPA) for helping in infographic work; Ms. Medha Narayan (Harvard University, Cambridge, MA, USA) for her assistance; and Shri Sunil Sethi (Senior Personal Assistant, IIPA) for assising in computer setting of the work. My sincere thanks to Prof. Vinod K. Sharma, Executive Vice-chairman, Sikkim State Disaster Management Authority and Faculty, IIPA for helping in publishing this volume. Finally, deep sense of gratitude to the family and friends for the understanding, patience and support for this endeavour.

CONTENTS

4. Conclusions and Observations. 115

Part 1

Part 3

Annexure – II : List of Figures

Part 2

Part 3

1. INTRODUCTION

Background

Globally, emphasis is increasingly being given to research in Higher Education (HE) as a key motor to advance in the knowledge society and economy. India, with a vast network of higher education institutions should aspire to have at least a few of its universities in the global top-league to make a mark in the global research landscape.

A data driven approach will effectively enable guiding investments through gap analysis, policy making for funding, and monitoring, evaluation and assessment. Evidence based, scientific data will enable all stakeholders, in government as well as academia, to work out strategic options and guide investments and initiatives that will drive the quality of higher education and research.

Currently the state of HE research in India is mosaic and heterogeneous. India lacks in many ways in terms of availability of meaningful data, analysis, and periodic assessments. Ranking systems are globally considered important indicators for research, and universities seeking to brand themselves as world-class entities.

Thus, mapping and analysis of HE research systems have become essential to acquire an understanding of their functioning and consequently of future requirements.

Present Research Study

The objectives of this study solely rests on a comprehensive undertaking of research and research systems on the state of HE and research in India that can be fed to all stakeholders so as to enable decision making and guide strategies as well as investment.

Thus, the prime focal point herein is to focus on accelerating research performance and ranking; to encourage peaks of excellence in HE research, in an endeavour to make India competitive on the global map; to increase collaborations and harmonize institutional excellence across levels, while facilitating constructive competition; and finally, providing policy makers, institutions and individuals with access to data, mapping and ranking of the higher education system.

The scope of the document attempts projecting research landscape of Indian higher education system through multipronged approach parameters to find meaningful performance metrics and procedure in Indian HE system. Alongside, endeavour has also been made to focus on creation of suitable frameworks for profile building, assessments and ranking for the Indian HE landscape using research profile tools to identify key performance metrics and measure results for driving success mandate.

The present evidence based study is a comprehensive undertaking of research and research systems on the state of HE and research in India.

Structure of the Document

The study comprises major three parts besides introduction, conclusion, and recommendations with limitations identified. The discussions are corroborated with schematic representations while the hard core data presented in tabulations is appended at the end. The document is also complimented with bibliography for further reading.

Part 1: Research Performance Measurement

Worldwide, research output itself, and recording, communicating and reading research output has become enormous and complex. Evaluation of researchers, research output, research areas and institutions are seen as key in decision-making and policy setting including focus setting, recruitment and promotion, funding; and promoting excellence of research output and higher education institutions. Advancements in technology and increased global interaction both demand and facilitate a higher order of evaluation and assessment practices.

The first part of the document deals with the various facets of the evaluation and institutional excellence. Historical context on ranking and assessment systems in higher education and detailed discussion of globally prevalent ranking systems.

Discussions of merits and demerits of peer review, bibliometric indicators and ranking systems with a detailed and systematic understanding of current systems, pitfalls, and suggested directions for the Indian context vis-à-vis the global scenario are presented here. Existing appraisal systems for award/reward, and also focus on the current systems in India for award, reward with a critical analysis has been dealt herewith in this part of the document.

Part 2: Reading the Research

Research performance measurement (RPM) has become an increasingly critical area for those in decision-making positions. Institutions, research councils, funding institutions and governments have turned their attention to creating robust and relevant structures linking research to policy.

Research policy operators including research councils that have previously focused primarily on prospective evaluation (assessment of applications) are increasingly directing their attention towards the retrospective evaluation of activities and results.

Evaluation experts argue that good evaluation practice provides more than just indicators. They recommend the use of "mixed methods" and characterize good evaluation practice as processes that create a dialogue between the parties involved about prioritization, organization, for both immediate and long-term effects. A performance measurement model for research by taking in a multi-pronged approach combining both qualitative and quantitative perspectives offered by a combination of methodologies demands an understanding of the culture, context, advantages and limitations within which these systems may work. A universe of research assessment tools created by the appropriate combination of these, can significantly improve and impact the way we read research and consequently direct us (India) to a systematic superiority in higher education and research.

This part while deliberating on the importance of research measurement systems for policy makers also throws light on how governments world-over are looking at these metrics. A study of how internationally, governments and policy makers read research on research and the relationship between policy and research measurement is presented in this context. Detailed representation and analysis of case studies of international practices in ranking and assessment such as, systems followed in Brazil and the UK Research excellence framework are focused specifically in this part.

Part 3: Health of Indian Research Landscape

The third part begins with an analysis projecting a comprehensive scenario of the health of Indian research landscape, taking into account a time period of the over four decades to give a holistic picture. However, top 1% publication output has been analyzed for a ten year period. Assessing research quality and impact and finding meaningful performance measures for higher education based on the research publication output being high on agenda and detailed here.

A brief overview of the current higher education research landscape in India as represented by currently available data; including data analysis, suitable comparisons and categorizations; and global contrasts are some of the highlights. A focus on research performance versus research excellence and a detailed examination, based on available data, of the top research producing and highly cited research of individuals/institutions at national level with global comparisons, while identifying key institutions and the pivotal people has been an interesting revelation and learning.

In the Indian context, analysis of data based on classifications (geographical/disciplines/type), comparing elements within each sub segment and also as a contrast across segments have been dealt with. Furthermore, a detailed mapping and analysis with emphasis on top research-focused institutions and top 1% of highly cited global research output has been included.

Focus on key trends with a global and Indian perspective, including (but not limited to) overview of collaboration (clustering, alliances and networking) across levels – that are, International, National, and Institutional have been analysed. Besides the journal preference aspect has been detailed, with some indices being applied to evolve evidence based metrics, reflecting health of Indian research landscape. The key findings from this part of the study for each sub-section have been focussed on.

Conclusions and Observations

Impact of global rankings on Higher education research and the production of knowledge have been catching up, of late. International research assessment has emerged as a key component of strategy, planning, management and implementation and as such, a meaningful systematized route to maximizing research output and guiding excellence in higher education is necessitated. However a comprehensive, effective and efficient research assessment model is a complex and context-sensitive system, which assists and enables all levels of stakeholders (government, institutions, individuals and society) that needs to be put in place.

Based on the experiences and analyses contained in this study, some indicative suggestions and cautions to mark the way forward has been included here. The specific issues that need to be mentioned in the discourse of projecting health of Indian research landscape based on evidence based metrics such as data management; collection, accuracy, accessibility; metrics and methodologies; parameters for institutional profiles; usability; strategic planning, funding etc; internationalization; competition and collaboration; and fulfilling the third Mission.

Bibliometric data are increasingly being used to assess the impact of research, to track and evaluate research activity. Classic indicators, linked to publication and citation, used in ever more sophisticated ways and stimulated by new technological possibilities, are being supplemented by indicators oriented towards the societal impact and quality of the research.

Various issues and concerns crop up as rapidly as advances are made. Attempts to accurately remove bias, attention to micro-level variations, assigning weightage and identifying parameters are in progress to fine-tune bibliometric processes. Simultaneously controversies remain in broad, conceptual areas such as, applying bibliometrics to areas in social sciences and cross-disciplinary fields. Practical aspects of consistency, benchmarking and data reliability further add to the nuances that require careful attention.

Nevertheless, with all its limitations, bibliometrics has come into its own and have gained rapid acceptability among governments and academic community. The need is ever more strong (which is being reflected in governments' and institutions' priorities), to create better systems, methodologies and also increase communication, awareness and education such that, better tools may be continuously developed and appropriately used to benefit the research landscape and community.

World over, many institutions are exceedingly showing concern for excellence and creating an environment that portrays them in an avatar of being world class universities or centres of excellence in higher learning. Securing higher position in the global scenario by significantly participating and identifying and setting-up for themselves benchmarking parameters to be the best of the rest in actions and activities. Projecting and dissemination of institutional potency in public domain is of prime importance. Consequently, India can be strongly placed and projected globally in HE system.

Existing scenario in India with respect to profiling and measurement systems and case for creating profiles to build/strengthen systems including suggested template profiles have also been included in the study. Finally, limitations of particularly the analytical part have been indicated.

2. COLLECTION OF EVIDENCE AND METHODOLOGY

2.1 Introduction

Compiling both quantitative and qualitative evidences to identify trends have been based on the extensive desk work, data analysis and drawing inferences from various sources. This Section outlines key concepts and methodology used in preparing this document.

2.2 Methodology

2.2.1 Research Performance Measurement (Part 1)

The report while presenting the evidences looks at the (global) ranking systems of Higher Education Institutions (HEI) and then draws an argument in favour of both objective and subjective systems of research evaluation.

2.2.2 Reading the Research (Part 2)

This looks at input-output mechanisms for deciphering the impact (if any) of inputs facilitated for promoting research excellence and the evidence of its reciprocation. The main focus being on relevance of access to research resources (as an input) to research promotion (as an output). The data based on empirical studies incorporating as to how government agencies are seeking to interpret impact metrics based on the research output analysis, giving trends, objectives and mechanisms, while including well established international experiences.

2.2.3 Health of Indian Research Landscape (Part 3)

Being the major part of this study, it focuses on various analytical parameters for deciphering evidence based metrics regarding health of Indian research landscape using bibliometric data and citation analysis.

Bibliometrics is the analysis of data derived from publications and their citations. Research evaluation increasingly makes use of bibliometric data and analyses. Research publications accumulate citation counts and are (believed to be) reflective of the value placed on a work by later researchers. However, some papers get cited frequently, yet many remain uncited. For instance, in social sciences there is also a diversity of publication modes and citation rates are typically much lower than in natural sciences. Indicators derived from publication and citation data should always be used with caution, though empirical studies have shown that high citation rates correlate with other qualitative evaluations of research performance, such as peer review.[3]

Since some fields publish at faster rates than others and citation rates also vary, accordingly citation counts must be carefully normalised to account for the variations across fields. Normalisation is usually done by reference to the relevant global average for the field and time period of the publication. As such, in the present study the normalisation has basis on the average scores.

Bibliometrics works best with large data samples. This study is based upon analysis of indicators for whole-country and disaggregated by the fields used in Thomson Reuters *Essential Science Indicators*. The analysis in Part 3 takes into account the following parameters.

2.2.4 Production – India Vs World

This looks at the comparative international analysis of India's world share of output and citation impact and disciplinary shifts. India's changing annual share of world output over a 10-year time frame (2002-2012) has been analyzed. India is compared with a selection of top 20 research producing and emerging research economies. India's citation impact is evaluated over the same time-frame. Data are sourced from Thomson Reuters National Science Indicators Database 2012 (ESI edition). Again the comparative disciplinary shifts studied for over a 40 year period, and discipline-wise aggregated by 10 year moving

---

[3] Adams J, *et al.*, Maintaining Research Excellence and Volume: A report by Evidence Ltd to the Higher Education Funding Councils for England, Scotland and Wales and to Universities UK (2002).

averages rather than annual data. Data are sourced from Thomson Reuters Web of Knowledge. Global baselines for field rankings for top ranking fields are also looked at.

2.2.5 Impact Analysis

This details the distribution of citation impact in the Indian research base across subject disciplines, while taking into consideration field weighted average impact. Impact analysis reveals uncitedness, clustering the research profile into impact categories, and indicates the disciplines of research which are (both) below and above world average citation impact. These impact sets are based on 2002-2012 data and the data are sourced from a customised National Citation Report for India upto June 2012 (ESI).

2.2.6 Highly-cited Paper Analysis

This flows from the impact analysis and assesses the percentage of research which falls into highly-cited impact categories of the impact analysis distribution, where citation impact is at least four times world average. The analysis considers the data for the period 2002-2012. Data are sourced from a customised National Citation Report for India June 2012 (ESI).

2.2.7 Key Institutes & Pivotal People

Depicting the analysis based on the decade-wise distribution of institutional research contribution in various subject disciplines, while covering a time frame of over 40 year data from 1971 to 2012 (June 2012). Data are sourced from Thomson Reuters *Web of Knowledge*. The author affiliation qualifies for the institution identification. The analysis clearly indicating the disciplinary emphasis and excellence in different decades. This also focuses on the decade-wise and subject-wise distribution of publication contributing institutions; number of highly prolific institutions; and, proportion and contribution of productive institutions. The identification of key performing institutions across over 4 decades (1971-2012) and disciplines (based on Web of Knowledge data) and institutions contributing to high impact publications (based on ESI data 2002-2012) that includes significant contribution of India's research power is detailed in this section.

The analysis also focuses on the pivotal people across disciplines who have contributed to the highly cited papers (HCPs). In cases where the author has more than one affiliation, the Indian affiliation has been credited for. Data are sourced from a customised National Citation Report for India June 2012 (ESI). It also looks at the authorship pattern of these HCPs in terms of (i) solo vs collaborative, (ii) Indian vs Foreign Collaboration, (iii) proportion of collaboration, and (iv) identifying the trends relating to Indian contributors and the proportion of Indian contribution in each of the HCPs.

2.2.8 Journal Analysis

This analyses the percentages of Indian research which are authored in journals linked to the *Journal Citation Report* (2012) and average citations received by the journals (using citations per paper) where all the papers for a 10 year period (Jan 1, 2002 - June 30, 2012) were published and covered by Essential Science Indicators (ESI). The journals are divided in quartiles (based on Journal Impact factor and Journal Citations per paper) into which Indian research falls by Journal Impact Factor wherein the top quartile covers top 25% of journals by impact, the top two quartiles covers top 50% and the top three quartiles cover top 75% respectively. This is then split into two data sets (i) all publications considered for a 10 year period based on Impact Factor score and average citations and (ii) similar analysis for the HCPs, with the same time frame (2002-2012). Data are sourced from a customised National Citation Report for India up to June 2012 (*ESI*).

2.2.9 International Collaboration Analysis

This analyses the percentages of Indian research which are internationally co-authored and the countries that are India's most frequent international partners. The selection focuses on the most frequently collaborating countries and collaborating Indian and foreign organisations, and analyses how the shares of international co-authorship have changed since 1971-2012. Data are sourced from *Web of Knowledge.*

2.3 Data Collection

The data for the present document has been gleaned from various sources and the same has been meticulously acknowledged. The main focus of the present study is Part 3 of the document relating to impact of research output using evaluative bibliometric techniques. Global share of publications in journals indexed by Citation Indexes has become one of the output indicators for assessing the competitiveness of National R&D systems.

Unless otherwise specified, primarily all the data for generating evidence based metrics using research publication output and its impact has been taken from Thomson Reuters databases - *Web of Knowledge* and *Essential Science Indicators*. Inputs sourced from various agencies in the preparation of this analysis report are gratefully acknowledged.

2.3.1 Data Source

In the present study, as indicated above, the data for the Part 3 has been sourced from Thomson Reuters database and used for presenting data on global share of India with respect to scientific publications as captured by *Web of knowledge (WoK)* and *Essential Science Indicators (ESI)* data-bases. The bibliometric data has been sourced from Thomson Reuters databases underlying the *WoK*, which gives access to conference proceedings, patents, websites, and chemical structures, compounds and reactions in addition to journals. The *Web of Science is* part of the *Web of Knowledge*, and focuses on research published in journals and conferences in science, medicine, arts, humanities and social sciences. It has a unified structure that integrates all data and search terms together and therefore provides a level of comparability. The authoritative, multidisciplinary content covers over 11,500 of the highest impact journals worldwide, including Open Access journals and over 110,000 conference proceedings. Coverage is both current and retrospective in the sciences, social sciences, arts and humanities, in some cases it dates back to 1900.

The Essential Science Indicators (ESI) of Thomson Reuters which is an analytical tool, offering data for ranking scientists, institutions, countries, and journals. Essential Science Indicators can determine the influential individuals, institutions, papers, publications, and countries in their field of study as well as emerging research areas that could impact their work. This unique and comprehensive compilation of science performance statistics and science trends data is based on journal article publication counts and citation data from Thomson Scientific databases. Broadly the tool helps to:

(i)   Analyze research performance of companies, institutions, nations, and journals.

(ii) Identify significant trends in the sciences and social sciences.

(iii) Rank top countries, journals, scientists, papers, and institutions by field of research.

(iv) Determine research output and impact in specific fields of research.

(v)   Evaluate potential employees, collaborators, reviewers, and peers.

The tool has an in-depth coverage, providing access to approximately ten million articles in over 11,000 journal titles from around the world. For facilitating a solid basis for comparison of research performance, it includes baselines, which are the benchmarks for assessing research impact. Available as a 10-year rolling file, the tool is updated every two months.

As per the Thomson Reuters source journal categorisation, Journals are mapped to one or more subject categories, and every article within that journal is subsequently assigned to that category. However the papers from prestigious, 'multidisciplinary' and general 'biomedical' journals such as Nature, Science, British Medical Journal (BMJ), The Lancet, New England Journal of Medicine and the Proceedings of the National Academy of Sciences (PNAS) are assigned to specific categories based on the journal categories of the citing and cited references in each article.

The bibliometric analyses presented in this report does not cover all research output covered by the *Web of Knowledge* and *Essential Science Indicators* over the period for deciphering trend analysis at national and global levels.

2.3.2 Terminological Usage

In order to benchmark India's performance to other countries, the analyses comparing India's research performance to comparative groups, based on the identified top performing countries, the names of the countries and /or institutions, though in many cases have been used in the full form, however wherever necessary (in some cases) have been abbreviated. In order to simplify the visual presentation of this information, abbreviations that clearly identifies the name have been used.

2.3.3 Timeline

Various time periods are used in this study that comprise primarily two time frames - long term and short term. The analyses rely on long-term time-frames, from (1971 to 2012), split into decadal periods of 1971-1980; 1981-1990; 1991-2000; 2001-2010 and 2011-2012 (June) (a historical time period one to contemporary). This was primarily used to focus on the decadal growth trends across disciplines; key institutions contributing to the growth and pivotal collaborating institutions; India's most favoured countries across disciplines and decades. The time period from 1970s was chosen for the simple reason, that by this time the established institutions after independence had matured and the tools for analyses had systematically been covering representative data, to portray a logical representative picture so as to arrive at meaningful evidences. The short term time frame 2002 – 2012 (June) (based on data from ESI) used in most of the sections of this study, besides dealing with the above indicated parameters, includes India's relative ranking in terms of publication production vis-à-vis world's top 20 ranking countries, coupled with impact profiles, Highly-Cited Papers (HCPs), authors and institutions contributing to these HCPs. To enable meaningful comparisons, the citations accrued by papers published during the time period up to the end of that time period are accounted for. The journals where the publications for this 10 year time frame (2002-2012) were published, have been analysed across disciplines to decipher journal preference pattern.

2.3.4 Research Fields

Standard bibliometric methodology uses journal categories as a proxy for research fields or areas. Using one such mapping scheme in this study to associate published research with research areas, the subject mapping has been done. The subject disciplines of the *Essential Science Indicators* (ESI) fields, which aggregate data at a higher level than the *Web of Science* journal categories have been adopted. There are 22 *ESI* fields compared to 254 *Web of Science journal* categories. For the short term time frame the data categorisation done by *ESI* has been directly used. However the data taken from *Web of Knowledge* listing 254 categories was re-categorised into 22 categories using the same scheme as used in *ESI* to ensure data consistency. *ESI* fields are defined by a unique grouping of journals with no journal being assigned to more than one field. Articles in journals such as *Science* and *Nature*, are assigned to Multidisciplinary field on the basis of an article-level classification. Analyses using ESI fields are useful to gain a headline understanding of the strengths and weaknesses of a research system, whereas *Web of Science* journal categories analyses are useful to identify strengths and weaknesses in more specific research areas. Though the prime motive of this study was to decipher trends, the domain of knowledge categorised in broad 22 subject categories have been used for feasible analyses to generate evidence based metrics.

Thus the relative performance of various sub sectors and disciplines of science with respect to citation index's publications has been studied for 22 disciplines. The 22 searchable fields of research includes: agricultural sciences, biology and biochemistry, chemistry, clinical medicine, computer science, economics and business, engineering, environment and ecology, geosciences, immunology, materials science, mathematics, microbiology, molecular biology and genomics, multidisciplinary, neuroscience and behaviour, pharmacology and toxicology, physics, plant and animal sciences, psychiatry and psychology, social sciences, and space science.

2.4 Core Parameters for Data Evaluation, Interpretation and Analyses

2.4.1 Papers/Publications

Often used interchangeably, the terms 'paper' and 'publication' are to refer to printed and electronic outputs of many types. Papers are the subset of publications for which citation data are most informative and which are used in calculations of citation impact. Publications refer to all document types. The minimum number of papers suitable as a sample for quantitative research evaluation is a subject of widespread discussion. Larger samples are always more reliable, but a very high minimum may defeat the scope and specificity of analysis. In the present study, India's contribution reflected in Thomson Reuters databases - *Web of Science* and *Essential Science Indicators* for the period 1971-2012 and 2002-2012 have been accounted for respectively.

2.4.2 Output and world share

Research papers are not the only output of the research process and some fields publish fewer research papers than others, nevertheless, they are universally important. The volume of research papers produced by an individual, research organisation or country can, therefore, be used as an indicator of research activity. Because publication behaviour differs between fields, the share of world output is a more useful indicator for comparing across disciplines.

2.4.3 Citations

The citation count is the number of times that a citation has been recorded for a given publication since it was published. Not all citations are necessarily recorded since not all publications are indexed.

2.4.4 Citation impact

'Citations per paper' is an index of academic or research impact (as compared with economic or social impact). It is calculated by dividing the sum of citations by the total number of papers in any given dataset (so, for a single paper, raw impact is the same as its citation count). Impact can be calculated for papers within a specific research field such as Chemistry, or for a specific institution or group of institutions, or a specific country. Also citation count declines in the most recent years of any time-period as papers have had less time to accumulate citations. For instance, papers published in 2012 are likely to have less citations then the ones published in 2009.

The Average Citation Rates table displays data on the average citation rates of papers within the scientific fields over each of the past 10 years. The calculation is *number of citations / number of papers*, where papers are defined as regular scientific articles, review articles, proceedings papers, and research notes.

The Average Citation Rates table (sample as given in *ESI* included below) lists all the scientific fields in rows, as well as an All Fields row combining all data. Data for the average citation rates are given for the last 10 years with the All Years column using all data for those years. A value of 3.16, for instance, in the field of Chemistry in 2001 indicates that the average paper in that year has since been cited 3.16 times. It may be highlighted that all papers from all the journals covered in the *Web of Science* product are used in the average citation rates calculation - not just the papers chosen for inclusion in *Essential Science Indicators*.

Average Citation Rates
for papers published by field, 2002–2012

Fields	2002	2003	2004	2005	2006	2007	2008	2009	2010	2011	2012	All Years
All Fields	20.83	19.34	18.09	16.04	13.65	11.75	8.88	6.55	3.93	1.42	0.20	10.50
Agricultural Sciences	15.33	14.77	13.86	11.96	10.61	8.58	5.96	4.25	2.40	0.82	0.12	7.16
Biology & Biochemistry	31.67	29.46	26.90	23.16	19.48	16.73	13.04	9.44	5.49	2.00	0.25	16.37
Chemistry	20.60	19.23	18.47	16.94	14.59	12.72	10.44	8.05	5.01	1.91	0.23	11.43
Clinical Medicine	24.77	23.55	22.12	19.93	16.87	14.27	10.49	7.52	4.58	1.53	0.22	12.56
Computer Science	9.42	6.56	5.01	4.74	3.73	5.43	4.30	3.04	1.77	0.54	0.08	4.01
Economics & Business	15.35	14.07	13.07	11.05	9.07	7.39	4.92	3.24	1.77	0.63	0.12	6.40
Engineering	9.17	8.74	8.59	7.58	6.63	6.15	4.58	3.61	2.06	0.73	0.10	4.96
Environment/ Ecology	23.73	22.28	20.59	17.90	15.30	13.11	9.72	6.86	3.89	1.42	0.23	11.24
Geosciences	19.10	18.01	16.52	14.52	13.40	10.01	7.96	6.00	3.51	1.40	0.27	9.55
Immunology	37.98	35.28	33.86	29.81	25.36	22.13	17.26	12.70	7.20	2.58	0.27	20.57
Materials Science	12.75	13.22	12.13	11.02	9.93	8.85	7.10	5.59	3.57	1.36	0.16	7.48
Mathematics	7.27	6.69	6.15	5.60	4.82	3.95	3.12	2.26	1.27	0.47	0.09	3.49
Microbiology	29.94	28.08	26.43	24.38	19.67	16.14	12.42	8.88	5.33	1.85	0.22	14.90
Molecular Biology & Genetics	48.82	44.33	40.67	34.94	29.87	24.98	19.28	13.99	8.13	2.92	0.34	23.48
Multidisciplinary	7.79	7.42	6.96	13.23	13.06	11.45	9.51	6.81	5.39	2.23	0.56	7.54
Neuroscience & Behavior	36.84	33.15	30.91	27.69	23.55	19.59	14.80	10.56	6.09	2.14	0.28	18.54
Pharmacology & Toxicology	24.59	21.99	21.83	18.29	17.10	14.14	10.94	7.58	4.25	1.49	0.21	11.97
Physics	15.16	14.26	13.84	12.51	10.86	8.39	6.65	5.92	3.88	1.57	0.23	8.47
Plant & Animal Science	15.58	14.46	13.52	11.58	9.90	8.19	6.10	4.34	2.55	0.92	0.15	7.67
Psychiatry/ Psychology	23.34	22.94	20.75	17.68	15.11	12.20	9.00	5.99	3.24	1.13	0.21	11.16
Social Sciences, general	10.34	9.76	9.50	8.58	7.17	5.86	4.13	2.82	1.56	0.57	0.14	4.71
Space Science	21.91	24.18	21.87	20.92	18.18	17.26	11.93	10.59	6.91	3.07	0.58	14.35

2.4.5 Field Weighted Relative Average Impact (FWRAI)

With output from India showing an upward trend, the question remains as to whether research quality has been able to keep up? The citations and Citations per paper (CPP) does throw light on this question. It will not be fair to compare, since the figures are not corrected for different citation levels from discipline to discipline (for instance one cannot compare an institute in Medical Sciences with that of Engineering Sciences or Department of Oncology with Department of International Studies). In order to have a fair comparison across subject domains, *Field Weighted Relative Average Impact* has been attempted that permits cross domain comparison.

2.4.6 Normalised Citation Impact (NCI)

Citation rates vary between research fields and with time, consequently, analyses must take both field and year into account. In addition, the type of publication will influence the citation count. As such values for individual papers vary widely and it is more useful to consider the mean NCI. This average can be at several granularities: field (either journal category or field), annual and overall (total output under consideration). The standard normalisation factor is the world average citations per paper for the year and journal category in which the paper was published. This normalisation is also referred to as 'rebasing' the citation count. The normalisation has been taken into account by using average values for India and the World to avoid discrepancies and showing the trend in the Indian research landscape vis-à-vis the world and thereby India's international recognition.

2.4.7 Highly-Cited Papers

Citation data are highly skewed, relatively large number of papers receiving none or very few citations and very few papers receiving many citations. There is no theoretical limit on the number of citations a paper could receive. Therefore, very highly-cited papers do occur and these can strongly affect average citation impact statistics. This effect is particularly exaggerated for fields that publish relatively small numbers of papers or countries with relative low outputs of research papers. The patterns can thus vary greatly depending on the type of paper, the field, and the nature of the finding reported. The present study has taken into account the HCPs as reflected and used in *Essential Science Indicators* (ESI) database of Thomson Reuters that account for the top1% of the world output. Percentage of highly-cited papers (those which received ≥4 world average citations) benchmark varies between subjects (and average varies between countries). In 2006-2010, 8.6% of UK papers were highly-cited compared to 2.7% for India.[4]

The selection procedure in *ESI* database while selecting highly cited papers takes into account the factors of variation in citation rates by field. The older papers typically receiving more citations than recent papers. The first step of the selection procedure is to count the number of papers cited at different levels of citation and construct distributions for each field and year. These distributions for each field/year are then used to set selection thresholds by taking the same fraction of papers.

The time period for counts is 10 years, plus partial year counts for the current year (data is updated every two to four months). This means that any paper in the 10+ year period can be cited by any items in that same period. Citations from all sources are counted, and are cumulated from the year of publication through the current year. Database years (the actual years when items are entered into the database, which is not necessarily the publication year) are used to define the time periods.

The selection criteria involves citation cutoffs specific to field and year which are applied to all papers in the journal set to select highly cited papers. Citation thresholds are based on the distribution of citations, picking the specified top fraction of papers for each year and field. The thresholds are based on the cutoffs given in the All Years column of the Baseline *Percentiles* table of *ESI*. The Percentiles table displays data on the minimum number of citations needed to meet six percentile breakdowns within the scientific fields.

The Percentiles table is divided into sections for each scientific field with the All Fields section combining all data. Within each section are rows for the six percentile breakdowns (0.01%, 0.10%, 1.00%, 10.00%, 20.00% and 50.00%). Data for the percentiles are given for the last 10 years with the All Years column using all data for those years. For instance in the Agricultural Sciences table, a value of 68 in the 0.01% column for 2001 indicates that the top 0.01% of papers in agricultural science journals in that year have been cited 68 times.

2.4.8 Types of Items Counted

Item types include papers which are defined as regular scientific articles, review articles, proceeding's papers, and research notes. Letters to the editor, correction notices, and

---

[4] *Department of Science and Technology Government of India, New Delhi*, (July 2012). Bibliometric study of India's Scientific Publication outputs during 2001-10 Evidence for Changing Trends.

abstracts are not counted. Only Thomson Scientific-indexed journal articles, or papers, are counted. Counts are based on a journal set categorized into *22 broad fields*. Fields are defined by a unique grouping of journals with no journal being assigned to more than one field. The Multidisciplinary field contains journals such as *Science* and *Nature*, which in an article level classification, would be assigned to specific fields. This should be taken into account when analyzing the field ranking of an individual scientist, institution, or country/territory. The HCPs here have been identified based on the country i,e; India and subjected to further analysis for identifying pivotal people, key institutions and collaboration (both international/national).

2.4.9 Relative Specialization Index (RSI)

Relative specialization Index looks at the global baselines top ranking fields for India viz-a-viz world scenario and their correlation therein. Identifies the top ranking fields for India and correlates their relative rank in the world, which determines the Relative Specialization Index.

2.4.10 Activity Index (AI)

This deals with figuring out the *Relative Specialization Index* (RSI) for a given country based on *Activity Index (AI)*. AI is usually applied to institutions or universities to figure out specialization index. However in the present case, it is projected as country specific and calculated using the following formula:

$$AI = \frac{\text{The share of a given field in the publications of a country}}{\text{The share of given field in the world total of publications}}$$

$$RSI = \frac{AI-1}{AI+1}$$

RSI takes the values in the range -1 to <1. The value indicates whether an institution or a country has a higher than average activity in the world in a given field (RSI>0) or a lower than world average activity (RSI<0). In situations where RSI=0, it reflects a completely balanced situation. As a benchmark an RSI=0 value is used for all research fields, which corresponds to the 'world standard case'. With unprecedented domestic growth and exceptional rates of change, this is a time of opportunity and activity for Indian research and Indian researchers. The conventional frames of reference, for activity, growth and impact are adjusting to these new dynamics.

2.4.11 Research Field

A problem frequently encountered in the analysis of data about the research process, is that of 'mapping'. For example, a funding body allocates money for a given field Biology, but this goes to researchers in Clinical Medicine and Engineering as well as to Chemistry and Biology departments. Similarly, Clinicians publish in Mathematics and Education journals. Publications in environmental journals come from a diversity of disciplines. This creates a problem while trying to define, for example, 'Biology research'. Is this the work funded under Biology programmes, the work of researchers in Biology units or the work published in Biology journals? For the first two options, tracking individual grants and researchers to their outputs, which is feasible but not the intention and scope of this study, nor for every comparator institution and this itself needs a further detailed study. Therefore, to create a simple and transparent dataset of equal validity across time and geography, we rely on the set of journals associated with the given subject say Biology as a proxy for the body of research reflecting the field. Hence the *ESI* 22 subject categorization was used for the 10 year data downloaded from *ESI* database, while the data for the years 1971-2012 downloaded from Web of Science scattered in 254 categories were reorganised into 22 subject categories for making analysis feasible based on the journal categorisation system used by Thomson Reuters. Thus, the classification of subjects is based on the journal in which the paper has been published.

2.5 Data Comprehensiveness and Constraints

The basic nature and flavour of this study was to study health of Indian research landscape. Thus it is imperative that the study was data intensive and hence it becomes inevitable to generate good metrics by minimizing error correction. The data are based on the Thomson Reuters Web of Knowledge and its analytical database Essential Science Indicators. Hence whatever inherent weaknesses (if any) are there would stay put. Though utmost care was taken to avoid any sort of error in identifying and crediting, institutions, people and the like, however, in some cases, its demarcation in situations as reflected below was difficult:

(i) **Comprehensiveness of Data:** Due to relying on third party database, it is but natural that all that has been contributed by Indian researchers and scholarly community does not find entry in the said sources. Hence (100%) comprehensiveness of the author publications emanating from India cannot be claimed.

(ii)   **Non-standardization of Institutional Affiliation:** The names of certain branded institutions whose demarcated identifier is either a place or an associated name is not standardized in many cases. For instance, examples of variation include – (a) in case of IITs, the place identifier was not in most cases included, so one does not know which IIT scores more. Hence in case of pivotal institutions these have to be clubbed. Nevertheless, the trend showed that earlier (old) IITs were mostly figuring as expected; (b) similarly in case of CSIR laboratories, many a times only CSIR was indicated, and while in some other cases the name of the concerned laboratory was mentioned. As such the projections for the key institutions have been based taking into account CSIR and the individual labs without clubbing them; (c) inconsistency in rendering of the name of the institution, for instance ISI, Indian Statistical Institution, Ind. Stat. Inst and the like; (d) change of names of the institutions have also resulted in the short coming. For instance the institutional affiliation data for NIT or REC (where in some NITs were earlier referred to REC).

(iii)   **Data Error Correction:** While downloading the data for past over four decades (1970s-), some institutions might have not been accounted for due to discrepancies in rendering and/or some records were out of range for unknown/unverified reasons. However the data error correction might be (utmost) to the tune of 5-10% which might not have any significant effect on the end result.

(iv)   **Subject-wise Distribution of Papers:** while deciphering the Highly Cited Papers for India, in some cases, the categorisation of papers appears to be somewhat skewed. For instance, a paper of Chemistry, published in the journal "Physical Review" has been placed in the field Physics. As indicated earlier, the reason being that the classification of subjects is based on the journal in which the paper has been published. Such a classification sometimes does impact trends in deciphering disciplinary shifts. However, such an error correction may not have debilitating impact on the end result.

(v)   **Subject Categorization:** The classification of the subjects has been based on the journals (by the Thomson Reuters) in which the paper has been published.

(vi)   **Author Affiliation:** In case of an author having two affiliations (Indian and foreign), the credit has been given to Indian affiliation.

(vii)   **Database Limitations:** Such databases do not adequately cover other content streams such as patents, books (book chapters) papers in conference proceedings and the like, which is a constraint while evaluating research productivity.

These key issues and limitations have surfaced due to use of a third party database in absence of an indigenous resource base/s at institution/national level. Unquestionably, mandating creation/regular updation of such (comprehensive/representative database of Indian contribution with multi-variate content streams) resource-base/s at the national level is necessary to derive evidence based metrics regarding research excellence and building institutional profiles. Therefore, while such an initiative would obviate the current inherent data documentation deficits, flavour and bridge the gap, thereby comprehensively projecting Indian research contribution on the global platform alongside other key parameters, important for international standing of the institution. The third party databases have to be used for data analytics purposes, since these databases can complement and supplement our indigenous resource-bases. These key issues and limitations can to a great extent be addressed, if suggestions in the recommendation section are implemented and brought in vogue.

Since the up-gradation of research activities is on the cards of the science policy formulations in the country, such an assessment would provide impetus to bridge the limitations by augmenting and appropriation of the resources thereby enhance the research impact assessment.

3. RESEARCH LANDSCAPE IN THE GLOBAL CONTEXT

Part 1: Research Performance Measurement

Part 2: Reading the Research

Part 3: Health of Research Landscape in India

PART 1: RESEARCH PERFORMANCE MEASUREMENT (RPM)

3.1 Systems and Issues in RPM

Worldwide, research output has become enormous; and due to its exponential growth, multi-disciplinarity and the emergence of new hybrid disciplines resulting in recording, communicating and reading research output as complex. Evaluation of researchers, research output, research areas and institutions are increasingly being adopted as key issues in decision-making and policy setting including in-focus setting, recruitment and promotion, funding, and promoting excellence of research and of higher education institutions. It is also observed that advancements in technology and increased global interaction, both demand and facilitate a higher order of evaluation and assessment practices.

Peer review is a time-tested and continue to be the most popular method of evaluation, but now increasingly bibliometrics, ranking and other (more objective) tools have gained popularity alongside though they are in an evolutionary phase. Where Peer review represents a subjective, bottom-up view and can be impacted, among other things, by individual bias and long-windedness; Bibliometric analysis offers a top-down and somewhat an objective perspective on a body of research and is impacted, for example, by oversimplification and lack of personal knowledge.

While for Bibliometric evaluation of the research output of an institution/department/individual, one observes that some (Bibliometric) tools are available, but for building institutional profiles for interpretation and analysis for the institutional rankings at local/regional/national and international levels, parameters other than Bibliometric indicators, must be taken into consideration. Alongside, such endeavours being data centric, the availability of data on various parameters is thus key resource for initiation of such activities and projects. There is a considerable deficit of data and tools and caution must be exercised in interpretation. Comparisons at an institutional level thus have to be done with care, and with the participation of all stakeholders. It is important, for instance, to ensure readiness to provide data for parameters on which institutions are judged internationally in rankings, for those rankings to be of value. In order to build up systems for deciphering recognition/reputational indices, it is essential, therefore, to build institutional datasets and context sensitive tools.

The discussion about existing methods is growing ever more intense with the importance of research performance measurement being viewed as an investment in quality - helping set strategic goals, allocate budgets, and communicate achievements to potential faculty, collaborators, funders, and students. As research evaluation is being integrated into management systems driving key decisions, research assessment policy trends have shifted from ex-ante i.e. focusing primarily on prospective evaluation (assessment of application) to ex-post i.e. retrospective evaluation, and rapid advancements (as well as controversy) have emerged as a result.

Even though existing evaluation systems are recognized to have various flaws, it is an evolving process. Many countries internationally, are engaging in research policy analyses to develop national research evaluation systems, seeking the right combination of data management and tools suitable to their research contexts and objectives. Good evaluation practice involves more than just indicators, and as no single approach standing alone will suffice, experts advocate the use of "mixed methods". Good evaluation practice involving, evolving systems and dynamic processes built through discussion and dialogue among all stakeholders'. A range of tools and reliable data can be selectively and subjectively applied, based on the objective or interpretation required can form, a powerful assessment system in the hands of an informed user/agency.

A performance measurement model for research combining both qualitative and quantitative perspectives demands a thorough understanding of the culture, context, advantages and limitations within which these systems may work. Such a system can significantly improve and impact the way we read research and consequently direct us (India) to a systematic improvement in higher education and research.

3.1.1 Peer Review

Peer review is one of the most entrenched yet controversial aspects of research assessment. Its processes are continuously evolving, not just in their methodology, but also in their basic objectives. Transitioning from its traditional role in publication of research, it has evolved to inform decision making in selection and promotion of scientists, resources allocation and the strategic planning. It has recently earned a role in ex-post evaluation and evaluation of institutions as well.

Peer review can be traced back in some form to at least the 18th century. It initially came into being as a process for selectivity in scientific journals, prior to which the '*burden of proof*' was considered to be on opponents and the editor usually, made publication decisions. The Royal Society of Edinburgh's *Medical Essays and Observations*, first published in 1731, was probably the first to introduce peer review as we would recognize it today, with submitted manuscripts being distributed by the editor to appropriate specialists for assessment[5]. Various forms of review were adopted by other journals over the next two centuries. Different editors employed varying styles of peer review and the development of peer review has therefore been gradual and somewhat haphazard.

Peer review's most undeniable value lies in its wide acceptance and prevalence among the academic community. It is considered to be a subjective method of evaluation and its subjectivity can be ascertained as the source of both its weaknesses and its strengths. Its legitimacy stems from the fact that it reflects the specialization of academic fields and underlines a widely held belief in the academic fraternity that it is impossible for outsiders to judge the quality of research in a particular discipline. Peer Review fortifies the academic fraternity's autonomy of academia, allowing them greater control over defining, assessing and monitoring research excellence.

With the introduction of quantitative metrics and impact oriented parameters, the design of any new robust research evaluation system will require, as a starting point, a careful consideration of the weaknesses, caution areas and strengths of the peer review process.

Peer review is under the burden of the expectations from research assessment systems based on contemporary demands. In the current global climate, there are pressures to make research more accountable to (funding) policy and to larger society (Third Mission). In addition there is an intensification of research-industry relationships, which is considered critical for the competitive position of states and regions in the context of economic globalization. Greater answerability and stronger public-policy-research linkages are sought with the increasing urgency to find research-driven solutions to global problems. In this scenario, peer review's biggest lacunae are perhaps the difficulty in defense and verification of its outcomes and the heavy and slow processes. Furthermore, it deals with the fact that it has conceptually never been subjected to

---

[5] Ray, Spier, August 2002. History of Peer Review Process. Trends in Biotechnology Vol.20 No.8, p.1.

a rigorous and systematic analysis. It remains a historically inherited practice, not a rationally constructed procedure.

A range of criticisms against peer review also stem from another widely accepted feature of peer review systems i.e. the inherent subjectivity brought on by the human element i.e. the 'reviewer'. 'The ideal reviewer,' notes Ingelfinger,[6] 'should be totally objective, in other words, supernatural.' In practice, reviewers operate within limited time and knowledge and are often greatly influenced by personal preferences. They also work within systems of professional relationships including hierarchy and status, and are part of the same systems as that, which is being reviewed. In addition, studies have pointed out that in some contexts there is a bias. Such bias can be a matter of the "Matthew effect" in the sense that "to those who have, more shall be given".[7] But bias can also be a matter of systematic unfair treatment as a result of an "old boys' network" or discrimination on the grounds of gender, age, race, institutional attachment and so on. Several authors, for instance, have investigated status or institutional bias, more recently a study by Ceci and Peters,[8] suggesting that researchers from prominent institutions are favoured in peer review. A study by Link[9] indicates bias in favour of US-based researchers, - 'strong' where the referees themselves are US-based, and 'weaker' (but still present) when the referees are not themselves based in the US.

As reviewers are at the same time researchers in their respective field, conflicts of interest may arise if researchers are tempted to hinder publication of findings contradicting their own work, or delay publications of similar research than their own. In an increasingly competitive academic environment, this raises important questions about accountability and publicity of the review process.

Yet another important issue about peer review concerns its possible detriment to heterodox or "path breaking" research. As peer review outcomes may reflect the bias of senior researchers, who might tend to discourage new paradigms that threaten the existing structure. Furthermore, epistemic communities that exercise influence serve to reinforce existing norms and can have an inhibiting effect on the emergence of methods, disciplines or viewpoints that are new or challenge conventional systems.

---

[6] Ingelfinger, Franz J., (1974). "Peer review in biomedical publication." The American journal of medicine 56(5), 686-692.

[7] Rigney, Daniel, (2010). The Matthew Effect: How Advantage Begets Further Advantage February, 176 pages, ISBN: 978-0-231-14948-8

[8] Ceci, Stephen J., and Douglas P. Peters, (1982). Peer review: A study of reliability. Change 14(6), 44-48.

[9] Link, Ann M., (1998). US and non-US submissions. JAMA: the Journal of the American Medical Association 280(3), 246-247.

In contrast to the peer review, Bibliometric tools present themselves, at first glance, as more objective, accountable and scalable. Notwithstanding their increasing popularity to track and evaluate research activity is undeniable, these methods have received their fair share of criticism.

3.1.2 Bibliometrics

Originally used in the field of library and information science as a set of methods to quantitatively analyze scientific and technological literature, bibliometrics have rapidly grown to wider applications in many research fields to explore impact of the field, researchers, or a particular paper. New indicators developed for the description of research activity and research results, supplement the innovative use of the more basic indicators and the advancements in technology, to create more and more sophisticated systems of analysis. Two of the simplest and longest-standing metrics have been number of papers produced (Publication Counts) and the number of citations received (Citation counts). Other popular indicators are the H-Index and the Journal Impact Factor (JIF), Crown Indicator. Indicators of journal quality, based on citations such as -Journal Citation Reports; SCImago journal & Country Rank; Journal eigenfactor and still more examples are indicators of journal quality, based on expert panels, such as European Reference Index for the Humanities (ERIH) and Norwegian register of scientific journals and publishers and the like. (Table 1.1)

Attempts to accurately remove bias, attention to micro-level variations, assigning weightage and identifying parameters are in progress to fine-tune Bibliometric processes. However, various issues and concerns crop up as rapidly as advances are made. In addition to these technical issues, controversies remain in broad, conceptual areas such as applying bibliometrics to social sciences and cross-disciplinary fields. Thirdly, practical aspects of consistency, benchmarking and data reliability further add to the nuances that require careful attention.

With a growing number of Bibliometric indicators around, a variety of evaluation models are used for classification and analysis - such as result models, process model, systems model, economic model and actor models. Using a systems approach, a distinction can be made between indicators of input, structure, process and results.[10]

---

[10] Hansen, Hanne Foss, and Birte Holst Jørgensen, (1995). Styring af forskning: kan forskningsindikatorer anvendes? [Anmeldelse]. Samfundslitteratur.

Input indicators relate, for example, to the ability of research systems and research organizations to attract, in competition, external research funding and to recruit recognized researchers and research groups. Structure indicators relate, for example, to indicators of reputation associated with the researchers' status in their network, including membership of editorial boards and international scientific committees. Process indicators relate, for example, to the researchers' participation in conferences and the like. Finally, result indicators relate to output and effect, including publications, citations, patents and contributions to research training in the form of PhD production.

Publication count and citation count result indicators are two of the most fundamental research indicators and ambiguity exists even at these most basic building blocks of Bibliometric assessment. Table 1.2 enumerates a series of theoretical perspectives on the meaning of citation count, throwing light on this ambiguity.

Citation count, for instance, considered an effective indicator, indicates how often publications and therefore, how often researchers are cited by other researchers. Since researchers cite each other for a variety of reasons,[11] what citation counts actually measure is a matter of debate. Garfield[12] enumerates 15 reasons why researchers may cite each other (Table 1.3). In the understanding where researchers cite each other to build on one another's work and results, citation counts are indicative of quality. However, citation behaviour can be argumentative (i.e. selective as support for the researcher's own viewpoint). It can be used to disagree, flatter or show command of a subject area.

It is clear that Bibliometric tools are sensitive to context and interpretation. Not surprisingly a wide range of issues are associated with poor application. Misrepresentations arise, for instance, when comparisons are sought and nuances and variations are neglected. Several index numbers have been developed that can easily be accessed and it can be tempting to make use of these, but unfortunately they are not always used with caution. For instance both publication and citation patterns can be expected to vary considerably between research fields. It is prudent only to compare "like with like". For example, while making a citation analysis related to research achievement at a university, it would be prudent to normalize data by calculating the average number of citations per article relative to the world average for individual research fields to show which subjects have greater or lesser impact than we could expect.

---

[11] Adler, Robert, Ewing John, and Taylor Peter, (2009). "Citation statistics." Statistical Science 24.1, 1.
[12] Garfield, Eugene, (1965). Can citation indexing be automated, Statistical association methods for mechanized documentation, symposium proceedings.

The H index[13] has been developed in recognition of the limitations of other citation measurements, however it comes with limitations of its own. A second-rate researcher can have a high total number of citations because he/she has published one highly cited article, popular with other researchers, however this anomaly will reflect in his or her H index. To achieve a high H index demands continuous achievement at a high level over a period of years. But this also means that the use of the H index only makes sense after some years of research. The H index varies according to number of years of employment, subject and collaboration patterns. The H index does not, then, solve the problem mentioned above as regards comparison.[14]

In addition, there are many challenges as regards technical measurement with respect to availability of data. Any metric requires objective, reliable data to deliver accurate, actionable results. Data sources therefore become important at multiple levels. Common to the international academic community, commercial (and competing) general databases are available. The most important are Thompson Reuters ISI Web of Science,[15] Scopus[16] and Google Scholar.[17] The Royal Society, London in its recent report on Knowledge Networks and Nations released in 2011 has made extensive use of publication and citation data from SCOPUS to indicate trends in research collaboration globally.[18] The Organization for Economic Co-operation and Development (OECD) annually publishes a Science, Technology and Industry Scoreboard looking at major trends in knowledge and innovation, the tenth edition of which (2011) builds on 50 years of indicator development.[19] The European Commission also has a well-established publication record in this area and draws on Thomson Reuters data. The Australian Research Council chose Elsevier to provide citation information for the next round of the Australian Government's Excellence in Research for Australia (ERA) initiative.[20] Also,

[13] Hirsch, Jorge E, (2005). An index to quantify an individual's scientific research output. Proceedings of the National Academy of Sciences of the United states of America 102(46), 16569.

[14] van Leeuwen, Thed, (2008). Testing the validity of the Hirsch-index for research assessment purposes. Research Evaluation 17(2), 157-160.

[15] *http://wokinfo.com/*

[16] *http://www.scopus.com/home.url*

[17] *http://scholar.google.com/*

[18] The Royal Society, London, (2011). Knowledge, networks and nations: Global scientific collaboration in the 21st century" The Royal Society, London.

[19] OECD Science, Technology and Industry Scoreboard, (2011). Innovation and growth in knowledge economies. Atailable at: *http://www.oecd.org/sti/oecdsciencetechnologyandindustryscoreboard2011 innovationandgrowthinknowledgeeconomies.htm* (Accessesd on 1-12-2013)

[20] Australian Government's - Australian Research Council. Excellence in Research for Australia (ERA). Available at: *http://www.arc.gov.au/era/* (Accessesd on 1-12-2013)

Elsevier has been awarded the contract for provision of citation data services in the 2014 Research Excellence Framework in the UK. UK Research Council commissioned a report on Bibliometric study of India's research output and international collaboration to Evidence.[21] The report explores India's research in terms of amount of activity taking place and the impact on the global research community. It also focuses on the India's collaboration at the international level particularly with the UK.

The major databases that primarily include and facilitate such Bibliometric tools thus are, Thomson Reuters Web of Science; SCOPUS; and GOOGLE Scholar.

Thomson Reuters Web of Science: *(*generally known as ISI Web of Science or ISI). This is the traditional source of citation data, established by Eugene Garfield in the 1960s. Many universities still use this as their only source of citation data. It has complete coverage of citations in over 10,000 journals that are ISI listed, going back to 1900. It is generally updated once or twice a week. Although its worldwide coverage has been improving recently, it still has a North American bias in many disciplines. It charges commercial rates for access.

Scopus: Introduced by Elsevier in 2004, aims to be the most comprehensive Scientific, Medical, Technical and Social Science abstract and citation database containing all relevant literature, irrespective of medium or commercial model. It covers nearly 18,000 titles from more than 5,000 publishers. It also claims worldwide coverage and more than half of Scopus content is said to originate from Europe, Latin America and the Asia Pacific region. The database is updated daily and charges commercial rates for access.

Google Scholar: Introduced by Google in 2004, Google Scholar has become a very popular alternative data source, not least through the fact that access is free and citation analysis programs such as Publish or Perish make bibliometric analysis easy. Some academics are skeptical about its wider coverage. However, studies (e.g. Vaughan and Shaw - 2008) have found most of the citations to be scholarly. After a relatively slow start Google Scholar coverage is increasing, although Google still does not provide a list of its sources. Google Scholar is updated several times a week.

Transparency in the coverage of the database and the criteria that form the basis for the material that is included or excluded leave something to be desired. Due to the

---

[21] UK Research Council, (2010). Bibliometric study of India's research output and international collaboration: A report commissioned to Evidence by UK Reseach Council. Available at: *http://www.rcuk.ac.uk/documents/ india/BibliometricstudyIndiaresearchoutput.pdf.* (Accessesd on 1-12-2013)

differences in the degree of coverage of these databases, searches on different databases can often give different results. Adding to this, the reality that the databases are prone to errors and that the degree of coverage varies from one research area to another.

Apart from the nuances of systematization and implementation, more conceptual drawbacks remain. Strong objections have been raised relating to the use of research indicators in the field of social sciences and humanities. Standard Bibliometric techniques are unfavorable, for instance, to monographs, and anthologies, or regional language publications. As a result, Bibliometric analyses provide an incomplete picture of research activity, as far as social sciences and humanities are concerned.[22]

Another interesting point arises regarding the unintended impact the assessment methodology may have on the reviewer or the possible abuse of the measurement tools. As with any assessment, there is danger that the test may become the ultimate objective rather than the activity being tested. In this case, where Bibliometric tools play an important role in all aspects of evaluation and assessment, the research process itself may be modified as researchers try to adapt to the assessment system[23].

Bibliometric tools require skillful interpretation and a nuanced approach, and can be effective when used in combination with subjectivity and specialist domain knowledge. Weingart[24] opines, the unique contribution of bibliometrics to the collective communication process in science and their greatest value to the scientific community itself as well as to policy makers and the public in providing this `greater picture'. It can `inform' the process about macro-patterns in scientific communication, for example about the unsuspected connection between research fields that are not yet institutionally connected. The interpretation of these patterns of unexpected contradictions to the common wisdom of the community or other irregularities must be left to the experts in the respective fields or at least assisted by them. Peer review must `correct' Bibliometric analyses wherever necessary.

---

[22] Center for Science and Technology Studies (CWTS) Leiden University The Netherlands (May, 2011). Research Report for KWR Watercycle Research Institute, Nieuwegein, the Netherlands: Bibliometric Study on KWR Watercycle Research Institute (2005-2009). Available at: *http://www.kwrwater.nl/uploadedFiles/ Website_KWR/Nieuws/2012/Bibliometric%20study%20of%20 the%20research%20performance%20of%20 KWR%20_2005-2009_%20versie%202.pdf*. Accessed on 5.12.2013).

[23] Butler, Linda. Assessing university research: A plea for a balanced approach, (2007). *Science and Public Policy*, 34(8): 565-574.

[24] Weingart, Peter, (2005). Impact of bibliometrics upon the science system: Inadvertent consequences? Scientometrics 62(1), 117-131.

Nevertheless, with all its constraints and the cautionary note, Bibliometrics have come in to their own as evidenced by their growing acceptance among governments and academic communities, as well as their prominence in internationals research assessment systems. There is a strong need to create better systems and methodologies, increase education and awareness, and initiate dialogue and research, in order that these tools may be effectively used to the benefit of the research landscape and community in India.

3.1.3 Ranking Systems

Ranking of higher education system and Universities has gained importance recently but rapidly. With the internationalization of higher education and increasing global competitiveness, rankings have gained prominence and comparisons are made not just at a national but also at an international level. Ranking system's increase in prominence and acceptability has been accompanied by a focus on the conceptual and methodological parameters and processes. While many of the ranking systems have overlapping parameters and indicators, even minor variations in weightage of isolated uniqueness is of interest in assessing their value and has attracted skepticism and criticism. As controversial as these rankings may be in their ability to accurately create a hierarchy among institutions, the fact remains that they answer to a demand in contemporary higher education and their increasing acceptability (and consequently their importance) is undeniable. It is worthwhile to review rankings from a systemic perspective, to understand how and to what extent they may impact the Higher Education landscape. While rankings might have had an innocuous beginning as a publicity promoter, they have now grown in stature to influence policy makers, funding, researchers and students and as such a critical understanding of ranking systems is essential - conceptually, methodologically and contextually.

The commonality between ranking systems is in the overall methodology used to develop ranking. They depend on a complex system of indicators, parameters and assigned weights, combining and using various mathematical models and methodologies, into a common score. This score is then used to provide overall ranking. The reduction of the multi-dimensional nature of higher education into a resultant linear scale causes much distortion. One of the broader conceptual criticisms leveled at ranking systems is that despite the complex methodologies involved, they are accused of over simplifying and as a consequence, lead to only superficial and unreliable analysis. Additionally, they are restricted because they measure what can be measured rather than what should be measured.

For instance, an analysis offered by Guarino[25] reveals that the simplistic ranking of individual institutions suggests an exaggerated difference between institutions where none may exist. According to the THE rankings for 2013-2014,[26] the University Imperial College London is at no. 10 and Yale University is at no. 11 (with overall score 87.5 and 87.4 respectively, thus their overall scores differing by .1). This suggests to the user that The Imperial College London is better than the Yale University, since they differ by a number. However what this might at best imply, is that both universities are better than university rank 50 and not as good as university rank 2. An extension of this, as Locke, William et all[27] suggests that 'The difference in scores between institutions placed several positions apart may not be statistically significant, even though the difference in positions suggests a disparity in quality or performance.'

Guarino further highlights the extent to which ranking systems recycle existing reputations, revealing the cyclical nature of the parameters used in ranking systems. Thus questioning whether they reflect any real difference or are merely reflective of a popularity contest serving according to Roberts and Thompson, 'to reinforce stereotypes and market stratification.[28]

Other highly controversial areas are concerned with the choice of parameters and the weightage assigned to them. There is, for instance the difficulty with measuring quality of teaching, under-represented as a parameter in many ranking systems. This leads to a bias in favour of research oriented schools as opposed to teaching centric schools and can lead thus to unfair comparisons. For instance, the Shanghai rankings are exclusively focused on research output and thus the neglect of strength or otherwise of teaching area for respective institutions. Where there is difficulty measuring, it is often observed that parameters are replaced with proxies. While the Times Higher Education includes a combination of 5 indicators (to a total weightage of 30%) to place focus on teaching, they represent the 'teaching environment', not to be confused with 'teaching

[25] Guarino, Cassandar, *et al.*, (2005). Latent variable analysis: A new approach to university ranking. *Higher Education in Europe* 30(2), 147-165.

[26] THE World University Rankings 2013-2014. Available at: *http://www.timeshighereducation.co.uk/world-university-rankings/2013-14/world-ranking* (Accessed on 3-12-2013)

[27] Locke, William, *et al.*, (2008). Counting what is measured or measuring what counts? League tables and their impact on higher education institutions in England.

[28] Roberts, David, and Lisa Thomson, (2007). Reputation Management for Universities: University league tables and the impact on student recruitment. *Reputation Management for Universities*.

quality' for which there is currently no accepted metric and therefore it does not find a place in rankings, merely an approximation through a measure of the environment. Furthermore, many studies have now emerged that suggest that the weights assigned are arbitrary and have neither empirical nor theoretical basis.[29] [30] [31]

Another aspect where misgivings arise is the collection and accuracy of data on which rankings are based. Data gathering varies for each ranking system and is usually a combination of independent surveys (from students, prospective employers, universities etc). The data provided by the universities themselves and in general statistical analysis of primary data is not carried out. According to Van Vught and Ziegele,[32] 'It seems that availability of quantitative data has precedence over their validity and reliability'.

The objectives and motivation based on which each ranking system operates gives interesting insight into their methodology, focus and usability. For instance QS (Quacquarelli Symonds) and THE (Times Higher Education), which parted ways in 2010, are geared towards different audiences. THE catering to a more academic readership, aims to influence government policy and to shape institutions' reputations, while QS is aimed to aid student decisions. According to Danny Byrne[33] of QS, "We've always been clear we're aimed at prospective international students. Our rankings are easy to understand and of direct relevance to students'. Accordingly, THEs tends towards a more stringent, verifiably methodology, while QS tends to rely heavily on quantity of survey data to ensure quality. The 'Shanghai Ranking', as a third example, originated in 2003 with Chinese government backing. It was designed for Chinese Universities to assess and plan their progress through international benchmarking. Their bias towards research output at the neglect of teaching, for instance, reflects China's strategic priority to catch up in 'hard scientific research' with a strong focus on long-term parameters such as Nobel prizes and publications in reputed scientific journals.

[29] Clarke, Marguerite, (2002). What can US News & World Report rankings Tell us about the quality of Higher Education? Education policy analysis archives 10(16).

[30] Van Dyke, Nina, (2005). Twenty years of university report cards. *Higher Education in Europe* 30(2), 103-125.

[31] Usher, Alex, and Massimo, Savino, (2006). A world of difference: A global survey of university league tables. Toronto, ON: Educational Policy Institute. Available at: *www.educationalpolicy. org.*

[32] Van Vught, Frans, and Frank, Ziegele. (2011). Design and testing the feasibility of a multidimensional global university ranking. Final report.

[33] How accurate are university rankings? Available at: *http://www.topix.com/forum/world/singapore/ T2VQMMMP55G4P6I7N* [Accessed on December 26, 2013]

League tables and ranking systems influence decision making in aspects such as strategy, personnel, recruitment, and public relations.[34] Student decision-making is also impacted by global rankings.[35] Rankings also affect the assessment of institutional reputation by faculty and institutional leaders. In fact, according to Bastedo and Bowman,[36] considering the difficulty in assessing quality parameters and making comparisons between universities (especially those in close competition) over time, the rankings become the reputation, rather than reputations being a parameter in determining rank. The influence of global benchmarking can be noted in various policy reforms in many countries. As noted by Davil Dill,[37] 'the sharp increase in R&D investment among a number of the Nordic countries, the adoption of performance-based funding for academic research, the reforms in doctoral education sweeping across Europe, and the new German Excellence Initiative would be difficult to understand without reference to debates about the relative standing of the world's universities'. Their growing impact holds the danger of creating an ill-motivated prestige-race. While competition and prestige have always been powerful motivators for excellence among higher education institutions, rankings threaten to become a measure for excellence. There is a danger of valuable efforts and resources being diverted to gaining supremacy in Global Rankings replacing the quest for quality with an effort to 'play the ranking game'

Despite all the discrepancies and conceptual issues, it is clear that ranking systems have come to have an influence on most, if not all players in Higher Education. Currently, despite India having the third largest education system, India is conspicuous with its absence among top institutions in international ranking. This may partly be attributed to a disadvantageous representation of Indian institutions and researchers. This underlines the importance to address the data deficit and highlights the importance of a concerted effort in building accurate institutional profiles. With globalization and massification, it is

---

[34] Hazelkorn, Ellen, (2007). The impact of league tables and ranking systems on higher education decision making. Higher Education Management and Policy 19(2), 87.

[35] UNESCO, UNESCO Forum on Higher Education, Research and Knowledge, (2009). *Occasional Paper* No. 15 - Impact of Global Rankings on Higher Education Research and the Production of Knowledge Ellen Hazelkorn. Available at: *http://unesdoc.unesco.org/images/0018/001816/181653e.pdf* (Accessed on 3-12-2013)

[36] Bastedo, Michael N. and Bowman, Nicholas A., (2011). College rankings as an interorganizational dependency: Establishing the foundation for strategic and institutional accounts. *Res High Educ* 52:3–23 DOI 10.1007/s11162-010-9185-0.

[37] Dill, David D., (2006). Convergence and diversity: The role and influence of university rankings. Keynote Address presented at the Consortium of Higher Education Researchers (CHER) *19th Annual Research Conference.* Vol. 9.

imperative to embrace the international ranking trend and contextualize it strategically for the betterment of the Indian Higher Education Landscape. In a brief overview of their systems, four of the more popular rankings of world universities are looked at here that include, QS, THE, SHANGHAI and U-MULTI Rank.

3.1.31 QS (Quacquarelli Symonds)[38]

Academic ranking of world universities compiled by the QS World University Ranking is a ranking of the world's top 500 universities by Quacquarelli Symonds since 2004. The QS rankings were originally published in collaboration with Times Higher Education, and known as the THE-QS World University Rankings. QS assumed sole publication of the existing methodology and Times Higher Education split in order to create a new ranking methodology in 2010, which became the 'THE' World University Rankings. QS World University Rankings uses Academic Peer Review (40%), Recruiter Review (10%), Faculty-Student Ratio (20%), Citations per Faculty (20%), International Orientation (10%) and International students (10%) as criteria for ranking universities.

3.1.32 Shanghai Jiao Tong University's Academic Ranking of World University (ARWU)

Popularly known as Shanghai Ranking, the Academic Ranking of World Universities (ARWU) was first published in June 2003 by the Centre for World-Class Universities and the Institute of Higher Education of Shanghai Jiao Tong University, China updated on an annual basis. The ranking compared 1,000 higher education institutions worldwide and published 500 of them. The methodology is spelled-out in an article by its originators, N.C. Liu and Y. Cheng.[39] Liu and Cheng explain that the original purpose of doing the ranking was "to find out the gap between Chinese universities and world-class universities, particularly in terms of academic or research performance". The Shanghai rankings draw on six different criteria to measure academic performance –

Alumni: alumni winning Nobel Prizes and Fields Medals	(10%)
Award: staff winning Nobel Prizes and Fields Medals	(20%)
HiCi: highly-cited research articles in 21 broad subject categories identified by Thomson Reuters	(20%)
N&S: articles published in the journals Nature and Science	(20%)

---

[38] *http://www.topuniversities.com/university-rankings/*
[39] Liu, Nian Cai, and Ying, Cheng, (2005). The academic ranking of world universities. *Higher education in Europe* 30(2), 127-136.

PUB: number of articles indexed in Science Citation Index Expanded and Social Sciences Citation Index	(20%)
PCP: per capita academic performance - weighted average of the scores obtained in the previous five categories, divided by the number of current full-time equivalent academic staff members	(10 %)

In each category the best performing university is given a score of 100 and becomes the benchmark by which the scores of all the other universities are measured. Universities are then ranked according to the overall score they obtain, which is simply a weighted average of their individual category scores. The Shanghai rankings have come under some criticism regarding both their methodology and choice of variables.[40-41] With respect to the choice of variables, Shanghai University uses only a limited set of criteria, which measure academic performance solely in terms of research excellence, to rank a wide range of universities. This "one size fits all approach" fails to capture the specific characteristics of a university and ignores the objectives an institution pursues outside of research, such as education and a social mission. In terms of its criteria, the ranking is biased in favour of science and technology and almost totally disregards other fields such as the arts and humanities. Thus, schools with strong scientific departments fare much better in the rankings than schools that specialize in the arts, humanities or social sciences. The ARWU also favours English-speaking universities as English is the predominant language of academic publications. Van Raan[42] points out these biases and warns against the misuse of overly simple bibliometric indicators. Finally, the ARWU does not take into account the effect of size on performance. Zitt, Michael et al[43] noted that ninety percent of criteria used in rankings are size-dependant. Indeed, the Shanghai rankings essentially measure overall production and not efficiency, an approach that favors large universities. And while they do include one variable to this effect ("PCP"), it is rendered almost useless as it is only computed for the universities which survive pre-selection based on their performance with respect to the other criteria.

[40] Vincke, Philippe, (2009). University rankings. *Ranking*.

[41] Van Raan, A. F.J., (2005). Challenges in ranking of universities." Invited paper for the First International conference on World Class University, Shanghai Jaio Tong University, Shanghai, June 16-18, 2005.

[42] Van Raan, A. F.J., (2005). Fatal attraction: Conceptual and methodological problems in the ranking of universities by bibliometric methods. *Scientometrics* 62(1), 133-143.

[43] Zitt, Michel, and Ghislaine, Filliatreau, (2007). Big is (made) beautiful: Some comments about the Shanghai - ranking of world-class universities. World-class university and ranking: Aiming beyond status, 141-160.

3.1.33 *Times Higher Education World University Rankings*

The Times Higher Education World University Rankings is an international ranking of the world's top universities published by Times Higher Education (THE). THE split from its original partner Quacquarelli Symonds in 2010, creating a new ranking methodology whose citation database information is compiled in partnership with Thomson Reuters. THE uses - Research (30 %), Citations (32.5 %), Teaching- Learning Environment (30%), Industry Income (2.5 %) and International Diversity (5 %) as criteria for ranking universities.[44]

The quality of the education correlates with the research productivity in terms of scholarly output of international repute, one of the important indicators used for ranking world-class universities. A cursory look at the indicators and criteria used for ranking of world universities by three different rankings reveals that the "Research" gets maximum weightage in ranking of world universities represented by criteria namely "Research", "Research Productivity", "Research Impact", "Research Excellence", "Academic Peer Review", "Citations" and "Citation per Article". In order to reflect a good ranking according to THE, therefore, academic institution must channelize their resources towards conducting high-quality research so as to obtain a good ranking in the world-class universities. Moreover, it is also important for a university to demonstrate its impact, cost-benefit and Return on Investment (ROI) to their administrators.

3.1.34 *U-Multirank*

U-Multirank[45] which is a new initiative, claims to provide an alternative to systems relying purely on ranking scores based on the understanding that while the ranking scores do capture an important aspect of each university's overall quality, they don't speak to a diverse range of other issues. The system designed and tested by the Consortium for Higher Education and Research Performance Assessment, and supported by the European Commission, aims to increase transparency in the information available to stakeholders about universities, and encourage functional diversity of the institutions. According to Richardson,46 unlike traditional university rankings such as the THE, ARWU and QS, U-Multirank features separate indicators that are not collapsed into an overall score.

---

[44] *http://www.timeshighereducation.co.uk/world-university-rankings/*

[45] *http://www.u-multirank.eu/*

[46] *http://www.researchtrends.com/issue24-september-2011/a-"democratization"-of-university-rankings-u-multirank/*

U-Multirank is a multi-dimensional, user-driven ranking tool, addressing the functions of higher education and research institutions across five dimensions: research, education, knowledge exchange, regional engagement and international orientation. In each of these dimensions, it offers indicators to compare institutions. In this sense it certainly focuses on the goals institutions set themselves. But unlike most current rankings, U-Multirank does not limit itself to one dimension only (research). It allows institutions to show whether they are winners or improvers over a range of dimensions. For the selection of U-Multirank's indicators it is reported that the team made use of a long and intensive process of stakeholder consultation, which included a broad variety of stakeholders, including the higher education and research institutions themselves. This stakeholder consultation reflected the criterion of 'relevance' in the process of indicator selection. In addition they reportedly used the criteria of validity, reliability, comparability and feasibility. For 'feasibility', they focused on the availability of data and the effort required to collect extra data. Also tried to ensure that data availability would not become the most important factor in the selection process. However, the empirical pilot test of the feasibility of U-Multirank indicators showed that, particularly in the dimension of 'knowledge exchange' and 'regional engagement' data availability is limited.[47] The team has designed a tool that allows users to select the institutions or programs they are interested. U-Map[48] is a mapping instrument that allows the selection of institutional activity profiles. As per the availability of literature, only certain comparisons are permitted, in order to avoid unfair comparisons.

---

[47] U-Multirank, (2011). The design and testing the feasibility of a multi-dimensional global university ranking. Draft version for distribution at the U-Multirank conference, Brussels, Thursday 9 June 2011.
[48] *http://www.u-map.eu/*

PART 2: READING THE RESEARCH

3.2 Reading the Research – The Insights

Research Performance Measurement has become an increasingly critical area for those in decision-making positions. Institutions, research councils, funding institutions and governments have turned their attention to creating robust and relevant structures linking research to policy. Through these systems decision makers can help more accurately plan and drive the quality of research to create peaks of excellence, while also being able to guide and monitor strategic direction. A universal and accessible system, also helps broaden the base while maintaining quality standards, not only from the point of view of policy makers but also by giving institutions and individuals a chance to benchmark and evaluate their progress and hindrances. A good system, therefore, serves the needs of decision makers while gaining approval of the academic community. At these various levels- government, policy, institution, departments, etc, funding and resource allocation is the most obvious and primary manner in which research performance is linked with decision making. However, recognition (motivation through credit/award/reward), recruitment, promotion (attracting, retaining and developing talent) an opportunity creation (collaboration, internationalization) are important mechanisms of influence. The relationship between research and policy is being recognized as complex and dynamic and systems are being sought to maximize its effectiveness. A few trends can be observed in the evolution of such systems internationally, and the same are briefly outlined here. One key shift has been in the role of research assessment. While decision-makers previously focused on the assessment of future applications (prospective evaluation), now they are focusing equally on the retrospective evaluation of activities and results. Evaluation of research programmes, *ex ante,* to distribute limited resources selectively, and *ex post,* to provide accountability over the value of investment, are both aimed for in combination. Weaving together the subjectivity of expert (peer) review and quantitative analysis (Bibliometric) has been (Promoting mixed method With sole objectives of aspiring for Excellence, Boosting quality, facilitating Strategic direction, Broadening base, and maintaining quality) the greatest challenge and goal of research assessment systems. These are only few examples of how modern research assessment systems favour mixed models, contextualized, holistic and nuanced approaches. Much contradiction and diversity requires to be addressed by a common system and therefore the parameters, the processes and the implementation is rigorous and feedback as well as flexibility is crucial. The evaluation of 'impact' as a parameter within the model of research assessment has also emerged as an area of research, evaluation and naturally conflict. Defining impact in terms of social, cultural, economic and environmental benefits, has underlined both the criticality of subjectivity in the process as well as the indispensible perspective and contextualization offered by objective methodologies. The global nature of research and the increasing focus on collaboration means that no system can exist in isolation. A successful system must necessarily take this into account and must facilitate

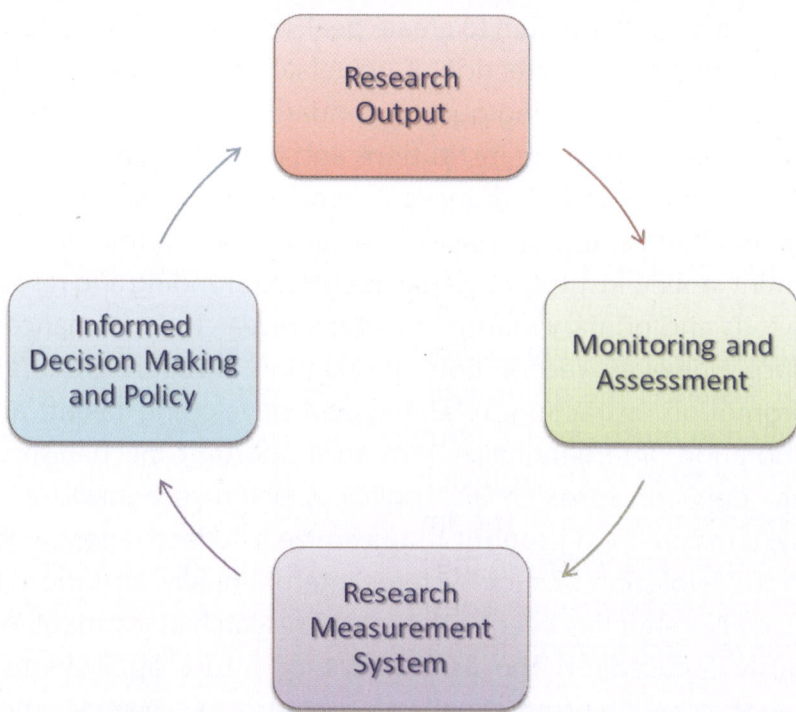

Figure 2.1: *Popularity of 'Mixed Method': Argument for Systemic View of Various Stages*

and recognize global, collaborative research. Competitiveness between countries and institutions in terms of research output is on the other side of the collaboration coin that reflects this increasingly global nature of research. To stay competitive as well as collaborative is a key issue that modern research assessment systems seek to serve.

Clearly, there is no single, correct methodology, and a multi-level approach seems prudent. An actionable assessment of research must consider the multi-functional, layered and diverse nature of the research landscape. Among others, this includes different disciplines (high tech, science to social sciences), different universities; levels at which research assessment are being conducted (individual, departmental, national, etc.) or the notion with which research is undertaken (from blue sky to practice based research). The key issue then becomes designing evaluations that fit their purpose. This needs an understanding of who will engage in research assessment, the reliability of data and processes used, a clear understanding of the objective the analysis seeks to serve and the acceptability to all stakeholders involved. A deeper realization of the relationship between policy, innovation and progress on the one hand and research activity on the other has highlighted this more holistic definition of research assessment, as governments and institution leaders seek to leverage their country/institution's research performance as relevant and accountable to the progress and status of the country/institution itself.

3.2.1 Interpreting Impact Metrics: Government Agencies

3.2.11 National and International Experiences

There are instances that show that governments around the world are increasingly using performance-based funding to allocate resources for research in higher education institutions. For instance, the introduction of the Performance-Based Research Fund (PBRF) has resulted in much greater scrutiny of the research activities of New Zealand universities, and there has been positive impact of this greater scrutiny on the research productivity of the universities.[49] The impact shows that most universities exhibit a significant increase in productivity in the period following the introduction of the PBRF. The new Government has expressed a desire to introduce performance-based funding of tertiary education research for more than a decade. The PBRF was designed to improve the

---

[49] The Impact of the Performance-Based Research Fund on the Research Productivity of New Zealand Universities. Available at: *http://www.educationcounts.govt.nz/__data/assets/pdf_file/0005/49901/ Social-Policy-Journal-paper.pdf*

average quality of the research in New Zealand tertiary education organizations through linking government funding directly to research performance. Greater scrutiny PBRF has placed on the research activities, most of the New Zealand universities have been associated with a significant increase in research productivity measured by the number of articles and reviews listed in the Web of Science per FTE research staff. The study shows that linking government funding directly to institutional research performance and ensuring the publication of that performance has been associated with significant changes in institutional behavior.[50]

Evidence shows a similar trend in Government agencies in India especially in relation to strengthening of funding. The Department of Science & Technology (DST), Ministry of Science and Technology, Government of India is at the forefront of this. The DST has identified 29 potential and richly contributing Universities in the country amongst 450 Central and State Universities, with an initiation of incentive support of Rs.9 crores (approx. USD 2 million) for a period of 3 years, under a new and innovative program namely PURSE (Promotion of University Research & Scientific Excellence) for incentivizing R&D activities in the Universities based on the scientific publications made by University faculty, research scientists and scholars (as per the SCOPUS Database). Similarly the development of a Results framework Document (RFD) and 'Bibliometric Study of India's Scientific Publication output from 2001-10' was entrusted to Thomson Reuters which primarily uses Bibliometric research indicators.[51]

DST's Consolidation of University Research for Innovation and Excellence (CURIE) is a special initiative to improve the R & D infrastructure of 'Women Universities'. These indicators endeavour to reflect on performance indicators to help policy makers arrive at certain conclusions. The DST is currently discussing with agencies for third party audit of India's performance as evident from S&T output indicators. This exercise is being explored for developing an evidence-based budgeting system for the science sector in the country. From the above, it clearly reflects that since the award of research grants is a competitive process and spreading this uniformly over the given time frame of the financial year and thus making full utilization of resources in making investment for developing the country's knowledge and technology base is high on Department's agenda and rightly

---

[50] The Impact of the Performance-Based Research Fund on the Research Productivity of New Zealand Universities. Available at: *http://www.educationcounts.govt.nz/__data/assets/pdf_file/0005/49901/ Social-Policy-Journal-paper.pdf*

[51] Results Framework Document (R F D) for Department of Science and Technology (2010-2011). Available at: *http://www.dst.gov.in/Results-Framework_Document__2010-11__DST_Final.pdf*

so. The National Institute of Communication and Information Resources (NISCAIR), a constituent establishment of CSIR has been using Bibliometric techniques for building national science indicators. They have been conducting Bibliometric analysis based on publication output using available tools for citation and impact analysis since 1986. These studies project the trends in research output, research excellence, highly cited - papers, authors and the institutions.

At a systemic level, many governments are creating/innovating/revamping frameworks and models for research assessment. One of the most discussed is the Research Excellence Framework[52] (REF) in UK. This new system for assessing the quality of research in UK higher education institutions (HEIs) replaces the Research Assessment Exercise (RAE). Similarly, The ERA[53] (Excellence in Research in Australia) initiative assesses research quality within Australia's higher education institutions using a combination of indicators and expert review by committees comprising experienced, internationally recognized experts. In 2008, the European Commission, Directorate-General for Research set up the Expert Group on Assessment of University- Based Research[54] to identify the framework for a new and more coherent methodology to assess the research produced by European universities. The Brazil and UK system cases are taken as a typical example and being briefly deliberated here for clarity of the case.

3.2.12 *Trends, Objective and Mechanism – The Case of Lattes Platform and REF*

In Brazil, The *Lattes* Platform is an information system (integrated data-base, web-based query interface, etc.) maintained by the Federal Bureau responsible for funding the scientific and technological research efforts, Conselho Nacional de Desenvolvimento Científico e Tecnológico (CNPq, National Counsel of Scientific and Technological Development). It is designed to be a one-stop source of information on science, technology and innovation related to individual researchers and institutions working in Brazil. *Lattes* is the main gate for ST&I (Science, Technology & Innovation) information used by all Brazilian researchers to update and access curricula, projects and other data. The key features facilitated by this system includes:

---

[52] *http://www.hefce.ac.uk/research/ref/*
[53] *http://www.arc.gov.au/media/releases/media_09september11.htm*
[54] *http://ec.europa.eu/research/era/docs/en/era-partnership-expert-group-external-funding-final-report-2008.pdf*

Free admission	Any person can log in and offer his/her CV
Mandatory	Registration required for funding, *Lattes* is disseminated throughout the National ST&I system
Information exchange standards	Defined in collaboration with universities and other S&T Agencies
Data certification	By the CV of the author and by institutions
Data quality	DOI, Audit reports of false information and mostly by social control
Interoperability and data extraction	*Lattes* interacts with 132 institutions
Alignment with other e-Gov initiatives	e.g, Innovation Portal, Coleta CAPES
Unique identification for researchers	Ready for ORCID
Public access to data and services	
Privacy policy	No personal information is made available through the Internet
Cross sectional views	It allows thematic views (e.g. Environment, Health, Engineering).

Citing an example, the following information gives a bird's eye view of the key parameters, the participating population and the broad usage statistics of this platform.

Category	CVs	R&D Groups	Institutions	Projects
Entries	2 million	23000	4000	20000 per year
Details	Personal profile Professional address Education background Publications Technical production Thesis Competencies Relationships	Name Institution Leaders Researchers Students Research areas Partnerships	Structure Department Course	Title R&D areas Researchers Students Grant

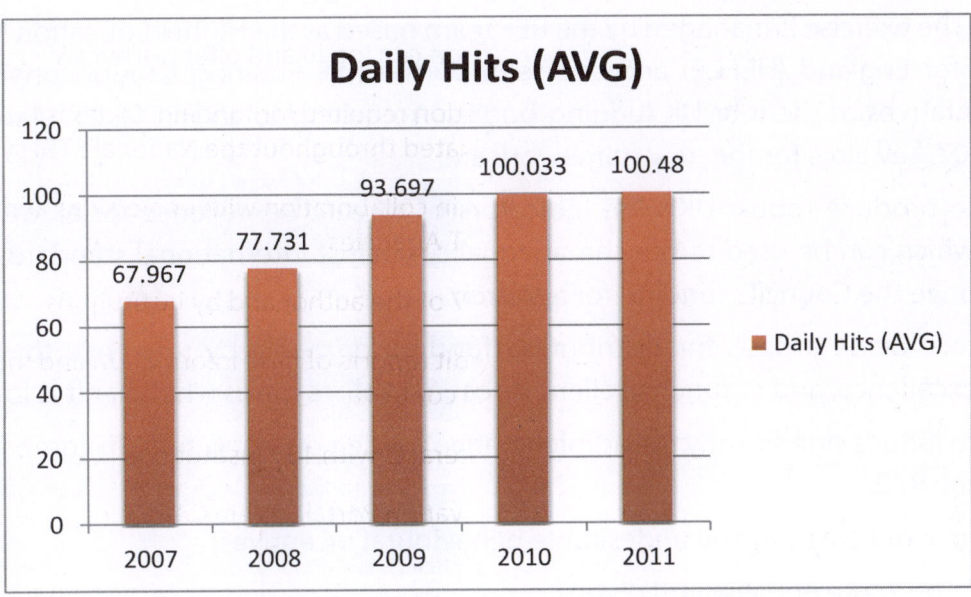

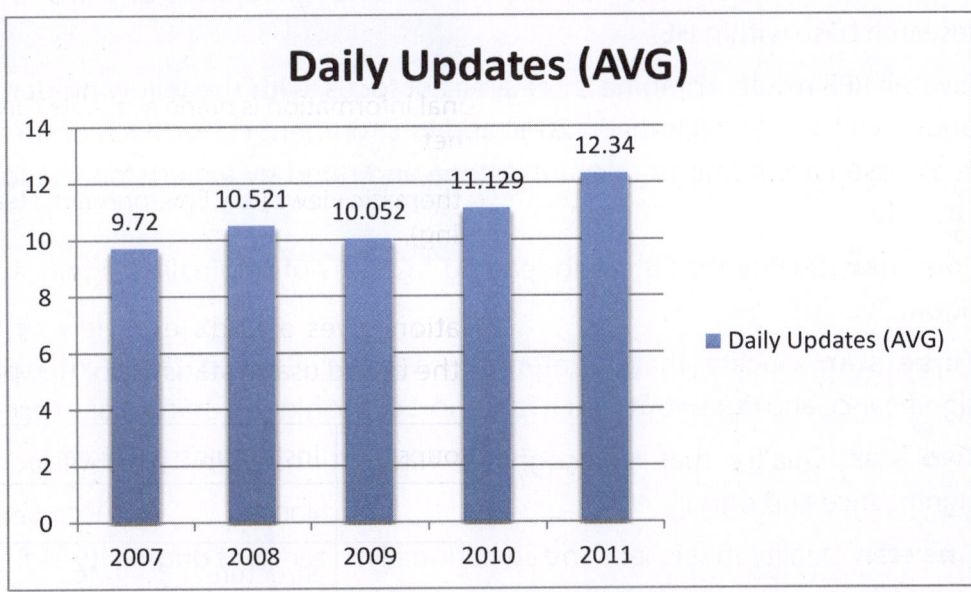

Research Excellence Framework[55] (REF) replaces the Research Assessment Exercise (RAE), as a method of assessing the research of UK higher education institutions (HEIs). It is currently planned to take place in 2014 to assess research that has taken place during the period 2008-2013 inclusive. The four UK higher education funding bodies undertake

---

[55] *http://www.ref.ac.uk/pubs/2011-02/*

the REF. The exercise is managed by the REF team based at the Higher Education Funding Council for England (HEFCE) and overseen by the REF Steering Group, consisting of representatives of the four UK funding bodies. According to an HEFCE circular[56] dated June 2007, key aims for the new framework are:

- to produce robust UK-wide indicators of research excellence for all disciplines which can be used to benchmark quality against international standards and to drive the Council's funding for research;

- to provide a basis for distributing funding primarily by reference to research excellence, and to fund excellent research in all its forms wherever it is found;

- to reduce significantly the administrative burden on institutions in comparison to the RAE;

- to avoid creating any undesirable behavioural incentives;

- to promote equality and diversity;

- to provide a stable framework for our continuing support of a world-leading research base within HE.

The overall REF results combine three areas of focus, with the following percentage contribution - Output (65%), Impact (20%) and Environment (15%). Each of these three areas are assessed according to relevant criteria and rated by experts on the following five-point scale.

- **Four star:** Quality that is world-leading in terms of originality, significance and rigour.

- **Three star:** Quality that is internationally excellent in terms of originality, significance and rigour but which falls short of the highest standards of excellence.

- **Two star:** Quality that is recognized internationally in terms of originality, significance and rigour.

- **One star:** Quality that is recognized nationally in terms of originality, significance and rigour.

- **Unclassified Quality:** that falls below the standard of nationally recognized work. Or work which does not meet the published definition of research for the purposes of this assessment.

---

[56] *http://webarchive.nationalarchives.gov.uk/20100202100434/http://www.hefce.ac.uk/pubs/circlets/2007/cl06_07/*

Based on submissions from Higher Education Institutes (HEIs), the quality of research outputs will be assessed by expert review, corroborated by citation analysis where data is available and reliable.

Impact assessment, which has attracted the most controversy, includes various types of impact parameters – economic, social, cultural, health, environmental, quality of life, public policy and services.

'Environment' parameter seeks to assess the extent to which the research environment supports the continuing flow of excellent research and its effective dissemination. The assessment is to be made based on (mostly qualitative) evidence in resourcing management and engagement. Much of the controversy and criticism of the REF has come around the nature of "impact" assessment. This caused a year's delay in the rolling out of REF and a reduction in its weighting to 20% from the 25% initially proposed.

3.2.2 Lessons Learnt

It is clear from the above deliberations that many nations around the world are keenly gauging the research scenario. Based on this, formulating requisite policies to harness the research outcomes for societal good on one side and promoting domains of research endeavours through judicious funding mechanisms for fostering R&D in core research areas.

The focus on the quality and quantity (as well) of the research and its impact is also being closely garnered in India. This is evident from the 75% increase in extramural Research and Development (R&D) Projects in the country from 3,336 in 2006-07 to 5,855 in 2010-11. The monthly emoluments of fellowships for research fellows have increased by 100% from Rs. 8,000 per month in 2007 to Rs. 16,000 per month in 2010. This was stated by the Union Minister of State for Science and Technology and Ministry of Earth Sciences (Independent Charge) Dr. Jitendra Singh in a written reply in Lok Sabha on 9 July 2014. (http://pib.nic.in/newsite/PrintRelease.aspx?relid=106310). As an indicator of development of technology and new products, the rate of commercialization of patents emanated from CSIR laboratories is above 9% while the global average is 3-4%. This was stated by Dr. Jitendra Singh, Union Minister of State for Science and Technology and Ministry of Earth Sciences (Independent Charge) in a written reply in Rajya Sabha on 17 July 2014 (http://pib.nic.in/newsite/PrintRelease.aspx?relid=106860).

Thus there is a clear trend that whether or not to measure is not the issue, but the big thing is how to measure, so that the robust system for extracting meaningful indicators can be evolved for assessing the research outcomes. This will help the in formulating a research assessment or excellence framework (RA/EF) to facilitate policy makers and funding bodies, interpreting the data for recognition and incentivizing institutions/ researchers through appropriate reward system, funding, etc.

PART 3: HEALTH OF INDIAN RESEARCH LANDSCAPE

3.3 Monitoring Global Research Landscape

The global research landscape is changing at a fast pace. As a natural consequence, the volume of (scientific) research, measuring to "know something," and recording and communicating that knowledge through publications, has itself become enormous and complex. Science research in the last few decades, evolved as a large enterprise and the substance of scientific research is so complex and specialised that personal knowledge and experience have to be rooted in data and its analysis for understanding trends or for making decisions.[57]

Global competitiveness among the community of Nations in the emerging knowledge economy of the world is often assessed on the basis of the research outputs originating from the country. The global share of publications reflected in the international databases has become one of the major output indicators for assessing the competitiveness of National R&D systems. As such national commitments to strengthen the R&D systems are partly drawn by such research findings.

Increasingly, the health of the research enterprise of various countries is regularly monitored to track the key trends in research productivity and to assess the impact in various subject domains at both national and global levels. Such research assessments require mapping of the research landscape with the aid of scientometric tools. These exercises allow appraisal of competencies/strengths in various domains, the recognition of emerging research trends and early recognition of inadequacies in specific subject domains.

---

[57] Pendlebury, David A., (2008). White paper: Using bibliometrics in evaluating research. Thomson Reuters, Philadelphia, USA.

The present study aims to gather evidence for analysing the research trends in India. While doing so, it also projects the position of India viz-a-viz other developed and emerging Nations, with the objective of assisting all stakeholders (policy-makers, academic segment and hardcore researchers) in evidence-based future planning and development of R and D.

3.3.1 Trends and Developments in R&D

Science is growing globally, which is evident from almost double spend on research and development, growth in the number of researchers, and the multifold growth in the number of publications since the beginning of the present century.[58] The social sciences domain has also shown growth in terms of the publications output. The Social Science Citation Index (SSCI) reported a fourfold increase from 1995 to 2007. Brazil, Mexico, Argentina and the Caribbean countries, and other developing countries including emerging economies (India, China) have shown significant increases. China has shown the most advances, moving from being fourth of the five countries in 1995 to the first position in 2007. Relatively India has shown the smallest increase and dropped from the first position amongst these five countries in 1995 to third, behind China and Brazil, at the end of the period.[59]

There is a paradigm shift in the global influence marked by an unprecedented rate of economic growth (1996-2007) that fuelled a spending spree in R&D between 2002 and 2007. The world GDP has risen by 43%; world expenditure on R&D by 45%; while global R&D intensity (Gross Expenditure on R&D (GERD)/GDP ratio) has remained stationary,1.7%.[60] Behind this apparent stability in the R&D establishment, a lot has changed, driven by Asia, largely attributed to China (+3.9% world share), India (+ 0.6%) and the Republic of Korea (+0.8%). Asia's share of world GERD rose from 27% to 32% between 2002 and 2007. This came at the expense of Europe, North America and Japan, the traditional leaders of science (Japan's share falling from 13.7% to 12.9%).[61]

---

[58] Spend on research and development data from UNESCO Institute for Statistics published in UNESCO Science Report 2010; Number of researchers data from UNESCO Institute for Statistics Data Centre, UNESCO Institute for Statistics, Montreal, Canada. Number of Publications Data from Thomson Reuters Web of Science.

[59] UNESCO World Social Science Report 2010, p 156.

[60] UNESCO Science Report 2010.

[61] *Ibid.*

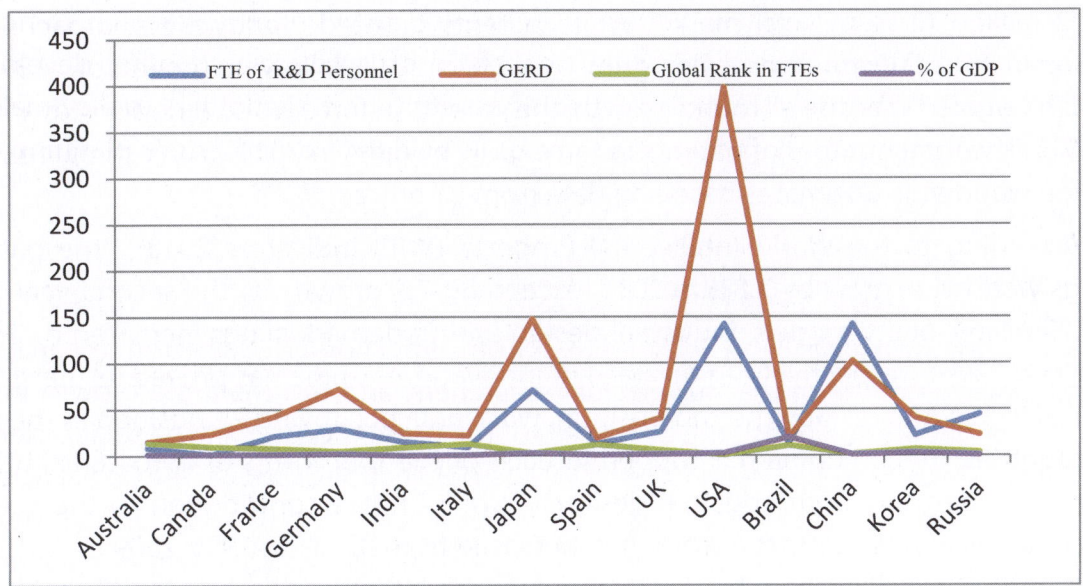

Figure 3.1 *Comparison of Investments and Number of Full Time Equivalents (FTE) of R&D Personnel for a Select Group of Nations*[62]

[Note: GERD relevant to 2007; GERD in billion dollar adjusted to PPP, FTE of R&D Personnel in Tens of Thousands, for visibility purposes]

While USA is dominating in research output with investment of US$ 400 billion per year in public and private research and development, Japan, Germany, France, UK are also relatively strong amongst the leading research nations. These five countries together are responsible for 59% of all spending on science globally.[63]

It is evident from Figure 3.1, that all countries depicted therein have better relative ranking, invest more, employ larger number of Full Time Equivalents (FTE) of R&D professionals (except Australia, Canada, Italy and Spain) compared to India.

3.3.2 Innovation vs Economic State

Innovation has been by and large judged by the patents filed. If this parameter is to be taken as a key yardstick, then in spite of the fact that the global economy continued to underperform, the Intellectual Property filings have shown an encouraging growth. For the first time in 2011, the total number of patent applications filed worldwide exceeded

---

[62] UNESCO, (2010). UNESCO Global Science Report 2010.

[63] UNESCO, (2010). Data from UNESCO Institute for Statistics, published in UNESCO Science Report 2010 (p.2, Table 1)

the 2 million (2.14 million) mark[64]. While patents granted worldwide approached 1 million in 2011.[65] Grants worldwide grew by 9.7% in 2011, following growth of 12.3% in 2010.[66] Coupled with the all round growth, the patents in force worldwide also advanced in 2011. The total number of patents in force grew by 6.9% in 2011 to an estimated 7.88 million worldwide (estimates based on data from 81 offices).[67]

According to the World Intellectual Property (WIP) indicators 2012[68] "the patent filings worldwide grew by 7.8% in 2011, exceeding 7% growth for the second year in a row. Similarly, utility models, industrial design and trademark filings increased by 35%, 16% and 13.3%, respectively. In this race of innovations, China has proved its mettle by overtaking USA to occupy the top position, with India occupying 7th position in the top 20 countries. In fact, China has topped in each of the four forms of IP (patents, utility models, trademarks and industrial designs). While China's contribution to the rise in patent applications across the world has increased from 37.2% in 1995-2009 to 72.1% in 2009-2011, the Indian share has increased over the last few years by more than 10%.[69] Though this is a positive sign but more needs to be done. According to WIP director general Francis Gurry.[70] "Sustained growth in IP filings indicates that companies continue to innovate despite weak economic conditions. This is good news as it lays the foundation for the world economy to generate growth and prosperity."

---

[64] World Intellectual Property Indicators, (2012). Available at: *http://www.wipo.int/pressroom/en/articles/ 2013/article_0025.html* (Accessed on April 5, 2013).

[65] World Intellectual Property Indicators, (2012). Available at: *http://www.wipo.int/pressroom/en/articles/ 2013/article_0025.html* (Accessed on April 5, 2013).

[66] World Intellectual Property Indicators, (2012). Available at: *http://www.wipo.int/pressroom/en/articles/ 2013/article_0025.html* (Accessed on April 5, 2013).

[67] World Intellectual Property Indicators, (2012). Available at: *http://www.wipo.int/pressroom/en/articles/ 2013/article_0025.html* (Accessed on April 5, 2013).

[68] Srivastava, Vanita (2013). China Patent Office World's Biggest, India at 7th Rank Srivastava Hindustan Times, January 1st, 2013.

[69] World Intellectual Property Indicators, (2012). Available at: *http://www.wipo.int/pressroom/en/articles/2013/ article_0025.html* (Accessed on April 5, 2013).

[70] Srivastava, Vanita (2013). China Patent Office World's Biggest, India at 7th Rank Srivastava Hindustan Times, January 1st, 2013.

Top 10 Patent Filers [71]

1.	China	526,412
2.	US	503,583
3.	Japan	342,610
4.	Republic of Korea	178,924
5.	European Patent Office	142,793
6.	Germany	59,444
7.	India	42.291
8.	Russian Federation	41,414
9.	Canada	35,111
10.	Australia	25,526

IP Fillings by Office and Income Group [72]

Office and Income Group	Share in World total (%)						Average Annual Growth(%)		
	2008	2011	2008	2011	2008	2011			
	Patents		Marks (Class count)		Designs (design count)		Patents	Marks	Designs
China	15.1	24.6	12.8	22.8	43.6	53.1	22.0	26.6	18.6
European Patent Office	7.6	6.7	n.a.	n.a.	n.a.	n.a.	-0.8	n.a.	n.a.
Japan	20.4	16.0	3.7	3.0	4.7	3.1	-4.3	-2.1	-2.8
OHIM	n.a.	n.a.	4.6	4.9	11.3	8.9	n.a.	6.7	2.4

---

[71] Srivastava, Vanita (2013). China Patent Office World's Biggest, India at 7th Rank Srivastava Hindustan Times, January 1st, 2013.
[72] WIPO Statistics Database, October 2012.

Republic of Korea	8.9	8.4	3.7	2.8	8.2	6.0	1.6	-4.8	-0.2
United States of America	23.8	23.5	7.3	6.6	3.9	3.1	3.3	0.9	3.1
World	100.0	100.0	100.0	100.0	100.0	100.0	3.8	4.3	11.0
High –income	74.8	67.0	52.8	45.1	44.9	37.2	-0.3	-1.0	4.2
Upper middle-income	22.2	29.8	35.5	43.9	52.0	59.5	14.2	12.1	16.0
Lower middle-income	3.0	3.2	10.4	9.9	2.8	3.1	5.2	2.7	15.9
Low-income	0.1	0.0	1.3	1.0	0.3	0.2	-38.5	-2.4	-7.4
World	100.0	100.0	100.0	100.0	100.0	100.0	3.8	4.3	11.0

[Note: OHIM= office for Harmonization in the Internal Market; trademark data refer to class counts, i.e. the number of classes specified in applications. Industrial design data refer to design counts, i.e., the number of designs contained in applications; n.a.=not applicable]

3.3.3 Trends in Publication Output: Global Share of India

The USA leads the world in research producing 23.69%[73] of the world's authorship of research papers, dominating world university league tables, Germany, Japan, UK, and France individually also command strong positions in the global league tables, producing high quality publications and attracting researchers to their world class universities and research institutes.

However, the traditional scientific leaders have gradually lost their relative share of published articles. Meanwhile, China has increased its publications to the extent that now it is the second highest producer of research output in the world.

The data analysis[74] merits careful readings; there is a great deal of useful and interesting information. The statistics (10 years data) projects noticeable trends, especially with regard to China-India comparisons. In terms of total published articles, China (996,935) is in the second position with 7.38% of the world share, behind the

---

[73] UNESCO Science Report, (2010).
[74] Based on Thomson Reuters database – Essential Science Indicators – 10 years database (January 1, 2002–June, 2012).

United States (3,199,249), while India (327,924) 2.42% is in the eleventh place, trailing behind Germany, Japan, England, France, Canada, Italy, Spain and Australia.

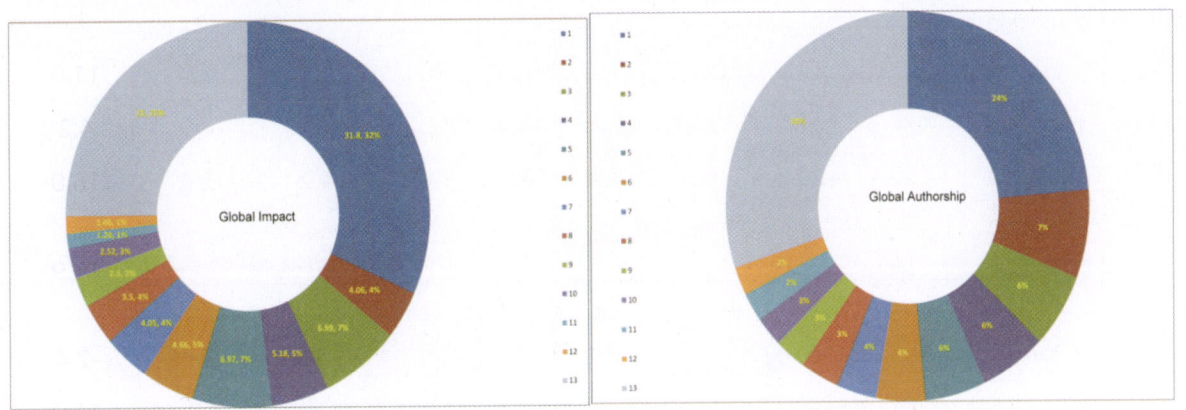

Key: 1. USA   2. China   3. Germany   4. Japan   5. England   6. France   7. Canada   8. Italy   9. Spain   10. Australia   11. India   12. South   Korea   13. Others

Figure 3.2 *Proportion of Global Publication & Impact Authorship by Country* [75]

As evident from Figure 3.2, only 12 countries account for 70% of the proportion of the global publication authorship, and in terms of impact represented by the proportion of global citations received by the country. These 12 countries shows supremacy by aggregating about 75% of the total citations. A large number of countries are not producing substantial number of papers, which may either be attributed to the small size of the country or relatively limited R&D endeavours. The statistics shows (Table 3.1), around 20 countries have greater than 1,50,000 of research publications. The break-up shows that 5 countries fall between 1,00,000-1,50,000; 13 countries between 50,000-1,00,000; 24 countries between 10,000-50,000; 12 between 5000-10000; 43 between 1000-5000, while 30 countries are with less than 1000 papers for a 10 year block period (January 1, 2002 – June 2012). This portrays that only a few nations contribute substantially to the map of R&D landscape. In the top 20 countries (Figure 3.3, Table 3.2), India ranks 11th in terms of the number of papers.

[75] Based on Thomson Reuters database – Essential Science Indicators – 10 years database (January 1, 2002–June, 2012)

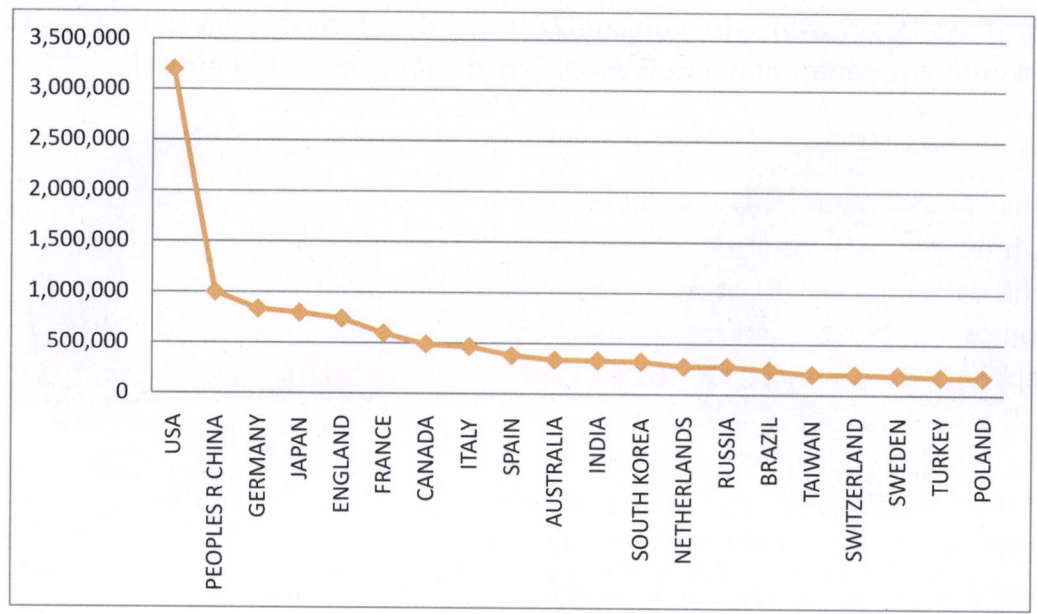

Figure 3.3 *Proportion of Global Publication : Top 20 Countries [Based on Publication output]*

India has replaced the Russian Federation in the top ten during 1996-2004[76], climbing from the 13th rank during the period to 11th rank between 2002 and 2012, besides moving ahead of Netherlands. Further down in the list are South Korea, Brazil, Turkey, South East Asian nations such as Singapore, Thailand and Malaysia and some European countries like Switzerland and Sweden.

As far as innovations are concerned, India has shown an upward trend in patent filing marking its strong presence amongst the innovatively developing countries. This is evident from the recent National Intellectual Property Organisation (NIPO) study[77]. According to the study "the number of patents issued for new pharma products in India have jumped more than three-fold from 765 in 2004-05 to 2,373 in 2008-09 after the Patents Act was introduced in the country. The total number of patents for companies across all segments has leaped from 1,911 to a whopping 18,230 during the period".

The study titled 'Patent Protection and Innovation' also points out that in 2004-05, as many as 40 % of the patents issued went to Indian companies and 60 % were given to foreign companies. In 2007-08, the number of patents issued to the Indian companies fell

[76] University Grants Commission, (2005). Research Handbook UGC, p63.
[77] James, T.C. Director, NIPO. NIPO Study Report. Available at: *http://www.nipoonline.org/Section-report.doc.*

to 21 % while those given to foreign companies went up to 79 %. This is a clear indication that the Indian Patents Act is a welcoming innovation for the nation.

3.3.31 Assessing Research Quality & Impact: Quantitative Vs Qualitative Parameters

In terms of the growth rate of publications, China outstrips India and shows a meteoric trend, however, both countries show a much higher growth rate in published papers than the developed world. In terms of publications (Figure 3.4), India occupies the 11th position on the global publication map, 16th in terms of citations. In terms of citations per paper (CPP) its rank is 112th (6.08 CPP) which needs notional reflection, with China on 106th rank (6.44 CPP). Based on this parameter, small countries like Bermuda (23.21), Gambia (16.59), and Switzerland (16.08). Panama (16.01), Iceland (15.39), Denmark (15.22) are on the top, with even the United States (15.15) trailing behind. It would therefore be inappropriate, to use overall statistics to compare the performance of small, low producing nations with that of highly prolific countries with large research domain base.

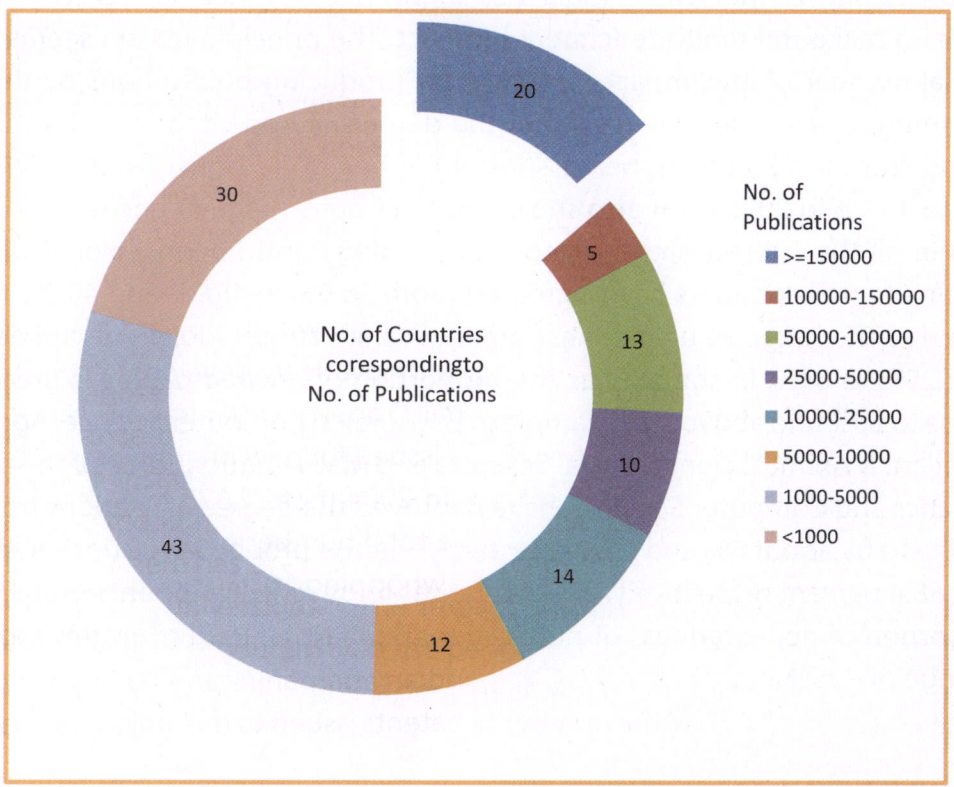

Figure 3.4 *Number of Top Publishing Nations*

This may not be a true reflection of the impact, since for evidence based metrics one needs to make comparisons amongst similarly sized research base. As reflected in Figure 3.5, only 20 countries contribute to 82.65% of the world's proportion of global publications with greater than 150000 publications each produced during a ten year period (January 1, 2002 - June 2012). This means on an average a minimum of 15000 papers have been produced per year by each of these 20 countries. Therefore only the top publishing nations have been considered here for the purposes of comparison to decipher the impact that the publications of these nations have made on the `impact geography' of research output. While China is far ahead in quantitative parameters, there is one parameter in which India appears at par with China - a reflection of `quality', viz., citations per paper (CPP). India scores 6.08, while China is at 6.44, leaving behind Russia and Turkey (4.93 & 5.51 respectively) in the top 20 publication producing nations.

3.3.32 *Impact Analysis (Uncitedness)*

This statistics shows that while India has improved in numbers, yet we have not been able to make the requisite (citation) impact. The principle reason seems to be, a substantial number of low impact or no impact producing publications by the Indian R&D community. The study[78] reveals that the degree of non-citedness of publications emanating from India despite being substantial has gone down from 51.9% during 2001-05 to 44.9% in 2006-10. It also shows that this percentage of uncited papers have dropped in all the subject areas, except Microbiology and Pharmacology, where the proportion of uncited papers have increased (from 36.9% to 45.9% and 40.9% to 42.1% respectively). The range of uncitedness amongst the various subject disciplines varies between 25% to 64%. In some areas the proportion of uncited papers is greater than 50% (close to 60% and above), for example in Economics, Computer Science, Agricultural Science, Plant & Animal Science, Social Sciences and Mathematics. Some of the areas like Mathematics and Computer Sciences, have improved the degree of citations from 2001-05 to 2006-10 by about 6% and 9% respectively, yet the proportion of uncited papers is 63.8 and 58.4 percent respectively, which requires to be reflected upon. Contrary to this, the proportion of non-citedness of publications from developed countries generally is in the range of ~25%.

---

[78] Department of Science & Technology, Government of India, (July 2012). Evidence Report of Thomson Reuters. Bibliometric Study of India's Scientific Publication Outputs during 2001-2010.

In general, there is a relatively high level of uncitedness, and also proportionately more research falls into impact categories where citation impact is below world average.

3.3.33 *Voicing Concerns*

India needs to critically examine the research output with low impact or no impact figures, strengthening the high impact research patches and developing reward strategies for low impact research areas. Efforts are required to further reduce the proportion of un-cited papers and make it comparable to those of the developed countries, if not less. This would call for taking strategic initiatives by the encouragement of publishing in high impact journals and thereby increase visibility, as well as discourage contributions to low or no impact journals which increases the total count of papers but lowers the quality of the research enterprise. Publishing in high impact journals requires processes that encourage and nurture creativity in academic and R&D institutions and discourage the 'numbers' syndrome. A parallel valuable step would be to increase the impact profile of Indian journals by applying stringent measures and restricting the proliferation of local and home journals in our institutions and universities.

A study of journal profile of un-cited publications would enable advisories to authors in journal selection.

This could be coupled with drawing up of a master list of journals by subject experts in each subject domain with impact analysis for consideration by the prospective authors. The anomaly that is prevalent in the higher education institutions regarding the gradation of journals in terms of assigning weightage to each journal publishing the institution's papers could be thus checked. Besides, undoubtedly, such an activity would facilitate somewhat standard list of quality journals to be considered by the prospective authors for publishing across institutions that could act as a benchmark for assessment purposes.

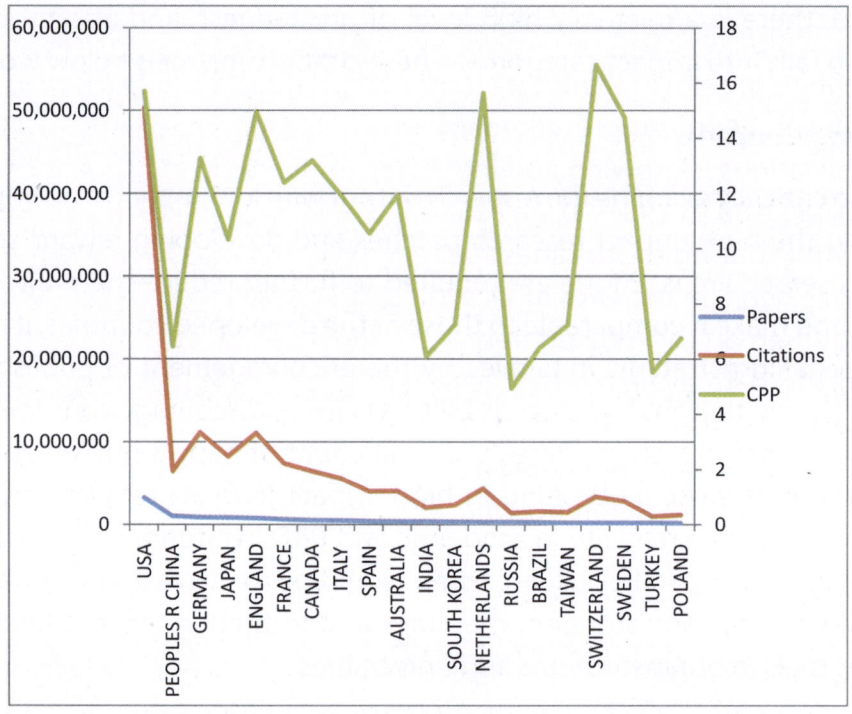

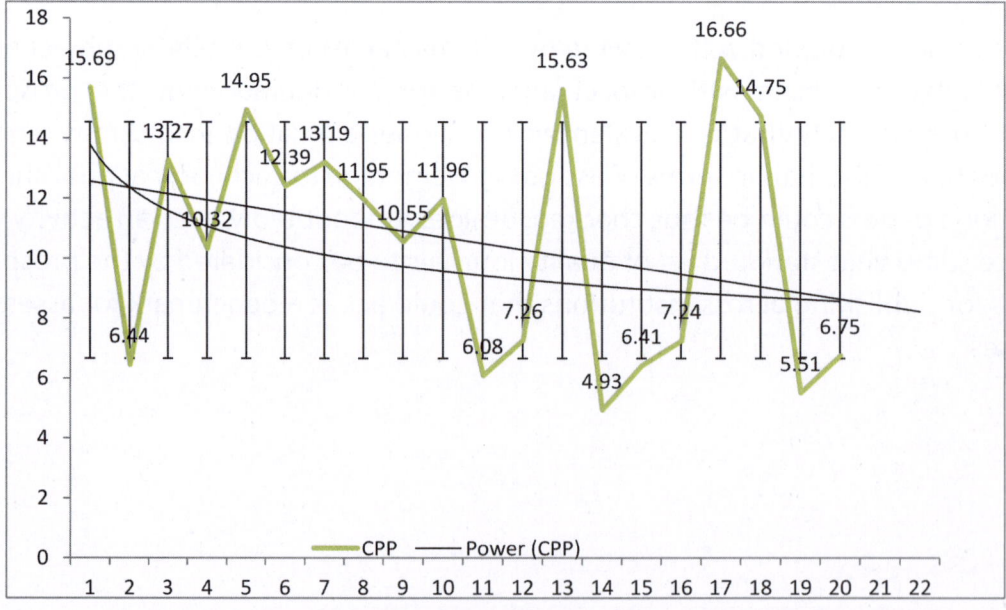

Figure 3.5 *Proportion of Global Publication, Citations and Citations per Paper (CPP): Top 20 Countries*

3.3.34 *Global Disciplinary Research Profile: Case for India*

There is a definite growth in the total number of publications from the 1970s to 2000s, as is evident from the trends presented in Figure 3.6 and Table 3.3. The volume share of research publications is growing and India's recent increase in publications is striking. Moving from 1971-80 to 1981-90, an estimated 42.87% increase in publication output is observed, while during the decades 1981-1990 to 1991-2000 and 1991-2000 to 2001-2010, there has been an increase of 24.96% and 51.48%respectively in the number of publications. Based on the moving trends in proportion of growth during the past few decades and the data available for the years 2011 and 2012, the future projections for India (Figure 3.6 (b)) for the decade 2010-2020 may increase by 38.70%. If this trajectory continues, then India's productivity will be on par or even may overtake most G8 nations by 2020. References have been made in the science policy literature to India's potential as 'sleeping giant'. This would require planning of research policy and preferential support to those disciplines/areas that have delivered high impact research in the past, and the seeding of research consortia in emerging inter and trans disciplinary research areas.

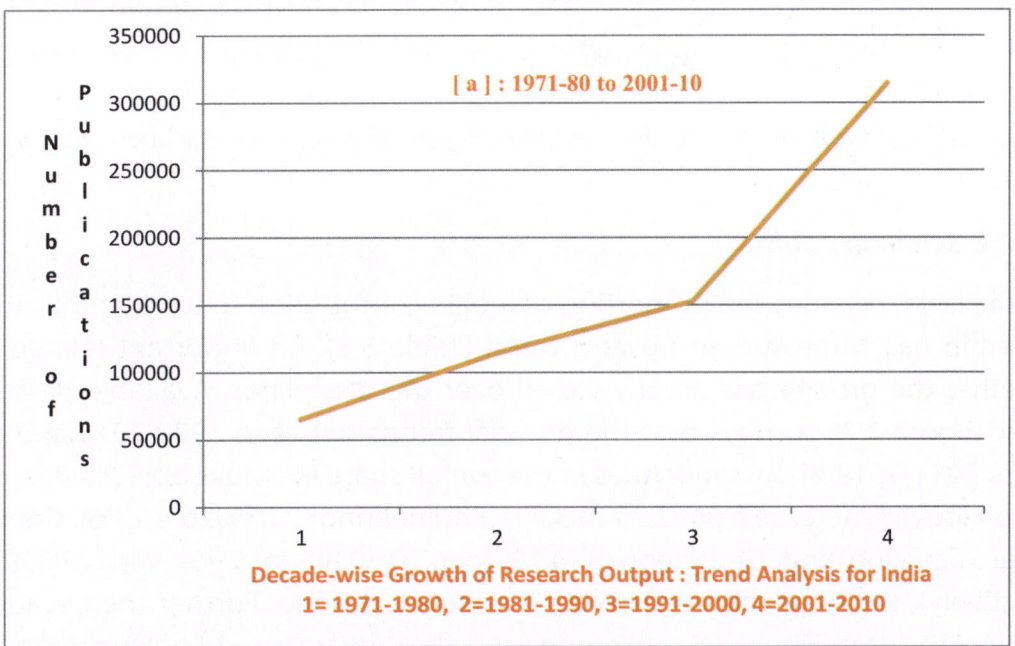

Decade-wise Growth of Research Output : Trend Analysis for India
1= 1971-1980, 2=1981-1990, 3=1991-2000, 4=2001-2010

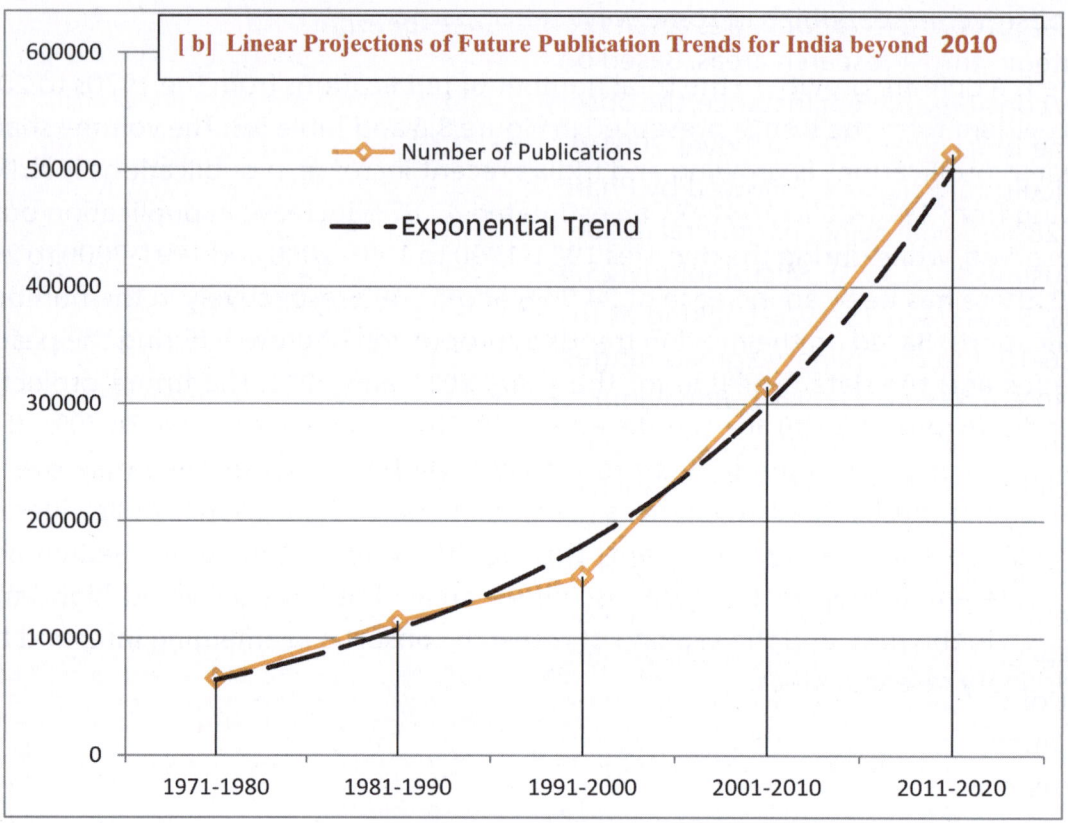

Figure 3.6 *Trends Analysis: Decade-wise Changing Trends in Indian Publication Contribution*

3.3.35 *Disciplinary Shifts*

Over the last 5 decades the proportion of India's contribution to world publication by authorship has witnessed an upward trend (Table 3.3). An important related query is whether the growth has greatly varied over the disciplines. Available data (Table 3.4 and Figure 3.7) portrays that the decadal trends between 1970-2010 and recent data for 2011-2012, show similarities in the overall share in many subject areas, except Microbiology (%increase from 2.25 to 5.47), Engineering (2.83% to 4.13%), Chemistry, Clinical Medicine and Geosciences (4.18 to 6.20; 0.89 to 1.70; and 2.90 to 3.74 respectively), with Agriculture showing a downward trend. Further the decade-wise analysis of the data (Figure 3.8a&b) emphasises the upward trend. Looking more closely at the data for the recent years for inter-annual comparisons (Table 3.5; Figure 3.9a&b),

we observe trends, which in some ways reflect either emerging or more appropriately strategic/thrust research areas. Based on these data,[79] the disciplinary spread seems to vary considerably. For instance, the observed values show increase in various disciplines in the block years 2006-10 over 2001-05 from 0.79 in Computer Science to 34.75% in Agricultural Sciences, followed by Pharmacology & Toxicology (31.99%), Mathematics (30.28%), Social Sciences general (27.86%), Environment/Ecology (27.80%), Economics & Business (26.96%), Microbiology(25.48%) while 7 subjects fall in the range of 20-25%, 5 between 15-19%. Looking at the scenario of expected values of 2011-12 over 2006-10, the disciplinary spread ranges between 3.82% in Biology & Biochemistry to 24.30% in Neuroscience & Behaviour. The spread is not as increasingly divergent in 2006-10 five year block over the previous one. There are not glaring variations across disciplines, except in a few cases, such as Biology & Biochemistry, Geosciences, Physics, etc. (Figure 3.10 a&b). However, comparisons across two blocks, shows that in the later block, while there is an upward trend in values (%) for the fields like Computer Science, Multidisciplinary, and Neuroscience & Behaviour ; fields like Space Science, Molecular Biology & Genetics have almost comparable proportion across two blocks; rest of the fields show downward trend with Geosciences having substantially larger variation. Perhaps the expected values based on a five year period block may be a better comparison.

---

[79] Based on the Thomson Reuters Data from Web of Science – deciphering percentage increase for the block period 2006-2010 from 2001-2005 and based on average of 5 years data (2006-2010) multiplied by 1.5 years to project the percentage increase or decrease in 2011 and 2012 (up to June 2012 for one and a half year period).

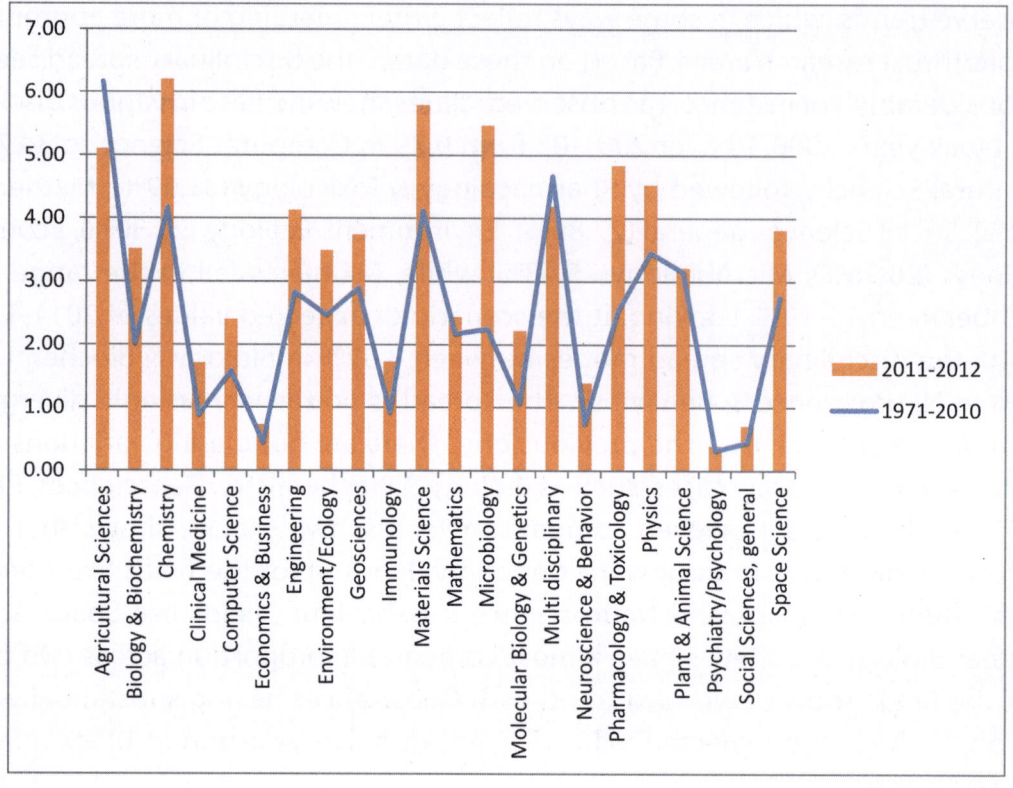

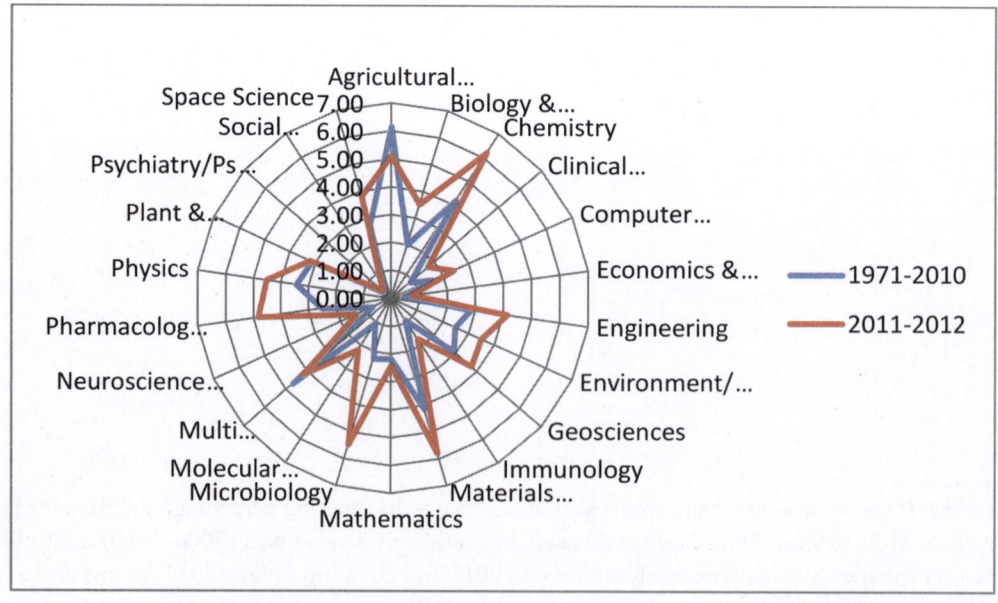

Figure 3.7 *Discipline-wise Share of Indian Publications viz-a-viz World Output*

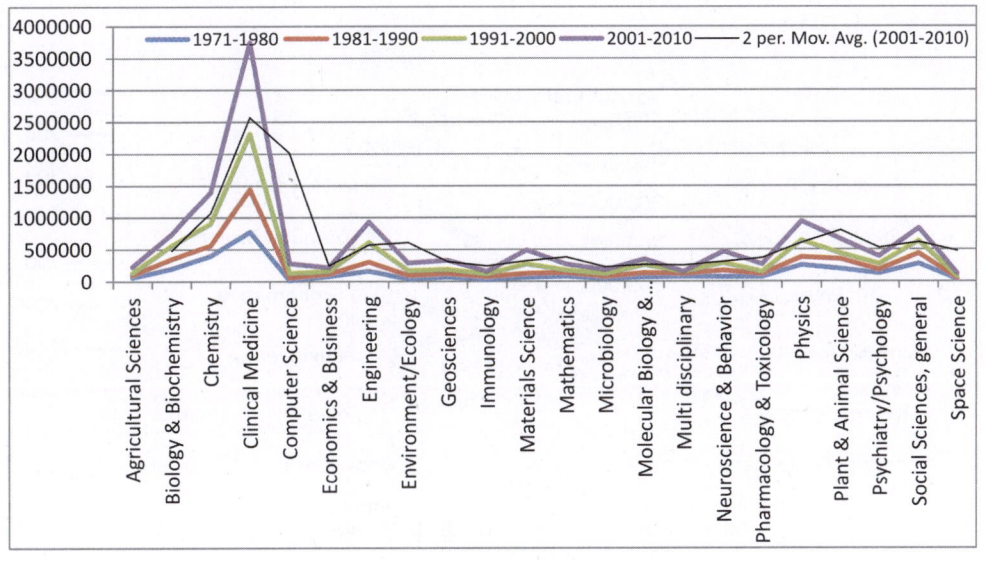

(a) *World*

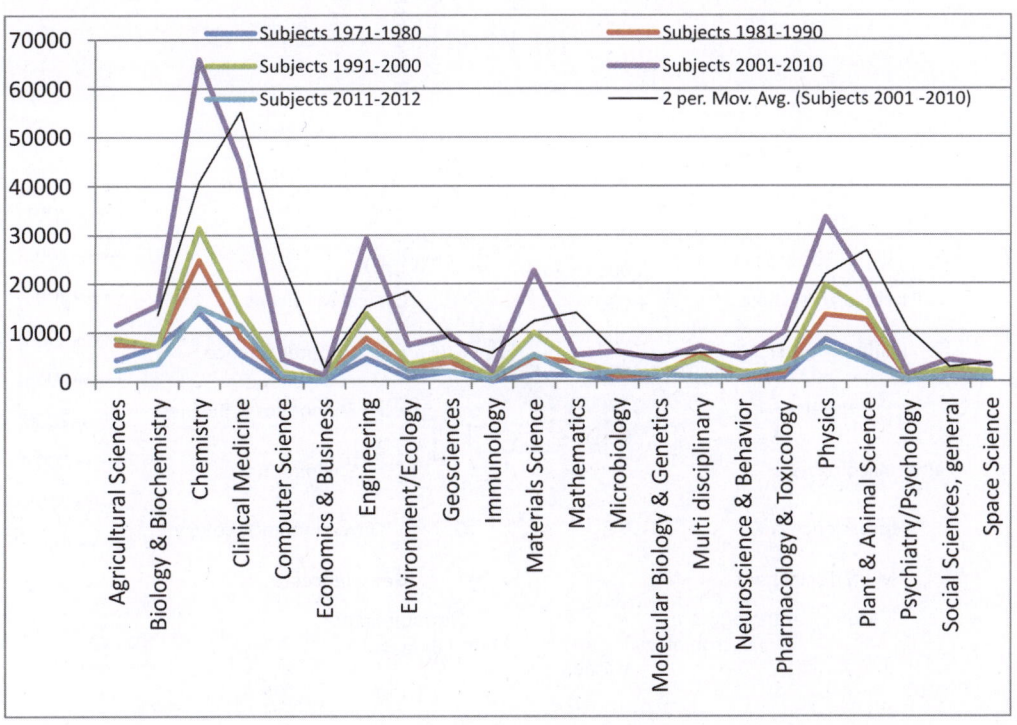

(b) *India*

Figure 3.8 *Trends in Global Research landscape Profile in Various Subject Domains and India's Share : Decade-wise Scenario*

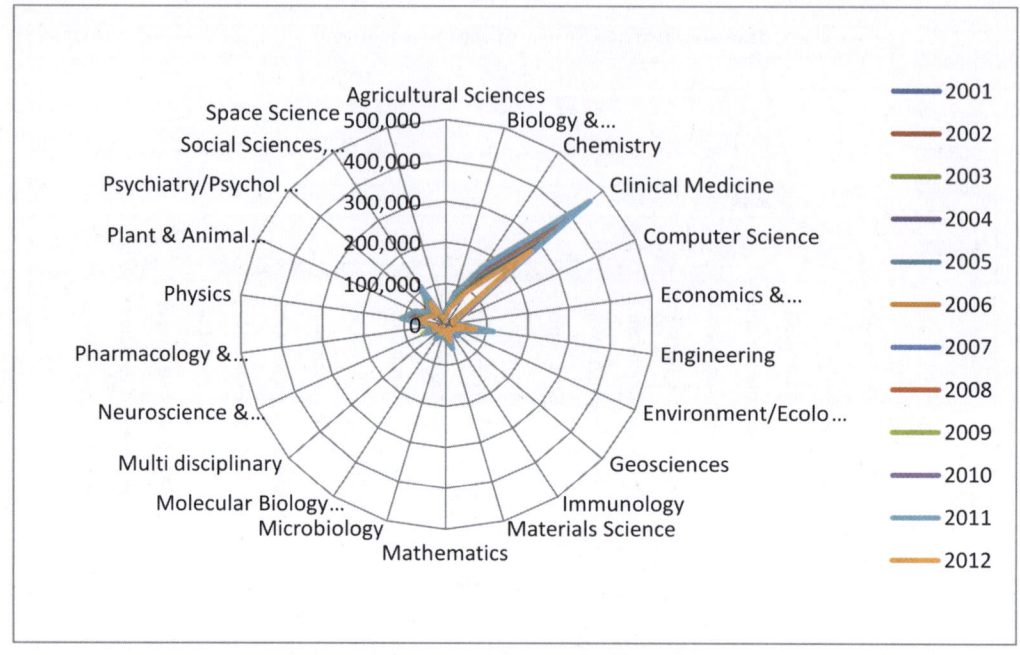

(a) *World*

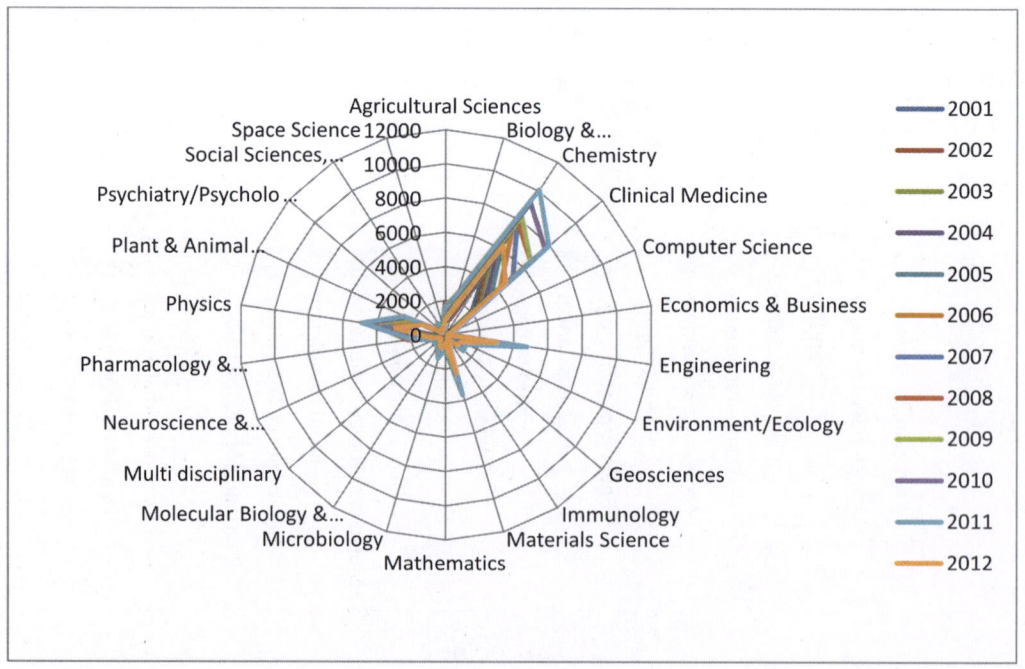

(b) *India*

Figure 3.9 *Year-wise Trends in Disciplinary Shifts in the Recent Years*

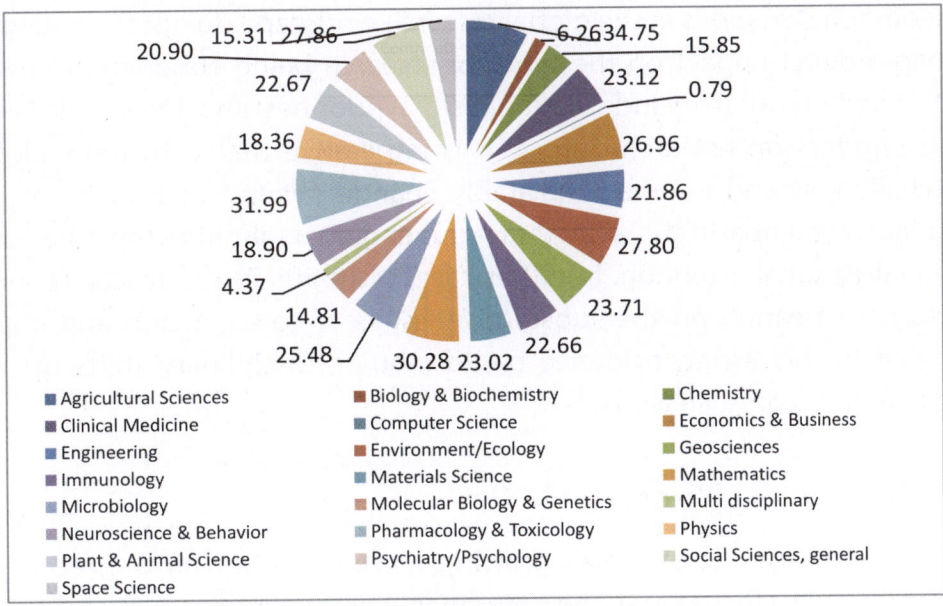

(a) *From 2001-05 to 2006-10 (Observed Values)*

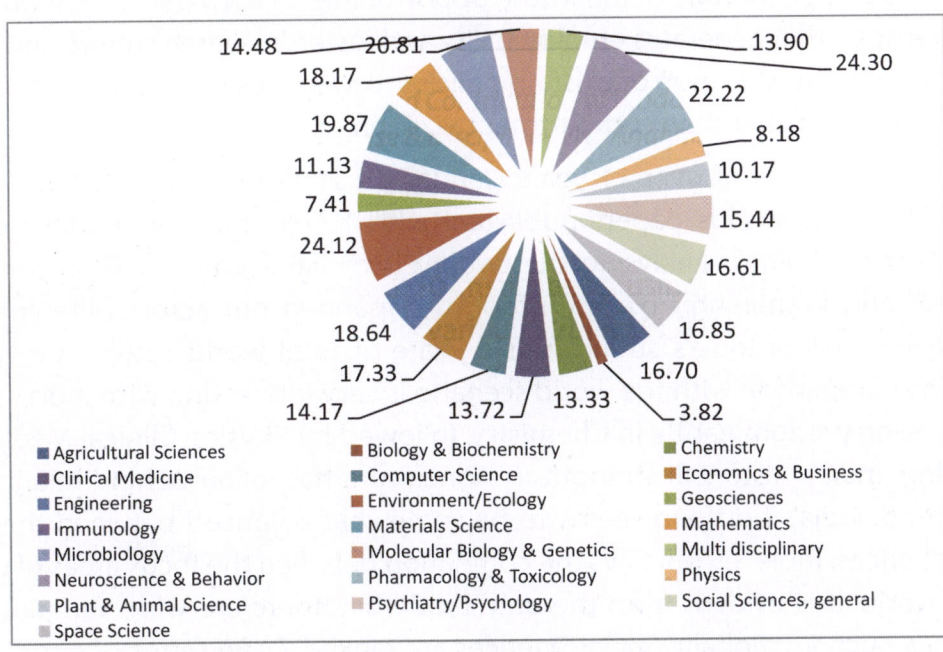

(b) *From 2006-10 to 2011-12 (Expected Values)*

Figure 3.10 *Disciplinary Spread : Proportion of %tage Increase*

The geographic changes in the global research landscape do not themselves appear to have had a direct impact on the types of research being conducted. However the domestic conditions that include national/state priorities and resources might have distinctive imprints on research output. It is imperative that with time, globally the research challenges and interests inevitably change. While some such changes that prioritise issues are global in character that requires global solutions, the others may need local/regional/national attention. In either case, the shifts in the research endeavours would also have bearing on the substantive issues of core concern and importance, besides other factors. Using evidence based metrics, disciplinary shifts by and large endeavours to focus on such issues.

3.3.4 Research Footprint: Global Impact

Mapping of the footprints of India's growing research output in the recent decades includes scrutiny of its focus areas and determining areas of research strength/weakness as compared with other countries. These issues are important to identify core research areas so as to take advantage of the future opportunities in such domains. Innovation in higher education that generates employability and fosters research culture and research synergies requires re-direction of existing policies towards research networks that foster inter-disciplinarity rather than compartmentalization.

In the recent block of over 11 years (January 1, 2002-June 30, 2012) India's share is about 3% of the worlds papers published with 1.77% share of citations captured in all subject disciplines. Analysing the twenty two major areas in Thomson Reuters databases[80] and examining how the proportion panned out across different subject areas, the scenario of India's strength as a share of total world activity proves to be diverse. The comparison with the world scenario is very interesting with India's research footprint being predominantly in Chemistry, followed by Physics, Clinical Medicine and Engineering. India's historical strength in Agriculture has fallen behind relative to the earlier period. India's portfolio seems to be somewhat balanced between the life and physical sciences. Table 3.5 portrays the correlation between the focus areas of India viz-a-viz the world. It is evident from the data that while there are wide variations in the thrust areas, such as, globally Social Sciences are ranked 7th (in terms of density) in the world while data for India shows it at 17th place. Similarly, a few areas such as Engineering

---

[80] Thomson Reuters. Essential Science Indicators data for the years January 1 2002 to June 30, 2012 publications published in journals indexed by Thomson Reuters.

(4), Geology (10), Environment /Ecology (11), Mathematics (13), Computer Science (14) have a similar rank at the national and global levels. However, by tracking the numerous subfields[81] at the micro level instead of broad subject category classification, one can fine tune the analysis to examine more closely the specific fields in which Indian research is focused. Such an analysis could facilitate decisions regarding guiding investments in higher education and research and also sensitise the research community at large regarding visibility of their research endeavours.

The correlation coefficient between India and World rank in various research focus areas is 0.73, which implies that both India and the world have similarity in the relative rankings and importance of different research disciplines.

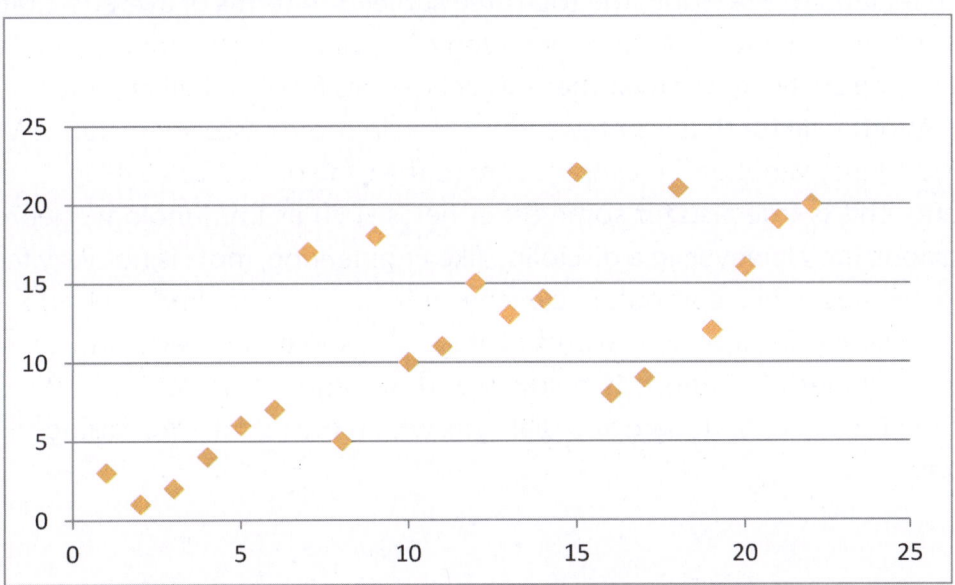

Figure 3.11 *Scatter Diagram Depicting a Positive Correlation Between India and the World in Various Subject Domains*

3.3.41 *Field Weighted Relative Average Impact (FWRAI)*

With output from India showing an upward trend, the question remains as to whether research quality has been able to keep up? The citations and Citations per paper (CPP) do throw light on this question. For instance, the average citations for India as a country is

---

[81] Field at the micro level by reorganising the 22 broad subject categories in the 250 categories, classified by the Web of Science.

6.08 (world average being 11) and this accounts for 1.77 % share of total citations received by papers in various subject disciplines.[82] However it will not be fair to compare, since the figures are not corrected for different citation levels from discipline to discipline. Thus one cannot compare an institute in Medical Sciences with that of Engineering Sciences or department of Oncology with department of International Studies. The data on FWRAI do permit such a comparison.

The FWRAI for various subject disciplines indicates that (Table 3.6 & Figure 3.12) there is diversity, average citations per paper across disciplines are ranging from around 3.5 to 23.5 and 2 to 9 for the world and India respectively. It is interesting to note that, though the relative impact has high variations across disciplines while assessing India's average in relation to the world, the top three subjects in terms of average citations per paper are in the same order - Molecular Biology & Genetics, Immunology, and Biology & Biochemistry respectively. For the other subjects it does follow a similar pattern with little variation. Another factor that is noteworthy is that in some cases, for instance Molecular Biology & Genetics variations in averages are to the tune of 9.05 to 23.48 in case of India & the world. This is true also for some other fields such as Immunology, Neuroscience and Behaviour, etc. However, in a discipline like Engineering, India is not very far behind the world average, while computer science is next in line after Engineering in relation to closer proximity for attaining world average threshold. It is revealing to note that the Indian averages of citations is below world averages in all subject categories, re-emphasising the need for debate and dialogue within the country for raising the quality of research.

---

[82] Based on the average of the data for the years January 1, 2002 to June 30, 2012 of Thomson Reuters Essential Science Indicators.

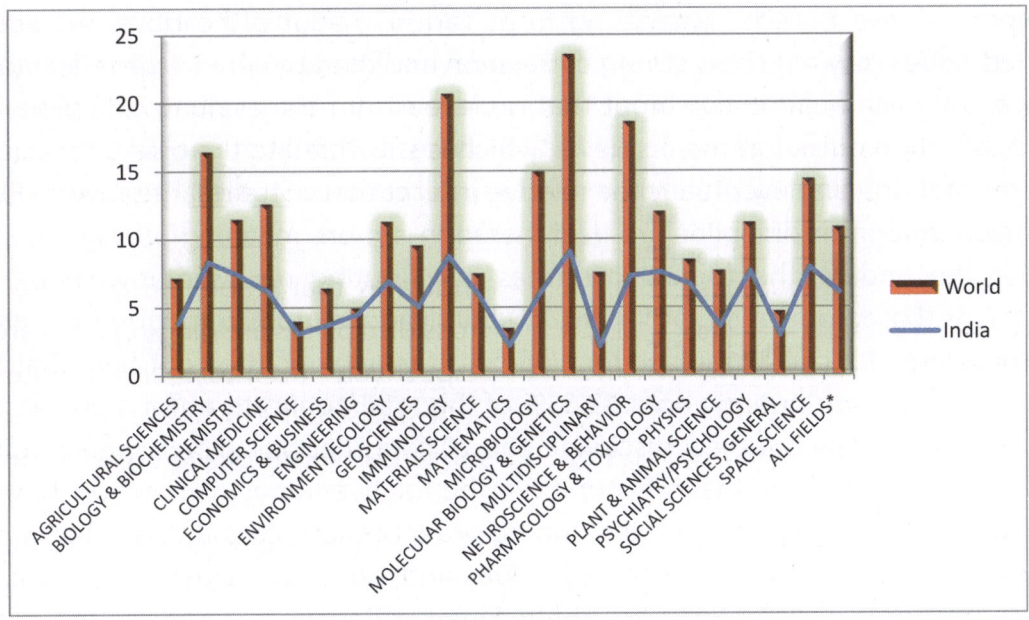

Figure 3.12 *Field Weighted Relative Average Impact (FWRAI)[83]*

The fields of Engineering and Computer Sciences show relatively less variations so far as the Average citation rates (Table 3.7) are concerned, while large variations are observed as in all other disciplines as one moves up from 2002.

A further impact analysis based on the top scoring nations and India's relative position to these nations reveals that while India is 11th in the publications (Table 3.8),[84] (out of 147 countries), the relative impact does not match in any way this number of quantitative research output. In order to focus on impact analysis of the publications across various countries, the quantitative parameters need to be of representative sample size. Since the quantifiable tangible entities are the ones subjected to impact study, unless there

---

[83] The Average Citation Rates table displays data on the average citation rates of papers within the scientific fields over each of the past 10 years. The calculation is number of citations / number of papers, where papers is defined as regular scientific articles, review articles, proceedings papers, and research notes. The Average Citation Rates table lists all the scientific fields in rows, as well as an All Fields row combining all data. Data for the average citation rates are given for the last 10 years with the All Years column using all data for those years. A value of 3.16, for instance, in the field of Chemistry in 2001 indicates that the average paper in that year has since been cited 3.16 times. All papers from all the journals covered in the product are used in the average citation rates calculation -- not just the papers chosen for inclusion in Essential Science Indicators®.

[84] Data from Thomson Reuters - Essential Science Indicators® for a 10 year period from January 2002 to Jun 2012.

is a representative number emanating from various nations, the observed and the expected values may not show strong correlation and thereby give a true reflection. For instance, the main quantitative input that is the basis for the evaluative bibliometrics, is the publication output at the first level, which is subjected to the qualitative analysis based on citations, and eventually the relative impact that these quantitative measures have made in a given discipline are deciphered in a more meaningful way. Hence for studying the impact, the top 20 countries contributing representative number of publications (based on cut off, of >=150,000 publications produced during the period) were considered for overall ranking for all fields. It may be pointed out here that these 20 countries are contributing about 82.65% of the total publications for a period of ten years. While within each subject discipline, top 20 countries based on papers published have been considered for analysis purposes. The relative impact is however based on the average citations per paper in descending order of these 20 countries, that projects somewhat true picture as the observed values and expected values does apparently show correlation. Thus the Field Weighted Relative Impact is based on the average citation per paper in each subject discipline, based on which the top 20 countries have been identified.

3.3.42 *Disciplinary World Share of Output and Citation Impact*

While charting qualitative parameters (citation impact) with the quantitative ones (publication count), interesting trends are being observed. With regard to the publications and Citations in relation to the overall ranking in all countries, India's position is 11 and 16 respectively, and in terms of average citations per paper (CPP), India's rank trails down to 111 (Table 3.9 a&b and Figure 3.13). However what is interesting to note that Switzerland which is 3rd in terms of CPP taking all the countries into consideration, is ranked 17th way behind India in terms of number of publications, but in citations it ranks 15th, just above India.. In top 20 countries based on CPP, Switzerland is on the top followed by USA, Netherlands, England, Sweden, Germany with India holding 18th position. India with 6.08 CPP, is slightly behind China (6.44) and Brazil (6.41), however it is ahead of Turkey (5.51) and Russia (4.93).

In the 22 subject disciplines, the comparisons have been made to determine the position of India viz-a-viz the world on two counts; firstly, to figure out what is the rank in terms of the publications, citations and CPP on a much wider scale taking all the countries (where all Countries including least producing countries that are listed in Thomson Reuters Essential Science Indicators data for Jan 2002 - June 2012) into consideration.

Simultaneously, the top twenty countries in each discipline were also ranked on the basis of number of publications in given disciplines. It may be noted here that in most of the subject disciplines, India has figured in the top 20 countries, but in 3 areas such as, Psychiatry/Psychology, Economics and the Social Sciences General? India's position is 35, 29 & 26 respectively, in the number of papers. Hence, while the rank in terms of Citation and CPP are based on the top 20 for the top 3 countries, for India the relative rank (35, 29 & 26) have been considered in each of these 3 disciplines respectively.

In the overall field weighted relative impact for India in various fields (Table 3.9 c), while comparing with all the countries in terms of number of publications, India's position is fairly sound with 3 fields in top 5 positions, 5 fields each in top 10, and > 10 positions and 3 & 1 fields in top 20s, and 30s respectively. So far as total citations in each discipline is concerned India's ranking scenario show 4 fields in top 10 positions, 8 fields between 11-19th position, and 8 and 2 fields occupying positions in 20s and 30s respectively. However in terms of CPP, the position is not very encouraging, the ranking spread of 22 fields is between 30s and 90s. For instance while only 1 field is having rank in 30s, 2 each in 40s & 90s, 3 in 60s, 4 in 70s, and 5 each in 50s and 80s. This means India has to strategise its policies that further foster quality research.

Yet, in another case field weighted relative impact based on top 20 highly research producing countries have also been analysed and the data has been projected (Table3.9 d) for top 3 countries viz-a-viz India's position.

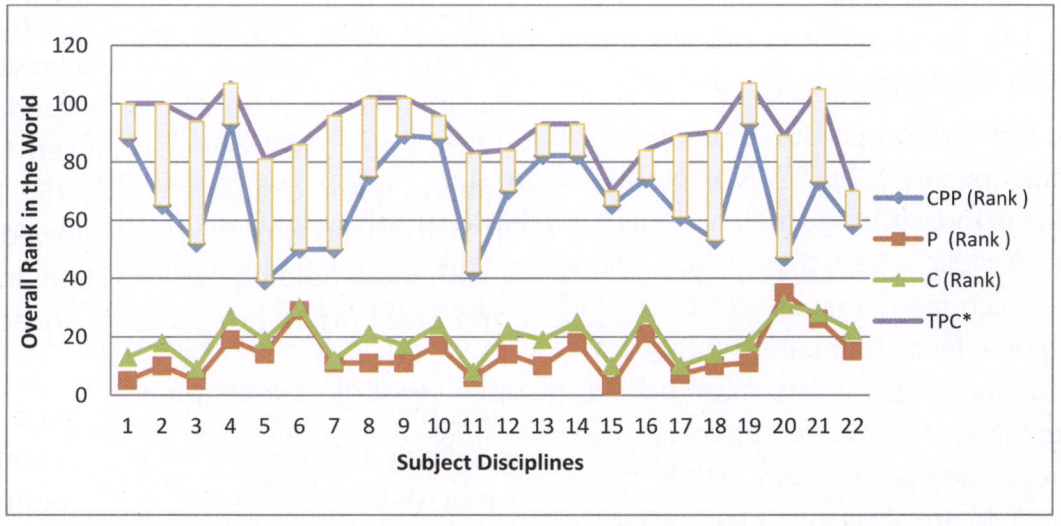

Figure 3.13 *Overall Field Weighted Relative Impact for India in Different Fields Viz -a-Viz All Countries*

[*TPC : Total participating countries]

3.3.43 *Relative Specialization Metrics*

In the research landscape, it is obvious that certain areas are more vibrant in contributing to the research landscape. Looking at the disciplinary spread, in terms of relative ranking of disciplines, the number of fractionalised publications contributed in India in general and in top 1%, it is observed that over 60% contribution comes from 5 disciplines (Chemistry, Physics, Clinical Medicine, Engineering, & Materials Science) and in top 1% only 4 disciplines (Engineering, Physics, Chemistry, and Clinical Medicine) -contribute to over 60% of the total proportion of publications respectively. (Table 3.10)

3.3.431 *Relative Specialization Index (RSI)*

The relative specialization indices for India viz-a-viz world scenario reveals a very interesting and revealing phenomenon. The picture that emerges by looking at the top subject disciplines in India and that of the world (Table 3.11 & 3.12) shows that there is a skewed correlation observed between India and the world so far as the relative subject specialization index is concerned. For instance, while the top 4 specialised areas in both the cases (India and rest of the world) are same with variation in rank, there are some areas having the similar rank, while a few areas showed disparities in order. Disciplines like Agricultural Sciences, Pharmacology and Toxicology, greatly vary in order of ranks. While for India these disciplines have a higher position than the world (8:16 & 9:17 respectively). For disciplines like Social Sciences (general) and Psychiatry/ Psychology India's rank order figure lower (17:7 & 22:15 respectively) then the world specialization order. This does point towards the important aspect of the core disciplines of Indian research landscape.

In terms of proportion of contribution in relation to the world, the disciplines like Multidisciplinary (19.55%), Biology & Biochemistry (12.98%) and Geosciences (12.38%) are on top three (though their national specialization position is 20, 7 and 10 respectively), while discipline Geosciences has the same rank at the national and international level, which infers that specialization index is same for Geosciences, both at national and global level. Comparing India's position in various disciplines taking all countries into account, India is fairly well placed (discipline wise) in papers published in some subject areas like Multidisciplinary, Agricultural Sciences, Chemistry, Material Science, Pharmacology & Toxicology with 3rd position, 5th in Agricultural Sciences & Chemistry, 6th and 7th in Material Science, Pharmacology & Toxicology, respectively, the remaining fields have ranks between 10 to 35. In terms of citations, fields like Material Science, Chemistry, Multidisciplinary, Pharmacology & Toxicology, are at 8th, 9th and last two at 10th position each respectively, with Engineering being at 12th position and Agricultural Sciences at 13th position, while the remaining fields are positioned between14 to 31st rank in the global sphere. The situation is bleak when the CPP scenario is projected, India's best on average citations per paper (which focuses on the impact made on the global platform) is Material Science, Psychiatry/Psychology, Engineering, Economics, Chemistry, Physics with 42, 47, 50,(Engineering & Economics both at 50th rank in respective areas) 52, 53 rank respectively, while the remaining fields fall between 58 to 93 positions.

In the second scenario while taking the top 20 countries in each of the 22 fields, the following scenario emerged. The citation impact represented in terms of CPP except for the fields Material Science, Engineering, Chemistry, Computer Science, Biology & Biochemistry which stands at 12,15,16,16,17 positions (out of top 20) respectively, the rest of the fields are at 18-20 position, while Psychiatry/Psychology Social Sciences and Economics are at 22,23,and 29th position respectively. Position in terms of papers may be better as reflected in (Table 3.9 d).

This indicates that though India may be improving in terms of growth in number of publication output which does not augur well with the impact that these publications are making at the global level. This is a matter of concern and thus needs to be addressed by the policy makers.

3.3.44 Activity Index

To figure out the *Relative Specialization Index (RSI)* for a given country the *Activity Index (AI)* is being attempted here to comprehend RSI based on Activity Index. It may be indicated that generally AI is applied to institutions or universities to figure out specialisation index. However in the present case, it is country specific.

$$AI = \frac{\text{Share of a given field in the publication of a country}}{\text{Share of given field in the world total of publications}}$$

$$RSI = \frac{AI-1}{AI+1}$$

RSI takes the values in the range -1 to <1. The value indicates whether an institution or a country has a higher than average activity in the world in a given field (RSI>0) or a lower than world average activity (RSI<0). In situations where RSI=0, it reflects a completely balanced situation.

As a benchmark an RSI=0 value is used for all research fields, which corresponds to the 'world standard case' as graphically presented by a regular radar.

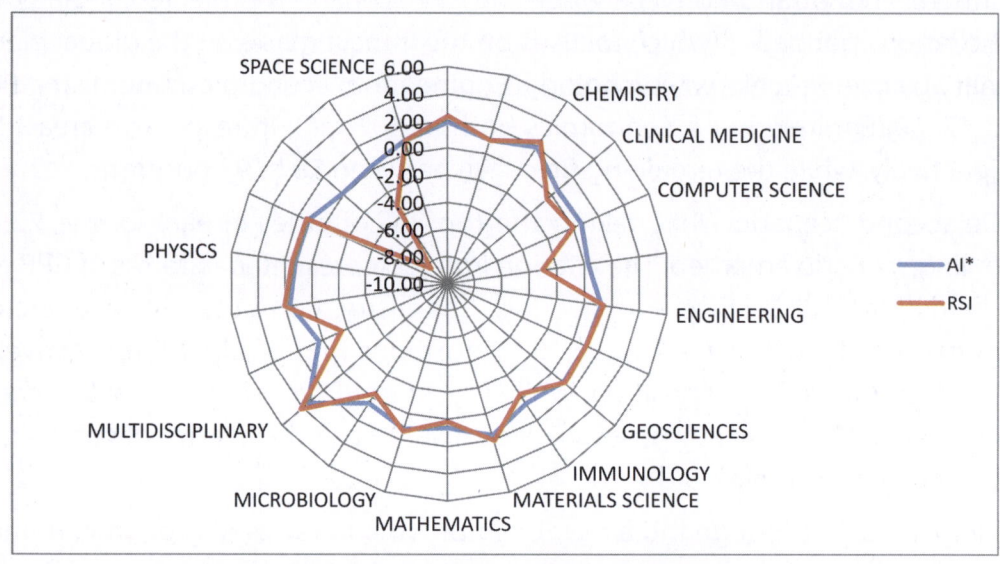

Figure 3.14 *Relative Specialisation Index and Corresponding Activity Index for India: Discipline-wise*

It is evident from the Table 3.13 (Figure 3.14) that there are 14 disciplines with value >1, 8 with value <1 and 3 disciplines (Biology & Biochemistry, Environment/Ecology and Space Sciences) have values closer to 1 (0.96, 0.99, and 0.93 respectively).

3.3.45 *Citation Publication Matrix*

While presenting the citation publication matrix (Table 3.14 & Figure 3.15), the fractionalised publication count is based on the proportion of each discipline calculated as the average of country's total contribution where in publication base for a period of 10 years have been considered. Besides the citations per paper for each discipline have been taken into account. The similar data-sets for the world as a whole have been taken into consideration for plotting the two sets of averages (fractional publication counts and relative citation impact) so as to figure out the disciplinary position of India viz-a-viz the world. It is evident from the Figure 3.14, that while on the fractionalised publication count front, India is above the world average in a number of disciplines (9 subject disciplines) and almost comparable in a few (3 disciplines) and lower than the world in remaining cases. On the citation impact front, India is nowhere near to the world average in any of the disciplines. Thus, wherein India may be fairly positioned so far as average publications in relation to the world is concerned, India is abysmally low in terms of impact its research publications have made on the world stage. Thus, what is important is to raise the bar of high impact publications and just not count on mere quantity but also quality by bridging the yawning gap between cited and un-cited publications.

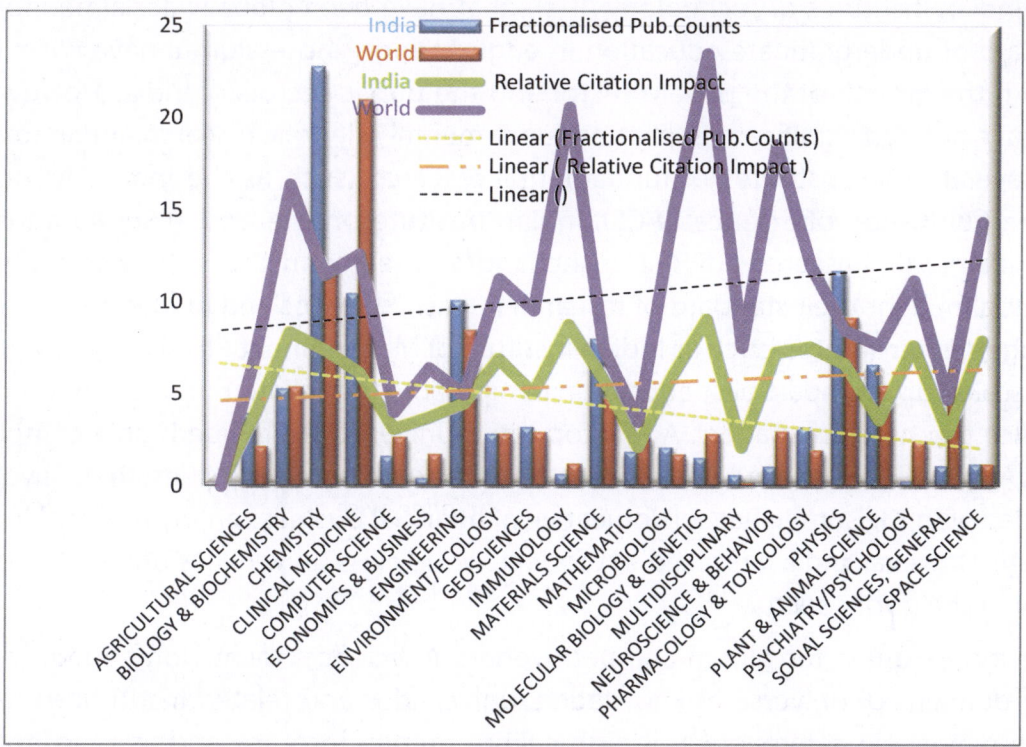

Figure 3.15 *Fractionalised Publication Counts Vs. Relative Citation Impact : India Viz-a-Viz World*

3.3.5 Key Institutes and Pivotal People

India's higher education system is the third largest in the world, after China and the United States.[85] The main apex institution at the tertiary level is the University Grants Commission (India), which enforces standards, advises the government, and helps coordinate between the centre and the state.[86] The higher education system in India

---

[85] World Bank. India Country Summary of Higher Education. Available at: *http://siteresources.worldbank. org/EDUCATION/Resources/278200-1121703274255/1439264-1193249163062/India_CountrySummary.pdf* (Accessed on 27-12-2012)

[86] India 2009: A Reference Annual (53rd edition), 237

though being one of the largest education networks in the world is highly fragmented and sub-optimally organised.

For instance, some institutions in this education system network are large, while some are relatively small with a few hundred students. Distance learning institutions are also an integral feature of the Indian higher education system. There are some institutions of India (e,g; Indian Institutes of Technology (IITs)), that have been globally acclaimed for their standard of undergraduate education in engineering, whose alumni have contributed to both the growth of the private sector and the public sectors of India. However, the IIT's have not had significant impact on fundamental scientific research and innovation. Yet, several other institutes of fundamental research (such as the Indian Association for the Cultivation of Science (IACS), Indian Institute of Sciences (IISc), Tata Institute of Fundamental Research (TIFR), Harishchandra Research Institute (HRI) and the like), are acclaimed for their standard of research in basic sciences and mathematics. Besides management institutes (such as Indian Institute of Management (IIMs) and a few others) have scaled up to impart best education comparable to the international standards and produce top notch managers. A few top rated universities provide highly competitive world class education, but India is also home to many universities that have been founded with the somewhat sole objective of financial business enterprise (making easy money). However, India has failed to produce world class universities both in the private sector or the public sector.[87-88]

In India's quest for global competitiveness in science, technology, medicine and other domains of universe of knowledge, universities and related institutions have to play a major role as providers of high calibre human resource and as repositories of national intellectual wealth in the R&D sector. In order to meet these diverse challenges, strengthening the institutional capacities in R&D through various initiatives, it becomes imperative to have evidence based metrics on the existing institutions for policy makers to take stock of existing scenario in a holistic way. It is hoped that the analysis based on the institutions and the personnel may assist in portraying somewhat realistic picture to aid institutions to do better.

---

[87] India doesn't figure in world top-100 universities, Press Trust of India via timesofindia.com, 2010-09-12
[88] Essential Science Indicators, Web of Science, Thomson Reuters database (Data covers 10 years period from Jan 1, 2002 to June 31, 2012)

3.3.51 Research Patterns

Keeping in view the above mixed kitty for Indian Higher Education system, an analysis to decipher the key institutes and the people whose contribution put Indian research on the world map of research landscape and article excellence was taken up using bibliometric analysis based on the publication count and the impact analysis.

The analysis based on the decade-wise distribution of institutional research contribution in various subject disciplines have been taken into account with data from 1971- 2012 (June 2012). The general trend observed from the data (Table 3.15 and Figure 3.16 a) which depicts that the number of institutions in almost all disciplines contributing to the research landscape are increasing as we move from 1970s to the recent years 2011-12. Only 6 subject disciplines (Chemistry, Clinical Medicine, Engineering, Multidisciplinary, Physics and Plant & Animal sciences) comprise over 50% (between 50-60%) of institutions engaged in research whose contributions have been reflected in various disciplines during 1971/81 - 2011/12. The institutions in these six areas are (Table 3.15 and Figure 3.16 b) contributing to about 65% of article output (ranging between 63%-70% in different decades).

The Figures 3.17 [a-v] are clearly indicating the disciplinary emphasis and excellence in different decades. For instance it is evident from the Figure 3.17c, that Chemistry has been doing well all along which is also corroborated from what we have seen in preceding paragraphs and analysis as well. Similarly, in the area of Clinical Medicine, while India was hardly in scene in 1970s, and partly in 1980s, the research activity has began to show R&D influx in 1990s and received a thrust in 2000-2010 which seems to be continuing, partly may be due to thrust at the national level and of course the field is on the top in the global arena in terms of R&D outputs. Similarly looking at the field of immunology, the analysis (Figure 3.17j) brings out a skewed picture, the decade of 1980s have shown active R&D endeavours as reflected by the research output of the institutions, followed by 1990s where one observes spurt in R&D activities, and unlike most of the other subject disciplines, there is limping phase in the decade 2001-2010, and again rise in activities in the recent years 2011-2012. This could be attributed again to the thrust given to the field.

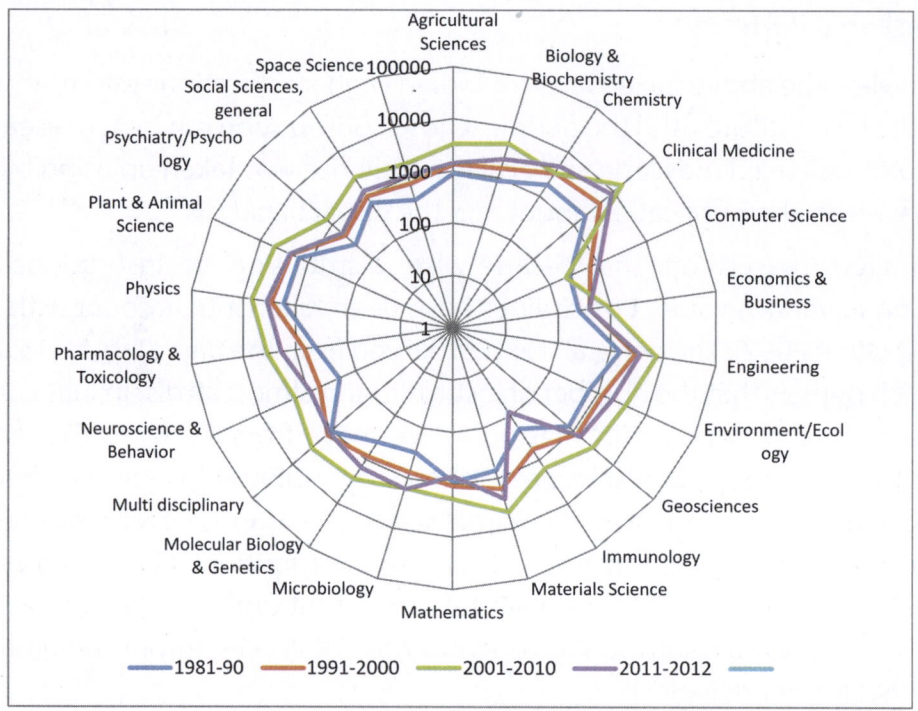

Figure 3.16(a) *Publication Contributing Institutions (No. of Institutions):*
Subject-wise & Decade-wise Analysis

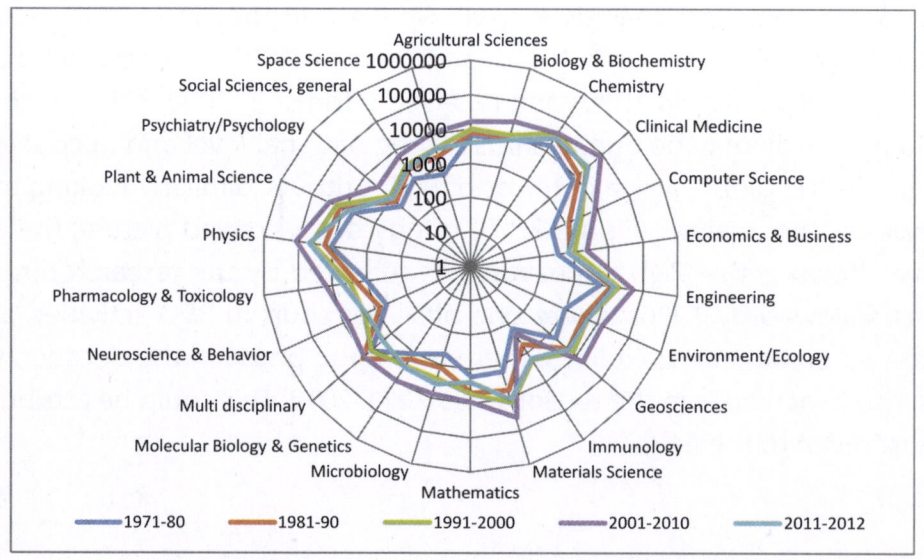

Figure 3.16(b) *Instititutional Productivity (No. of Publication Records):*
Subject-wise & Decade-wise Analysis

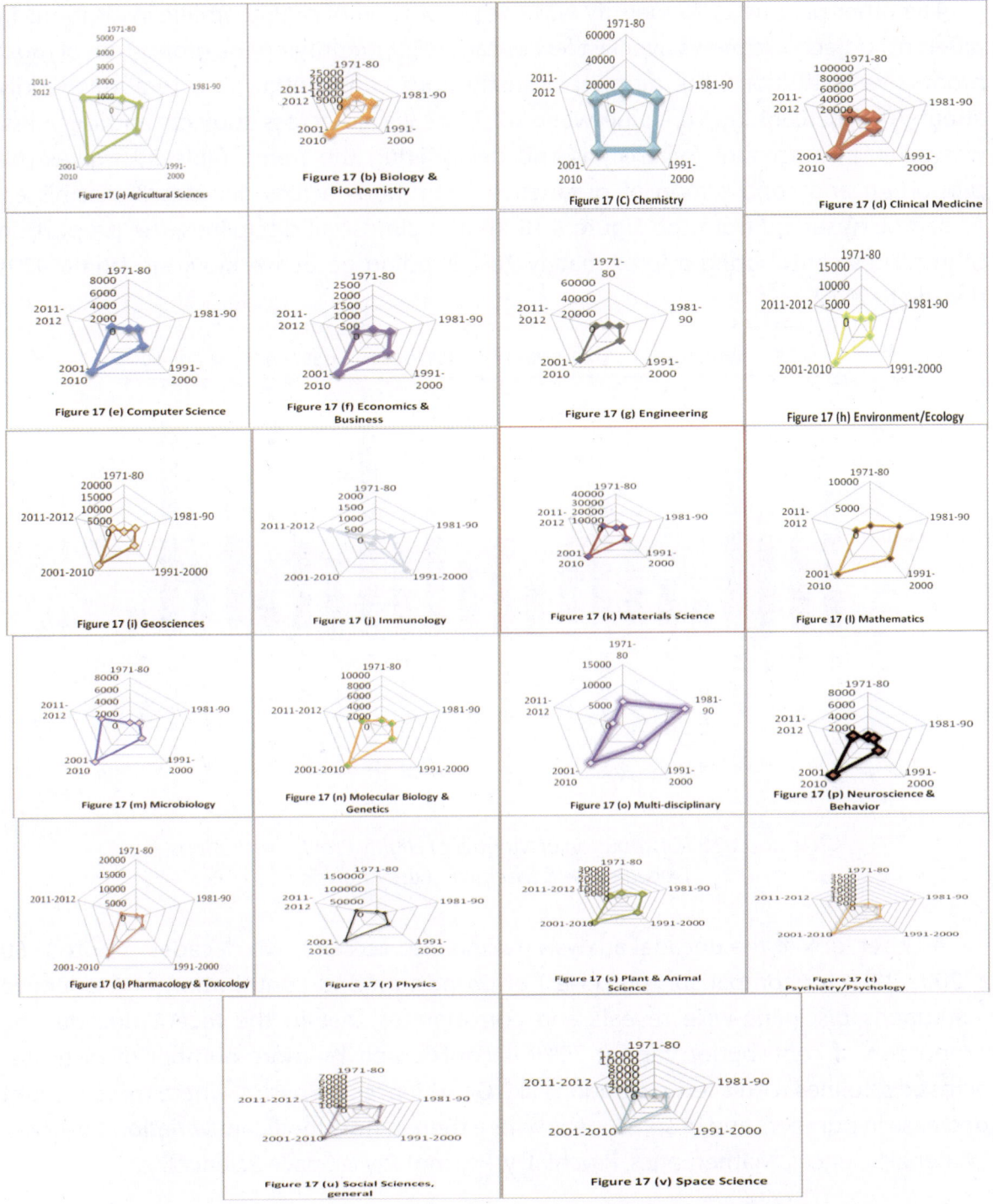

Figure 3.17(a-v) *Decade-wise Distribution [number] of Publication Contributing Institutions : Subject-wise Analysis*

The other parameter to identify was the proportion of highly prolific institutions in different subject disciplines over last few decades. For identifying the proportion of most productive institutions, the number of institutions contributing to a total of roughly about 70% (percentage varies between 67-70 as we go across subjects and decades) was taken into account for analysis and deciphering the trend. Table-3.16 gives the proportion and contribution of productive institutions across decades & disciplines. As can be observed from the Figure 3.18, that in almost all disciplines, the proportion of institutes contributing approximately 70% is going up as we move up from 1970s decade to 2012.

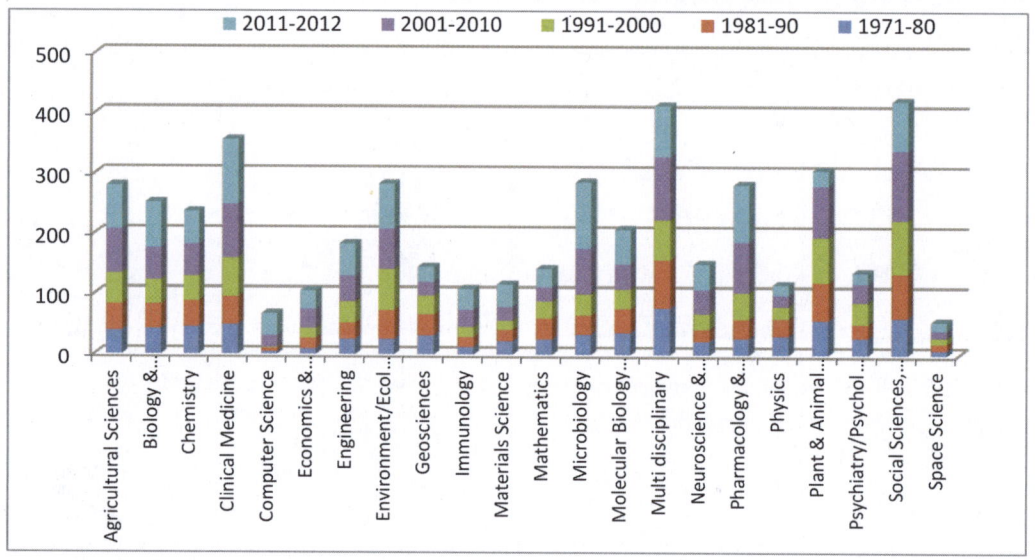

Figure 3.18 *Distribution of Number of Highly Prolific Institutions Decade-wise & Discipline-wise Analysis*

A closer look at the decadal analysis (taking into account two decades – 1971-1980 & 2001-2010 for comparison purposes) of proportion and contribution of productive institutions discipline-wise reveals and corroborates that in the recent decade, the proportion of contribution (about 70%) is contributed by more number of institutes across disciplines with a few exceptions like Geosciences & Physics, where there is slight decrease in numbers and in some areas where there is not significant variation observed (Material Sciences, Mathematics, Psychiatry/Psychology & Space Science).

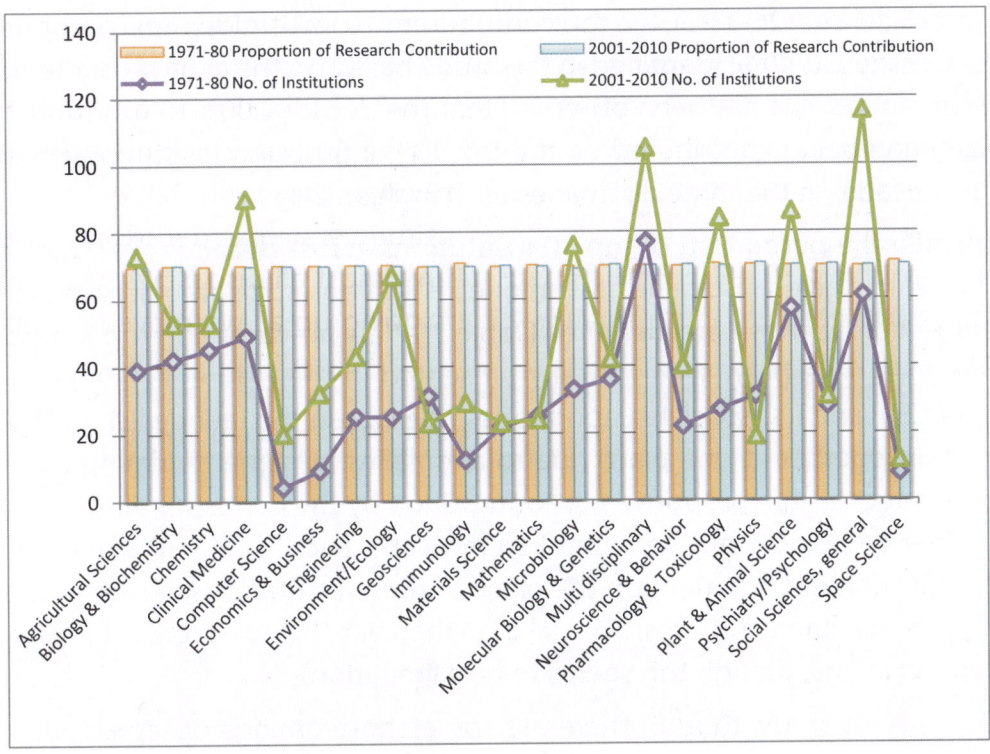

Figure 3.19 *Proportion and Contribution of Productive Institutions Decade-wise & Discipline-wise Analysis*

3.3.511 *Key Institutions*

A further analysis based on the top 10 institutions using the data for the decades 1971-1981 to 2001-2010 & 2011-12 in various subject disciplines reveals the decadal contribution of the prolific institutions whose contribution has been substantial. The analysis (Table 3.17) shows that in various subject disciplines and across decades (1971/81 -2001/10 and 2011-12) on an average top 10 institutions are contributing to roughly around over 30-40% with a few exceptions of some research fields doing better in terms of cotribution of top 10 institutions, while a few disciplines have lower proportion for the same number of institutions. For instance in research disciplines like Computer Science (50-80%), Economics & Business (45-70%), Engineering (45-60%), Immunology (>30-68%), Material Science (40-60%), Mathematics (47%), Neuroscience & Behaviour (38-54%), Physics (47—53%) and Space Science (58-75%). With some disciplines like Economics & Business, Engineering, Immunology, Space Science, showing upper limit in the initial decades (1970s & 1980s). While research disciplines like Multidisciplinary (25-

30%), Social Sciences (26-33%), the top contributing 10 institutions are contributing less than the average (30-40%) identified in this study based on the analysis of the observed data. In certain cases, it has been observed that the decade 2001-10 has comparatively less percentage being contributed by the top 10 contributing institutions then in the rest of the decades in the same disciplines used in this study.

To identify the porportion of top 10 institutions across decades (1971/80 -2001-10 and 2011/12) in a given discipline, the unique institutions across the decades in each of the 22 subject disciplines were identified (Table 3.17(a)), however only top 10 institutions across decades have been schematically presented in the Figure 3.20 (a-v) (except in the case of CSIR and its associate Laboratories, the labs in the unique list and CSIR have been plotted together at one place due to anomalies in name rendering). The size of the bubble indicates the proportion of contribution by the corresponding institution. It may however be pointed out here that in a few cases such as IIT's, IIMs and rarely CSIR labs, the affiliation did not mention the place of these institutions, hence in such cases clubbing was inevitable. (for instance all IITs wherever they occurred, their values have been clubbed in one, though this seems to be a limitation).

In the present study, though there are some shortcomings observed, viz, IIT's are not clearly specified as which IIT is doing better due to incomplete rendering of author affliations. However since the purpose of present study is to throw light on the status of health of Indian research landscape, wherein the notion of a few institutions contributing for the research excellence of the Indian research landscape can be deciphered even with the current shortcomings. There is thus a need for standardization of the names alongwith working on data deficit by doing the gap analysis.

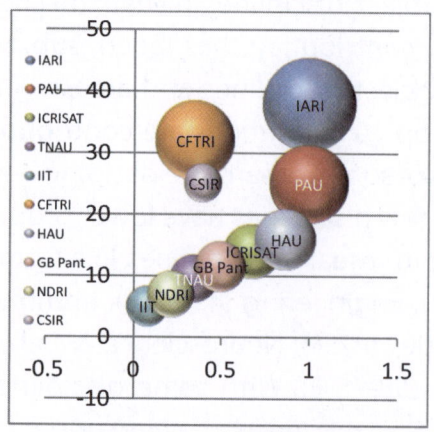

(a) Agricultural Sciences

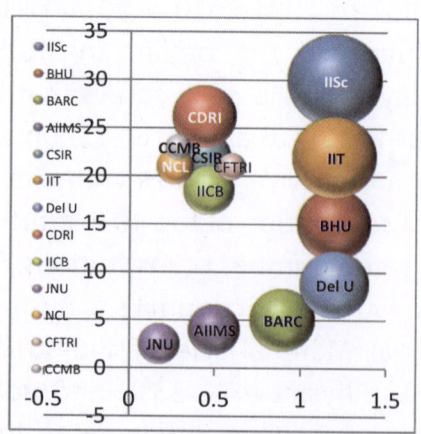

(b) Biology and Biochemistry

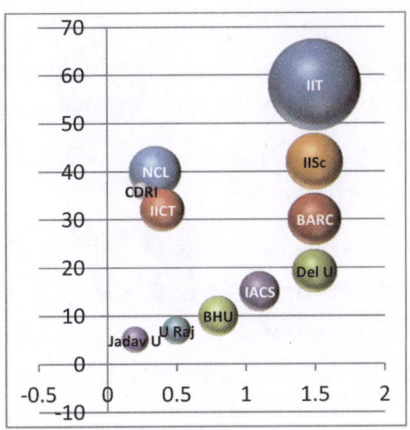

(c) Chemistry

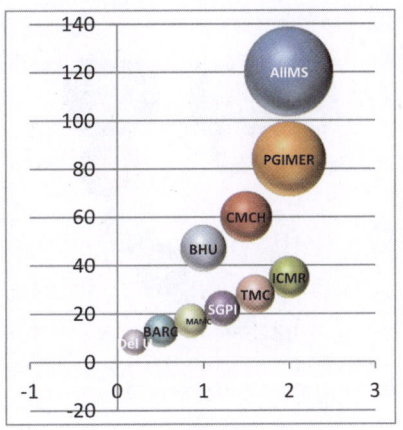

(d) Clinical Medicine

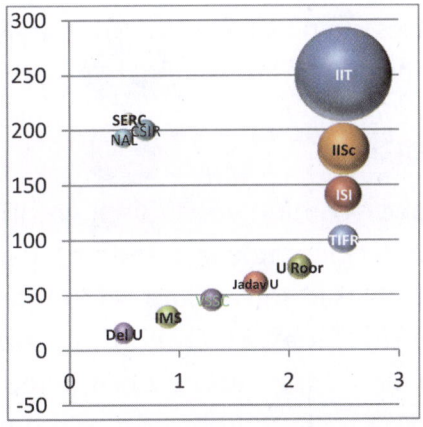

(e) Computer Science

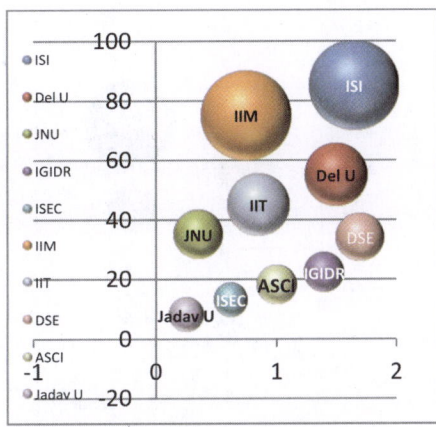

(f) Economics and Business

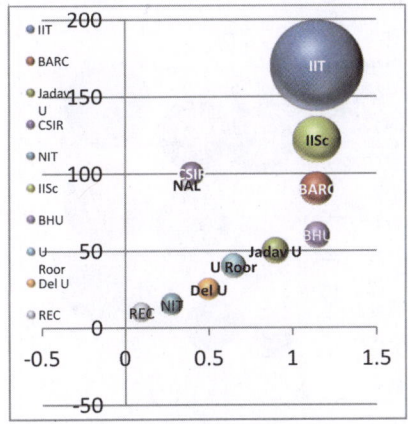

(g) Engineering

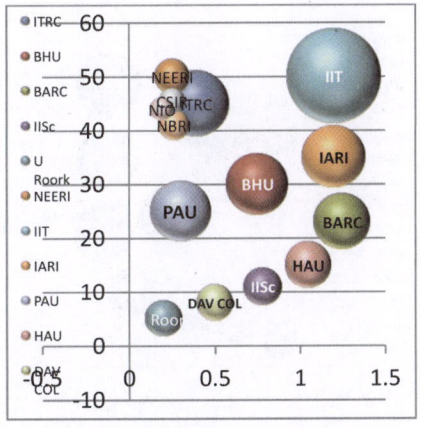

(h) Environment / Ecology

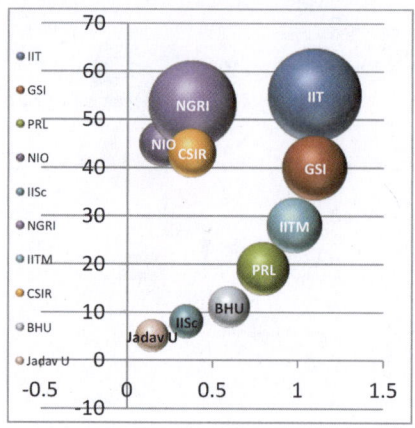

(i) Geosciences

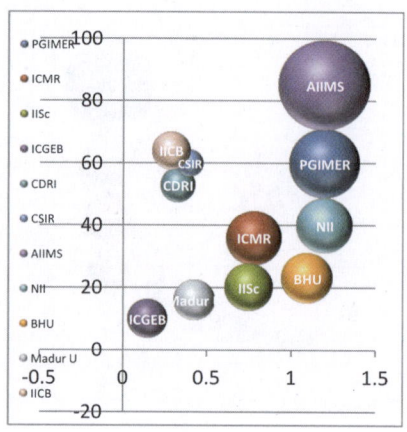

(j) Immunology

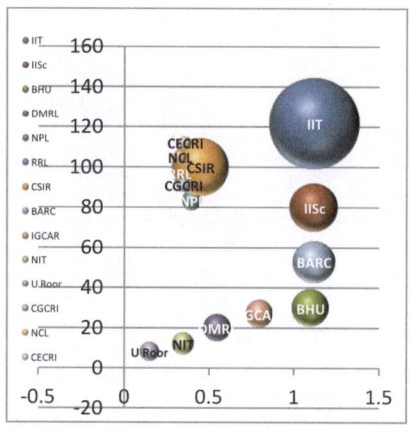

(k) Materials Science

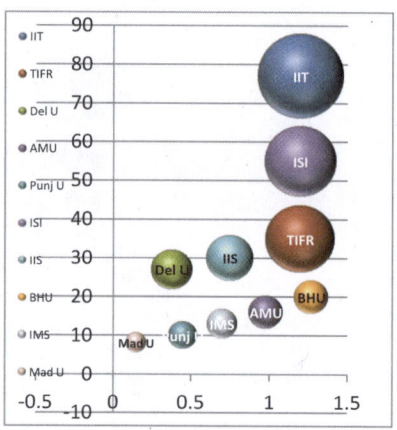

(l) Mathematics

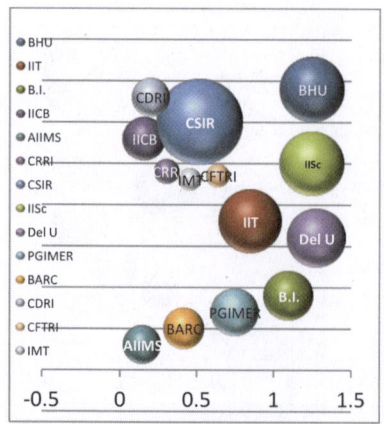

(m) Microbiology

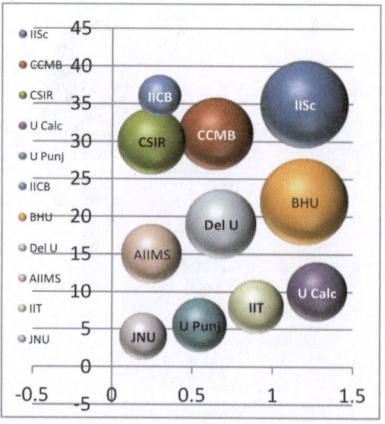

(n) Molecular Biology & Genetics

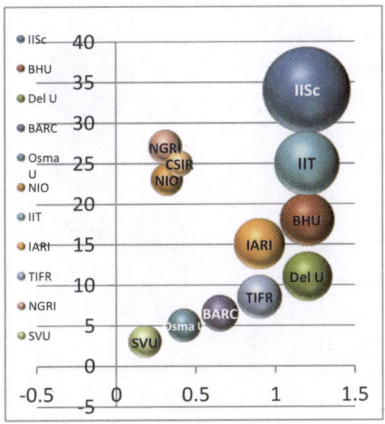

(o) Multidisciplinary

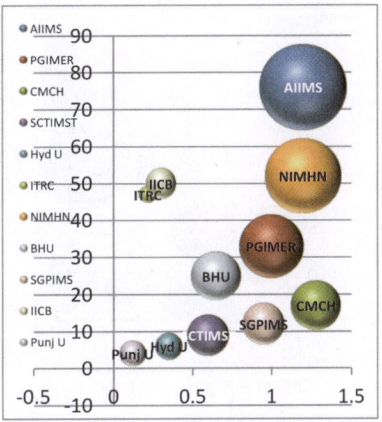

(p) Neuroscience & Behaviour

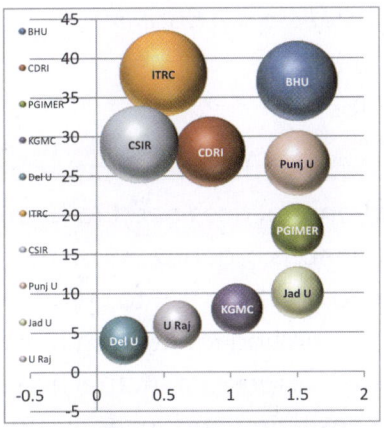

(q) Pharmacology & Toxicology

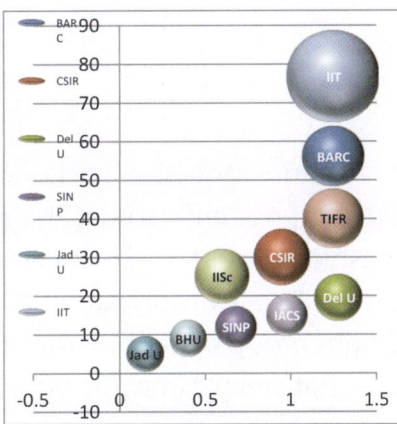

(r) Physics

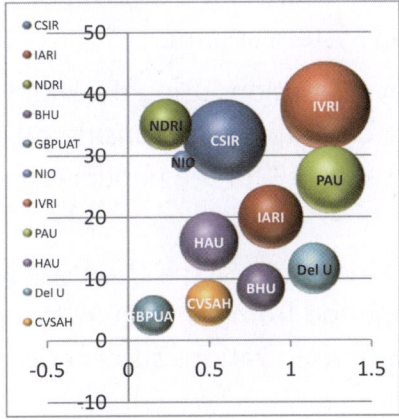

(s) Plant & Animal Science

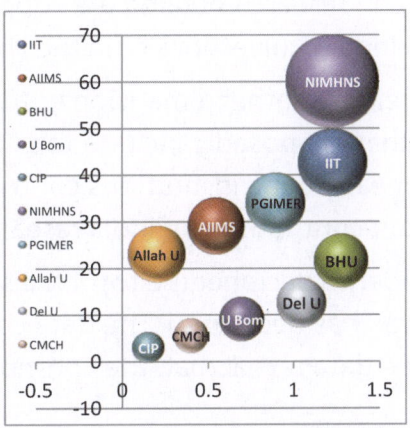

(t) Psychiatry / Psychology

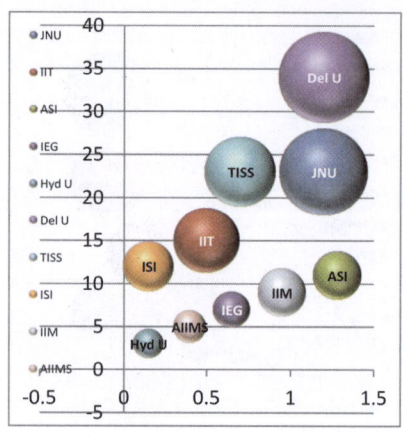

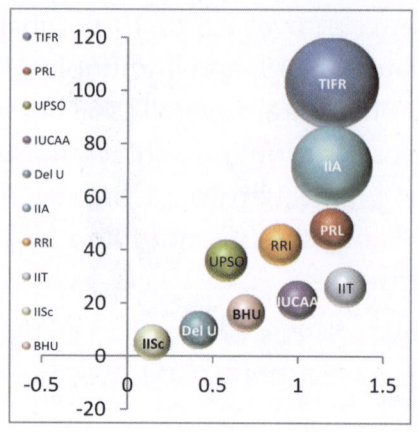

(u) Social Sciences, General   (v) Space Science

Figure 3.20   *Institutions with Strong Focus on 22 Research Disciplines*

3.3.512 Assessment of Science & Scientific Institutions

While we are addressing issues of key performing institutions, the identification becomes rather difficult if not impossible. The national innovation practices need to be modeled in such a fashion so that the characteristics of heterogeneous organizations, multifarious activities do not pose unsolvable challenges of whom to or whom not to credit or are we crediting right group of people. The statistics for evaluation and identifying for instance, top 1 or 2% that creates (may be 90%) maximum value is important. Actual need comes for institutions performing top level (80% or so) for sensitizing and incentivizing them. Innovations in higher education (that supports employability) fostering a research culture and research pooling are supposed to be at the heart of the innovations taking place in Indian universities or in the Higher Education System at large.

Therefore another dimension used to focus on key institutions using the Highly Cited Papers that supposedly include top 1% of the research output was considered. Hence by any measure the institutions contributing for these top 1% of the papers represent a significant contribution of India research power.

In order to decipher the top 1%, Essential Science Indicators of the Thomson Reuters database[89] has been used. The data covers 10 years period from Jan 1, 2002 to June 31, 2012. The data reveal that, the contribution of institutions in various subject disciplines

---

[89] Moed, Henk F. The use of big datasets in bibliometric research. Research Trends : Special issue on Big Data. Issue 30, p31-33 (Elsevier)

varies between over 2.5 to 13 per institution with Physics in lead (13.01), followed by Mathematics (12.13) and Immunology (3.73) and Multidisciplinary (2.58) at the tail end. (Table 3.18). There are overall 4963 unique institutions, out of which 637 are unique Indian institutions contributing a total number of 1531 highly citied papers (HCPs), and having about 2.4 publications per institute. Thirty institutions are contributing for 50% of the total HCP and 70% is contributed by 110 institutions, while 80% has been contributed by 206 institutes.

Further analysis of these 1531 highly cited papers from India revealed that 1039 (where each institution within a subject has been counted only once, however across the fields, the institution has been counted as many times as the number of fields it appears, although unique institutions publishing these papers are 637 in number) institutions from India are overall contributing partners. Engineering discipline with higher number of institutions, followed by Clinical Medicine & Chemistry, while Psychiatry/Psychology and Economics & Business subject discipline have least number of institutions with none in the Multidisciplinary field (Table 3.19 and Figure 3.21. The other inference drawn that needs to be mentioned is that, though there are number of institutions engaged in multidisciplinary research, yet there is none in the top 1% of Indian papers included in ESI.

Table 3.20 lists top 5 institutions under each subject discipline. There are though some universities listed in top 5, however, in certain fields like, Computer Science, Economics, Immunology and Psychiatry/Psychology, there is no university in top 5 list of institutions. This is exceptional as in most countries worldwide, universities are prominent in research output and also its impact. However, Table 3.21 focuses on an interesting point that there are 8 universities listed in top 20 Indian institutions (contributing to HCPs) with 23.7% proportion of contribution. This indicates that though in terms of number proportion (40%), universities have decent presence, but contribution wise, their proportion to other institutes is only 1/4th. This can also be corroborated with the fact that Indian premier institutes occupy high ranks in the top institute listing (Table 3.21). The level of collaboration (national and international) is also relatively high for these institutes that could be a contributory factor to the article excellence and resultant competencies.

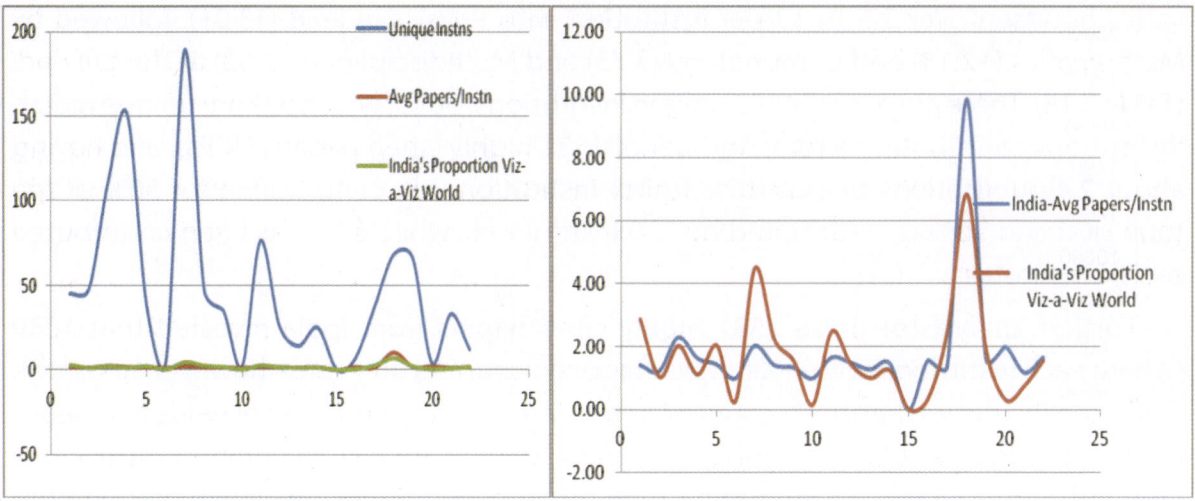

Figure 3.21 *Institutions (Unique) Corresponding to the Proportion of Contribution to HCP (Subject-wise Segregation)*

1. Agr.Scs   2. Bio & Bioch   3. Chem   4. Clin Med.   5. Comp. Sc   6. Eco & Bus.   7. Eng
8. Env & Ecol   9. Geosc.   10. Immunol   11. Mat. Sc.   12. Maths   13. Microbiol.   14. Mol.Biol & Gen
15. Multidisc.   16. Neuro & Behav   17. Pharm & Tox.   18. Phys.   19. Pl & Anim Sc.
20. Psy. & Pshchol.   21. Soc. Sc.   22. Space Sc

Table 3.22 and Figure 3.22 depicts the scenario of India's contribution to the top 1% of research productivity in relation to the world. The subject wise analysis of the highly cited papers focuses on the relative ranking of disciplines contributing to publications in top 1% viz-a-viz citation impact. Proportion of India's contribution to the world varies from 0.16 to 3.38 in various subject disciplines with Engineering at the top (3.38%) and Immunology at the bottom (0.16%) and multidisciplinary subject area with no contribution in top 1% publications. The disciplines like Engineering, Chemistry, Computer Science, Environment/Ecology, Materials Science, Pharmacology & Toxicology, Physics, Plant & Animal Science are above the world average in terms of cut off citations for the top 1% publications, and the remaining ones are below the world average.

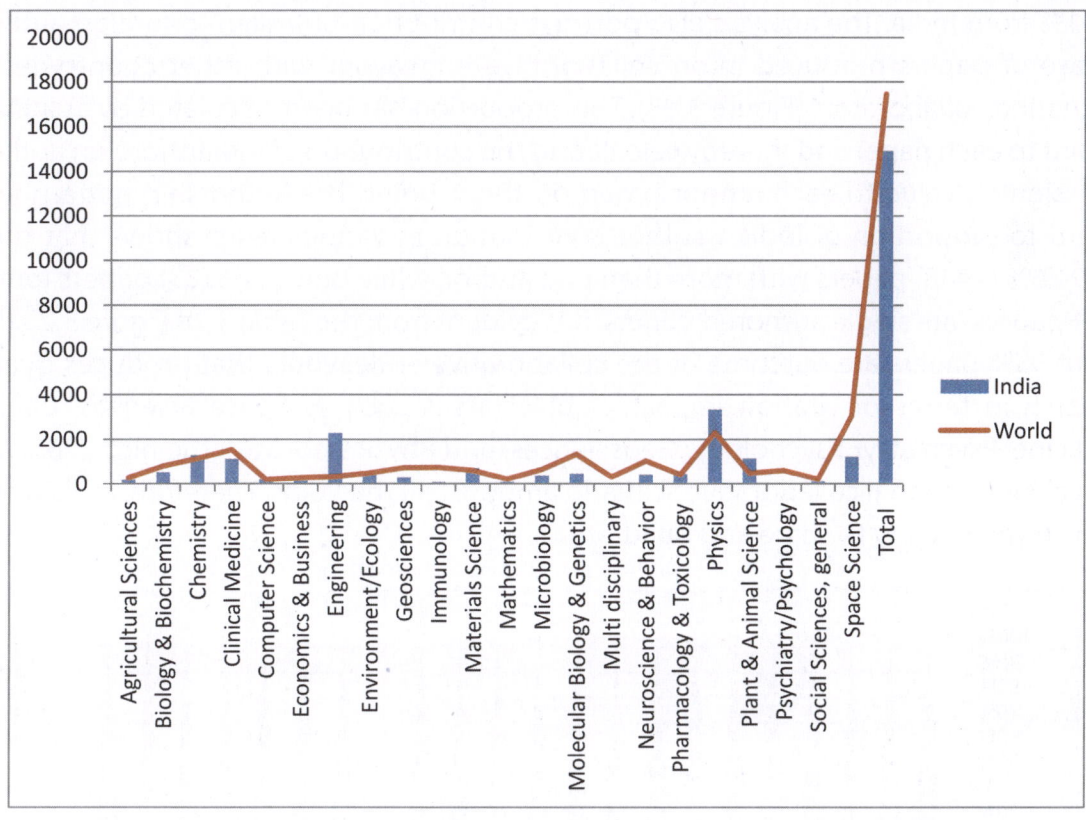

Figure 3.22 *Relative Position of India viz-a-viz World Average in Terms of Citations*

3.3.513 Pivotal People

The pivotal people contributing to these HCPs affiliated to the institutions as indicated above were further analyzed by taking into account the affiliation of the authors contributing to these HCPs. The contribution of Indian authors were deciphered and are presented in the Table 3.23 which revealed the following scenario. Out of 1,16,903 total authors contributing to the 1531 HCPs, 6609 are Indian authors (by Indian, it is inferred that the corresponding authors are affiliated to an Indian institution). In the cases where the author has two affiliations, the credit has been given to Indian affiliation. Except the fields like Materials Science, Mathematics, Pharmacology & Toxicology, rest all other fields have lower proportion of Indian Contributors; while fields like Clinical Medicine, Physics and Space Science have very low proportion of Indian contributors (.09, .03, & .02 respectively), and Multidisciplinary field is not represented at all with no paper in top 1% from India. The analysis also portrays contribution of Indian to foreign authors in case of papers produced in an collaborative endeavour with other country/ies as partnering collaborators (Figure 3.23). This proportion has been calculated by assigning 1 point to each paper and thereby calculating the contribution of Indian/foreign authors by assigning value to each author based on this 1 point. The authorship pattern with regard to proportion of Indian author contribution in various fields shows that there are 94.6% (1448) papers with more than one author, while only 5.4% (83) papers (out of 1531 papers) are single authored papers. It is evident from the Table 3.24 Figure 3.23 that about 95% papers are outcome of the collaborative endeavours that produces quality research in terms of citation impact. Subject areas such as Space Sciences, Clinical Medicine, Psychiatry/ Psychology, Geosciences, and Physics portrays the high (70% and above) international collaborative scenario amongst all the fields. Therefore, India needs to nurture its ability to share and build.

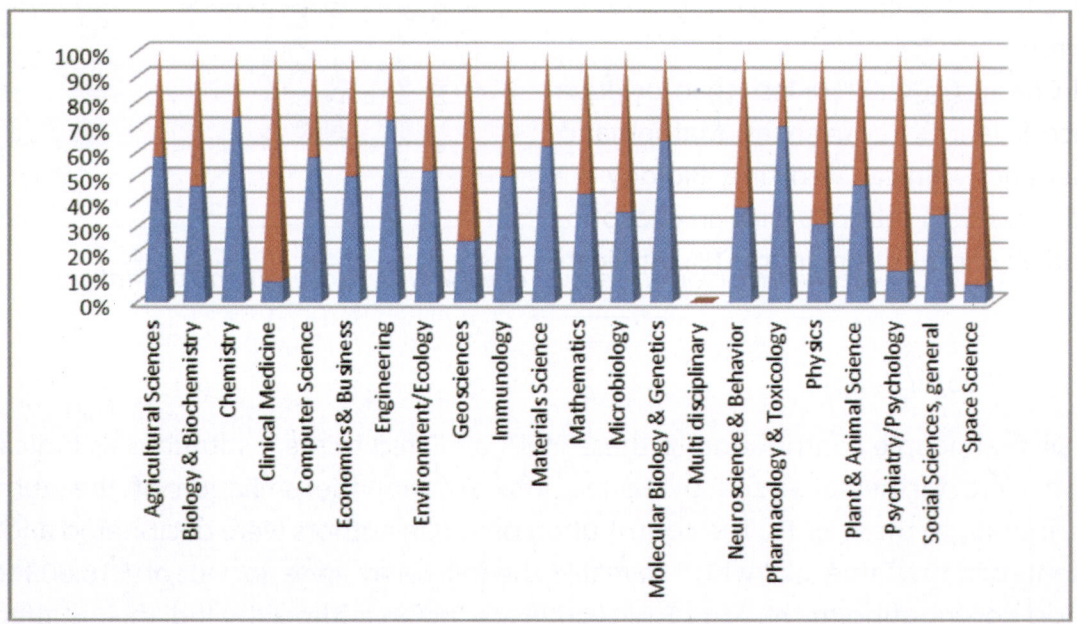

Figure 3.23 *Proportion of Indian Contribution in Collaborative (HCP) Papers*

Further authorship pattern shows that (out of 1531 papers) while there are (Table 3.25) about 5% single authored papers, the proportion of papers contributed by multiple authors by and large declines as we move in ascending order of author blocks (B1-B18). For instance, collaboration pattern shows while the observed proportion of about 60%

contribution comes from block 1 that is from 2-5 authors; over 13% by block 2 authors (6-10 authors); 8% by block 3 (11-20authors); around 2.5% by authors in the range of 301-400; over 1% by authors in the range of 21-30, 31-40, 101-200, 501-1000 & >2000 authors. While the remaining blocks of authors have contributed less than 1%. In all, only the first three multi-authored blocks constitute over 80% of the publications. It can also be deciphered that in almost all the disciplines (except multidisciplinary area which is not represented in highly cited paper section as there is no paper from India in top 1%), the collaborative clustering/alliances is glaringly visible (Figure 3.24). Therefore, it can be stated here that global problems need global solutions.

The Indian Scientists (Table 3.26) contributing to top 5 papers of top 1% of Indian highly cited papers also infers that most of these papers in many disciplines are outcome of collaborative endeavours. The proportion of Indian contribution in each discipline is about 7% in Physics & 4.5% in Engineering Sciences; about 3% in Agricultural Sciences; over 2% in subjects like - Pharmacology & Toxicology (2.50%), Materials Science (2.49%), Environment/Ecology (2.24%), Space Science (2.2%), Computer Science (2.04%), and Chemistry (2.02%); less than or equal to 2% in fields like Plant & Animal Science, Space Science, Geosciences, Mathematics, Biology & Biochemistry, Molecular Biology & Genetics, Clinical Medicine, Biology & Biochemistry, Microbiology; while rest of the fields have less than 1% with an exception of Multidisciplinary field. Overall proportion of India's contribution (in top 1% publications) viz-a-viz world is 2.19%.

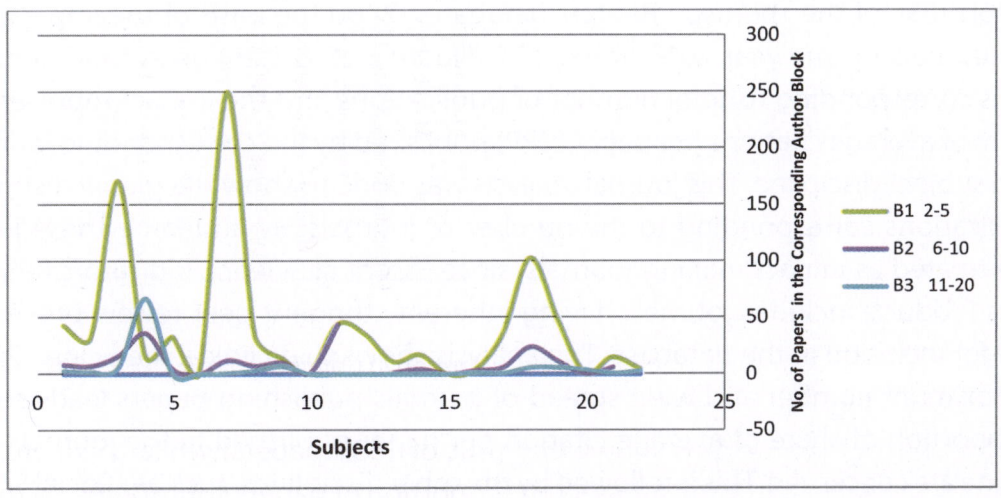

Figure 3.24 *Authorship Pattern: Trend Analysis of Number of Authors Contributing to Maximum Papers*

Finally, while we are arguing the point about the most prolific institutions and people who placed India on the map of global research landscape, wherein we treat the publication count with the impact metrics (where impact is directly related to citations and where no citations is taken as of poorer quality). It may be worthwhile to mention that there is a debatable question about papers published vs citations received. In a study by Moed[90] regarding correlation between a journal's downloads and its citations reveals large differences in the degree of correlation between disciplines. The author argues that the distinction between author vs. readers, where both can be publishing authors in one case; but in situations where readers are not necessarily authors can make the hypothesis of looking at downloads as the possible parameter for added citation impact unclear. However this merits further detailed investigation.

3.3.6 Journal Preference

The relative importance of the journal in which author chooses to publish his/her research work is determined by the Impact Factor score. Impact factor is a measure of the relative importance of a journal within its field, with journals of higher journal impact factors deemed to be more important than those with lower ones. In the present study, the journal analysis was done using primarily two criterion – (1) Average citations received by the journals (using citations per paper), where all the papers for a 10 year period (Jan 1, 2002 - June 30, 2012) were published and covered by Essential Science Indicators (ESI) of the Thomson Reuters database; (2) on the basis of the Impact Factor of the journals for the year 2012. Table 3.27 (Figure 3.25 & 3.26) gives total number of journals corresponding to total number of publications and the impact represented in the form of average citations per paper (CPP) published by the corresponding journals in a given subject discipline. This journal analysis was done to show the global distribution of publications corresponding to the number of journals. By and large these journals can be treated as impact making journals, since Essential Science Indicators (Thomson Reuters Product) includes journals having inherent stringent peer review processes to qualify for inclusion in the database. The analysis shows that clinical medicine (20.73%) with maximum number and wide spread of journals publishing papers leads so far as the proportion of share of average citation per paper published in the journals in this discipline are concerned. This is followed by the other disciplines, such as - Social Sciences

---

[90] Department of Science and Technology, Government of India. Bibliometric study of India's Scientific Publication outputs during 2001-10: Evidence for Changing Trends.

(7.60%), Biology & Biochemistry (7.24%), Psychiatry/Psychology (6.11%), Chemistry (5.98%), Molecular Biology & Genetics (5.88%), Plant & Animal Science (5.86%) with multidisciplinary discipline in the last (0.26%). It is however important to note that only 5 subject disciplines (out of 22) are capturing about 60% (59.40%) of citations., while next 7 disciplines account for 23% (22.81%), which means the remaining 10 disciplines account for only 17% (17.79%) citations.

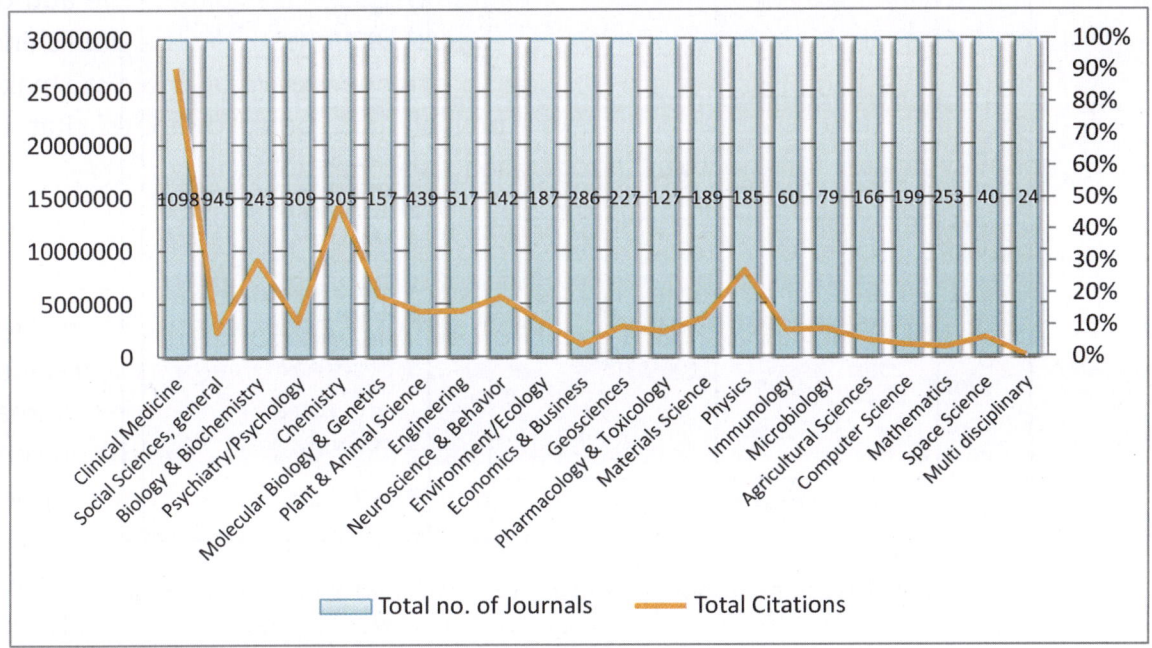

Figure 3.25 *Journal Preference Analysis: Field-wise Number of Journals Vs Citations*

Alternatively based on the average CPP, all the 6177 journals in various subject disciplines were arranged and grouped in five groups, each group referring to the level of impact making journals in which these papers were published. The number of journals in top impact making journals that is group 1 generally is between 1 – 2% The Figure 3.27 (Table 3.28 and Table 3.29) portrays the scenario, from which it can be interpreted that journals in six subject disciplines have none in group 1, 7 disciplines have less than 1% journals, 4 disciplines have slightly more than 1% journals, while 5 disciplines have slightly more than 2% journals, with an exception of 2 disciplines with over 3% journals (Molecular Biology & Genetics (4.46%); Immunology (8.33%) respectively) in the Group 1.

This also infers that very low proportion of journals are in the top bracket of high impact making journals.

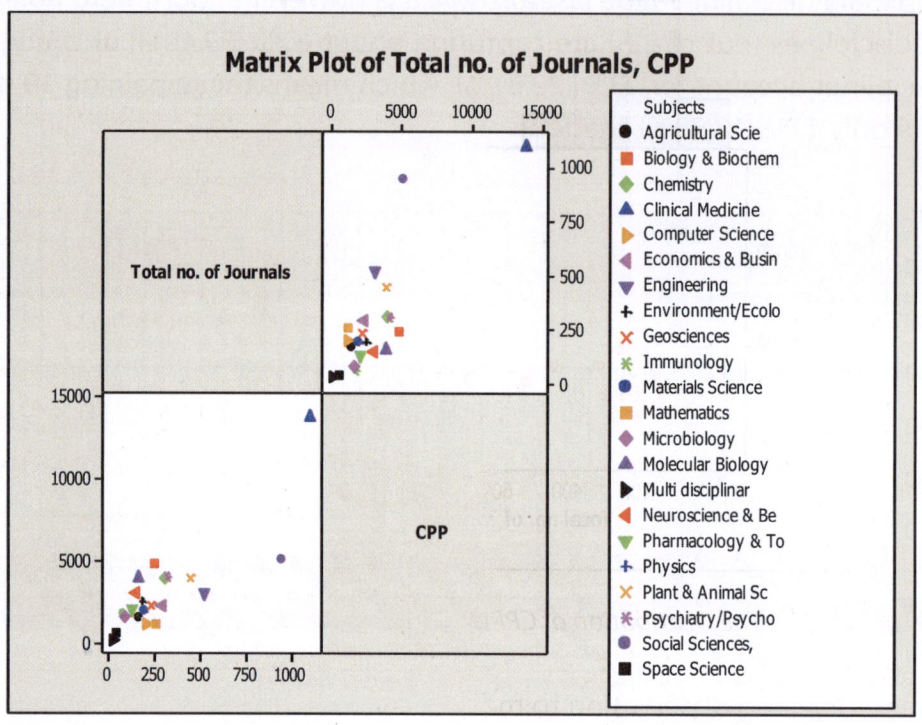

Figure 3.26 *The Interrelationship Between CPP and Total Number of Journals*

Figure 3.26 exhibits a positive trend between CPP and total number of journals, which implies that as total number of journals is increasing, CPP is also increasing. This can be considered as a healthy sign in terms of research and the growing acceptance of journals being added in the existing body of research.

Continuing with the earlier graphical display, the next figure (Figure 3.27) echoes the same view. Even if number of journals is very less as compared to other areas, the CPP is growing consistently.

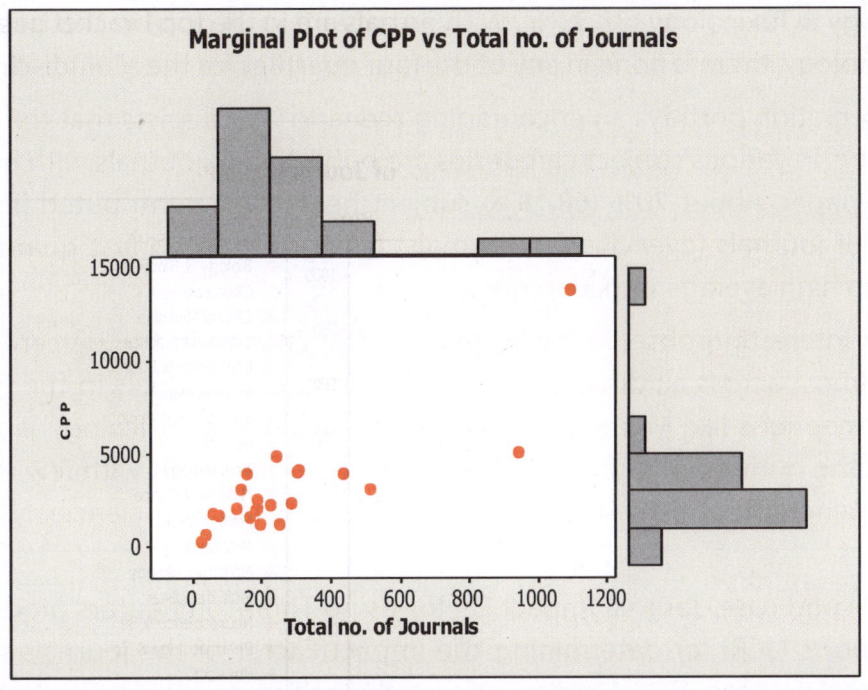

Figure 3.27 *The Distribution of CPP and Journal Alongwith Their Interrelation*

Another interesting observation to make, is the change in the percentages. This was deciphered based on the unique journals where these high impact papers have been published in various subject fields (subject categorization based on ESI 22 categories). The average of citations per paper (CPP) of each of these journals was taken into account for all the journals covered by the Essential Science Indicators (ESI) (Thomson Reuters database) and arranged in descending order of CPP score. Within the subject field, the entire list of journals has been divided into four quartiles based on the CPP score. Looking at the percentage of the journals in the top quartile by Journal average impact based on the average of citations per paper (CPP), fields which had a comparatively high percentage of journals (in which Indian high impact papers have been published) are placed in this top quartile. The top quartile included 15 fields with over 50% that includes 2 fields with 100% - Economics & Business, Psychiatry/Psychology; 3 fields with over 70% - Neuroscience & Behaviour(83.33%), Space Science (75%) clinical medicine (74.55%); 4 fields in the range of 60-70% that includes Material Science (68.75%), Geosciences (65%), Computer Science (62.50%), Molecular Biology & Genetics (60%); 3 fields with over 40% (Between 40-50%) like Engineering (49.30%), Chemistry (47.76%), Microbiology (41.67%); 2 fields with over 30% (Between 30 -40%) that includes Mathematics (33.33%),

Pharmacology & Toxicology (33.33%). With an exception of none in top quartile for the field Immunology, there is none in any of the four quartiles for the Multidisciplinary field.

The observation portrays an encouraging scenario that depicts that the Indian high impact papers in various subject categories are published in journals with high average impact per paper. About 70% (68.18%) subject fields have contributed in substantial proportion of journals (over 50-80% journals) falling in the top first quartile covering journals with high average impact score.

Another interesting observation to make is the change in the percentages across quartiles. Almost across all disciplines there is increase of journals in the top quartile, (with few exceptions like Mathematics and Pharmacology & Toxicology) and a parallel decrease in the number of journals in the lowest quartile (again with few exceptions – Biology & Biochemistry, Clinical Medicine, Computer Science, Geosciences, Microbiology) (Table 3.32).

In the second case, Journal Impact Factor using Thomson Reuters product, Journal Citation Reports (JCR) for determining the Impact Factor of the journals for the data sample comprising top 1% of Indian papers christened as Highly Cited papers (HCP) were taken into account. The journals publishing Indian HCPs were ranked based on the Impact Factor Score and grouped into 4 groups. The total of the top 1% of papers for India are 1531, downloaded from Essential science Indicators of Thomson Reuters for a ten year period (Jan 1, 2002 to June 30, 2012). In all there were 523 journals (unique titles) in various subject domains publishing these 1531 Indian highly cited papers. The analysis portrays that of all the papers, 7 disciplines are contributing for over 75% of HCPs that are published by 272 (50%) journals. (Table 3.30 & 3.31) and (Figure 3.28).

Alternatively, these journals were also checked viz-a-viz CPP in each subject discipline and the following was revealed. More than 60% of publications in high impact journals are from just four disciplines such as Engineering, Physics, Chemistry and Clinical Medicine.

In disciplines with comparatively large number of (top 1%) published papers, it has been observed that the spread of concentration of papers is small (Table 3.32). For instance in the subject disciplines of Engineering, Physics, Chemistry, and, Clinical Medicine, over 50% of (top 1% Indian papers) are published in just 5, 3,12, & 5 journals respectively. While, in many disciplines with less number of papers such as Space Science, Social Science, Psychiatry/Psychology, Microbiology, etc., the wide spread of papers in different journals is observed. The impact of these journals was deciphered using two-pronged approach: i) using average citations per paper (CPP) for each of the journals in which these high impact papers have been published for a 10 year period

(Jan 2002 - June 2012); ii) identification of the Impact Factor (IF) score for each of these journals (Tables 3.30 (a -v)). In some cases CPP/IF could not be deciphered. The CPP was taken for all the journals covered in the Essential Science Indicators (ESI, Thomson Reuters Product). The CPP and also the Impact Factor Score is fairly impressive, hence it can be presumed that these are fairly quality journals. As such combining such a list with some quality Indian journals drawn from a good secondary indexing source (e,g; Indian Science Citation Index) can form basis of the list of journals that Indian researchers can consider for publishing their scholarly communications. Since the visibility factor of the journals is also important while selecting the journals for submitting the papers, this can act a guide to the prospective researchers [Figure 3.28 exhibits cross section view of journal and the relative position of their rank and CPP. For instance, in row 1, panel 5, the rank of the journal is very high (the smaller the rank, better is the positioning of the journal) and it has a high CPP too as the indicator for CPP is on higher side. However, there are few journals, which were not included in the analysis as the minimum benchmark value for CPP was 100.]

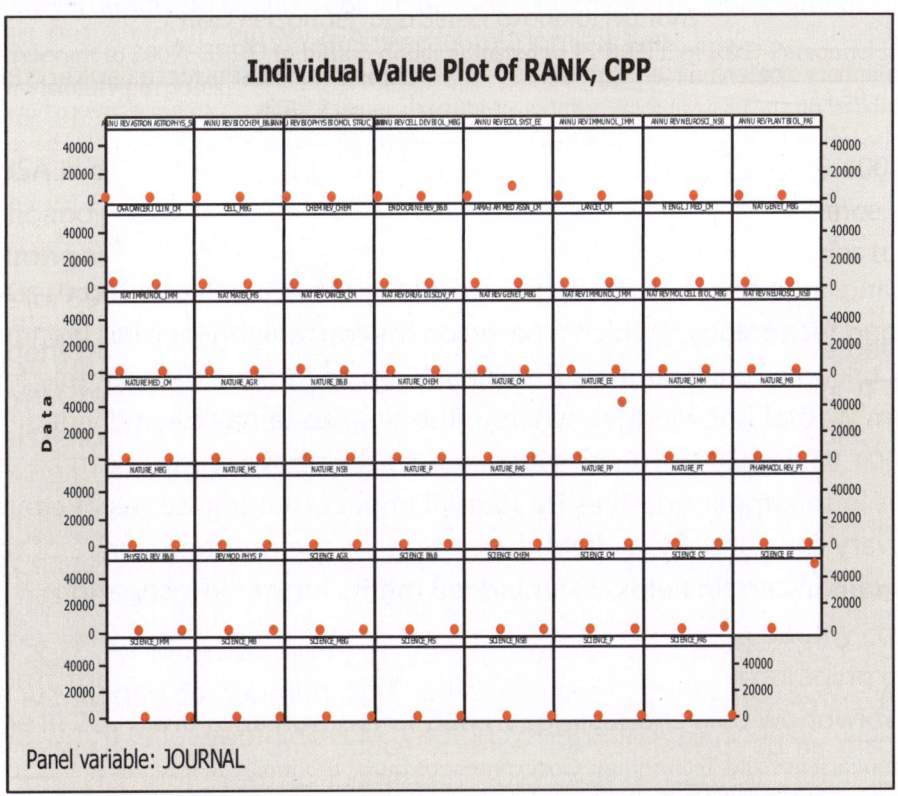

Figure 3.28 *Distribution of CPP and Rank for Individual Journals*

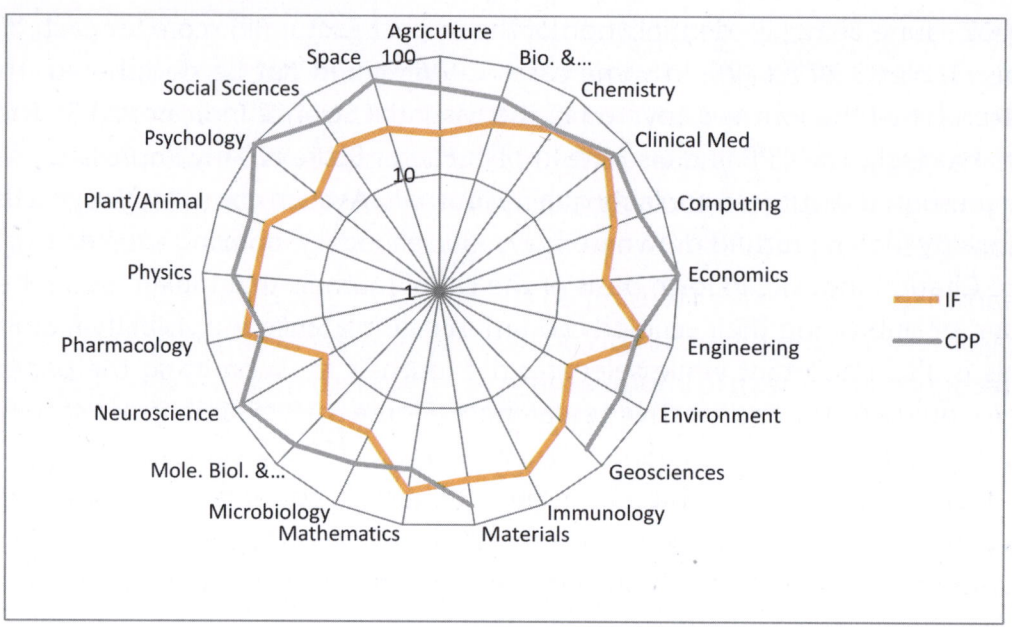

Figure 3.29 *Journal Analysis: Percentage Papers in the Top Quartile by Journal Impact Factor (JIF) and Journal Citations per Paper (CPP)*

[Time period January 2002 - June 2012, Ordered by percentage papers, ESI fields, India. The CPP quartiles has been calculated on the basis of total number of journals in each subject category covered by ESI and arranged on the basis of CPP/Journal in the descending order. The Impact Factor quartiles have been calculated on the basis of the journals in each subject category for the total (HCP) papers. In both the cases sample size includes total HCP's with Indian Authorship]

This by and large corroborates with the study by Thomson Reuters for the Department of Science and Technology,[91] which is based on the top ranking journals in various subject disciplines, taking into account a substantially representative data sample for a 10 year period (from Web of Knowledge), wherein the time zone has been divided in two time periods (2001-05 and 2006-10). As suggested by the above study, it can be argued that the journals in the upper quartiles by Journal Impact Factor in terms of citation impact is likely to vary substantially by field depending on the preponderance of certain high impact journals in certain fields.[92] This indeed merits further investigation.

---

[91] Department of Science and Technology, Government of India. Bibliometric study of India's Scientific Publication outputs during 2001-10: Evidence for Changing Trends. p. 181-182

[92] Department of Science and Technology, Government of India. Bibliometric study of India's Scientific Publication outputs during 2001-10: Evidence for Changing Trends.

Again, as stated in the DST report,[93] the relatively low share of India in top 1% journals has been a matter of concern. This report was released in 2011 and has coverage from 2001-10, where as another study[94] reports that the Indian publication share in top 1% impact making journals has been as low as 0.54. Thus the share in top 1% of the high impact journals could be treated as one of the parameters for assessing the global impact of national research output landscape. This has also been corroborated by DST report.[95]

Meanwhile, while we are focusing on the impact of the journals, many have argued and reasoned on laying the emphasis on the default value of impact factor.[96,97,98,99,100,101] While recognizing the need to improve the ways in which the outputs of scientific research are evaluated, the San Francisco Declaration on Research Assessment[102] (DORA's) focus lies on, over emphasizing Impact Factor parameter for research assessments. Of late, many journal metrics are available such as SNIP (Source Normalized Impact per Paper) and SCR (SCImago Journal Rank), ensuring by and large balanced treatment of journals from different fields and are primarily based on average and sources of citations to a given journal.

However rather than taking Impact Factor as a surrogate measure of quality of research output by individuals/Departments/institutions, a normalized impact score may be more appropriate. Thus considering CPP may be a better alternative to the existing Impact Factor Score in absence of a more viable solution. Many of the problems inherent in using Impact Factor can perhaps be obviated.

---

[93] Report of Science Advisory Council to the Prime Minister, 2010.

[94] Department of Science and Technology, Government of India. Bibliometric study of India's Scientific Publication outputs during 2001-10: Evidence for Changing Trends.

[95] *Urid.* Alberts, B. (2013). Impact Factor Distortions. (editorial) Science, 340, 17 May, 2013.

[96] Simmons, K. (2008). The misused Impact Factor. Science 322(5899), 165, 10 October, 2008.

[97] Alberts, B, Hanson, B, and Kelner, KL. (2008). Science, 321(5885), 15, 2008. Reviewing peer review.

[98] Seglen, P.O. (1997) Why the impact factor of journals should not be used for evaluating research. *BMJ*, 314, 498-502.

[99] Vanclay, J.K., (2012) Impact Factor: Outdated artefact or stepping-stone to journal certification. *Scientometrics 92*, 211-238.

[100] The PLoS Medicine Editors, (2006). The impact factor game. *PLoS Med* 3(6): e291 *doi:10.1371/journal.pmed.0030291.*

[101] Rossner M., Van Epps H., and Hill E. (2008). Irreproducible results: A response to Thomson Scientific. J. *Cell Biol.* 180, 254–255.

[102] San Francisco Declaration on Research Assessment: Putting science into the assessment of research. Available at: *http://am.ascb.org/dora/files/SFDeclarationFINAL.pdf* (Accessed on July 3, 2013)

3.3.7 Collaboration

Creating an eco-system conducive to seek inter-disciplinary (research) problem solving, in an endeavour for promoting innovation and (academic/research) entrepreneurship, it is important to nurture our ability to share and build. There are studies[103] that project two instinctive core issues in relation to research endeavours – difference (clash) in motives of the national (governmental) level and the individual level. While the former is concerned with overall gaining competitive edge over other nations in multivariate aspects (prestige, economic aspect, superiority), the individual researcher is driven by competition and curiosity. In either cases, creating and nurturing mechanism to build and enable academic partnerships with public/private agencies/institutions for seeking excellence in research endeavours is important.

The Royal Society study[104] on collaboration notes that the rapidly growing scientific nations that includes India are collaborating less than their developed counterparts, with over 70% of the research coming from the endeavours of national researchers. This low proportion of Indian collaborative endeavours has been further corroborated by various other studies[105–106]. These studies also observed that India is amongst the countries with lowest percentages of international collaboration. The reason for this, as reported in[107] states that this is due to Indian Law that restricts foreign universities to enter into wedlock of partnerships with their Indian counterparts.. However changes are being envisaged that would facilitate international collaboration thereby benefitting the Indian scientific system, reports the study.[108]

3.3.71 Collaboration Pattern

Keeping above in perspective, Thomson Reuters data has been used to examine India's international links. Nobody would argue in the present globalised scenario, that international collaboration is beyond an iota of doubt, an important marker of the

---

[103] The Royal Society, (2011). Knowledge, Networks and Nations: Global Scientific collaboration in the 21st Century. The Royal Society, London, p. 47.

[104] Tijssen R, Waltman L, and Van Eck N J. 12 May, 2011. Collaborations span 1,533 kilometres. *Nature*, 473, 154.

[105] Kamalski J, (2009). Small countries lead international collaboration. *Research Trends*, 14. Available at: *http://www.researchtrends.com/issue14-december-2009/country/* (Accessed on December 12, 2012)

[106] Nayar A, April, 2011. Educating India. *Nature*, 472(7), 24-26.

[107] *Ibid.*

[108] Department of Science and Technology, Government of India. Bibliometric study of India's Scientific Publication outputs during 2001-10: Evidence for Changing Trends.

significance of research activity to all the partners including others engaging with the domestic research base.

The data as indicated has been downloaded from Thomson Reuters database using author affiliation as the country of origin of research. The data thus analyzed have been presented in the Tables 3.33 - Table 3.41 and also represented schematically in Figures 3.30 – Figure 3.31.

In terms of volume, collaboration is increasing. Collaboration generally represents an increasing contribution to India's research output. The decadal collaboration scenario in general in all the subject disciplines shows an upward trend (Table 3.33).

On an average, the average rate of increase in collaboration in co-authored papers (during last four decades i,e;1971/80 – 2001/10) is 44.27% which means about 4.4 % per year. Alongside the average growth in number of papers, the average number of collaborating institutions during the same period has increased by 33.44% (approximately by 3.3% per year). By way of an international comparison, whilst international collaboration varies substantially by country and by field, the USA and China like India have much lower levels of international collaboration compared to the UK due to their larger internal geographies as reported in.[109]

Having said that, analysis to focus on discipline-wise collaboration pattern showed that during 1971-2012, (for the last over four decades) the highest collaboration has been observed for disciplines like Space Sciences, followed by Physics; Psychiatry/Psychology; Economics & Business; Microbiology; and Mathematics; while the lowest collaborating disciplines are Biology & Biochemistry; Plant & Animal Science; Pharmacology & Toxicology; Chemistry; Multidisciplinary, and with Agricultural Sciences having the least share of international collaborative endeavours. This corroborates to the above statement that the different subject disciplines show variations in collaboration patterns. Some research disciplines may have exception of touching base with localised research problems, which cannot be generalised for all disciplines.

There have been evidences that the average citation impact of internationally co-authored work is significantly higher than the overall average.[110]

A further analysis was carried out to decipher proportion of India's total international contribution corresponding to the number of countries. This was done by identifying

---

[109] Adams J, et al Patterns of international collaboration for the UK and leading partners (2007)
[110] Department of Science and Technology, Government of India. Bibliometric study of India's Scientific Publication outputs during 2001-10: Evidence for Changing Trends.

the unique countries (representing each country only once) and calculating the average based on the discipline wise total share (%) of India's collaboration with other countries for the years 1971-2012 (~ 4 decades). Based on the average, the countries whose proportion was above that average were taken into account for each discipline. The data reveals that on an average >=80% of the collaboration is as a result of alliances with limited number of countries in each discipline (number varies between 20-29 countries), with exception of fields like Clinical Medicine (42 countries with 88.92% share); Plant & Animal Science (32 countries with 85.74%) and Physics (31 countries constituting 85.72%) on one hand. On the other hand, the exceptions are disciplines like Geosciences (19 countries contributing 82.07% of share); Materials Science (18 countries with 83.05%); Computer Science (16 countries with 83.32%); Economics & Business (14 countries with 83.44%); while in the field of Agricultural Sciences, only 13 countries (crossing the threshold average) are contributing 64.5% of the total of India's share of international collaboration endeavours. This shows that India's major collaborative endeavours are with a limited group of countries (although the level of collaboration may vary from time to time) with sporadic efforts (where proportion of share is very less and the dispersion of this proportion is wide with a number of participating countries) with a wide range of number of countries (Table 3.33). Looking at the overall trend of the decadal growth over the last few decades, it is observed that with growth in number of countries almost by over 65% per decade, the share of collaboration is also witnessing comparatively upward trend. (Figure 3.30).

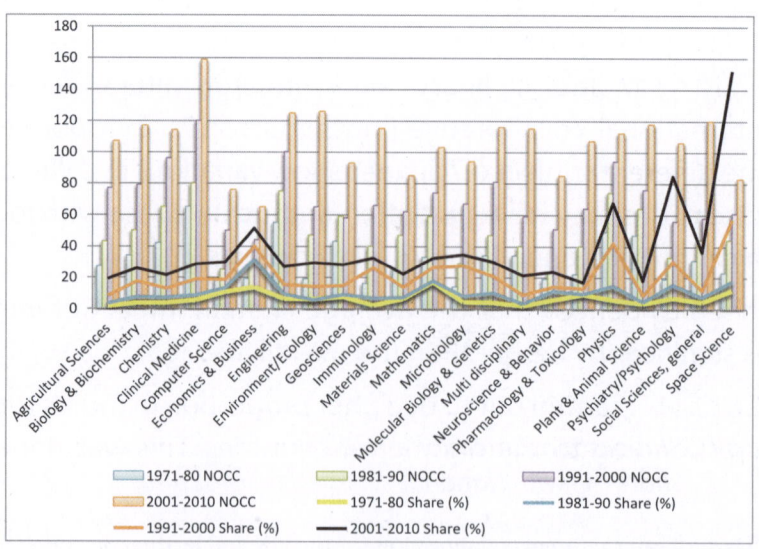

Figure 3.30 *Collaboration Pattern: India's International Collaboration Links*

3.3.72 *Whom is India Collaborating with?*

India's partners as the Tables 3.34-3.39 indicates, have largely remained stable (so far as the top collaborating nations with India are concerned), although there are some discernible changes

The USA was India's most frequent collaborating country throughout the decadal period under consideration in all the block years (1971-80 ;1981-90; 1991-2000;2001-2010; or 2011-12) period, representing 5.166% of India's total research output. There is a constant upward trend as move from decades 1971-80 to 2011-12 with an exception of couple of fields in one decade as observed in the case of Mathematics and Pharmacy. In these fields, there was a notable fall in 1981-90 when expressed as an overall percentage contribution to India's research output compared to the 1971-80 period. However, thereafter there has been again rise in the subsequent years (1991-2000; 2001-2010 and 2011-12). UK was the second most frequent collaborating country with India in the same period (0.786%) and the Australia and Germany at the third (.583%), and fourth place (.557%) respectively The USA stands apart in terms of its frequency of co-authorship with India-based institutions. However even with this rising trajectory, as reported in.[111] India appears to have been less well connected to international networks than some other countries.

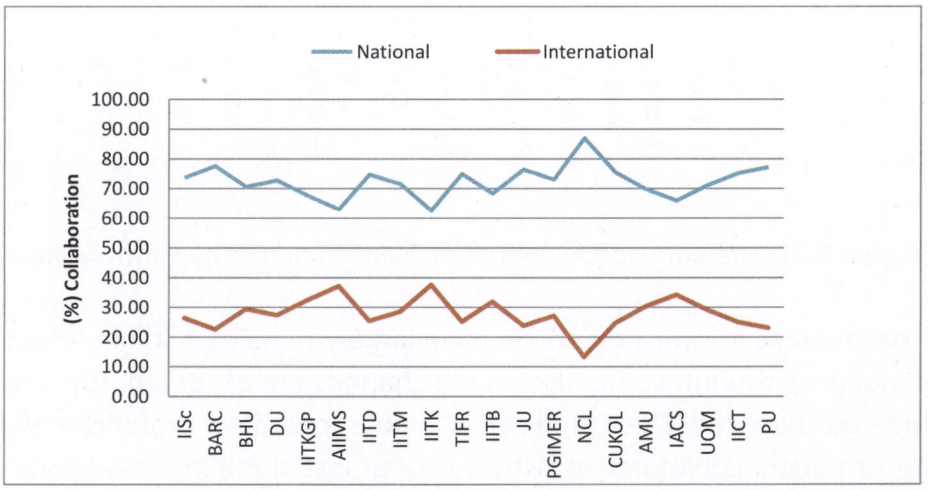

Figure 3.31 *Overall Collaboration Pattern of Top Collaborating Indian Institutions: Collaboration Share (%) with National/Internatinal Organisations*

---

[111] San Francisco Declaration on Research Assessment : Putting science into the assessment of research. Available at: *http://am.ascb.org/dora/files/SFDeclarationFINAL.pdf* (Accessed on July 3, 2013)

In an assessment of India's top 20 research partners over the last 42 years, the USA stands apart in terms of its frequency (60%) of co-authorship with India-based institutions, while remaining 40% with other 6 countries (with China & France about .5% and Germany, Japan, Korea and UK about .25% each) (Table 3.37). However, the level of collaboration as a fraction of national domestic output is lower for India, than it is for other emergent nations. Perhaps, this coordinates to India being less connected to international networks than some other countries. Having said that, Table 3.34 clearly indicates the brighter side of the scenario, that is rising trend over the decades. In a further investigation the time series analysis shows that there is encouraging growth in terms of collaboration over the last few decades 44.27%. Therefore, it retains a significant capacity to expand its collaborative links as the upward trajectory indicates.

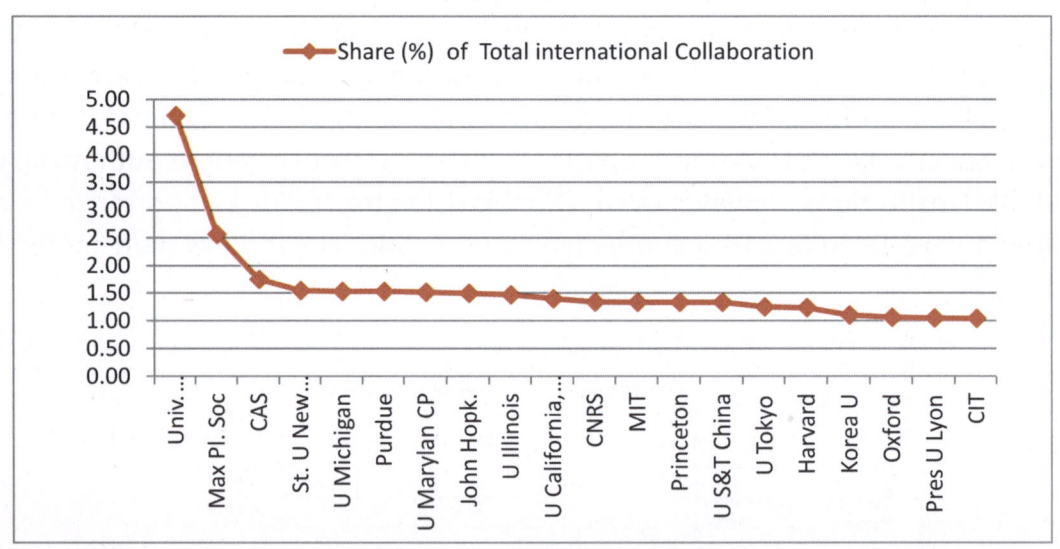

Figure 3.32 *International Organisations Collaborating Frequently with India*

India's partners, as is indicated above have largely remained stable (especially with developed nations), though some discernible changes are observed. Top 3-5 countries are by and large the same. Over the years the countries have (significantly) increased percentage of papers collaborative with India, almost doubling in volume of Indian collaborative output with some countries.

This infers that India is making constant endeavours to reach out to the international partners for research excellence. For the subject-wise analysis, top 10 countries collaborating with India have been considered. However in instances where several countries have the same proportion (%) of India's total, in such cases all the countries are being accounted for.

The subject-wise analysis (Table 3.34) of top 10 countries in each subject discipline reveals features which are common to most (though not all) fields and highlights several key themes in these analyses, though, there are few variations by field.

The USA is India's most frequent collaborating partner in all fields, often by a significant margin, while the UK and Germany are often in India's top three most frequent collaborating partners. There is capacity for further international collaboration particularly in certain high-growth fields. However, there are few variations by field, which we discuss here.

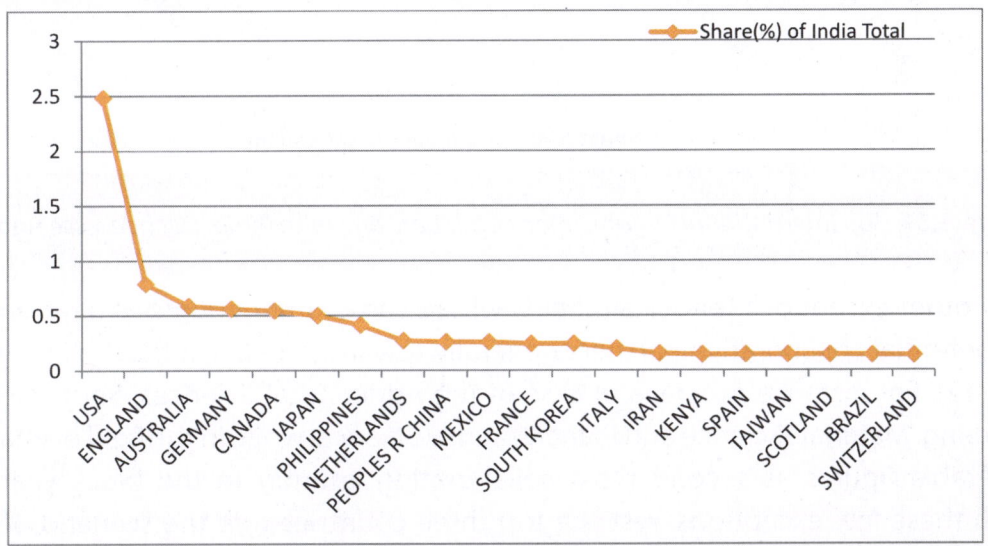

Figure 3.33 *Top 20 Countries Collaborating with India (1971-2012)*

In certain subject areas, Canada is also frequently collaborating partner only next to USA mostly in some block years and particularly in initial decades. For instance, the fields like Economics & Business (1971-80 & 1991-2000), Environment/Ecology (1971-80), Engineering (1981-90), Geosciences (1981-90), Psychiatry/Psychology (1971-80), Mathematics (1971-80, 1981-90, 1991-200), while for computer science, second most collaborative partner for all the block years have been Canada.

Similarly in the field of Microbiology, Japan has been observed as the second favoured collaborative country for the block years 1971-80 and 1991-2000.

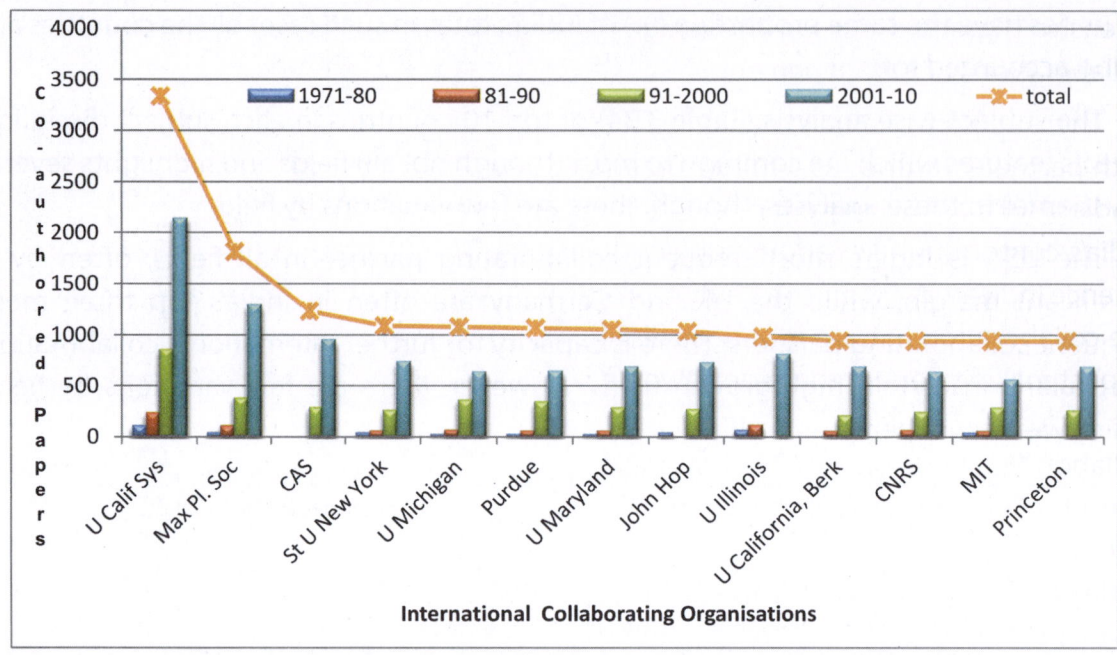

Figure 3.34 *Top International Organisations Collaborating with India: Decade-wise Analysis*

Few other exceptions for certain fields where some other countries have emerged as the potential collaborative partners generally towards the latest block of years (that is 2011-12). For instance Australia in case of the Agricultural Sciences, South Korea for Engineering, Materials Science and Plant & Animal Science, while in the field of Pharmacy, Saudi Arabia figures as second most collaborating country in the block year 2011-12. With these few exceptions, rest the top three countries rule the scenario of India's international research partners.

The diversity of India's research ties is elaborated in Table 3.37, which presents a selection of international organizations (Top 20) that have made especially numerous collaborative endeavours with Indian institutions in the last four decades. The relatively sparcity of the European partners, and absence of UK institutions in top 10 will be a surprise to policy observers.

Table 3.39 portrays the top Indian organizations that are mostly collaborating with international partners. Of the 20 top collaborating Indian organizations, there are 7 Universities, with a few surprises (Jammu University, and Aligarh Muslim University) leaving some other anticipated universities behind. While, the remaining 13 organizations (with all the old 5 IITs in the top 20 list) are as expected, having research active and intensive characteristics. On an average (Table 3.40), the ratio of national to international collaboration is 72 to 28 percent.

There could be value in exploring the citation impact gain of collaboration for India with partner countries as a follow-up analysis which would bring value addition to the studies, such as the present one. This has been tried on a limited sample of top 1% of India's contribution in various subject disciplines. The data downloaded from Essential Science Indicators (Thomson Reuters product) was arranged subject disciplines wise and the collaboration pattern in various disciplines was studied. Since these papers are high impact making papers, therefore it would be worthwhile to project whether such papers are exclusively Indian contribution or they have emerged as a result of International collaborative links. The study shows (Table 3.41), by and large, top impact research is a result of collaborative endeavours. Particularly with the fields like Space Science and Chemistry, not even a single paper is from indigenous efforts (resulting from research endeavours of the same institute), while for the fields like Clinical Medicine and Plant and Animal Sciences, very small proportion of high impact papers are brought out without international collaboration.

It may be worthwhile to mention that the analyzed observed data revealed that the collaborating authorship pattern in top 1% of publications in a given discipline are by and large having same group of people as collaborating partners (with little variation so far as additional members are concerned). This needs further merit. Such studies that focus on key people are important. If the above trend is observed on a larger data sample of high impact co-authored publications, it divulges that group of people leading the research front, nurturing second in command can bring around excellence in research. Thus nurturing our ability to share and built through alliances, thereby promoting innovation and entrepreneurship. As research in all domains of knowledge is an enterprise in itself, while researchers are entrepreneurs indulging in an act of collateral academic/knowledge entrepreneurship.

4. CONCLUSIONS AND OBSERVATIONS

4.1 Implications for Reputation, Research and Production of Knowledge

Impact of global rankings on higher education research and the production of knowledge have been catching up, of late. International research assessment has emerged a key component of strategy, planning, management and implementation and as such a meaningful systematized route to maximizing research output and guiding excellence in higher education. However a comprehensive, effective and efficient research assessment model is a complex and context-sensitive system, which assists and enables all levels of stakeholders (government, institutions, individuals and society). As a nation, India had been the country with the highest population of S & T qualification since the seventies.

While higher education landscape reveals encouraging advancements, there is yet immense potential for systemic improvements that can create a transformational impact and place India at a competitive position on the international landscape, as well as create a quality balance within the nation.

A major theme underlying the current higher education and research context is the internationalization of higher education and research, whether it is represented by increased competition or collaboration. Systems and initiatives need to work well with this global perspective, and promote a culture and spirit of excellence towards personal, national, and international goals. There are myriad international experiences to take a cue from, but key lies in innovating and encouraging internally generated solutions that cater specifically to the unique Indian context.

It is also significantly desirable to create awareness and education regarding research assessment and ensure that the emergent systems are universal in their acceptability, access and adoption throughout the academic community.

While research assessment is an important tool towards realizing strategic vision, it by no means, must be confused with the end goal of research excellence itself. As such it must always be valued only in terms of the purpose served within a given context.

A huge demand for quantitative studies and indicators for impact of research exists now. This demand does not come only from the governments, but also from researchers for different purposes. For instance evaluating performance as return on investment and for theoretical purposes respectively.

There are various key components that need detailed attention including (but not limited to) data management; assessment methodologies, metrics and parameters; resource and fund allocation and performance assessment. This requires a comprehensive exercise taken forward in conjunction with various stakeholders. Based on the experiences and analysis contained in this document, below are some indicative suggestions and cautions to mark the way forward.

➢ Data Management: Collection, Accuracy and accessibility

➢ Metrics and Methodologies

➢ Parameters – Institutional profiles

➢ Usability: Strategic Planning, Funding, etc.

➢ Internationalization : Competition and Collaboration

➢ Third Mission

Each of the above indicated issues are interlinked and thus have bearing on one another. While data (authentic) is in place, the metrics and methodologies can be evolved for creating evidence, based on which feasibility of identifying features for building institutional parameters across higher education system could be made possible. This would ensure strategic planning based on evidence based metrics and promoting competition and collaboration for setting alliances with the best in the world, while addressing challenges for sustainable development, thereby fulfilling third mission.

4.1.1 Data Management: Collection, Accuracy and Accessibility

Innovation requires diversity of high quality research and development of standardized (normalised/rationalised) performance metrics that reliably reflect the complexity and societal expectations of contemporary research. Enabling the highest impact research requires observed and expected data to assess programs and evaluate key opportunities in a resource-constrained environment. Thus, for strategic decision-making at different levels, requires data that reflect a local, national and international scope.

The data is not required only for attracting sponsors in academia but more importantly making academic system more efficient and effective. Though a few institutions have the predictive and program-based comparative data to make informed decisions, many are out of this ambit. Even where data are available, the organisations lack the discipline to utilize the data appropriately (for guiding them). Added to this is the major hurdle of data inconsistency, it has been deciphered that, data are not available holistically about research programmes, faculty, grants, strategic collaborative alliances, and other related administrative data, critical for guiding decisions by policy makers and potential sponsors.

It is thus imperative that strategies need to be re-oriented for *open and consistent data infrastructure*. Re-inventing may not be necessary, existing international best practices can be adopted, such as to create something on the lines of *Lattes data base* in Brazil (http://lattes.cnpq.br/). The Lattes Platform is an information system comprising integrated data-base, web-based query interface, etc., maintained by the Brazilian Government to manage information on science, technology, and innovation related to individual researchers and institutions working in Brazil.[112] It may be thus worthwhile to put together a facility of evolving a *unique research identification numbers or Open Researcher and Contributor Number*. This will ensure creating a database of research and researchers at the local, state and national level in a *trusted environment*. Such a system needs to be mandated by the authorities at the national level, if accountability, accessibility and authenticity parameters are to be factored for creating a repository of research resource-base in a trusted environment. Once such an initiative is in place, gap analysis is feasible to be undertaken, many subsequent data deficit functions and processes could be put in place and gap analysis made to bridge the gap. For instance, the *Lattes database* is made interoperable with Elsevier's SCOPUS, one of the world's largest abstract and citation database of peer-reviewed research literature, *Curriculo Lattes* are mapped to matching Scopus author IDs allowing users to access relevant profiles easily between the two databases. Thus the link to *Curriculo Lattes* greatly enhances Brazil's science research community by allowing more efficiency, visibility, and connectivity amongst Brazilian researchers Therefore, such a facility will truly support data management in terms of its collection, accuracy and accessibility.

[112] Mena-Chalco, J.P., (2009). ScriptLattes: An open-source knowledge extraction system from the Lattes platform. *Journal of the Brazilian Computer Society*15 (4): 31–39. *doi:10.1007/BF03194511. ISSN 0104-6500.*

4.1.2 Metrics and Methodologies

The overarching aspiration of higher education institutions (HEIs) is to offer most highly recognized education in its areas of expertise, and perform highest impact research. Optimizing the potential impact of nation's academic research, in order to accelerate research excellence, provision to evaluate, benchmark & strategize is necessitated to be factored in the academic and research landscape system.

In devising research methods for evaluating the impacts and outcomes of research projects or initiatives, research tools and instruments must be engaging and designed with a clear idea of the proposed end user in mind. They must be able to be practically implemented and appropriate to the individuals or groups concerned; otherwise there is a risk that the tools themselves will undermine the process of evidence gathering.

There are a number of research indicators that can be used for the description of research activity and results, and more are being developed. If we adopt a systems' approach, a distinction can be made between indicators of *input, structure, process* and *results*. Input indicators relate, for example, to the ability of research systems and research organizations to attract, in competition external research funding and to recruit recognized researchers and research groups. Structure indicators relate, for example, to indicators of reputation associated with the researchers' status in their network, including membership of editorial boards and international scientific committees. Process indicators relate, for example, to the researchers' participation in conferences and the like. Finally, result indicators relate to output and effect, including publications, citations, patents and contributions to research training in the form of PhD production.[113]

4.1.21 Metrics: Challenges

While setting the tone for assessments/measurements, the question is not whether or not to measure but how to measure. The assessment is based on universities evaluation taking into consideration multi-pronged approach to academic and research evaluation, besides administrative parameters. Academic evaluation encompasses assessment of scholarly activities, achievements, and return on investment (outcome of research investment), and the like; while the research evaluation is more specific than the academic evaluation. Thus besides research, teaching, services, stakeholders and overall

---

[113] Cave, Martin; (1997). The Use of Performance Indicators in Higher Education: The Challenge of the Quality Movement. Jessica Kingsley Publishers, 1997, ISBN 1853023450, 9781853023453, 276 pages *Higher Education Policy Series*, ISSN 0954-3716

administration defines the higher education institutions' (HEIs) evaluation. The ranking and evaluation through similar concepts differ in purpose and outcome. For instance, evaluation sets a benchmark against which a university performance in certain aspects could be assessed with a goal to surpass the basic level of requirement. While institution ranking shows its relative strength or otherwise in relation to its peer institutions in the areas represented by the indicators.

There are several systems (indicated in previous discussions) those currently rank universities in relation to their research intensity and, to a lesser degree, performance. These involve both public and self-reported data. Other national measures of research activity related to Government/public total funding obligations and self-reported total research expenditures are more subjective rankings and reward size rather than productivity, quality, or impact.

At the same time, there are subjective and objective approaches to research evaluation complimenting each other. While subjective approach (peer evaluation) continues to be trusted by the research community, the objective approach (using bibliometric/scientometric analytical tools) are subjected to many a time to criticism. Nevertheless, not withstanding antagonism, factoring normalisation (using average citations/paper and its derivatives thereof) does reveal evidence based metrics in a trusted environment.

Debating the health of any environmental domain (any aspect/phenomenon that is relevant in a given situation/institution/region/etc.) pre-empts an informed debate which necessitates measuring. For any intuitive predictions, mere right to know is not enough, it is duty to know to be able to accurately predict thereby avoiding being perilous.

In order to do so it is important that the nature of research is focussed on and made clear. For instance, university research by and large is investigator oriented focussing on fundamental research, problems narrowly defined, with education as the primary goal and teaching/mentoring as the core activity. On the other hand, public funded research is generally team driven, multi-disciplinary (interdisciplinary) with (comparatively) very high end problems needing solutions, addressing long term needs of stakeholders (industry and society). While the industrial research though multi-disciplinary and team oriented has business flavour with focus on small or medium term. Research enterprise is thus a complex fame game. For instance research favourites–solo/collaborative, mission driven/leading to IP (intellectual property), and/or products and prototypes, aimed at societal needs and teaching/mentoring/communication purposes.

The research assessment has number of challenges with focus on - both quantity and quality, consistency in performance, not occasional peaks, crediting authors (first and subsequent), anxiety of career graphs (early/late stage) and the like. Currently, individual and institutional assessment by and large seems to be fragmentary with skewed distribution of performers.

4.1.22 *Metric Parameters*

While for some, measurements are necessarily evil, for others it is an angel. In either case, the methodology followed needs to be coherent in nature – encompassing tangible/intangible rewards, capturing all aspects of research and developmental activities that support/facilitate and transmit ideas for further innovation and to a greater extent should endeavour to obviate the hazards of assessment.

Poor or misuse of metrics or over emphazing and linking them to personal rewards leads to perverse and devious human behaviour, besides impact funding decisions, thereby sidelining good researchers. Good metrics must be grounded in theory, data driven and owned by the research (scholarly) community. Thus (good) metrics that are fair and acceptable to the R&D community is necessary, though difficult to develop, yet not impossible, keeping national priorities and flavour in view. Moreover, strong incentives are required to use them uniformly across all decision making.

A multidimensional view of research landscape is required wherein one needs to move beyond economic measures to measure the quantifiable and there are very few indicators that link domains of research to these economic pay-offs. The systematic measurements and indicators having impact on research landscape should also thus focus on social, cultural, political and organizational dimensions. The social responsibility of higher education institutions fulfilling 'Third Mission' (3 missions -teaching, research and service as third mission generating new knowledge to pressing problems) have been emphasised by many in one or the other context.[114,115,116]

---

[114] Tandon, Rajesh, and Hall, Budd (Co-Chairs). UNESCO Chair on Community based research & social responsibility in higher education: A framework for action 2012-2016. Available at: *http://www.livingknowledge.org/livingknowledge/wp-content/uploads/2012/09/UNESCO-Chair-on-Framework-for-action.pdf* (Accessed on May 21, 2014).

[115] Godin, Benoit and Dore, Christian, (2005). Measuring the impacts of science: Beyond the economic dimension, urbanicatia INRS, Culture at Societe. Helsinki, Finland Helsinke Institute for Science and Technology Studies. Available at: *http://www.csiic.ca/Godin_Dore_Impacts.pdf*

[116] Reeves, Michelle, (2002). Measuring the economic and social impact of the arts: A review. London: Arts Council of England.

A far sighted action can ensure that metrics goes beyond identifying a few researchers, nation or ideas; to capturing the essence of what it means to be a core research centric and research intensive researcher/institution and the nation. Its essence lies in its impact 'a dynamic concept' which pre-supposes a relationship of cause and effect. It can be measured through the evaluation of the outcomes of particular actions, be that an imitative, a set of initiatives forming a policy or set of policies which form a strategy.[117]

Therefore statistics of evaluation has to be inclusive, stakeholder centric and development oriented. In the absence of a productivity-based assessment system, each institution (university and sponsor) believes that they have the best system or at least the means to discover what changes should be made that would be an improvement over their current processes.

In the present case of health of research landscape in India, the *performance ranking* of institutions need to cater to both the quality and quantity of research and current research performance and other factors crucial to the research assessment excellence and for framing research excellence framework.

A carefully designed quantitative data based ranking offers objective information for policymaking, besides setting a benchmark for the HEIs to strive to be the best of the rest, thereby engendering more peaks of excellence.

Using multipronged approach for finding meaningful performance metrics and procedures in the Indian higher education system, there is a need to go for a comprehensive policy framework which channelizes entrepreneurial energy. This will eventually assist in unlocking the potential for productivity improvements by *securing institutional success*, and by *driving research outcomes* through *improved information discovery*.

Clearly, it may be pertinent to devise and develop local multipliers for the sector or national benchmarking metric parameters for evaluating (part of it available in the NAAC system) and ranking at the national level, taking clue from the existing international ranking systems. Together with this, developing metric based analysis of the HEIs/ university's research output in order to "form a world class educational & research base", is necessitated, thereby accelerating growth in an increasingly competitive global academic market. Identification of attributes (for assessing & benchmarking) relevant to the national context may be important, because of its uneven & diverse character.

---

[117] Landry, C, Bianchini, F, Maguire, M and Worpole, K, (1993). The Social Impact of the Arts A Discussion Document, Comedia, Stroud.

Prestige Factors of a University or a higher education institution include:

(a)	**Teaching-Learning Environment**	• Academic Programmes • Faculty • Students • Research and Research impact (measured using metrics) • Awards/ Honours, alumni and staff winning international prestigious Prizes and Field Medals
(b)	**Collaborations**	• International diversity/mix
(c)	**Institutional Administration**	• Administrative Structure & Functioning • Financing & Funding • Industry Income
(d)	**Infrastructural Facilities /Usage**	• Building & Infrastructure • Technology Usage
(e)	**Others**	• Stakeholder participations

It is important to feature these prestige factors while devising methodology for metrics and measurements for evaluation and ranking.

It may be worthwhile to consider setting up of an independent centre - *Centre for Excellence for Academic and Research Performance & Evaluation*. The centre can facilitate (besides other things) the following:

• undertake research to identify parameters

• assess periodically methodology used &

• bring out white papers (sensitize stakeholders & capacity building)

The reliability of the methodology requires that the system generates consistent results in replication, and its validity concerns how well its indicators conduct the evaluation taking into account normalisation/rationalization of the interacting variables and entities.

4.1.3 Parameters – Institutional Profiles

Research institutions and universities are expected to have contributed repeatedly to the development and economic prosperity of the nation. Higher education has evolved in response to internal and external pressures and will continue to do so. Thus the foundation of success of academic research rests on a high degree of programmatic self-direction, a competitive environment that rewards success, and an entrepreneurial approach to attract the resources necessary to be successful.

To enable impact oriented research that addresses significant social challenges, higher education institutions and research sponsors must work together in providing flexible and adaptive strategies, tactics and most importantly operational structures.

Innovation requires diversity of high quality research and development of standardized performance metrics that reliably reflect the complexity and societal expectations of today's research. This should be initiated by the academic research community, in partnership with key stakeholders.

For obtaining a holistic view of the health of Indian research landscape, measuring scientific, economic and societal impact needs to be considered in a cohesive platform. Developing and managing a research portfolio is not easy for the individual researcher, and it is equally, if not more difficult, at the institutional level. There are many points of failure and the benefits are often not easy to measure or immediately obvious. On a broader level, universities are heavily regulated and scrutinized by government, sponsors, and the public who seek transparency, integrity, and measurable value for their investment. Competitive academic environments require efficient and highly responsive evidence-based decision-making that depends on data rich information systems.

Institutions are increasingly becoming aware of necessity of performance data. Strategically, timely research performance data informs an institution of its performance and competitiveness and allows it to make decisions based on facts rather than instinct or opinion.

Within the research university community there is a growing recognition of the need for research intelligence and well-established performance and risk management systems. These can help focus institutional strategies on research quality, raise the profile of the institution nationally and internationally, manage talent and resources and build a high-quality research environment. Yet, there is considerable dissatisfaction with current systems and a lack of coordination within and between institutions, as each implement their own solution to what are often common needs.

Broadly speaking, the future aim is to develop well-defined solutions while exploring how these can be implemented to create synchronization of demands and aspirations of various stakeholders. For building an evidence base, nobody would argue a consistent theme emerging from such debate on assessment and ranking, including agreement around the need for (i) common definitions and concepts to underpin measurement of research outcomes and consistency in their use; (ii) the need for systematic evaluation and more robust methodologies and evidence; (iii) embracing a multi-value approach to impact/outcome measurement which recognises both quantitative data, qualitative description and administrative narrative; (iv) the need to establish the relationship between intermediate and strategic outcomes accruing from research projects; and (v) specified degree of standardisation of methodologies to enable comparison between different levels of intervention, and between different scales of organisations and initiatives, but which allows flexibility to address specific contexts, sector/sub-sector needs and priorities, and target groups.

For building institutional profiles on the identified parameters across HEIs on the basis of deliberations in point 1 (data management) and 2 (metrics and methodologies) would enable to generate persistent yet interoperable system that would facilitate unified approach across institutions to be assessed and ranked using similar datasets. Once such a system of cohesive institutional profiles for majority of Indian HEIs are in vogue, the third party services for furthering and nurturing the assessment system particularly for research component could be put in place which seems very much feasible. For instance, the *Lattes database in Brazil*, is collaborating and made interoperable with Elsevier's SCOPUS database, while Times Higher Education draws evidence based metrics for the research component based on the big data analytics from Thomson Reuters database. Such a tie up could facilitate identification of research impact (based on citations), research excellence (based on highly cited papers and high impact journals) and normalization of impact scores to help overcoming distortions.

4.1.4 Usability: Strategic Planning, Funding, etc.

Assessment reflects increased competition to be best of the rest, increase prestige and ranking, funding and productivity. Governments are increasingly focusing on research performance measurement (RPM). Various governments attempt to have a more mature assessment process for research excellence that does not aim only to enable the higher education funding bodies to distribute public funds for research selectively on the basis of quality, but also reduce the clear distinction between the arrangements for science-based subjects and those for all other subjects. Thus factoring evidence based metrics evaluation across all domains of knowledge.

Research assessment is becoming more transparent, comprehensive and systematic. There is a need for a data driven scientific approach for guiding investments using gap analysis, policy making for funding education and research in the country, while assessing outcomes for future support. These guidelines are important to attract and retain talent through well defined strategies for creating centres of research excellence, and formulation of research policies that drives the quality of higher education and research. These challenges are to an extent taken note of by the stakeholders.

Within the higher education institutions research community, there is a growing recognition of the need for research intelligence and fairly well established research performance management systems. These systems can factor multi-variate strategic measures conducive for quality research environment, such as - facilitate in focusing on institutional strategies for research quality, enhance institutional profile (at national or global level), and manage resources (both retaining talent and other resources) and the like. Currently, institutions are going their own way, based on the individual institution's basic requirements. However, the demand for competitive advantage coupled with restrictive resources, the sponsors are expecting that for the strategic planning, the identification of peeks of excellence and subsequent cadre levels of institutions is a must to nurture the excellence and support to raise few additional peeks of excellence. Thus seeking to develop an understanding of shared institutional profiles (for maintaining research portfolios) as discussed in point 3 above that could sustain current efforts and give boost to their continued success. An important element of any strategy will be to enhance evidence based decision making with reliable and performance based data focused on maximizing the scholarly potential and research productivity of faculty and research community. Usability of such solutions undoubtedly helps all stakeholders including the sponsors for funding cutting edge research and maintains a robust research mission. For instance many Governments in various countries along the globe are taking these indicators very seriously to fund research in their countries. At the same time, such metrics also brings with it the visibility and other achievements (honours and awards). The usability of such systems on one hand is required for facilitating adequate funding from sponsors/funding agencies, besides making academic system more effective, efficient and responsive to the needs. Thus the research assessment and relative ranking are important to universities and higher education institutes because they are the basis of institutional reputation, signal a presumption of excellence and capacity to sponsors, and serving to attract talented faculty, good students and funding donors.

4.1.5 Internationalization: Competition and Collaboration

It is crucial to drive international collaborations, synergizing beyond boundaries so as to factor competition, choose and work with best minds in the world and be party to provide global solutions to critical global problems. The existing set of indicators such as number of scientists, engineers, corporate and government R&D, intellectual property (productivity and trade), etc. on one hand, and indicators relating to societal issues for sustainable development are quantising the pace of change in research that is reflected in the competitiveness for leveraging resource advantage and rewarding collaboration.

The increasing tendency towards a multipolar scientific world demands concentration of development activities to widely dispersed research hubs addressing (both) local and global issues of sustainability.

So the real attributes of a scientific/research team can be viewed in terms of (multivariate issues ranging from) problem solvers, integrator, implementers, (to) problem finders, predictable/unpredictable, simple/complex roles, ranging from creative generators of new ideas to coordinators who keep everyone working together.

Keeping in touch is the key thing, and that means meeting as many people working in the field. This is because science is at heart a progressive evolutionary subject and a collective endeavour. However in many cases, the evidence suggest that while the domain of art and artists have always fathomed collective endeavours (musical orchestra, opera….), science and scientists by and large have solo solace landfills. But on the contrary now, strong alliances are sought, research activities in the world (more specifically for scientific ones) are becoming increasingly interconnected. International collaboration is on rise, which is evident from the fact that over 35% of research publications in international journals are through international collaborative endeavours.[118] Global Challenges have received great attention of late, and all countries (developed/developing) have a role to play to engage in global efforts for arresting these challenges.

As such, need to nurture research by building strong alliances are a must for sustenance and sustainability.

---

[118] Royal Society, London. Knowledge, networks and nations: Global scientific collaborationin the 21st century, (March 2011). Science Policy Centre Report 03/11, p.6.

4.1.6 Third Mission

Research is less and less assessed on scientific impact alone – we should aim to quantify the increasingly important contributions of science to society.

Lutz Bornmann[119]

Institutions of Higher Education are institutions in perpetuity. If GE, Microsoft and IBM can do cutting edge research in India, so can our institutions of higher learning and research. At a time when, Corporate Social Responsibility (CSR) – the societal responsibility of companies, a smart approach to sustainability is being promoted all around, where mandatory spending on CSR is brought in vogue. The companies have to spend at least 2% of its average net profit during the three preceding years on CSR activities. The amount has to be preferably spent near or around the areas company operates with stringent auditing procedures.

This being the case the social shaping of the national research (science) base becomes all the more important. Because where private sector is expected to contribute for the societal good, the public sector needs to chip in its own way to fulfil the *Third Mission*. Addressing questions as to what societal problems have been solved by the given piece of research produced by the institutions in public sector is important. Thus higher education institutions have to factor social responsibility in their research mission.

However, the roles of Higher Education Institutions (HEIs) and universities as public institutions, and knowledge as a common public good, can only be advanced in the perspective of knowledge society. In this perspective, knowledge is seen to serve the larger public purpose for human and social well-being, even though for the most part the nature of knowledge and the origins of knowledge are not brought into question.

The Final Communiqué from the UNESCO World Conference on Higher Education of July 8, 2009[120] opens with a section on the "Social Responsibility in Higher Education". While *Item 1 notes*: "Higher education is a public good and the responsibility of all stakeholders." *Item 2 notes*: "Higher education has the social responsibility to advance our understanding of multifaceted issues...and our ability to respond to them... It should lead society in generating global knowledge to address global challenges, inter alia,

---

[119] Bornmann, Lutz., (2012). Measuring the societal impact of research. EMBO Ref. 2012 Aug; 13(8), 673-676 *doi:10.1038/embor.2012.99*.

[120] Tandon, Rajesh, and Hall, Budd (Co-Chairs). UNESCO Chair on Community based research & social responsibility in higher education: A framework for action 2012-2016. Available at: *http://www. livingknowledge.org/livingknowledge/wp-content/uploads/2012/09/UNESCO-Chair-on-Framework-for-action. pdf* (Accessed on May 21, 2014)

food security, climate change, water management, intercultural dialogue, renewable energy and public health." Therefore it seemingly is time to review and reconsider the interchange of values between HEIs (university) and society and thus the felt imperative need to rethink the social relevance of universities and other HEIs.

On September 23, 2010, eight international networks[121] supporting community–university engagements across the globe gathered to issue a call for increased North-South cooperation in community–university research and engagement. They called for "all higher education institutions to express a strategic commitment to genuine community engagement, societal relevance or research and education and social responsibility as a core principle."[122]

In the current scenario, HEIs are expected to serve three missions, viz. teaching, research and service. While mission of teaching (or education) and research (or knowledge) have dominance, the third or the service mission seems to be independent though emerging as equally (if not more) important. Community engagement may sometimes actually contribute to improvements in HEIs, especially to their teaching and research functions.

Broadly defining the mandate of the UNESCO Chair[123] in three distinct, yet inter-related arenas, the suggested framework for action must include:

➢   Research and Knowledge Mobilization

➢   Capacity Enhancement

➢   Policy Development

Research & Knowledge Mobilization again has two broad streams of research to focus on – (i) Innovations in Community based Research Methodologies. (ii) Synthesizing local experiences with professional expertise, and eventually the synergy of such interactions leads to the co-creation of new knowledge on a variety of issues facing the people and communities (variously emphasized under the rubric of MDGs, Human Development, Sustainable Livelihoods, etc), that has practical and theoretical relevance and resonance.

---

[121] The eight networks were the Centro Boliviano de Estudios Multidisciplinarios (CEBEM), Commonwealth Universities Extension and Engagement Network, Global Alliance on Community Engaged Research (GACER), Global Universities Network for Innovation (GUNI), Living Knowledge Network, PASCAL International Observatory, Participatory Research in Asia (PRIA), and the Talloires Network.

[122] Tandon, Rajesh, and Hall, Budd (Co-Chairs). UNESCO Chair on Community based research & social responsibility in higher education: A framework for action 2012-2016. Available at: *http://www. livingknowledge.org/livingknowledge/wp-content/uploads/2012/09/UNESCO-Chair-on-Framework-for-action. pdf* (Accessed on May 21, 2014).

[123] *Ibid.*

The approaches in social responsibility lie in the demand for public accountability and local relevance of higher, post-secondary education which is growing rapidly in many societies. This demand is being responded in many different ways by different types of institutions through service learning and student internships, by co-production of knowledge (with local communities acting as partners), or by bringing in the experiences of communities and practitioners in designing curricula and teaching new problem and issue-centred courses. This social responsibility is expressed both, inside and outside the institutions.

For the capacity enhancement, community based research has emerged over the past 35 years in a variety of discourses and practices. While easy to define as an approach to knowledge construction based on themes, issues and questions coming from the "community," the reality is that this is a complex and value based process.

As a policy development, the capacity of institutions of higher education to sponsor, support and promote initiatives that deepen social accountability practices internally and externally, needs to be strengthened in an action-learning mode. Practical experiences and insights gained from actual efforts will form the basis for such dialogue and sharing. Special attention needs to be given to the institutional policies, structures and leadership.[124]

Ranking has several advantages, while it makes it easy to compare and contrast the performances of the universities being evaluated, it also focuses on relative achievement of each HEI and facilitate in identifying problems/ suggesting directions of development, brings in accountability and transparency in making HEIs data public (open information) about a university's performance and above all makes provision for objective information for policymaking (based on a carefully designed quantitative data based ranking system).

Functionally, there is a lack of means to objectively evaluate the impact of policy alternatives, organizational structure, or different administrative approaches. Consequently the result is an academic research enterprise that is recalcitrant to change or experiment, even in the face of impending decline. Such attitudes have long-term implications upon healthy competitive growth and hence the core of the academic enterprise.

---

[124] Tandon, Rajesh, and Hall, Budd (Co-Chairs). UNESCO Chair on Community based research & social responsibility in higher education: A framework for action 2012-2016. Available at: *http://www. livingknowledge.org/livingknowledge/wp-content/uploads/2012/09/UNESCO-Chair-on-Framework-for-action. pdf* (Accessed on May 21, 2014)

Many studies are pointing towards identification of parameters for ranking institutions in India.[125] However many of these are aligned towards publication output. There is a need to devise mechanisms to have a broader framework for obtaining indicators at a glance for policy makers, researchers /funding bodies etc. Perhaps, devising and working out parameters based on various systems could be one way of looking at this larger issue.

It thus resonates that an Indian monitoring system should be established to address important aspects of reform and performance in higher education systems, being in constant flux. An Indian scoreboard for higher education touching on the trade-off between autonomy and accountability could integrate and further develop important indicators for performance and for the characteristics of higher education systems and their reform. Such a monitoring system would also provide a valuable foundation for the analysis of national systems and the development of tailor-made recommendations for further reform.

4.2 Way Forward and Recommendations

4.2.1 Way Forward

System of accountability needs to be factored in the Higher Education System. Unarguably, there is a pressing need to put all domains of knowledge into assessment of research framework. The question is not whether or not to measure, but how to measure. Thus, while research assessment is important for bringing research excellence, equally important are the ways in which the research output is evaluated by various stakeholders. In the recent San Francisco Declaration on Research Assessment[126], the emphasis has been laid on the ways in which the output of scientific research is evaluated by funding agencies, academic institutions, and other parties and a number of important (18) recommendations have been made focussing on various players. Most of the recommendations are in sync with what is being highlighted in this study. As a suggestive measure, as indicated in analytical portion in Part 3, it is highlighted that the research evaluation mechanisms need further merit. As a first step in this direction, the corrective measures to obviate biasness and inherent weakness in evaluation mechanisms need to be worked out and put in place. For this, the following suggestive measures

---

[125] Pratha, Gangan P and Gupta, B. M. (2009). Ranking of Indian engineering and technological institutes for their research performance during 1999–2008. Current science, vol. 97, no. 3, 10 August 2009, p304-06.

[126] Putting science into the assessment of research. Available at: *http://am.ascb.org/dora/files/ SFDeclarationFINAL.pdf* (Accessed on July 3, 2013).

included here could be considered – (i) rationalisation of the Impact Factor score across disciplines; (ii) citations per paper (CPP) for an individual may be important than just H Index or Impact Factor; (iii) since H index varies, it is important to fix thresholds across disciplines and classification of these thresholds is equally important; (iv) an in-depth mechanism of identifying who is citing and where the citations are coming from may also be an important factor (for instance, a single citation from a Noble Laureate or from an emerging area of importance may be worth 10 or 100 citations); (v) the yawning gap between cited and uncited publications needs to be bridged. Finally, mere head counting is not enough, the qualitative factors in bringing about excellence is necessitated, for which research assessment and excellence framework is a must. More so, since global rankings have started having impact on higher education research and the production of knowledge, wherein innovatively developing or developed (successful) economies develop and exploit new knowledge for competitive advantage and performance.[127] In this competitive environment, it is reputation fame game and right or wrong; the reputation factor is primarily piggy backing on the research component. As higher education is viewed as critical to international competitiveness and individual opportunity, its quality and status have become vital indicators. It may not be an overarching statement that India needs a Centre of Excellence to bring in force such a framework and carry out such studies in a professional manner and feed the information to all stakeholders to create world class institutions through collective measures.

4.2.2 Recommendations

Based on the above, it becomes imperative to create a system that brings in framework for HEIs assessment in a holistic way (including the core research assessment and excellence framework). Keeping the above parameters on board would facilitate, transparency, accountability and excellence. This initiative will lay the foundation for the *Third Mission*. Thus the study broadly recommends that national strategies for reforming and assessing higher education, the system of accountability needs to be factored in the (HE) system. For this introduction of *measurement and transparent reporting system* together with positioning of a well designed performance related incentive/reward

---

[127] Brinkley, I., 2008. The Knowledge Economy: How knowledge is reshaping the economic life of nations. The Work Foundation, London, pp. 17-18. Available at: *http://www.workfoundation.com/assets/docs/ publications/41_KE_life_of_nations.pdf*

system is important for further strengthening the research base of the country. In doing so, the following needs to be accounted for :

(i) **Data-Planning & Management:** For facilitating evolving of evidence based metrics for research excellence framework, the comprehensiveness of data, bridging data gaps and fostering data reliability and accuracy is the need of the hour. Thus mandating creation of resource-base at institutional and national level by the competent authority comprising the total intellectual output of the institutions/nation for subsequent analysis is necessitated.

Failure to provide accurate or incomplete information might give a misleading impression of our institutions that could result in exclusion or misrepresentation in comparisons or groupings of leading institutions. Thus need for central governance structure to mandate standards, only then balanced comparisons can take place.

(ii) **Comprehensive Content Streams:** (While) so far the conduit for the research communications have primarily been journal articles, now the intellectual output need to focus on all content streams, besides journal articles including patents, books, research reports, etc. Thus for balanced assessment and excellence framework, accounting not just publications, but technological innovations as well is important.

Point (i) and (ii) can thus help quantifying the quality of research in our extremely mosaic higher education system by creating evidence.

(iii) **Other Parameters:** The intellectual (academic) output to be supplemented by other indicators such as administrative/recognition/teaching-learning and others, that would facilitate in evolving and building institutional profiles in more holistic way.

Indigenous Model: *Identification of Benchmarking Measurement Indicators (National Level):* While the above indicated data will help in measuring the unmeasured, there is a need for appropriate methodology. The requisite parameters (such as but not limited to – research funding; infrastructure; mission driven research; research leading to - IP, products and prototypes; publications; awards; journal metrics; patents; Doctoral/post-doctoral statistics; consultancy; collaboration, etc.) for evaluating like entities in a more objective manner needs to be identified/ evolved and factored in the framework at the national level, that will help in deciphering quality gaps for excellence. Besides, this would also help (to some extent) in filling the pitfalls in metrics based measurement of research. The indigenous model is important to give a national context to excellence. One cannot seed world institutions of excellence, they evolve. Although indigenous model will work, but learning experiences from other countries is also important.

(iv) **Using International Benching Marking Systems for International Foothold:** At the international level, the existing ranking systems may thus be able to include many of our institutions for evaluation purposes thereby, projecting the international standing (reputation/foothold) viz-à-viz world HEIs.

(v) **Creation of a Centre of Excellence:** It may not be an overarching statement that India needs a centre of excellence to bring in force such a framework. The centre thus can carry out such studies in a professional manner and eventually feed the information to all stakeholders to create world class institutions through collective measures.

(vi) **Innovative Indicators for Evaluative Metrics:** Innovatively improved indicators are required to facilitate factoring accuracy while evaluating global research outputs and evolving mechanisms for capturing trends for quantification, and benchmarking so as to determine the vitality of the research landscape.

For accelerating the research excellence, building institutional profiles for evolving measures for ranking and evidence based metrics ought to be in place.

BIBLIOGRAPHY

1. Abbott, C. (2001). Some young male website owners: The technological aesthete, the community builder and the professional activist. *Education, Communication and Information*, 1, 197-212.

2. Adkins, D., and Budd, J. (2006). Scholarly productivity of U.S. LIS faculty. *Library & Information Science Research*, 28(3), 374-389.

3. Aguillo, I. F., Ortega, J. L., and Fernández, M. (2008). Webometric ranking of world universities: Introduction, methodology, and future developments. *Higher education in Europe*, *33*(2-3), 233-244.

4. Aina, L. O., and Mooko, N. P. (1999). Research and publication patterns in Library and information Science. *Information Development*, 15(2), 114-119.

5. Aldridge, J. M., Fraser, B. J., and Huang, T. (1999). Investigating classroom environments in Taiwan and Australia with multiple research methods. Journal of Educational Research, 93, 48-57.

6. Ale Ebrahim, N. (2013). Introduction to the Research Tools mind map. *Research World*, *10*, Article A10.4.

7. Alemna, A. A. (1996). The periodical literature of library and information in Africa: 1990-1995. *International Information & Library Review*, 28, 93-103.

8. Alemna, A. A. (2001). The periodical literature of library and information in Africa: 1996-2000. *Information Development*, 17(4), 257-261.

9. Allen, E, S. (1929). Periodicals for mathematicians. Science 70, 552.

10. Almind, T. C., and Ingwersen, P. (1997). Infometric analyses on the World Wide Web: Methodological approaches to 'Webometrics'. *Journal of Documentation,* 53(4), 404-426.

11.   Arasu, A., Cho, J., Garcia-Molina, H., Paepcke, A., and Raghavan, S. (2001). Searching the Web. *ACM Transactions on Internet Technology*, 1(1), 2-43.

12.   Arunachalam S.(2001). Mathematics research in India today: What does the literature reveal?, *Scientometrics*, 52(2), 235-259.

13.   Arunachalam S., and Gunasekaran S. (2002). Tuberculosis research in India and China: From bibliometrics to research policy, *Current Science*, 82(8), 933-947.

14.   Arunachalam S., and Manorama K. (1989). Are citation-based quantitative techniques adequate for measuring science on the periphery? *Scientometrics*, 15(5-6), 393-408.

15.   Arunan E., Brakaspathy R., Desiraju G.R., and Sivaram S. (2013). Chemistry in India: Unlocking the potential. *Angewandte Chemie - International Edition*, 52(1), 114-117.

16.   Bailey J., Zhang C., Budgen D, Turner M., and Charters S. (2007). Search Engine Overlaps : Do they agree or disagree? In *Second International Workshop on Realising Evidence-Based Software Engineering (REBSE '07)*, p.2.

17.   Bala A., and Gupta B.M. (2010). Mapping of Indian Neuroscience Research: A Scientometric Analysis of Research Output during 1999-2008. *Neurology India*, 58(1), 35-41.

18.   Barabási, A. L. (2002). *Linked: The new science of networks*. Cambridge, Massachusetts: Perseus Publishing.

19.   Bar-Ilan, J. (2007). Which h-index? - A comparison of WoS, Scopus and Google Scholar. *Scientometrics*, 74(2), 257-271.

20.   Bar-Ilan, J. (1999, online). Search Engine Results over Time - A Case Study on Search Engine Stability. *Cybermetrics*, 2/3(1), http://www.cindoc.csic.es/cybermetrics/articles/v2i1p1.html.

21.   Bar-Ilan, J. (2001). Data collection methods on the Web for informetric purposes - A review and analysis. *Scientometrics*, 50(1), 7-32.

22.   Bar-Ilan, J. (2004). A microscopic link analysis of academic institutions within a country - The case of Israel. *Scientometrics*, 59(3), 391-403.

23.   Bar-Ilan, J., and Peritz, B. C. (2004). Evolution, continuity, and disappearance of documents on a specific topic on the Web: A longitudinal study of 'informetrics'. Journal *of the American Society for Information Science and Technology*, 55(11), 980-990.

24. Basu, A. (2013). Some differences in research publications of Indian scientists in India and the diaspora. 1986-2010. *Scientometrics*, 943, p.1007-1019.

25. Basu, A., and Aggarwal, R. (2001). International collaboration in science in India and its impact on institutional performance. *Scientometrics*, 52(3), 379-394.

26. Bayer, A.E. and Folger, J. (1966). Some Correlates of a Citation Measure of Productivity in Science. *Sociology of Education*, 39, 381-390.

27. Beel J. and Gipp B. (2009). Google Scholar's Ranking Algorithm: The Impact of Articles' Age (An Empirical Study), in *Proceedings of 6th International Conference on Information Technology: New Generations (ITNG'09)*, IEEE.

28. Beel J.and Gipp B. (2008) The Potential of Collaborative Document Evaluation for Science, in *11th International Conference on Digital Asian Libraries (ICADL'08)*, ser. Lecture Notes in Computer Science (LNCS), G. Buchanan, M. Masoodian, and S. J. Cunningham, Eds., vol. 5362. Heidelberg (Germany): Springer, December 2008, pp. 375-378.

29. Beel J.and Gipp B. Collaborative Document Evaluation: An Alternative Approach to Classic Peer Review, in *5th International Conference on Digital Libraries (ICDL'08)*, ser. Proceedings of World Academy of Science, Engineering and Technology, vol. 31, August 2008, pp. 410-413.

30. Beel, J., & Gipp, B. (2009, July). Google Scholar's ranking algorithm: an introductory overview. In *Proceedings of the 12th International Conference on Scientometrics and Informetrics (ISSI'09)* (Vol. 1, p. 230-241).

31. Bernier, C.L., Gill, W.N., and Hunt, R.G. (1975). Measures of Excellence of Engineering and Science Departments: A Chemical Engineering Example. *Chemical Engineering Education*, 9:94-97.

32. Beumer, Koen. Bhattacharya, Sujit. (2013, October). Emerging Technologies in India: Developments. Debates and Silences about Nanotechnology. *Science and Public Policy*, Volume 40(5).

33. Bhat, M. H. (2009). Open access publishing in Indian premier research institutions. *Information Research*, 14(3), 4.

34. Bhatia, K., & Mavalankar, D. V. (1990). Output of Medical Research From India. *British Medical Journal*, 872-873.

35. Björneborn, L., and Ingwersen, P. (2001). Perspectives of Webometrics. *Scientometrics*, 50(1), 65-82.

36.   Bollen, J.; Van de Sompel, H., Smith, J. and Luce, R. (2005). Toward Alternative Metrics of Journal Impact: A comparison of download and citation data. *Information Processing and Management*, 41(6), 1419–1440, *arXiv:cs.DL/0503007, doi:10.1016/j. ipm.2005.03.024.*

37.   Borgman, C. L., and Furner, J. (2002). Scholarly Communication and Bibliometrics. *Annual Review of Information Science and Technology*, 36, 3-72.

38.   Börner, K., Chen, C. M., and Boyack, K. W. (2003). Visualizing knowledge domains. *Annual Review of Information Science and Technology,* 37, 179-255.

39.   Börner, K., Sanyal, S., and Vespignani, A. (2007). Network Science. *Annual Review of Information Science and Technology,* 41, 537-607.

40.   Bornmann, L., and Daniel, H. (2008). What do citation counts measure? A Review of Studies on Citing Behavior. *Journal of Documentation*, 64(1), 45-80.

41.   Bornmann, L., de Moya-Anegón, F., and Leydesdorff, L. (2012). The new excellence indicator in the World Report of the SCImago Institutions Rankings 2011. *Journal of Informetrics*, 6(3), 333-335.

42.   Bornmann, L., Leydesdorff, L., and Van den Besselaar, P. (2010). A Meta-Evaluation of Scientific Research Proposals: Different Ways of Comparing Rejected to Awarded Applications. *Journal of Informetrics,* 4(3), 211-220.

43.   Bouabid, H. (2011). Revisiting citation aging: A model for citation distribution and life-cycle prediction. *Scientometrics,* 88(1), 199-211. *doi:10.1007/s11192-011-0370-5.*

44.   Bradford, S. C. (1985). Sources of information on specific subjects. *Journal of Iinformation Science*, (10), 173-180.

45.   Brody, T., Harnad, S., and Carr, L. (2006). Earlier web usage statistics as predictors of later citation impact. *Journal of the American Society for Information Science and Technology*, *57*(8), 1060-1072.

46.   Harnad, S., and Brody, T. (2004). Comparing the impact of open access (OA) vs. non-OA articles in the same journals. *D-lib Magazine*, *10*(6).

47.   Bruce, G. Charlton and Peter Andras. *(2007).* Evaluating Universities using simple scientometric research-output metrics: Total citation counts per university for a retrospective seven-year rolling Sample. *Science and Public Policy,* 34(8), 555-563. *doi:10.3152/030234207X254413.*

48.   Callon, M., Law, J., and Rip, A. (Eds.) (1986). *Mapping the Dynamics of Science and Technology.* London, Macmillan.

49. Cano, V. (1999). Bibliometric Overview of Library and Information Science Research in Spain. *Journal of the American Society for Information Science*, 50(8), 675-680.

50. Cao, Y., Zhou, S., and Wang, G. (2013). A bibliometric analysis of global laparoscopy research trends during 1997–2011. *Scientometrics*, *96*(3), 717-730.

51. Carter, G.M. Peer Review, Citations, and Biomedical Research Policy: NIH Grants to Medical School Faculty. *Rand Report, R-1583-HEW* (Santa Monica, California: Rand Corporation, 1974). 90.

52. Cash, D., Clark, W. C., Alcock, F., Dickson, N., Eckley, N., and Jäger, J. (2002). *Credibility, legitimacy and boundaries: Linking, assessment and decision making.* Cambridge, MA, Kennedy School of Government.

53. Center for Science and Technology Studies (CWTS) (2005-2009). *Bibliometric Study on KWR Watercycle Research Institute.* Center for Science and Technology Studies (CWTS) Leiden University, The Netherlands, Research Report for KWR Watercycle Research Institute, Nieuwegein, the Netherlands.

54. Chen, C., Newman, J., Newman, R., and Rada, R. (1998). How did university departments interweave the Web: A study of connectivity and underlying factors. *Interacting with computers*, *10*(4), 353-373.

55. Chidambaram, R. (2005). Measures of progress in science and technology. *Current Science*, 88(6).

56. Ching, J. T. Y., and Chennupati, K. R. (2002). Collection evaluation through citation analysis techniques: A case study of the Ministry of Education, Singapore. *Library Review*, 51(8), 398-405.

57. Chubin, D., and Moitra, S. (1975). Content analysis of references: adjunct or alternative to citation counting? *Social Studies of Science*, 5, 423-441.

58. Chubin, D. E. (1987). Research evaluation and the generation of big science policy. *Science Communication*, 9(2), 254-277.

59. Claire, Donovan. (2007). Future pathways for science policy and research assessment: Metrics vs peer review, quality vs impact. *Science and Public Policy*, 34(8), 538-542. *doi:10.3152/030234207X256529.*

60. Claire, Donovan. (2007). The qualitative future of research evaluation. *Science and Public Policy*. 34(8), 585-597. *doi:10.3152/030234207X256538.*

61.   Cole, S.; Cole, J. R.; and Dietrich, L. (1978). Measuring the cognitive state of scientific disciplines. In Y. Elkana, J. Lederberg, & IC Merton (Eds.), *Toward a metric of science: The advent of scientific indicators* (p. 209-251). New York, Wiley.

62.   Cozzens, S. E. (1989). What do citations count? The rhetoric-first model. *Scientometrics,* 15(5), 437-447.

63.   Creswell, J., and Plano Clark, V. (2007). *Designing and conducting mixed methods Research*. London, Sage.

64.   Cronin, B. (2001). Bibliometrics and beyond: some thoughts on Web-Based Citation Analysis. *Journal of Information Science*, 27(1), 1-7.

65.   Cronin, B. (1984). The citation process. *The role and significance of Citations in scientific communication* (Vol. 1). London, Taylor Graham.

66.   Crosbie, G.C. and Heckel. (1976, September). Citation Criteria for Ranking Academic Departments. *Journal of Metals*, 28(9), 27-28.

67.   Aksnes, D. W. (2006). Citation rates and perceptions of scientific contribution. *Journal of the American Society for Information Science and Technology*, *57*(2), 169-185.

68.   Dandona, L., Raban, M. Z., Guggilla, R. K., Bhatnagar, A., & Dandona, R. (2009). Trends of public health research output from India during 2001-2008. *BMC Medicine*, *7*(1), 59.

69.   Dastidar, P. G., and Ramachandran, S. (2005). Engineering research in ocean sector: An international profile. *Scientometrics*, *65*(2), 199-213.

70.   Davenport, E., and Cronin, B. (2000). The Citation network as a Prototype for Representing Trust in Virtual Environments. In B. Cronin & H. B. Atkins (Eds.). The *Web of Knowledge: A Festschrift in Honor of Eugene Garfield.* (p. 517-534). Metford, NJ, Information Today Inc. ASIS Monograph Series.

71.   Davis, P. M., Lewenstein, B. V., Simon, D. H., Booth, J. G., & Connolly, M. J. L. (2008). Open access publishing, article downloads and citations: randomised trial. *British Medical Journal*, 337, a568.

72.   Davis, P. M. (2011). Open access, readership, citations: a randomized controlled trial of scientific journal publishing. *The FASEB Journal*, *25*(7), 2129-2134.

73.   Derrick, G. E., & Pavone, V. (2013). Democratising research evaluation: Achieving greater public engagement with bibliometrics-informed peer review. *Science and Public Policy*, *40*(5), 563-575.

74. Dhawan, S. M. (1998). Comparative study of physics research in India and China based on INSPEC-Physics for 1990 and 1995. *Scientometrics, 43*(3), 423-441.

75. Dosi, G. (1982). Technological Paradigms and Technological Trajectories: A suggested Interpretation of the Determinants and Directions of Technical Change. *Research Policy*, 11(3), 147-162.

76. Edge, D. (1979). Quantitative measures of communication in science: A critical review. *History of Science, 17*, 102-134.

77. Egghe, L. (2006). Theory and practise of the g-index. *Scientometrics*. 69(1), 131-152. *doi:10.1007/s11192-006-0144-7*.

78. Elkana, Y., Lederberg, J., Merton, R. K., Thackray, A., and Zuckerman, H. (1978). Toward *a Metric of Science: The advent of Science Indicators*. New York, etc., Wiley.

79. Ellman, I. A. (1983). Comparison of Law Faculty Production in Leading Law Reviews. *Journal of Legal Education*, 33(4), 681.

80. Eysenbach, G. (2006b). The Open Access Advantage. *J Med Internet Res,* 8(2), e8.

81. Eysenbach G. (2006a). Citation Advantage of Open Access Articles. *PLoS Biol.*, 4(5), e157.

82. Finkenstaedt, T. (1990). Measuring Research Performance in the Humanities. *Scientometrics,* 19(5-6), 409-417.

83. Florian, R. (2007). Irreproducibility of the results of the Shanghai academic ranking of world universities. *Scientometrics, 72*(1), 25-32.

84. Foot, K., Schneider, S., Dougherty, M., Xenos, M., and Larsen, E. (2003). Analyzing linking practices: Candidate sites in the 2002 US electoral web sphere. *Journal of Computer Mediated Communication,* 8(4). Available at: *http://www.ascusc.org/jcmc/vol8/issue4/foot.html*.

85. Fosu, V. K., and Alemna, A. A. (2002). Fifty years of library literature on Ghana: A Bibliometric Study. *Ghana Library Journal*, 14, 1-5.

86. Foucault, M. (1969). *The Archaeology of Knowledge* (A. Sheridan, Trans.). London, Routledge.

87. Francisco M. Couto, Catia Pesquita, Tiago Grego, and Paulo Veríssimo. (2009). Handling self-citations using Google Scholar, *Cybermetrics*, 13(1).

88. Fujigaki, Y. (1998). Filling the Gap between Discussions on Science and Scientists' Everyday Activities: Applying the Autopoiesis System Theory to Scientific Knowledge. *Social Science Information*, 37(1), 5-22.

89.   Garfield, E. (1972, November). Citation analysis as a tool in journal evaluation. American Association for the Advancement of Science.

90.   Garfield, E. (1977). Caution urged in the use of citation analyses. *Trends in Biochemical Sciences*, *2*(4), N84.

91.   Garfield, E. (1955). Citation Indexes for Science: A New Dimension in Documentation through Association of Ideas. *Science*, 122(3159), 108-111.

92.   Garfield, E. (1973). Citation Frequency as a Measure of Research Activity and Performance in *Essays of an Information Scientist*, 1, 406-408, 1962-73, *Current Contents*, 5.

93.   Garfield, E. (1988). Can Researchers Bank on Citation Analysis? *Current Contents*, (44), October 31, 1988.

94.   Garfield, E. (1998). The use of Journal Impact Factors and Citation Analysis in the Evaluation of Science. *41st Annual Meeting of the Council of Biology Editors*, Salt Lake City, UT, May 4, 1998.

95.   Garfield, E. (1979). Most-cited authors in the Arts and Humanities, 1977-1978. *Current Contents*, 32. In E. Garfield, *Essays of an information Scientist*, Volume 4 (p. 238-244). Philadelphia, Institute for Scientific Information.

96.   Garfield, E. (1986). The 250 most-cited authors in the Arts and Humanities Citation Index, 1976-1983. *Current Contents*, 48, 3-10.

97.   Garfield, E. (1977). Is the Ratio Between Number of Citations and Publications Cited a True Constant? *Essays of an Information Scientist*, Vol. 2 (Philadelphia: IS1 Press, 1977), (p. 419-421).

98.   Garfield, E. (1977, August 8). Will ISI's Arts & Humanities Citation Index Revolutionize Scholarship? *Current Contents*, (32), 5-9.

99.   Garfield, E. (1970).Citation Indexing for Studying Science. *Nature*, 227(5259), 669-671.

100. Garfield, E. (1977, December 19). The 250 Most-Cited Primary Authors, 1961-1975. Part III. Each Author's Most Cited Publication. *Current Contents*, (51), 5-20.

101. Garfield, E. (1977, April 25). Citation Analysis and the Anti-Vivisection Controversy. *Current Contents*, (17), 5-10.

102. Garfield, E. (1977, November 28,). Citation Analysis and the Antivivisection Controversy. Part II. An Assessment of Lester R. Aronson's Citation Record. *Current Contents*, (48), S-14.

103. Garfield, E. (1977, December 12). The 250 Most-Cited Primary Authors, 1961-1975. Part II. The Correlation among or between Citedness, Nobel Prizes, and Academy Memberships. *Current Contents,* (50), 5-15.

104. Garfield, E. (1979). *Citation Indexing: Its theory and application in Science, Technology, and Humanities.* New York, John Wiley.

105. Garfield, E. and Sher, I.H. (1963). Citation Indexes in Sociological and Historical Research. *American Documentation,* (14), 289-91.

106. Garfield, E. A. (1977, March 9). List of 100 Most Cited 'Chemical' Articles. *Current Contents,* (10), 5-12.

107. Garg K.C. Scientometrics of Laser Research in India and China. *Scientometrics,* 55(1), 71-84.

108. Garg K.C., Padhi P. Scientometrics of Laser Research in India during 1970-1994. *Scientometrics,* 55(2), 215-241.

109. Garg K.C., Padhi P. (1999). Scientometrics of Laser Research Literature as viewed through the Journal of Current Laser Abstracts. *Scientometrics,* 45(2), 251-268.

110. Gargouri, Y., Hajjem, C., Lariviere, V., Gingras, Y., Brody, T., Carr, L. and Harnad, S. (2010). Self-Selected or Mandated, Open Access Increases Citation Impact for Higher Quality Research. *PLoS one,* 5(10), e13636.

111. Gavel,Y. and Iselid, L. (2008). Web of Science and Scopus: a Journal Title Overlap Study. *Online Information Review,* 32(1), 8-21.

112. Geller, N.L., deCani, J.S., and Davies, R.E. (1975). Lifetime Citation Rates as a Basis for Comparisons within a Scientific Field. In Goldfield, E.D. (ed.) *American Statistical Association Proceedings of the Social Statistics Section.* (Washington, D.C.: American Statistical Association, p. 429-433.

113. Gerritsen, Alwin L., Stuiver, Marian, Termeer, Catrien J. A. M. (2013, October). Knowledge Governance: An Exploration of Principles, Impact, and Barriers. *Science and Public Policy,* 40(5).

114. Gilbert, G. N. (1977). Referencing as persuasion. *Social Studies of Science,* 7, 113-122.

115. Giles, C. L., Bollacker, K., and Lawrence, S. (1998). CiteSeer: An Automatic Citation Indexing System. *DL'98 Digital Libraries, 3rd ACM Conference on Digital Libraries:* 89-98. *doi:10.1145/276675.276685.*

116. Gingras, Y., and Larivière, V. (2011). There are neither king nor crown in scientometrics: Comments on a supposed alternative method of normalization. *Journal of Informetrics*, 5(1), 226-227.

117. Gipp, B., and Beel, J. Scienstein. (2009). A Research Paper Recommender System, in *International Conference on Emerging Trends in Computing*. IEEE, 2009, 309-315.

118. Greene, J. C., and Caracelli, V. J. (1997). Defining and describing the paradigm issue in mixed- method evaluation. In J. C. Greene & V. J. Caracelli (Eds.), *Advances in mixed-method evaluation: The challenges and benefits of integrating diverse paradigms* (New Directions for Program Evaluation, No. 74, p. 5-17).

119. Griffith, B., Small, H.G., Stonehill, J.A., and Dey, S. (1974). The Structure of Scientific Literature, II: Towards a Macro- and Micro-Structure for Science. *Science Studies*, 4, 339-365.

120. Gruhl, D., Guha, R., Liben-Nowell, D., and Tomkins, A. (2004). Information diffusion through Blogspace. Paper presented at the *WWW2004*, New York. Avilable at: *http://www.www2004.org/proceedings/docs/1p491.pdf.*

121. Guba, E., Lincoln, Y., and Greene, J. (1989). Fourth Generation Evaluation.

122. Gunasekaran M., and Balasubramani R.(2012). Scientometric analysis of artificial intelligence research output: An Indian Perspective. *European Journal of Scientific Research*, 702, 317-322.

123. Gupta, B., Sharma, S., and Mehrotra, N. (1990). Subject-based publication activity indicators for medicinal & aromatic plants research. *Scientometrics*, 18(5-6), 341-361.

124. Gupta, B. M. (2011). Ranking of Indian institutions in agricultural & allied sciences for their research output during 1999-2008. *Annals Lib. Inf. Stud*, 58, 63-70.

125. Gupta, B. M. (2012). Measurement of Indian science and technology using publications output data during 1996-2010. *Indian Journal of Science and Technology*, 5(6), 2899-2911.

126. Gupta B.M., Bala A. Mapping of Asthma Research in India: A scientometric analysis of publications output during 1999-2008. *Lung India*, 284, 239-246.

127. Gupta B.M., Bala A. A Bibliometric Analysis of Malaria Research in India during 1998-2009. *Journal of Vector Borne Diseases*, 483, 163-170.

128. Gupta B.M., Bala A. A Scientometric Analysis of Indian Research Output in Medicine during 1999-2008. *Journal of Natural Science*, Biology and Medicine, 21, 87-100.

129. Gupta B.M., Kshitij A. Verma C.,(2011, February). Mapping of Indian Computer Science Research Output, 1999-2008. *Scientometrics*, 86(2), 261-283.

130. Gupta B.M., Kumar S., Khanna H.K.,and Amla T.K. (1989, July). Impact of Professional and Chronologcal Age on the Productivity of Scientists in Engineering Science Laboratories of CSIR,. *Malaysian Journal of Library and Information Science*, 41, 103-107.

131. Gupta V.K. Inventors' productivity in a publicly funded R&D agency - The case of CSIR in India. *World Patent Information*, 263, 235-238.

132. Gustafson, T. (1975, December 12.). The Controversy Over Peer Review.. *Science*, 190(4219), 1060-1066.

133. Hajjem, C., Harnad, S. and Gingras, Y. (2005). Ten-Year Cross-Disciplinary Comparison of the Growth of Open Access and How It Increases Research Citation Impact, *IEEE Data Engineering Bulletin*, 28(4), 39-47.

134. Harries, G., Wilkinson, D., Price, E., Fairclough, R., & Thelwall, M. (2004). Hyperlinks as a data source for science mapping. *Journal of Information Science*, 30(5).

135. Hemlin, S. (1996). Social Studies of the Humanities: A case study of research conditions and performance in ancient history and Classical Archaeology, and English. *Research Evaluation*, 6(1), 53-61.

136. Hemlin, S., & Gustafsson, M. (1996). Research production in the Arts and Humanities. A questionnaire study of factors influencing Research Performance. *Scientometrics*, 37(3), 417-432.

137. Henk F Moed. (2007). The future of Research Evaluation rests with an intelligent combination of advanced metrics and transparent peer review. *Science and Public Policy*, 34(8), 575-583. *doi:10.3152/030234207X255179*.

138. Henzinger, M. R. (2001). Hyperlink analysis for the Web. *IEEE Internet Computing*, 5(1), 45-50.

139. Hernández-Borges, A. A., Macías-Cervi, P., Gaspar-Guardado, M. A., Torres-Álvarez de Arcaya, M. L., Ruiz-Rabaza, A., and Jiménez-Sosa, A. (1999). Can examination of WWW usage statistics and other indirect quality indicators distinguish the relative quality of medical Web sites? *Journal of Medical Internet Research*, 1(1). Available at: *http://www.jmir.org/1999/1991/e1991/index.htm*.

140. Hicks, D., Tomizawa, H., Saitoh, Y., and Kobayashi, S. (2004). Bibliometric Techniques in the Evaluation of Federally Funded Research in the United States. *Research Evaluation*, 13(2), 78-86.

141. Hider, P., and Pymm, B. (2008). Empirical research methods reported in high-profile LIS Journal Literature. *Library & Information Science Research*, 30, 108-114.

142. Hirsch, J. E. (2005, 15 November). An index to quantify an individual's scientific research output. *PNAS,* 102(46), 16569-16572. *arXiv:physics/0508025, doi:10.1073/ pnas.0507655102.*

143. Hirsch, J. E. (2005). An index to quantify an individual's scientific research output. Proceedings of the National Academy of Sciences of the USA, 102(46), 16569-16572.

144. Hoang, D.; Kaur, J. and Menczer, F. (2010). Crowdsourcing Scholarly Data, *Proceedings of the WebSci10: Extending the Frontiers of Society,* On-Line, April 26-27th, 2010, Raleigh, NC, US.

145. Holmes, R. (2006). The THES university rankings: Are they really world class?

146. Hyland, K. (2000). *Disciplinary Discourses: Social Interactions in Academic Writing.* Harlow, Longman.

147. Ingwersen, P. (1998). The calculation of Web Impact Factors. *Journal of Documentation*, 54(2), 236-243.

148. Inter-University Committee of the Flemish Faculties of Law. (1996). *The Assessment of Performance in Juridical Research*. Final Report (in Dutch).

149. Irvine, J., and Martin, B. R. (1984). *Foresight in Science: Picking the Winners*. London, Frances Pinter.

150. Jain A., Garg K.C. (1992). Laser Research in India: Scientometric Study and Model Projections. *Scientometrics,* 233, 395-415.

151. Jang, E. E., McDougall, D. E., Pollon, D., Herbert, M., and Russell, P. (2008). Integrative mixed methods data analysis strategies in Research on School Success in challenging circumstances. *Journal of Mixed Methods Research*, 2(3), 221-247.

152. Jayashree B., and Arunachalam S. (2000). Mapping Fish Research in India. *Current Science*, 79(5), 613-620.

153. Jeremic, V., Bulajic, M., Martic, M., and Radojicic, Z. (2011). A fresh approach to evaluating the academic ranking of world universities. *Scientometrics, 87*(3), 587-596.

154. Johnson, A.A. and Davis, R.B. (1975, June). The Research Productivity of Academic Materials Scientists. *Journal of Metals*, 27(6), 28-29.

155. Johnson, R. B., & Onwuegbuzie, A. (2004). Mixed Methods Research: A Research Paradigm whose time has come. Educational Researcher, 33(7), 14-26.

156. Jossey-Bass. Greene, J. C., Caracelli, V. J., & Graham, W. F. (1989). Toward a conceptual framework for mixed- method evaluation design. Educational Evaluation and Policy Analysis, 11(3), 255-274.

157. Justiss, L. K. (1993). A bibliometric study of Texas Law Reviews. *Law Library Journal*, 85, 407-408.

158. Karki M.M.S. (1990, May). Environmental Science Research in India: An analysis of publications. *Scientometrics*, 18(5), 363-373.

159. Karki M.M.S., and Garg K.C. (1997). Bibliometrics of alkaloid chemistry research in India, *Journal of Chemical Information and Computer Sciences*, 37(2),157-161.

160. Karpagam R., Gopalakrishnan S., Natarajan M., and Ramesh Babu B. (2011). Mapping of nanoscience and nanotechnology research in India: A scientometric analysis, 1990-2009. *Scientometrics*, 89(2), 501-522.

161. Kaur H., and Gupta B.M.. (2010). Mapping of dental science research in India: A scientometric analysis of India's research output, 1999-2008. *Scientometrics*, 85(1), 361-376.

162. Kim, H. J. (2000). Motivations for hyper linking in scholarly electronic articles: A qualitative study. *Journal of the American Society for Information Science*, 51(10), 887-899.

163. Kleinberg, J. M. (1999). Authoritative sources in a hyperlinked environment. Journal of the ACM, 46(5), 604-632.

164. Koshy, G.P. (1976). The Cite ability of a Scientific Paper. *Proceedings of Northeast Regional Conference of American Institute for Decision Sciences* (Philadelphia, Pa., April/May 1976), p. 224-227.

165. Kostoff, R.N., Bhattacharya, S., Pecht, M. (2007a). Assessment of China's and India's science and technology literature - introduction, background, and approach. *Technological Forecasting and Social Change*, 74(9), 1519-1538.

166. Kostoff, R.N., Briggs, M.B., Rushenberg, R.L., Bowles, C.A., Pecht, M., Johnson, D., Bhattacharya,S., Icenhour, A.S., Nikodym, K., Barth, R.B., and Dodbele, S. (2007a).

Comparisons of the structure and infrastructure of Chinese and Indian Science and Technology. *Technological Forecasting and Social Change,* 74(9),1609-1630.

167. Kuhn, T. (1962). *The Structure of Scientific Revolutions.* Chicago, IL, University of Chicago Press.

168. Kumar, H. A., and Dora, M. (2012). Research Productivity in a Management Institute: An Analysis of Research Performance of Indian Institute of Management Ahmedabad during 1999-2010. *DESIDOC Journal of Library & Information Technology,* 32(4).

169. Kumar, Patra S., and Chand, P. (2005). Biotechnology Research Profile of India. *Scientometrics,* 63(3), 583-597.

170. Kurtz,, M. J.; Eichhorn, G.; Accomazzi, A.; Grant, C. S.; Demleitner, M. and Murray, S. S. (2004). The Effect of Use and Access on Citations. *Information Processing and Management,* 41 (6), 1395-1402.

171. Kyvik, S. (1989). Productivity differences, fields of learning, and Lotka's Law. *Scientometrics,* 15(3-4), 205-214.

172. Larson, R. R. (1996). Bibliometrics of the World Wide Web: an exploratory analysis of the intellectual structure of cyberspace. Paper presented at the *AISS 59th Annual meeting.*

173. Lawrence, S. (2001). Online or Invisible? *Nature* 411 (2001) (6837), 521.

174. Lewison G., Roe P. (2012). The evaluation of Indian Cancer Research, 1990-2010. *Scientometrics,* 93(1), 167-181.

175. Leydesdorff, L., and Milojević, S. (2012). Scientometrics. *arXiv preprint arXiv:1208.4566.*

176. Leydesdorff, L. (1998). Theories of Citation. *Scientometrics,* 43(1), 5-25.

177. Leydesdorff, L. (1995). *The challenge of scientometrics: The development, measurement, and self-organization of scientific communications.* Leiden: DSWO Press, Leiden University.

178. Leydesdorff, L. (1998). Theories of Citation? *Scientometrics,* 43(1), 5-25.

179. Leydesdorff, L., & Bornmann, L. (2011). Integrated Impact Indicators (I3) compared with Impact Factors (IFs): An alternative design with policy implications. *Journal of the American Society for Information Science and Technology,* 62(11), 2133-2146. *doi:10.1002/asi.21609.*

180. Leydesdorff, L., and Curran, M. (2000). Mapping university-industry-government relations on the Internet: the construction of indicators for a knowledge-based economy. *Cybermetrics,* 4. Avaialable at: *http://www.cindoc.csic.es/cybermetrics/articles/v4i1p2.html.*

181. Leydesdorff, L., and Opthof, T. (2010). Scopus' Source Normalized Impact per Paper (SNIP) versus the Journal Impact Factor based on fractional counting of citations. *Journal of the American Society for Information Science and Technology,* 61(11), 2365-2396.

182. Leydesdorff, L., and Van den Besselaar, P. (1997). Scientometrics and Communication Theory: Towards Theoretically Informed Indicators. *Scientometrics*, 38, 155-174.

183. Leydesdorff, L., and Wouters, P. (1999). Between Texts and Contexts: Advances in Theories of Citation? *Scientometrics*, 44(2), 169-182.

184. Linda, Butler. (2007). Assessing university research: A plea for a balanced approach. *Science and Public Policy,* 34 (8), 565-574. *doi:10.3152/030234207X254404.*

185. Lindsey, D. and Brown, G.W. (1977). Problems of Measurement in the Sociology of Science: Taking AC. count of Collaboration. (unpublished).

186. Lindsey, D. and Brown, G.W. (1977). The Measurement of Quality in Social Studies of Science: Critical Reflections on the Use of Citation Counts. (unpublished).

187. Liu, N. C., and Cheng, Y. (2005). The academic ranking of world universities. *Higher education in Europe, 30*(2), 127-136.

188. Liu, N. C., Cheng, Y., and Liu, L. (2005). Academic ranking of world universities using scientometrics-A comment to the "Fatal Attraction". *Scientometrics, 64*(1), 101-109.

189. Lotka, A. J., (1926). The frequency distribution of scientific productivity. *Journal of the Washington Academy of Sciences*, 16(12), 317-323.

190. Lutz Bornmann, Hans-Dieter Daniel, (2008), What do citation counts measure? A review of studies on citing behavior. *Journal of Documentation,* 64(1), 45-80.

191. Luwel, M.. Moed, H. F., Nederhof, A. J., De Samblanx, V., Verbrugghen, K., and van der Wurff, L. J.(1999). Towards indicators of research performance in the social sciences and humanities: An exploratory study in the fields of law and linguistics at Flemish Universities. Brussels: *Vlaamse Universitaire Raad, Depot nr. d/1999/2939/9.*

192. MacCallum, C.J. and Parthasarathy, H. (2006). Open access increases citation rate. *PLoS Biol,* 4(5), e176.

193. MacRoberts, M. H., and MacRoberts, B. R. (1996). Problems of citation analysis. *Scientometrics, 36*(3), 435-444.

194. Mahapatra, G. (2000). *Bibliometric studies on Indian library & information science literature* (1st ed.). New Delhi: Crest Publishing House.

195. Martino, J.P. (1971). Citation Indexing for Research and Development Management. IEEE Transactions on Engineering Management, EM-18(4),146-151.

196. Mavalankar, D.V. (1990,13 October). Output of Medical Research from India (II), *British Medical Journal*, 301(6756), 873.

197. McCain, K. W. (1990). Mapping authors in intellectual space: A technical overview. Journal *of the American Society for Information Science, 41*(6), 433-443.

198. McGrath, W. E. (1996). The unit of analysis (objects of study) in bibliometrics and scientometrics. *Scientometrics, 35*(2), 257-264.

199. McKechnie, L. E. F., Goodall, G. R., Lajoie-Paquette, D., and Julien, H. (2005). How human information behaviour researchers use each other's work: a basic citation analysis study. *Information Research-an International Electronic Journal*, 10(2), 12.

200. Meadows, A.J. (1974). *Communication in Science*, London, Butterworths, 1974, p. 45.

201. Meho, L. I. (2007). The rise and rise of citation analysis. *Physics World*, 20(1), 32-36.

202. Meho, L. I., and Spurgin, K. M. (2005). Ranking the research productivity of LIS faculty and schools: An evaluation of data sources and research methods. *Journal of the American Society for Information Science and Technology* 5 6(12), 1314-1331.

203. Meho, L. I., and Yang, K. ((2007). Impact of data sources on citation counts and rankings of LIS faculty: Web of Science vs. Scopus and Google Scholar. *Journal of the American Society for Information Science and Technology*, 58(13), 2105-25.

204. Meier J. J. and Conkling T. W. (2008). Google Scholar's Coverage of the Engineering Literature: An Empirical Study. *The Journal of Academic Librarianship*, 34(3), 196-201.

205. Merton, R. K. (1973). *The Sociology of Science: Theoretical and empirical investigations.* Chicago/ London: University of Chicago Press.

206. Merton, R. K. (1968, January).The Matthew Effect in Science, *Science*, 159(3810), 56-63.

207. Merton, R.K. (1968). *Social Theory and Social Structure*, (New York: The Free Press, 1968), 26-28.

208. Mertoa. R.K. (1965). *On the Shoulders of Giants: A Shandean Postscript* (New York: Harcourt Brace & Jovanovich, 1965), 218-219.

209. Michael, N., Bastedo, Nicholas A. (2010). *Bowman College Rankings as an Interorganizational Dependency: Establishing the Foundation for Strategic and Institutional Accounts*, Published online, 21 September 2010! Springer Science+Business Media, LLC 2010. Available at : *https://www.google.co.in/url?s a=t&rct=j&q=&esrc=s&source=web&cd=1&cad=rja&uact=8&ved=0CBwQFjAAah UKEwj69bzog6jlAhXTj44KHYkuC4E&url=http%3A%2F%2Flink.springer.com%2Fco ntent%2Fpdf%2F10.1007%252Fs11162-010-9185-0.pdf&usg=AFQjCNG6pSt1uH_ p7pBZLmPQM2hTZpb7JA.*

210. Middleton, A. (2005). An attempt to quantify the quality of student bibliographies. Performance Measurement and Metrics. *The International Journal for Library and Information Services,* 6(1), 7-18.

211. Milojević, S., and Leydesdorff, L. (2013). Information Metrics (iMetrics): A Research Specialty with a Socio-Cognitive Identity? *Scientometrics,* 95(1), 141-157.

212. Moed, H. F. (2005a). *Citation Analysis in Research Evaluation.* NY Springer.

213. Moed, H. F. (2005b). Statistical Relationships Between Downloads and Citations at the Level of Individual Documents Within a Single Journal. *Journal of the American Society for Information Science and Technology,* 56 (10), 1088-1097. *doi:10.1002/ asi.20200.*

214. Moed, H. F. (2002). The impact-factors debate: the ISI's uses and limits. *Nature,* 415, 731-732.

215. Moed, H. F. (2005). *Citation analysis in research evaluation.* Dordrecht. The Netherlands, Springer.

216. Moed, H. F. (2010). Measuring contextual citation impact of scientific journals. *Journal of Informetrics,* 4(3), 265-277.

217. Morgan, D. L. (2007), Paradigms lost and pragmatism regained: Methodological implications of combining qualitative and quantitative methods. Journal of Mixed Methods Research, 1(1), 48-76.

218. Morse, J. M. (2003). Principles of mixed methods and multi-method research design. In A. Tashakkori and C. Teddlie (Eds.), *Handbook of mixed methods in social and behavioral research* (pp. 189-208). Thousand Oaks, CA, Sage.

219. Myers, K., and Oetzel, J. (2003). Exploring the dimensions of organizational assimilation: Creating and validating a communication measure. *Communication Quarterly*, 51, 436-455.

220. Nader, Ale Ebrahim. (2013, June 14). Introduction to the Research Tools Mind Map Research Support Unit, Centre of Research Services Institute of Research Management and Monitoring (RMM). University of Malaya, Malaysia. *Research World*, 10, 2013 Published Online.

221. Narin, F. (1976). Evaluative Bibliometrics. The Use of Publication and Citation Analysis in the Evaluation of Scientific Activity. (Cherry Hill, N.J.: Computer Horizons, Inc., 1976). 500 p. *NTIS- PB252339/AS.*

222. National Science Board. (2012). Science and Engineering Indicators. Washington DC, National Science Foundation. Available at: *http://www.nsf.gov/statistics/seind12/.*

223. Nayak, Y., Mor, V. Unnikrishnan, M.K. (2010). Research in pharmacy schools of India: A study based on Scopus database. *Indian Journal of Pharmaceutical Education and Research*, 45(1), 1-7.

224. Nederhof, A.J., and Noyons, E.C.M. (1992). International comparison of departments' research performance in the humanities. *Journal of the American Society for Information Science*, 43(3), 249-256.

225. Nederhof, A. J., and Zwaan, R. A. (1991). Quality judgments of journals as indicators of research performance in the humanities and the social and behavioral sciences. *Journal of the American Society for Information Science*, 42(5), 332-340.

226. Nederhof, A. J., Luwel M., and Moed, H. F. (2001). Assessing the quality of scholarly journals in linguistics: An alternative to citation based journal impact factors. *Scientometrics,* 51(1), 241-265.

227. Nederhof, A. J., Zwaan, R. A., de Bruin, R. E., and Dekker, P. J. (1989). Assessing the usefulness of bibliometric indicators for the humanities and tile social behavioral sciences. *Scientometrics,* 15(5-6), 423-435.

228. Nightingale, Paul and Scott, Alister. (2007). Peer review and the relevance gap: Ten suggestions for policy-makers. *Science and Public Policy, 34 (8): 543-553. doi:10.3152/030234207X254396.*

229. Novak, J. D., and Gowin, J. B. (1984). *Learning how to learn.* Cambridge, UK, Cambridge University Press.

230. Noyons, E. (2001). Bibliometric mapping of science in a science policy context. *Scientometrics,* 50(1), 83-98.

231. Odumosu, Tolu. (2013, October). Making the World. One Recipe at a Time. *Science and Public Policy*, 40(5).

232. OECD. (1963, 1976). *The Measurement of Scientific and Technical Activities: Frascati Manual.* Paris.

233. Open University, Hobsons, and Higher Education Funding Council for England. (2008). Counting what is measured or measuring what counts? League tables and their impact on higher education institutions in England. Bristol, Higher Education Funding Council for England. Available at: *http://www.hefce.ac.uk/pubs/ year/2008/200814/*

234. Oppenheim, C. (2000). Do patent citations count? In B. Cronin and H. B. Atkins (Eds.), *The web of knowledge: A festschrift in honor of Eugene Garfield* (p. 405-432). Metford, NJ, Inormation Today Inc. ASIS Monograph Series.

235. Oppenheim, C., and Renn, S. (1978). Highly cited old papers and the reasons why they continue to be cited. *Journal of the American Society for Information Science,* 29(5), 225-231.

236. Palys, T. (1992). *Research decisions: Quantitative and qualitative perspectives* (3rd ed.). Toronto: Thompson Canada.

237. Pandya S.K. (1990, August 11). Why is the output of medical research from India low?. *British Medical Journal*, 301(6747), 333. PMC1663625.

238. Park, H. W. (2003). Hyperlink network analysis: A new method for the study of social structure on the web. *Connections*, 25(1), 49-61.

239. Park, H. W., and Thelwall, M. (2003). Hyperlink analysis: Between networks and indicators. *Journal of Computer-Mediated Communication,* 8(4). Available at: *http:// www.ascusc.org/jcmc/vol8/issue4/park.html.*

240. Persson, O., and Åström, F. (2005). Most cited universities and authors in Library & Information Science 1990-2004. *Bibliometric Notes,* 7(2). Available at: *http://www. umu.se/inforsk/BibliometricNotes/BN2-2005/BN2-2005.htm.*

241. Philipps, Axel. (2013, October). Mission Statements and Self-Descriptions of German Extra-University Research Institutes: A Qualitative Content Analysis. *Science and Public Policy*, 40(5).

242. Pouris, A. (2007). Nanoscale research in South Africa: A mapping exercise based on scientometrics. *Scientometrics, 70*(3), 541-553.

243. Prathap, G., & Gupta, B. M. (2011). Ranking of Indian medical colleges for their research performance during 1999-2008. *Annals of library and information studies, 58*(3), 203-210.

244. Price, D. J. de Solla. (1961). *Science Since Babylon.* New Haven: Yale University Press.

245. Price, D. J. de Solla. (1963). *Little Science. Big Science.* New York, Columbia University Press.

246. Price, D. J. de Solla. (1965). Networks of scientific papers. *Science,* 149(3683), 510-515.

247. Price, D. J. de Solla. (1976). A general theory of bibliometric and other cumulative advantage processes. *Journal of the American Society for Information Science,* 27(5), 292-306.

248. Price, D.J.D. and Beaver, D.B. (1966). Collaboration in an Invisible College. *American Psychologist,* 21, 1011-1018.

249. Pritchard, A. (1969). Statistical bibliography or bibliometrics? *Journal of Documentation,* 25(4), 348-349.

250. Protogerou, Aimilia, Caloghirou, Yannis, and Siokas, Evangelos. (2013, October). Research Networking and Technology Fusion through EU-Funded Collaborative Projects. *Science and Public Policy,* 40(5).

251. Radicchi, F., Fortunato, S., and Castellano, C. (2008), Universality of citation distributions: Toward an objective measure of scientific impact. *Proceedings of the National Academy of Sciences,* 105(45), 17268-17272.

252. Rafols, I., Leydesdorff, L., O'Hare, A., Nightingale, P., and Stirling, A. (2012). How journal rankings can suppress interdisciplinary research: A comparison between innovation studies and business & management. *Research Policy,* 41(7), 1262-1282.

253. Raja S., Balasubramani R. (2011) Plasmodium falciparum research publication in India: A scientometric analysis. *European Journal of Scientific Research,* 56(3). 294-300.

254. Rajagopal T., Archunan G., Surulinathi M., and Ponmanickam P. (2012). Research output in pheromone biology: A case study of India. *Scientometrics,* 94(2), 711-719.

255. Ramesh, R., and Ravi Chandra Rao, I. K. (2004). Modelling the growth of indian medical literature. In *Digital Information Exchange. Annual conference* (p. 162-171).

256. Rafols, I., Leydesdorff, L., O'Hare, A., Nightingale, P., and Stirling, A. (2012). How journal rankings can suppress interdisciplinary research: A comparison between innovation studies and business & management. *Research Policy, 41*(7), 1262-1282.

257. Rip, A. (1988). Mapping of Science: Possibilities and Limitations. In A. F. J. Van Raan (Ed.). *Handbook of Quantitative Studies of Science and Technology* (pp. 253-273). North: Elsevier.

258. Rorty, R. (1999). *Philosophy and social hope*. London, Penguin.

259. Ross, Beveridge, (2007). Smart and snappy, *Science and Public Policy, 34*(8), 603-604. *doi:10.1093/spp/34.8.603.*

260. Rousseau, R. (1997). Citations: an exploratory study. *Cybermetrics, 1*(1). Available at: *http://www.cindoc.csic.es/cybermetrics/articles/v1i1p1.html.*

261. Roy, R. (1976, June). Comments on Citation Study of Materials Science Departments. *Journal of Metals, 28*(6), 29-30.

262. Sadlak, J., and Liu, N. C. (2007). *The world-class university and ranking: Aiming beyond status.* Bucharest, Romania/Shanghai, China/Cluj-Napoca, Romania, Unesco-Cepes.

263. Sandelowski, M. (2001). Real qualitative researchers do not count: The use of numbers in qualitative research. *Research in Nursing and Health, 24*(3), 230-240.

264. Scharnhorst, A., Börner, K., and van den Besselaar, P. (2012). *Models of science dynamics: encounters between complexity theory and information sciences*. Springer Science & Business Media.

265. Schubert, A., and Braun, T. (1996). Cross-field normalization of scientometric indicators. *Scientometrics, 36*(3), 311-324.

266. Sekar, N., Shah, N. K., Abbas, S. S., Kakkar, M., (2011). Roadmap to Combat Zoonoses in India Initiative: Research options for controlling zoonotic disease in India, 2010–2015. *PLoS one, 6*(2), e17120.

267. Shapiro, F. R. (1992). Origins of bibliometrics, citation indexing, and citation analysis: The neglected legal literature. *Journal of the American Society for Information Science, 43*(5), 337-339.

268. Shapley, D. (1975, August 22). NSF: A 'Populist' Pattern in Metallurgy, Materials Research?. *Science, 189*(4203), 622-624.

269. Shapley, D. (1976, January 9). Materials Research: Scientists Show Scant Taste for Breaking Ranks. *Science, 191*(4222), 53.

270. Sharif, M. A., and Mahmood, K. (2004). How economists cite literature: Citation analysis of two core Pakistani economic journals. *Collection Building*, 23(4), 172-176.

271. Sher, LH., and Garfield, E. (1966). New Tools for Improving and Evaluating the Effectiveness of Research. In Yovits, M.C., Gilford, D.M., Wilcox, R.H., Stavely, E., and Lemer, H.D. (eds.). *Research Program Effectiveness* (New York: Gordon and Breach, 1966). (p. 135-146).

272. Shin, J. C., Toutkoushian, R. K., and Teichler, U. (Eds.). (2011). *University rankings: Theoretical basis, methodology and impacts on global higher education* (Vol. 3). Springer Science & Business Media.

273. Small, H.G. (1974). Characteristics of Frequently Cited Papers in Chemistry. *Final Report on NSF Contract*.

274. Small, H.G. and Griffith, B.C. (1974). The Structure of Scientific Literatures, I: Identifying and Graphing Specialties. *Science Studies*, 4, 17-40.

275. Smith, A. G. (1999). A tale of two web spaces; comparing sites using Web Impact Factors. *Journal of Documentation*, 55(5), 577-592.

276. Sorensen, J. R. (1994). Scholarly productivity in criminal justice: Institutional affiliation of authors in the top ten criminal justice journals. *Journal of Criminal Justice*, 22(6), 535-547.

277. Stefan,Carlstein. Evaluative Bibliometrics. Available at: *http://hj.se/bibl/en/ publishing/bibliometrics/evaluative-bibliometrics.html*.

278. Stefan, Carlstein. Journal evaluation. Available at: *http://hj.se/bibl/en/publishing/ bibliometrics/journal-evaluation.html*.

279. Stefan, Carlstein. Ranking lists. Available at: *http://hj.se/bibl/en/publishing/ bibliometrics/ranking-lists.html*.

280. Stefan, Carlstein. Tools for bibliometric analyses: Available at: *http://hj.se/bibl/en/ publishing/bibliometrics/tools-for-bibliometric-analyses.html*.

281. Stefan, Carlstein. Descriptive bibliometrics. Avaialble at: *http://hj.se/bibl/en/ publishing/bibliometrics/descriptive-bibliometrics.html*.

282. Stefan, Carlstein. Databases for bibliometric analysis, Available at: *http://hj.se/bibl/ en/publishing/bibliometrics/databases-for-bibliometric-analysis.html*.

283. Stefan, Carlstein. Definitions & Terminology. Available at: *http://hj.se/bibl/en/ publishing/bibliometrics/definitions–terminology.html*.

284. Stefan, Carlstein. Evaluative Biblimetrics: The Swedish model. Available at: *http:// hj.se/bibl/en/publishing/bibliometrics/evaluative-bibliometrics/the-swedish-model. html.*

285. Stefan, Carlstein. Evaluative Biblimetrics: The Norwegian model: Available at: *http:// hj.se/bibl/en/publishing/bibliometrics/evaluative-bibliometrics/the-norwegian-model.html.*

286. Stewart, J., Van Kirk, J., and Rowell, R. (1979). Concept Maps: A tool for use in biology teaching. *American Biology Teacher*, 41(3), 171-175.

287. Stillwell, W., Winterfeldt, D. V., and John, R. S. (1987). Comparing hierarchical and nonhierarchical weighting methods for eliciting multiattribute value models. *Management Science*, 33(4), 937-943.

288. Sujatha, Raman. (2007). Lack of balance *Science and Public Policy*, 34 (8), 601-602. *doi:10.1093/spp/34.8.601.*

289. Swygert, M. I., and Gozansky, N. E. (1985). Senior law faculty publication study: Comparisons of law school productivity. *Journal of Legal Education*, 35(3), 373-394.

290. Sylvia, M. J. (1998). Citation analysis as an unobtrusive method for journal collection evaluation using psychology student research bibliographies. *Collection Building*, 17(1), 20-28.

291. Tashakkori, A., and Teddlie, C. (1998). *Mixed methodology: Combining qualitative and quantitative approaches.* Thousand Oaks, CA, Sage.

292. Teddlie, C. B., and Tashakkori, A. (2009). *Foundations of mixed methods research: Integrating quan- titative and qualitative approaches in the social and behavioral sciences.* Thousand Oaks, CA, Sage.

293. Terttu, Luukkonen. (2007). A new European S&T governance. *Science and Public Policy*, 34(8), 599-601. *doi:10.3152/030234207X257825.*

294. Thelwall, M. (2002). The top 100 linked pages on UK university Web sites: High in link counts are not usually directly associated with quality scholarly content. *Journal of Information Science*, 28(6), 485-493.

295. Thelwall, M., (2003). What is this link doing here? Beginning a fine-grained process of identifying reasons for academic hyperlink creation. *Information Research*, 8(3). Available at: *http://informationr.net/ir/8-3/paper151.html.*

296. Thelwall, M., (2005). *Link Analysis: An Information Science Approach.* San Diego, Academic Press.

297. Thelwall, M., and Wilkinson, D. (2003). Graph structure in three national academic Webs: Power laws with anomalies. *Journal of American Society for Information Science and Technology*, 54(8), 706-712.

298. Thelwall, M., Vaughan, L., and Björneborn, L. (2005). Webometrics. *Annual Review of Information Science and Technology*, 39.

299. Turns, J., Atman, C., and Adams, R. (2000). Concept maps for engineering education: A cognitively motivated tool supporting varied assessment functions. *IEEE Transactions on Education*, 43, 164-173.

300. Van Den Berghe, H., de Bruin, R. E., Houben, J. A., Kint, A., Luwel, M., Spruyt, E., and Moed, H. F.(1998). Bibliometric indicators of research performance in Flanders. *Journal of the American Society for Information Science,* 49(1), 59-67.

301. Van Raan, A. F. J. (1996). Advanced bibliometric methods as quantitative core of peer review based evaluation and foresight exercises. *Scientometrics,* 36(3), 397-420.

302. Van Raan, A. F. J. (1998). In matters of quantitative studies of science the fault of theorists is offering too little and asking too much. *Scientometrics,* 43(1), 129-148.

303. Van Raan, A. F. J. (2000). The Pandora's Box of Citation Analysis: Measuring Scientific Excellence- The Last Evil? In B. Cronin & H. B. Atkins (Eds.), *The Web of Knowledge: A Festschrift in Honor of Eugene Garfield* (p. 301-319). Medford, NJ: Information Today, Inc. ASIS Monograph Series.

304. Van Raan, A. F.J. (2005). Fatal attraction: Conceptual and methodological problems in the ranking of universities by bibliometric methods. *Scientometrics*, *62*(1), 133-143.

305. Van Wesel, M.; Wyatt, S.; and ten Haaf, J. (2013). What a difference a colon makes: how superficial factors influence subsequent citation. *Scientometrics. doi:10.1007/ s11192-013-1154-x.*

306. Vickery, B.C. (1971). The administration of research in institutions. Paper presented at the *Seminar on Objectives and Administration of Library Research*, Nottingham, 20-21 September, 1971.

307. Vieira, E.S., and Gomes, J. A. N. F. (2009). A comparison of Scopus and Web of Science for a typical university. *Scientometrics,* (81)2, 587-600.

308. Vijayalakshmi S., and Iyer N.R. Mapping strategies and performance evaluation of research organizations, Recent Researches in Communications and IT - *Proc.*

of the 15th WSEAS Int. Conf. on Communications, Part of the 15th WSEAS CSCC Multiconference, Proc. of the 5th Int. Conf. on CIT'11, 50-53.

309. Virgo, J.A. (1977). A Statistical Procedure for Evaluating the Importance of Scientific Papers. *The Library Quarterly,* 47, 415-430.

310. Visalakshi, S., and Sandhya, G. (1997). An analysis of biotechnology and non-biotechnology R&D capabilities in the Indian pharmaceutical industry. *R&D Management,* 27(2), 177-180.

311. Vreeland, R. C. (2000). Law libraries in hyperspace: A citation analysis of World Wide Web sites. *Law Library Journal,* 92(1), 9-25.

312. Wade, N. (1975). Citation Analysis: A New Tool for Science Administrators. *Science,* 188(4187), 429-432.

313. Walters, W. H. (2007, July). Google Scholar coverage of a multidisciplinary field. *Information Processing & Management,* 43(4),1121-1132.

314. Waltman, L., Van Eck, N.J., Van Leeuwen, T.N., Visser, M.S., & Van Raan, A.F.J. (2011). Towards a new crown indicator: Some theoretical considerations. *Journal of Informetrics,* 5(1), 37–47. Available at: *http://dx.doi.org/10.1016/j.joi.2010.08.001.*

315. Waltman, L., Van Eck, N.J., Van Leeuwen, T.N., Visser, M.S., & Van Raan, A.F.J. (2011). Towards a new crown indicator: An empirical analysis. *Scientometrics,* 87 (3), 467-481. Available at: *http://arxiv.org/abs/1004.1632.*

316. Wert, C.A. (1975, December). The Citation Index Revisited. *Journal of Metals,* 27(12), 20-22.

317. Wheeldon, J. P. (2010). Mapping mixed methods research: Methods, measures, and meaning. *Journal of Mixed Methods Research,* 4(2), 87-102.

318. Wheeldon, J. P., and Faubert, J. (2009). Framing experience: Concept maps, mind maps, and data collection in qualitative research. *International Journal of Qualitative Methods,* 8(3), 68-83.

319. White, B. (2006).Examining the claims of Google Scholar as a serious information source. *New Zealand Library & Information Management Journal,* 50(1), 11-24.

320. Wickson, F. (2012). Researching standards: No recipe, no risk. *Science and Public Policy,* scs111.

321. Wilkinson, D., Harries, G., Thelwall, M., & Price, E. (2003). Motivations for academic Web site interlinking: Evidence for the Web as a novel source of information on informal scholarly communication. *Journal of Information Science,* 29(1), 49-56.

322. Wood, J. B. (1988). The growth of scholarship: An online bibliometric comparison of dissertations in the sciences and the humanities. *Scientometrics*, 13(1-2), 53-62.

323. Wouters, P. (1999). *The Citation Culture*. Amsterdam: Unpublished Ph.D. Thesis University of Amsterdam.

324. Wouters, P., and Leydesdorff, L. (1994). Has Price's Dream Come True: Is Scientometrics a Hard Science? *Scientometrics*, 31, 193-222.

325. Yang K. and L. I. Meho. (2006). Citation Analysis: A Comparison of Google Scholar, Scopus, and Web of Science, in *69th Annual Meeting of the American Society for Information Science and Technology*, Austin (US), (p. 3-8).

326. Yu, G., and Li, Y.J. (2010). Identification of referencing and citation processes of scientific journals based on the citation distribution model. *Scientometrics*, 82 (2): 249–261. *doi:10.1007/s11192-009-0085-z*.

327. Zhou, P., and Leydesdorff, L. (2011). Fractional counting of citations in research evaluation: A cross- and interdisciplinary assessment of the Tsinghua University in Beijing. *Journal of Informetrics*, 5(3), 360-368.

328. Zuckerman, H. (1968). Patterns of Name-Ordering Among Authors of Scientific Papers. *American Journal of Sociology*, 74, 276-291.

329. Zuckerman, H. (1987). Citation analysis and the complex problem of intellectual influence. *Scientometrics*, 12(5), 329-338.

INDEX

Table 3.1: Country Categorization Based on Number of Papers Published

No. of Publications	No. of Countries
>=150000	20
100000-150000	5
50000-100000	13
25000-50000	10
10000-25000	14
5000-10000	12
1000-5000	43
<1000	30

Table 3.2: Top 20 Countries Based on the Number of Papers
[Data for 10 year Period Jan. 2002 - Jun 2012]

Sl. No.	Country	Papers	Citations	CPP
1	USA	3,199,249	50,208,701	15.69
2	Peoples R China	996,935	6,420,348	6.44
3	Germany	831,676	11,038,202	13.27
4	Japan	793,163	8,188,434	10.32
5	England	735,916	11,001,142	14.95
6	France	593,631	7,354,022	12.39
7	Canada	485,580	6,406,015	13.19
8	Italy	462,968	5,534,428	11.95
9	Spain	375,263	3,960,297	10.55
10	Australia	332,527	3,976,977	11.96
11	India	327,924	1,994,057	6.08
12	South Korea	318,539	2,311,340	7.26
13	Netherlands	272,015	4,252,011	15.63
14	Russia	271,611	1,337,782	4.93
15	Brazil	239,055	1,533,297	6.41
16	Taiwan	199,135	1,441,074	7.24
17	Switzerland	197,062	3,283,869	16.66
18	Sweden	188,006	2,773,025	14.75
19	Turkey	171,883	946,496	5.51
20	Poland	167,132	1,127,867	6.75

Note: Top 20 contries with >=150000 Publication Ouput

CPP = citations Per Paper

Table 3.3: Trends in Global Research landscape Profile in Various Subject Domains and India's Share: Decade-wise Scenario

Fields	1971-1980		1981-1990		1991-2000		2001-2010		2011-2012		All Years 1971-2012	
	World	India	World	India	World	India	World	India	World	India	World	India
Agricultural Sciences	62676	4426	99452	7525	131978	8695	228930	11611	46354	2364	569390	34621
Biology & Biochemistry	202806	7232	347711	7376	560066	7222	743283	15654	104944	3684	1958810	41168
Chemistry	399964	14182	560882	24716	900983	31361	1399222	65978	244212	15140	3505263	151377
Clinical Medicine	774341	5529	1442614	8861	2310054	14416	3743289	44437	666071	11346	8936369	84589
Computer Science	18514	298	55479	874	127581	1900	278888	4526	45138	1083	525600	8681
Economics & Business	78898	234	115246	456	157353	573	217235	1269	41855	304	610587	2836
Engineering	162949	4658	303171	8837	603714	13788	927767	29330	175562	7245	2173163	63858
Environment/ Ecology	39987	803	91454	2770	166502	3472	289172	7444	60413	2108	647528	16597
Geosciences	66548	2044	121998	3869	186519	5159	332319	9437	55673	2081	763057	22590
Immunology	32123	106	73027	345	111726	901	156449	2092	27779	479	401104	3923
Materials Science	64082	1409	121475	4860	267627	9921	489566	22653	94784	5490	1037534	44333
Mathematics	80409	1283	132601	3753	172490	3790	268002	5382	53020	1291	706522	15499
Microbiology	43582	429	72984	985	114793	1722	180259	6126	36089	1974	447707	11236
Molecular Biology & Genetics	66330	936	131178	1400	254110	1788	348544	4428	58365	1293	858527	9845
Multi disciplinary	76373	5450	115977	5495	147788	4822	150423	7166	23747	995	514308	23928

Neuroscience & Behavior	70605	400	166546	585	286052	1751	468575	4646	72707	1007	1064485	8389
Pharmacology & Toxicology	70170	870	112852	1752	145646	2422	269566	10117	55941	2707	654175	17868
Physics	255826	8560	377454	13530	633806	19530	939200	33601	154156	6986	2360442	83207
Plant & Animal Science	197609	4888	347106	12569	436075	14380	660322	19846	112546	3615	1753658	55298
Psychiatry/ Psychology	125613	352	181737	521	267369	763	393050	1486	64815	253	1035584	3375
Social Sciences, general	277452	934	433663	1914	627153	2523	830672	4362	159737	1126	2328677	10859
Space Science	29672	502	44631	1656	72304	1901	121977	3344	21606	824	290190	8227
Total	3196529	65525	5449238	114649	8681689	152800	13436710	314935	2375514	73395	33142680	722304

Table 3.4: Discipline-wise Share of Indian Publications viz-a-viz World Output 1971-2010 & 2011-2012

Fields	India's share in	india's share in	India's Overall % of world share
	1971-2010	2011-2012	1971-2012
Agricultural Sciences	6.17	5.10	6.08
Biology & Biochemistry	2.02	3.51	2.10
Chemistry	4.18	6.20	4.32
Clinical Medicine	0.89	1.70	0.95
Computer Science	1.58	2.40	1.65
Economics & Business	0.45	0.73	0.46
Engineering	2.83	4.13	2.94
Environment/Ecology	2.47	3.49	2.56
Geosciences	2.90	3.74	2.96
Immunology	0.92	1.72	0.98
Materials Science	4.12	5.79	4.27
Mathematics	2.17	2.43	2.19
Microbiology	2.25	5.47	2.51
Molecular Biology & Genetics	1.07	2.22	1.15
Multi disciplinary	4.67	4.19	4.65
Neuroscience & Behavior	0.74	1.39	0.79
Pharmacology & Toxicology	2.53	4.84	2.73
Physics	3.45	4.53	3.53
Plant & Animal Science	3.15	3.21	3.15
Psychiatry/Psychology	0.32	0.39	0.33
Social Sciences, general	0.45	0.70	0.47
Space Science	2.76	3.81	2.84

Table 3.6: Global Baselines: Field Rankings for India & World

Rank	India Field	Papers	Citations Per Paper	Rank	World Field	Papers	Citations Per Paper
1	Chemistry	74,618	7.38	1	Clinical Medicine	2,243,290	12.56
2	Physics	38,238	6.65	2	Chemistry	1,277,663	11.43
3	Clinical Medicine	34,161	6.16	3	Physics	980,416	8.47
4	Engineering	33,010	4.47	4	Engineering	906,529	4.96
5	Materials Science	26,110	5.83	5	Plant & Animal Science	583,513	7.67
6	Plant & Animal Science	21,673	3.6	6	Biology & Biochemistry	574,795	16.37
7	Biology & Biochemistry	17,254	8.32	7	Social Sciences, General	512,563	4.71
8	Agricultural Sciences	12,849	3.84	8	Materials Science	505,252	7.48
9	Pharmacology & Toxicology	10,428	7.56	9	Neuroscience & Behavior	314,871	18.54
10	Geosciences	10,374	5.02	10	Geosciences	308,764	9.55
11	Environment/Ecology	9,142	6.9	11	Environment/Ecology	299,856	11.24
12	Microbiology	6,735	5.99	12	Molecular Biology & Genetics	298,779	23.48
13	Mathematics	6,044	2.13	13	Mathematics	286,862	3.49
14	Computer Science	5,235	3.07	14	Computer Science	282,804	4.01
15	Molecular Biology & Genetics	4,986	9.05	15	Psychiatry/Psychology	263,525	11.16
16	Space Science	3,849	7.89	16	Agricultural Sciences	227,351	7.16
17	Social Sciences, General	3,529	2.91	17	Pharmacology & Toxicology	204,138	11.97
18	Neuroscience & Behavior	3,425	7.26	18	Economics & Business	184,532	6.4
19	Immunology	2,122	8.75	19	Microbiology	179,670	14.9
20	Multidisciplinary	1,982	2.1	20	Space Science	129,538	14.35
21	Economics & Business	1,299	3.69	21	Immunology	127,424	20.57
22	Psychiatry/Psychology	861	7.63	22	Multidisciplinary	18,051	7.54
23	All Fields*	327,924	6.08	23	All Fields	10,710,186	11

* Includes data for all papers from ranked and unranked fields.

Table 3.7: Average Citation Rates for Papers Published by Field (2002 - 2012)

Fields	2002	2003	2004	2005	2006	2007	2008	2009	2010	2011	2012	All Years
All Fields	20.83	19.34	18.09	16.04	13.65	11.75	8.88	6.55	3.93	1.42	0.2	10.5
Agricultural Sciences	15.33	14.77	13.86	11.96	10.61	8.58	5.96	4.25	2.4	0.82	0.12	7.16
Biology & Biochemistry	31.67	29.46	26.9	23.16	19.48	16.73	13.04	9.44	5.49	2	0.25	16.37
Chemistry	20.6	19.23	18.47	16.94	14.59	12.72	10.44	8.05	5.01	1.91	0.23	11.43
Clinical Medicine	24.77	23.55	22.12	19.93	16.87	14.27	10.49	7.52	4.58	1.53	0.22	12.56
Computer Science	9.42	6.56	5.01	4.74	3.73	5.43	4.3	3.04	1.77	0.54	0.08	4.01
Economics & Business	15.35	14.07	13.07	11.05	9.07	7.39	4.92	3.24	1.77	0.63	0.12	6.4
Engineering	9.17	8.74	8.59	7.58	6.63	6.15	4.58	3.61	2.06	0.73	0.1	4.96
Environment/Ecology	23.73	22.28	20.59	17.9	15.3	13.11	9.72	6.86	3.89	1.42	0.23	11.24
Geosciences	19.1	18.01	16.52	14.52	13.4	10.01	7.96	6	3.51	1.4	0.27	9.55
Immunology	37.98	35.28	33.86	29.81	25.36	22.13	17.26	12.7	7.2	2.58	0.27	20.57
Materials Science	12.75	13.22	12.13	11.02	9.93	8.85	7.1	5.59	3.57	1.36	0.16	7.48
Mathematics	7.27	6.69	6.15	5.6	4.82	3.95	3.12	2.26	1.27	0.47	0.09	3.49
Microbiology	29.94	28.08	26.43	24.38	19.67	16.14	12.42	8.88	5.33	1.85	0.22	14.9
Molecular Biology & Genetics	48.82	44.33	40.67	34.94	29.87	24.98	19.28	13.99	8.13	2.92	0.34	23.48
Multidisciplinary	7.79	7.42	6.96	13.23	13.06	11.45	9.51	6.81	5.39	2.23	0.56	7.54
Neuroscience & Behavior	36.84	33.15	30.91	27.69	23.55	19.59	14.8	10.56	6.09	2.14	0.28	18.54
Pharmacology & Toxicology	24.59	21.99	21.83	18.29	17.1	14.14	10.94	7.58	4.25	1.49	0.21	11.97
Physics	15.16	14.26	13.84	12.51	10.86	8.39	6.65	5.92	3.88	1.57	0.23	8.47
Plant & Animal Science	15.58	14.46	13.52	11.58	9.9	8.19	6.1	4.34	2.55	0.92	0.15	7.67
Psychiatry/Psychology	23.34	22.94	20.75	17.68	15.11	12.2	9	5.99	3.24	1.13	0.21	11.16
Social Sciences, general	10.34	9.76	9.5	8.58	7.17	5.86	4.13	2.82	1.56	0.57	0.14	4.71
Space Science	21.91	24.18	21.87	20.92	18.18	17.26	11.93	10.59	6.91	3.07	0.58	13.84

Table 3.8: Global Publication Authors by Country

	Country/Territory	Papers	Citations	Citations Per Paper
1	USA	3,199,249	50,208,701	15.69
2	Peoples R China	996,935	6,420,348	6.44
3	Germany	831,676	11,038,202	13.27
4	Japan	793,163	8,188,434	10.32
5	England	735,916	11,001,142	14.95
6	France	593,631	7,354,022	12.39
7	Canada	485,580	6,406,015	13.19
8	Italy	462,968	5,534,428	11.95
9	Spain	375,263	3,960,297	10.55
10	Australia	332,527	3,976,977	11.96
11	India	327,924	1,994,057	6.08
12	South Korea	318,539	2,311,340	7.26
13	Netherlands	272,015	4,252,011	15.63
14	Russia	271,611	1,337,782	4.93
15	Brazil	239,055	1,533,297	6.41
16	Taiwan	199,135	1,441,074	7.24
17	Switzerland	197,062	3,283,869	16.66
18	Sweden	188,006	2,773,025	14.75
19	Turkey	171,883	946,496	5.51
20	Poland	167,132	1,127,867	6.75
21	Belgium	149,383	2,077,628	13.91
22	Israel	116,034	1,457,925	12.56
23	Scotland	114,954	1,794,156	15.61
24	Denmark	105,467	1,658,499	15.73
25	Austria	102,985	1,340,896	13.02
26	Iran	99,413	439,482	4.42
27	Finland	93,807	1,265,994	13.5
28	Greece	90,902	828,875	9.12
29	Mexico	82,439	586,964	7.12
30	Norway	79,237	1,001,427	12.64
31	Czech Republic	74,413	608,383	8.18
32	Singapore	74,396	760,677	10.22
33	Portugal	72,181	669,227	9.27
34	Argentina	62,561	518,233	8.28

35	New Zealand	62,536	687,018	10.99
36	South Africa	60,188	517,161	8.59
37	Hungary	53,009	542,418	10.23
38	Ireland	51,439	602,991	11.72
39	Ukraine	45,176	191,735	4.24
40	Romania	42,467	189,061	4.45
41	Egypt	41,376	216,134	5.22
42	Wales	39,013	493,286	12.64
43	Chile	38,465	336,370	8.74
44	Thailand	38,079	300,711	7.9
45	Malaysia	32,754	140,607	4.29
46	Slovenia	26,459	188,968	7.14
47	Slovakia	25,607	169,596	6.62
48	Pakistan	25,188	112,968	4.48
49	Saudi Arabia	24,806	103,104	4.16
50	Croatia	24,461	142,851	5.84
51	Bulgaria	20,060	134,018	6.68
52	Serbia	19,180	58,964	3.07
53	North Ireland	19,151	228,219	11.92
54	Tunisia	18,091	78,800	4.36
55	Colombia	16,192	104,754	6.47
56	Lithuania	13,863	70,683	5.1
57	Nigeria	13,259	59,296	4.47
58	Venezuela	11,889	84,317	7.09
59	Morocco	11,867	60,906	5.13
60	Algeria	11,534	47,453	4.11
61	Byelarus	10,281	46,258	4.5
62	Estonia	10,032	96,375	9.61
63	Vietnam	8,567	58,915	6.88
64	Jordan	8,429	40,711	4.83
65	Kenya	8,325	88,171	10.59
66	Indonesia	7,583	59,567	7.86
67	Cuba	7,487	46,146	6.16
68	U Arab Emirates	7,479	41,478	5.55
69	Bangladesh	7,460	49,413	6.62
70	Philippines	6,461	59,592	9.22

71	Kuwait	6,107	33,979	5.56
72	Iceland	5,991	95,730	15.98
73	Lebanon	5,933	40,067	6.75
74	Uruguay	5,294	47,973	9.06
75	Armenia	4,935	35,509	7.2
76	Peru	4,744	51,416	10.84
77	Cyprus	4,620	31,529	6.82
78	Yugoslavia	4,267	37,994	8.9
79	Tanzania	4,222	43,379	10.27
80	Cameroon	4,148	27,423	6.61
81	Ethiopia	4,068	24,811	6.1
82	Uganda	4,062	39,608	9.75
83	Latvia	4,048	27,467	6.79
84	Sri Lanka	3,612	25,603	7.09
85	Rep of Georgia	3,598	20,605	5.73
86	Costa Rica	3,592	43,675	12.16
87	Oman	3,406	17,793	5.22
88	Luxembourg	3,314	28,001	8.45
89	Uzbekistan	3,311	14,361	4.34
90	Azerbaijan	3,145	9,462	3.01
91	Ghana	3,062	22,760	7.43
92	Kazakhstan	2,535	9,704	3.83
93	Ecuador	2,533	25,323	10
94	Senegal	2,417	19,084	7.9
95	Nepal	2,378	17,595	7.4
96	Bosnia & Herceg	2,297	7,881	3.43
97	Zimbabwe	2,258	19,976	8.85
98	Moldova	2,175	13,736	6.32
99	Panama	2,175	35,834	16.48
100	Syria	2,090	11,445	5.48
101	Qatar	2,086	10,330	4.95
102	Serbia Monteneg	1,913	15,191	7.94
103	Malawi	1,849	18,929	10.24
104	Iraq	1,805	5,159	2.86
105	Macedonia	1,753	8,990	5.13
106	Burkina Faso	1,724	14,303	8.3

107	Jamaica	1,660	9,532	5.74
108	Sudan	1,657	9,431	5.69
109	Botswana	1,630	12,610	7.74
110	Bolivia	1,615	15,702	9.72
111	Cote Ivoire	1,601	14,476	9.04
112	Trinid & Tobago	1,538	8,997	5.85
113	Zambia	1,365	14,852	10.88
114	Benin	1,339	8,595	6.42
115	Madagascar	1,322	9,618	7.28
116	New Caledonia	1,140	10,603	9.3
117	Mongol Peo Rep	1,057	7,538	7.13
118	Libya	998	3,567	3.57
119	Reunion	995	7,945	7.98
120	Mali	990	10,591	10.7
121	Guadeloupe	977	7,103	7.27
122	Malta	944	8,577	9.09
123	Cambodia	902	8,710	9.66
124	Mozambique	836	7,882	9.43
125	Gambia	826	14,150	17.13
126	Gabon	801	10,129	12.65
127	Guatemala	760	6,336	8.34
128	Fiji	758	4,825	6.37
129	Yemen	755	3,301	4.37
130	Papua N Guinea	744	7,748	10.41
131	Namibia	733	7,989	10.9
132	Congo	689	5,143	7.46
133	Niger	665	4,865	7.32
134	Monaco	652	7,750	11.89
135	Laos	592	5,098	8.61
136	Barbados	578	5,858	10.13
137	Fr Polynesia	550	4,545	8.26
138	Kyrgyzstan	531	2,087	3.93
139	French Guiana	510	5,179	10.15
140	Nicaragua	493	4,507	9.14
141	Greenland	430	3,975	9.24
142	Liechtenstein	381	3,464	9.09

143	Honduras	344	3,745	10.89
144	Bermuda	272	6,480	23.82
145	Guinea Bissau	213	3,016	14.16
146	Chad	148	1,861	12.57
147	Vatican	92	816	8.87

Table 3.9(a): Overall Field Weighted Relative Impact: All Fields and All Countries India's Placement

		Top 3 Countries			
All Fields	**TPC**	**1st**	**2nd**	**3rd**	**India**
	147	**Bermuda**	**Gambia**	**Switzerland**	
CPP (Rank & Value)		CP= 1 (23.82)	CP= 2 (17.13)	CP=3 (16.66)	CP=111(6.08)
P (Rank & Number)		P =145 (272)	P = 147 (826)	P= 17 (197062)	P= 11(327924)
C (Rank & Number)		C=129 (6480)	C=103 (14150)	C=15(3283869)	C=16(1994057)

CPP= Citations per paper; P= Papers; C= Citations

The number within brackets indicates actual value or number

Note: For reference puposes based on average citations per paper for all the countries the list of all counties derived to show how un-purposeful metrics it can generate if proper sampling mechanism is not taken into account for evaluative bibliometrics

Table 3.9(b) Overall Field Weighted Relative Impact: All Fields and Top 20 Countries

Country/Territory	Papers	Citations	Citations Per Paper
Switzerland	197,062	3,283,869	16.66
USA	3,199,249	50,208,701	15.69
Netherlands	272,015	4,252,011	15.63
England	735,916	11,001,142	14.95
Sweden	188,006	2,773,025	14.75
Germany	831,676	11,038,202	13.27
Canada	485,580	6,406,015	13.19
France	593,631	7,354,022	12.39
Australia	332,527	3,976,977	11.96
Italy	462,968	5,534,428	11.95
Spain	375,263	3,960,297	10.55
Japan	793,163	8,188,434	10.32
South Korea	318,539	2,311,340	7.26
Taiwan	199,135	1,441,074	7.24

Poland	167,132	1,127,867	6.75
Peoples R China	996,935	6,420,348	6.44
Brazil	239,055	1,533,297	6.41
India	327,924	1,994,057	6.08
Turkey	171,883	946,496	5.51
Russia	271,611	1,337,782	4.93

Table 3.10: Relative Spacilization Metrics

Field	Total Papers	Top 1% Papers
Agricultural Sciences	12,849	59
Biology & Biochemistry	17,254	49
Chemistry	74,618	215
Clinical Medicine	34,161	160
Computer Science	5,235	40
Economics & Business	1,299	4
Engineering	33,010	290
Environment/Ecology	9,142	49
Geosciences	10,374	33
Immunology	2,122	2
Materials Science	26,110	88
Mathematics	6,044	35
Microbiology	6,735	14
Molecular Biology & Genetics	4,986	27
Multidisciplinary	1,982	0
Neuroscience & Behavior	3,425	8
Pharmacology & Toxicology	10,428	43
Physics	38,238	290
Plant & Animal Science	21,673	73
Psychiatry/Psychology	861	8
Social Sciences, General	3,529	26
Space Science	3,849	18
All Fields*	327,924	1531

Table 3.11: Relative Specialization Metrics (RSM): World & India (Based on 10 year period)

		India					World		
Rank	Field	Papers	Citations Per Paper	% contribution	Rank	Field	Papers	Citations Per Paper	% contribution
1	Chemistry	74,618	7.38	22.75	1	Clinical Medicine	2,243,290	12.56	20.95
2	Physics	38,238	6.65	11.66	2	Chemistry	1,277,663	11.43	11.93
3	Clinical Medicine	34,161	6.16	10.42	3	Physics	980,416	8.47	9.15
4	Engineering	33,010	4.47	10.07	4	Engineering	906,529	4.96	8.46
5	Materials Science	26,110	5.83	7.96	5	Plant & Animal Science	583,513	7.67	5.45
6	Plant & Animal Science	21,673	3.6	6.61	6	Biology & Biochemistry	574,795	16.37	5.37
7	Biology & Biochemistry	17,254	8.32	5.26	7	Social Sciences, General	512,563	4.71	4.79
8	Agricultural Sciences	12,849	3.84	3.92	8	Materials Science	505,252	7.48	4.72
9	Pharmacology & Toxicology	10,428	7.56	3.18	9	Neuroscience & Behavior	314,871	18.54	2.94
10	Geosciences	10,374	5.02	3.16	10	Geosciences	308,764	9.55	2.88
11	Environment/ Ecology	9,142	6.9	2.79	11	Environment/ Ecology	299,856	11.24	2.80
12	Microbiology	6,735	5.99	2.05	12	Molecular Biology & Genetics	298,779	23.48	2.79
13	Mathematics	6,044	2.13	1.84	13	Mathematics	286,862	3.49	2.68
14	Computer Science	5,235	3.07	1.60	14	Computer Science	282,804	4.01	2.64
15	Molecular Biology & Genetics	4,986	9.05	1.52	15	Psychiatry/ Psychology	263,525	11.16	2.46

16	Space Science	3,849	7.89	1.17	16	Agricultural Sciences	227,351	7.16	2.12
17	Social Sciences, General	3,529	2.91	1.08	17	Pharmacology & Toxicology	204,138	11.97	1.91
18	Neuroscience & Behavior	3,425	7.26	1.04	18	Economics & Business	184,532	6.4	1.72
19	Immunology	2,122	8.75	0.65	19	Microbiology	179,670	14.9	1.68
20	Multidisciplinary	1,982	2.1	0.60	20	Space Science	129,538	14.35	1.21
21	Economics & Business	1,299	3.69	0.40	21	Immunology	127,424	20.57	1.19
22	Psychiatry/Psychology	861	7.63	0.26	22	Multidisciplinary	18,051	7.54	0.17
23	All Fields*	327,924	6.08	100.00	23	All Fields	10,710,186	11	100.00

*Includes data for all papers from ranked and unranked fields.

Table 3.12 Rank Correlation: Discipline-wise India's Share to world output

Rank	Field		India's Share
	India	**World**	
1	Chemistry	Clinical Medicine	5.84
2	Physics	Chemistry	3.90
3	Clinical Medicine	Physics	1.52
4	Engineering	Engineering	3.64
5	Materials Science	Plant & Animal Science	5.17
6	Plant & Animal Science	Biology & Biochemistry	3.71
7	Biology & Biochemistry	Social Sciences, General	3.00
8	Agricultural Sciences	Materials Science	5.65
9	Pharmacology & Toxicology	Neuroscience & Behavior	5.11
10	Geosciences	Geosciences	3.36
11	Environment/Ecology	Environment/Ecology	3.05
12	Microbiology	Molecular Biology & Genetics	3.75
13	Mathematics	Mathematics	2.11
14	Computer Science	Computer Science	1.85
15	Molecular Biology & Genetics	Psychiatry/Psychology	1.67
16	Space Science	Agricultural Sciences	2.97
17	Social Sciences, General	Pharmacology & Toxicology	0.69
18	Neuroscience & Behavior	Economics & Business	1.09
19	Immunology	Microbiology	1.67
20	Multidisciplinary	Space Science	10.98
21	Economics & Business	Immunology	0.70
22	Psychiatry/Psychology	Multidisciplinary	0.33

Table 3.13: Relative Specialization Indices: India

Field	AI*	RSI
Agricultural Sciences	1.85	2.31
Biology & Biochemistry	0.98	0.96
Chemistry	1.91	2.38
Clinical Medicine	0.50	-0.51
Computer Science	0.61	-0.04
Economics & Business	0.23	-3.07
Engineering	1.19	1.35
Environment/Ecology	1.00	0.99
Geosciences	1.10	1.19
Immunology	0.55	-0.28
Materials Science	1.69	2.09
Mathematics	0.69	0.23
Microbiology	1.22	1.40
Molecular Biology & Genetics	0.54	-0.29
Multidisciplinary	3.53	4.25
Neuroscience & Behavior	0.35	-1.47
Pharmacology & Toxicology	1.66	2.06
Physics	1.27	1.49
Plant & Animal Science	1.21	1.39
Psychiatry/Psychology	0.11	-8.36
Social Sciences, General	0.23	-3.21
Space Science	0.97	0.93
All Fields	1.00	1.00

*AI= Activity Index

RSI= Relative Specialization Index

Table 3.14: Fractionalised Publication Counts Vs. Relative Citation Impact: India Viz a Viz World

	Fractionalised Pub.Counts		Relative Citation Impact	
	India	World	India	World
Agricultural Sciences	3.92	2.12	3.84	7.16
Biology & Biochemistry	5.26	5.37	8.32	16.37
Chemistry	22.75	11.93	7.38	11.43
Clinical Medicine	10.42	20.95	6.16	12.56
Computer Science	1.60	2.64	3.07	4.01
Economics & Business	0.40	1.72	3.69	6.4
Engineering	10.07	8.46	4.47	4.96
Environment/Ecology	2.79	2.80	6.9	11.24
Geosciences	3.16	2.88	5.02	9.55
Immunology	0.65	1.19	8.75	20.57
Materials Science	7.96	4.72	5.83	7.48
Mathematics	1.84	2.68	2.13	3.49
Microbiology	2.05	1.68	5.99	14.9
Molecular Biology & Genetics	1.52	2.79	9.05	23.48
Multidisciplinary	0.60	0.17	2.1	7.54
Neuroscience & Behavior	1.04	2.94	7.26	18.54
Pharmacology & Toxicology	3.18	1.91	7.56	11.97
Physics	11.66	9.15	6.65	8.47
Plant & Animal Science	6.61	5.45	3.6	7.67
Psychiatry/Psychology	0.26	2.46	7.63	11.16
Social Sciences, General	1.08	4.79	2.91	4.71
Space Science	1.17	1.21	7.89	14.35
All Fields*	100.00	100.00	6.08	11

Table 3.15: Distribution of Contributing Institutions and their Contributions: Subjectwise & Decadewise Analysis

Subjects	1971-80		1981-90		1991-2000		2001-2010		2011-2012	
	Institutions	Records	Institutions	Records	Institutions	Records	Institutions	Records	Institutions	Records
Agricultural Sciences	635	6494	915	7884	1375	10057	3326	16836	1451	4098
Biology & Biochemistry	765	7787	875	8243	1526	9597	4670	24927	2343	6912
Chemistry	1217	15666	1830	28051	3317	40139	3317	40139	4822	27331
Clinical Medicine	1235	6871	1859	11405	4308	21313	15292	80191	8329	26776
Computer Science	86	343	249	1138	649	2882	206	7793	1019	2220
Economics & Business	113	319	237	670	380	952	849	2203	378	637
Engineering	837	5386	1483	10847	2839	18650	7169	48201	3473	13599
Environment/Ecology	211	887	655	3270	1199	4576	3625	13411	1920	4664
Geosciences	465	115.552	735	4583	1201	6920	3010	17019	1442	4792
Immunology	70	146	193	506	565	1701	1527	215.423	81	1544
Materials Science	301	1554	688	5694	1598	13673	4736	38212	2557	10864
Mathematics	406	1647	878	4965	1060	5450	1935	9056	868	2420
Microbiology	152	464	310	1220	771	2615	1642	3770	578	1023
Molecular Biology & Genetics	281	1090	398	1688	865	2997	2750	9499	1522	3279
Multi Disciplinary	989	5893	1082	113.397	1370	6098	3385	11331	1208	2160
Neuroscience & Behavior	161	471	225	770	594	2323	1950	7511	920	2012
Pharmacology & Toxicology	244	1028	424	2071	711	3166	3619	15778	1930	5087
Physics	1047	11161	1733	18089	3056	44191	7081	124676	3648	50082
Plant & Animal Science	810	5287	1616	13873	2430	17391	5204	29950	1024	7095
Psychiatry/Psychology	193	460	270	749	497	1218	1422	3452	579	1038
Social Sciences, General	417	1105	705	2250	998	3043	2641	6999	1310	2474
Space Science	133	622	361	2235	765	3826	1849	11315	1104	3732

Table-3.16: Proportion and Contribution of Productive Institutions Decade-wise and Discipline-wise Analysis

Subjects	Decadewise Analysis									
	1971-80		1981-90		1991-2000		2001-2010		2011-2012	
	No. of Institutions	Propostion of Research Contribution	No. of Institutions	Propostion of Research Contribution	No. of Institutions	Propostion of Research Contribution	No. of Institutions	Propostion of Research Contribution	No. of Institutions	Propostion of Research Contribution
Agricultural Sciences	39	69.747	44	70.298	51	70.137	73	70.089	73	70.43
Biology & Biochemistry	42	70.135	41	69.972	40	69.953	53	70.187	76	70.146
Chemistry	45	69.949	43	69.814	41	70.204	37	61.813	55	70.181
Clinical Medicine	49	70.178	46	69.786	64	69.999	90	69.869	107	69.848
Computer Science	4	70.135	6	70.823	1	98.627	20	69.952	36	70.138
Economics & Business	9	70.085	18	69.734	16	70.507	32	69.896	30	69.757
Engineering	25	70.16	27	70.103	35	69.719	43	70.107	53	70.01
Environment/ Ecology	25	70.233	48	69.821	68	69.968	67	69.853	75	70.071
Geosciences	31	69.716	35	69.966	31	70.078	23	69.741	25	69.608
Immunology	12	70.755	17	69.565	16	70.258	29	69.739	34	69.784
Materials Science	22	69.767	19	70.203	15	70.105	23	70.007	37	69.789
Mathematics	25	70.146	35	69.837	28	70.107	24	69.767	30	70.104
Microbiology	33	69.933	32	69.745	35	70.152	76	69	110	69.877
Molecular Biology & Genetics	36	69.979	40	70.002	32	70.079	42	70.211	57	69.926
Multi disciplinary	77	70.034	80	69.832	67	70.1	105	70.01	84	70.136

Neuroscience & Behavior	22	69.75	20	70.084	26	69.849	40	70.126	42	70.09
Pharmacology & Toxicology	27	70.455	32	69.862	44	70.314	84	69.869	95	69.927
Physics	31	70.178	29	69.921	20	70.737	19	70.609	17	71.584
Plant & Animal Science	57	69.927	63	70.017	75	70.05	86	69.994	25	68
Psychiatry/ Psychology	28	69.887	23	70.252	36	70.243	31	70.208	18	71.003
Social Sciences, general	61	69.693	74	69.902	89	69.997	116	69.872	81	69.753
Space Science	8	71.115	11	70.048	10	70.698	12	70.188	14	70.442

Table 3.17(a): Publication Output of the Most Prolific Research Institutions

Top 10 Contributing Institutions: Subjecwise and Decadewise Analysis
Decadewise Analysis

Agricultural Sciences									
1971-80		**1981-90**		**1991-2000**		**2001-2010**		**2011-2012**	
Indian Agr Res Inst	14.031	Indian Agr Res Inst	7.362	Cent Food Technol Res Inst	5.554	Cent Food Technol Res Inst	6.373	Cent Food Technol Res Inst	6.228
Punjab Agr Univ	8.450	Haryana Agr Univ	7.124	Indian Agr Res Inst	4.656	Indian Agr Res Inst	6.141	Indian Agr Res Inst	5.696
Haryana Agr Univ	5.423	Punjab Agr Univ	6.429	Punjab Agr Univ	4.403	Punjab Agr Univ	4.022	Csir	4.638
Cent Food Technol Res Inst	3.479	Cent Food Technol Res Inst	6.172	Int Crops Res Inst Semi Arid Trop	4.383	Indian Inst Technol	3.462	Punjab Agr Univ	3.772
Govind Ballabh Pant Univ Agr Technol	3.479	Int Crops Res Inst Semi Arid Trop	2.844	Haryana Agr Univ	3.514	Int Crops Res Inst Semi Arid Trop	2.452	Indian Inst Technol	3.620
Univ Agr Sci Bangalore	2.988	Natl Dairy Res Inst	2.844	Govind Ballabh Pant Univ Agr Technol	2.989	Tamil Nadu Agr Univ	2.274	Int Crops Res Inst Semi Arid Trop	3.342
Natl Dairy Res Inst	2.824	Govind Ballabh Pant Univ Agr Technol	2.578	Gujarat Agr Univ	2.793	Govind Ballabh Pant Univ Agr Technol	1.852	Natl Dairy Res Inst	2.238
Chandra Shekhar Azad Univ Agr Technol	2.237	Gujarat Agr Univ	1.967	Tamil Nadu Agr Univ	2.345	Guru Nanak Dev Univ	1.791	Tamil Nadu Agr Univ	1.912
Banaras Hindu Univ	1.988	Andhra Pradesh Agr Univ	1.794	Cent Arid Zone Res Inst	2.265	Natl Dairy Res Inst	1.697	Univ Agr Sci	1.709

1971-80		1981-90		1991-2000		2001-2010		2011-2012	
Tamil Nadu Agr Univ	1.905	Tamil Nadu Agr Univ	1.743	Rajasthan Agr Univ	2.265	Csir	1.593	Banaras Hindu Univ	1.671
% Contribution	46.804	% Contribution	39.114	% Contribution	35.167	% Contribution	31.657	% Contribution	34.826

Biology and Biochemistry

1971-80		1981-90		1991-2000		2001-2010		2011-2012	
Cent Drug Res Inst	9.541	Indian Inst Sci	7.728	Indian Inst Sci	7.893	Indian Inst Technol	7.244	Indian Inst Technol	7.140
Indian Inst Sci	6.554	Indian Inst Chem Biol	4.598	Indian Inst Technol	5.168	Indian Inst Sci	5.756	Csir	5.755
Banaras Hindu Univ	4.632	Banaras Hindu Univ	4.447	Banaras Hindu Univ	3.974	Univ Delhi	3.486	Indian Inst Sci	3.205
Bhabha Atom Res Ctr	4.328	Cent Drug Res Inst	4.325	University of Delhi	2.908	Banaras Hindu Univ	2.702	Univ Delhi	3.026
Univ Delhi	4.187	Bhabha Atom Res Ctr	3.539	Bhabha Atom Res Ctr	2.714	Csir	2.574	Banaras Hindu Univ	2.701
Univ Calcutta	2.247	Univ Calcutta	3.115	Cent Food Technol Res Inst	2.672	All India Inst Med Sci	2.433	All India Inst Med Sci	2.062
All India Inst Med Sci	2.212	Indian Inst Technol	2.871	Natl Chem Lab	2.437	Ctr Cellular Mol Biol	2.096	Aligarh Muslim Univ	1.918
Sri Venkateswara Univ	1.979	Univ Delhi	2.766	Jawaharlal Nehru Univ	2.364	Jawaharlal Nehru Univ	1.980	Jawaharlal Nehru Univ	1.666
Aligarh Muslim Univ	1.922	Natl Chem Lab	2.210	Indian Inst Chem Biol	2.356	Bhabha Atom Res Ctr	1.878	Bhabha Atom Res Ctr	1.641
Indian Institute of Technology	1.836	Bose Inst	2.183	All India Inst Med Sci	2.188	Indian Institute of Chemical Biology	1.796	Panjab Univ	1.489
% Contribution	37.602	% Contribution	35.599	% Contribution	34.674	% Contribution	31.945	% Contribution	30.603

Chemistry

1971-80		1981-90		1991-2000		2001-2010		2011-2012	
Indian Inst Technol	10.527	Indian Inst Technol	11.351	Indian Inst Technol	11.333	Indian Inst Technol	12.430	Indian Inst Technol	12.252
Univ Delhi	4.767	Indian Inst Sci	4.916	Indian Inst Sci	5.252	Indian Inst Chem Technol	4.546	Indian Inst Chem Technol	3.983
Indian Inst Sci	3.624	Natl Chem Lab	3.856	Natl Chem Lab	4.649	Indian Inst Sci	4.297	Bhabha Atom Res Ctr	3.632
Bhabha Atom Res Ctr	2.980	Bhabha Atom Res Ctr	3.778	Bhabha Atom Res Ctr	4.270	Bhabha Atom Res Ctr	3.641	Indian Inst Sci	3.468
Natl Chem Lab	2.884	Univ Delhi	3.419	Indian Inst Chem Technol	4.069	Natl Chem Lab	3.281	National Chemical Laboratory	2.790
Univ Rajasthan	2.743	Banaras Hindu Univ	2.618	Indian Assoc Cultivat Sci	3.201	Indian Assoc Cultivat Sci	2.390	Indian Assoc Cultivat Sci	2.394
Banaras Hindu Univ	2.545	Indian Assoc Cultivat Sci	2.452	Banaras Hindu Univ	2.034	Banaras Hindu Univ	0.967	Banaras Hindu Univ	2.222
Panjab Univ	2.426	Univ Rajasthan	2.326	Univ Hyderabad	1.869	Univ Hyderabad	0.888	Mangalore Univ	1.904
Univ Allahabad	2.393	Osmania Univ	2.039	Jadavpur Univ	1.862	Jadavpur Univ	0.885	Univ Delhi	1.828
Aligarh Muslim Univ	2.179	Cent Drug Res Inst	1.995	Univ Delhi	1.859	Univ Delhi	0.884	Jadavpur Univ	1.753
% Contribution	37.068		38.750		40.398		34.209		36.226

Clinical Medicine

1971-80		1981-90		1991-2000		2001-2010		2011-2012	
All India Inst Med Sci	10.309	All India Inst Med Sci	13.114	All India Inst Med Sci	13.256	All India Inst Med Sci	12.241	All India Inst Med Sci	8.223
Postgrad Inst Med Educ Res	8.718	Postgrad Inst Med Educ Res	10.100	Postgrad Inst Med Educ Res	7.797	Postgrad Inst Med Educ Res	6.473	Postgrad Inst Med Educ Res	6.101

1971-80		1981-90		1991-2000		2001-2010		2011-2012	
Banaras Hindu Univ	4.956	Banaras Hindu Univ	3.788	Christian Med Coll Hosp	4.530	Christian Med Coll Hosp	3.699	Council of Scientific Industrial Research Csir India	4.371
Indian Council Med Res	3.752	Christian Med Coll Hosp	3.495	Sanjay Gandhi Postgrad Inst Med Sci	3.572	Sanjay Gandhi Postgrad Inst Med Sci	3.172	Christian Med Coll Hosp	2.703
Christian Med Coll Hosp	3.581	Indian Council Med Res	3.222	Tata Mem Hosp	3.184	Tata Mem Hosp	2.742	Tata Mem Hosp	2.318
Bhabha Atom Res Ctr	3.363	Bhabha Atom Res Ctr	2.630	Banaras Hindu Univ	2.756	Indian Council Med Res	2.041	Banaras Hindu Univ	1.825
Maulana Azad Med Coll	2.243	Maulana Azad Med Coll	2.505	Indian Council Med Res	2.738	Banaras Hindu Univ	1.747	Sanjay Gandhi Postgrad Inst Med Sci	1.728
Cent Drug Res Inst	1.971	Tata Mem Ctr	2.212	Bhabha Atomic Research Center	1.493	Maulana Azad Med Coll	1.667	Manipal Univ	1.574
University of Delhi	1.709	Cent Drug Res Inst	2.133	Gb Pant Hosp	1.367	Kasturba Med Coll Hosp	1.543	Univ Delhi	1.474
Goa Med Coll	1.139	Gb Pant Hosp	1.715	Kasturba Med Coll Hosp	1.367	Univ Delhi	1.336	Maulana Azad Med Coll	1.360
% Contribution	41.741		44.914		42.06		36.661		31.677

Computer Science

1971-80		1981-90		1991-2000		2001-2010		2011-2012	
Indian Inst Technol	42.987	Indian Inst Technol	40.160	Indian Inst Technol	34.147	Indian Inst Technol	32.722	Indian Inst Technol	24.028
Indian Inst Sci	12.081	Indian Inst Sci	12.471	Indian Inst Sci	9.018	Indian Inst Sci	9.700	Indian Inst Sci	6.537
Univ Delhi	6.040	Indian Stat Inst	5.939	Indian Stat Inst	6.680	Indian Stat Inst	6.297	Indian Stat Inst	4.770

1971-80		1981-90		1991-2000		2001-2010		2011-2012	
Univ of Roorkee	5.829	Vikram Sarabhai Space Ctr	5.721	Inst Math Sci	6.407	Council of Scientific Industrial Research Csir India	4.800	Jadavpur Univ	4.364
Vikram Sarabhai Space Ctr	4.698	Tata Inst Fundamental Res	5.492	Tata Inst Fundamental Res	6.407	Jadavpur Univ	2.673	Natl Inst Technol	4.038
Jadavpur Univ	3.020	Natl Aeronaut Lab	4.983	Jadavpur Univ	5.378	Anna Univ	2.475	Council of Scientific Industrial Research Csir India	3.793
Tata Inst Fundamental Res	2.349	Univ Roorkee	4.233	Univ Roorkee	3.890	Natl Inst Technol	1.900	Anna Univ	2.032
Univ Calcutta	2.013	Madras Christian Coll	2.403	Struct Engn Res Ctr	2.517	Inst Math Sci	1.657	Vit Univ	1.502
Natl Aeronaut Lab	1.342	Banaras Hindu Univ	1.945	Reg Engn Coll	2.288	Tata Inst Fundamental Res	1.370	Inst Math Sci	1.413
Indian Stat Inst	1.007	Reg Engn Coll	1.831	University of Delhi	2.261	Int Inst Informat Technol	1.149	Bengal Engn Sci Univ	1.325
% Contribution	81.366		85.178		78.993		64.743		53.802

Economics and Business

1971-80		1981-90		1991-2000		2001-2010		2011-2012	
Indian Inst Management	23.504	Indian Stat Inst	14.035	Indian Stat Inst	19.023	Indian Stat Inst	12.057	Indian Stat Inst	8.951
Indian Statistical Inst	14.530	Indian Inst Management	9.211	Indian Inst Management	8.768	Indian Inst Management	8.668	Indian Inst Management	8.642

1971-80		1981-90		1991-2000		2001-2010		2011-2012	
Univ Delhi	10.684	Inst Social Econ Change	8.991	Indira Gandhi Inst Dev Res	7.155	Indian Inst Technol	8.126	Indian Inst Technol	5.556
Shri Ram Ctr Ind Relations Human Resources	6.235	Indian Inst Technol	6.500	Univ Delhi	5.934	Univ Delhi	5.595	Indian Sch Business	4.630
Jawaharlal Nehru Univ	5.983	Delhi Sch Econ	6.140	Delhi Sch Econ	5.236	Jawaharlal Nehru Univ	4.255	Univ Delhi	4.321
Adm Staff Coll India	5.635	Univ Delhi	6.140	Indian Inst Technol	4.538	Indira Gandhi Inst Dev Res	2.600	Ctr Studies Social Sci	3.704
Delhi Sch Econ	5.128	Adm Staff Coll India	3.125	Jadavpur Univ	4.147	Indian Sch Business	2.304	Indira Gandhi Inst Dev Res	2.778
Indian Inst Technol	2.998	Univ Calcutta	2.250	Jawaharlal Nehru Univ	3.490	Ctr Studies Social Sci	2.285	Jawaharlal Nehru Univ	2.778
Indian Inst Sci	2.137	Aligarh Muslim Univ	2.193	Adm Staff Coll India	2.443	Reserve Bank India	1.891	Jadavpur Univ	2.160
Birla Inst Sci Res	1.709	Jawaharlal Nehru Univ	2.193	Indian Inst Sci	2.269	Jadavpur Univ	1.812	Univ Calcutta	2.160
% Contribution	78.543		60.778		63.003		49.593		45.680

Engineering

1971-80		1981-90		1991-2000		2001-2010		2011-2012	
Indian Inst Technol	29.820	Indian Inst Technol	30.021	Indian Inst Technol	28.372	Indian Inst Technol	28.941	Indian Inst Technol	25.753
Indian Inst Sci	10.198	Indian Inst Sci	8.114	Indian Inst Sci	7.818	Indian Inst Sci	6.229	Council of Scientific Industrial Research Csir India	8.756
Univ Roorkee	3.465	Bhabha Atom Res Ctr	4.221	Bhabha Atom Res Ctr	3.503	Bhabha Atom Res Ctr	3.263	Bhabha Atom Res Ctr	4.730
Univ Delhi	2.898	Univ Roorkee	3.689	Univ Roorkee	2.524	Natl Inst Technol	3.154	Natl Inst Technol	4.468

1971-80		1981-90		1991-2000		2001-2010		2011-2012	
Banaras Hindu Univ	2.705	Banaras Hindu Univ	2.252	Banaras Hindu Univ	2.393	Jadavpur Univ	2.431	Indian Inst Sci	4.455
Bhabha Atom Res Ctr	2.447	Univ Delhi	1.992	Indian Stat Inst	2.103	Anna Univ	2.230	Jadavpur Univ	2.927
Jadavpur Univ	2.426	Reg Engn Coll	1.777	Indira Gandhi Ctr Atom Res	1.973	Banaras Hindu Univ	1.715	Anna Univ	2.051
Reg Engn Coll	2.340	Jadavpur Univ	1.663	Reg Engn Coll	1.617	Univ Delhi	1.568	Banaras Hindu Univ	1.986
Natl Aeronaut Lab	1.868	Vikram Sarabhai Space Ctr	1.200	Jadavpur Univ	1.603	Indira Gandhi Ctr Atom Res	1.469	Aligarh Muslim Univ	1.898
Vikram Sarabhai Space Ctr	1.417	Aligarh Muslim Univ	1.166	Univ Delhi	1.588	Indian Stat Inst	1.377	Indira Gandhi Ctr Atom Res	1.829
% Contribution	59.584		56.095		53.494		52.377		58.853

Environemnt/Ecology

1971-80		1981-90		1991-2000		2001-2010		2011-2012	
Punjab Agr Univ	10.585	Ind Toxicol Res Ctr	6.101	Indian Inst Technol	8.603	Indian Inst Technol	9.897	Indian Inst Technol	13.842
Indian Agr Res Inst	7.721	Banaras Hindu Univ	5.379	Bhabha Atom Res Ctr	4.081	Natl Environm Engn Res Inst	3.424	Csir	2.868
Bhabha Atom Res Ctr	7.549	Indian Inst Technol	3.922	Univ Roorkee	3.786	Indian Agr Res Inst	2.726	Indian Agr Res Inst	2.515
Ind Toxicol Res Ctr	7.347	Punjab Agr Univ	3.610	Banaras Hindu Univ	3.772	Banaras Hindu Univ	2.551	Anna Univ	2.471
Haryana Agr Univ	6.600	Haryana Agr Univ	3.105	Ind Toxicol Res Ctr	3.743	Anna Univ	2.412	Banaras Hindu Univ	2.427
Dav Coll	5.978	Indian Agr Res Inst	2.780	Natl Bot Res Inst	2.303	Indian Inst Sci	2.404	Natl Inst Technol	2.336
Indian Inst Technol	4.709	Bhabha Atom Res Ctr	2.635	Punjab Agr Univ	2.102	Ind Toxicol Res Ctr	2.296	Natl Environm Engn Res Inst	2.030

1971-80		1981-90		1991-2000		2001-2010		2011-2012	
University of Delhi	3.139	Sri Venkateswara Univ	2.419	Indian Inst Sci	2.061	Natl Inst Oceanog	2.202	Aligarh Muslim Univ	1.898
Sri Venkateswara University	3.064	Univ Roorkee	2.093	Indian Agr Res Inst	1.929	Natl Bot Res Inst	1.920	Indian Inst Sci	1.898
Banaras Hindu Univ	2.989	Univ Delhi	2.022	Univ Madras	1.699	Univ Calcutta	1.867	Univ Calcutta	1.898
% Contribution	59.681		34.066		34.079		31.699		34.183

Geosciences

1971-80		1981-90		1991-2000		2001-2010		2011-2012	
Natl Geophys Res Inst	9.464	Natl Geophys Res Inst	13.342	Natl Geophys Res Inst	17.536	Indian Inst Technol	20.815	Indian Inst Technol	20.100
Geol Survey India	7.730	Geol Survey India	6.177	Indian Inst Technol	7.364	Natl Geophys Res Inst	7.449	Csir	15.434
Indian Inst Technol	5.626	Indian Inst Technol	5.299	Geol Survey India	6.512	Geol Survey India	5.022	Indian Inst Trop Meteorol	6.572
Andhra Univ	3.914	Banaras Hindu Univ	3.799	Natl Inst Oceanog	4.873	Phys Res Lab	4.567	Natl Geophys Res Inst	4.382
Univ Allahabad	3.914	Phys Res Lab	3.205	Phys Res Lab	4.477	Indian Inst Trop Meteorol	4.111	Indian Inst Sci	4.004
Indian Inst Trop Meteorol	3.425	Ctr Earth Sci Studies	3.153	Indian Inst Trop Meteorol	3.682	Natl Inst Oceanog	3.740	Phys Res Lab	3.606
Phys Res Lab	3.033	Natl Inst Oceanog	3.127	Banaras Hindu Univ	3.043	Indian Inst Sci	3.581	Indian Meteorol Dept	3.149
Osmania Univ	2.642	Univ Allahabad	2.895	Jadavpur Univ	2.868	Wadia Inst Himalayan Geol	3.242	Natl Inst Oceanog	2.784
Jadavpur Univ	2.202	Indian Inst Trop Meteorol	2.585	Wadia Inst Himalayan Geol	2.733	Banaras Hindu Univ	2.490	Geol Survey India	2.602

Institution	Value	Institution	Value	Institution	Value	Institution	Value	Institution	Value
Banaras Hindu Univ	2.055	Jadavpur Univ	2.326	Univ Roorkee	2.384	Vikram Sarabhai Space Ctr	2.437	Vikram Sarabhai Space Ctr	2.602
% Contribution	44.005		45.908		55.472		57.454		65.235

Immunology

1971-80		1981-90		1991-2000		2001-2010		2011-2012	
All India Inst Med Sci	25.472	All India Inst Med Sci	13.817	Natl Inst Immunol	11.099	All India Inst Med Sci	8.546	All India Inst Med Sci	6.628
PGIMER Chandigarh	9.686	Postgrad Inst Med Educ Res	9.900	All India Inst Med Sci	9.989	Indian Council Med Res	6.268	Banaras Hindu Univ	5.653
Madurai Univ	6.604	Indian Inst Sci	6.087	Postgrad Inst Med Educ Res	7.389	Postgrad Inst Med Educ Res	5.949	Council of Scientific Industrial Research Csir India	5.425
Jawaharlal Inst Postgrad Med Educ Res	5.660	Madurai Kamaraj Univ	5.797	Indian Council Med Res	6.664	Indian Inst Sci	4.398	Pgimer Chandigarh	4.939
Govind Ballabh Pant Hosp	4.717	Natl Inst Immunol	4.717	Int Ctr Genet Engn Biotechnol	6.548	Natl Inst Immunol	4.111	Indian Council Med Res	3.725
Banaras Hindu University	4.276	Indian Council Med Res	3.768	Indian Inst Chem Biol	5.438	Univ Delhi	3.776	Univ Delhi	2.955
King Georges Med Coll	3.774	Cent Drug Res Inst	3.188	Indian Inst Sci	4.218	Banaras Hindu Univ	3.537	Natl Inst Immunol	2.729
Maulana Azad Med Coll	3.774	Indian Inst Chem Biol	3.188	Natl Inst Cholera Enter Dis	4.218	Int Ctr Genet Engn Biotechnol	3.250	Christian Med Coll Hosp	2.534
Indian Council Med Res	2.830	King Georges Med Coll	3.188	Banaras Hindu Univ	3.663	Indian Inst Chem Biol	2.629	Ctr Dis Control Prevent	2.534

1971-80		1981-90		1991-2000		2001-2010		2011-2012	
Indian Inst of Sci	2.830	Banaras Hindu University	2.763	Cent Drug Res Inst	3.552	Cent Drug Res Inst	2.533	Int Ctr Genet Engn Biotechnol	2.144
% Contribution	69.623		57.783		62.778		44.997		39.266

Materials Science

1971-80		1981-90		1991-2000		2001-2010		2011-2012	
Indian Inst Technol	26.757	Indian Inst Technol	27.737	Indian Inst Technol	24.806	Indian Inst Technol	22.575	Indian Inst Technol	19.998
Council of Scientific Industrial Research Csir India	9.549	Indian Inst Sci	6.831	Indian Inst Sci	9.636	Csir	17.024	Csir	15.810
Bhabha Atom Res Ctr	7.030	Banaras Hindu Univ	4.621	Bhabha Atom Res Ctr	5.679	Indian Inst Sci	6.956	Indian Inst Sci	5.929
Banaras Hindu Univ	6.917	Bhabha Atom Res Ctr	4.858	Indira Gandhi Ctr Atom Res	4.473	Bhabha Atom Res Ctr	4.454	Bhabha Atom Res Ctr	4.401
Indian Inst Sci	4.116	Def Met Res Lab	3.818	Banaras Hindu Univ	4.050	Indira Gandhi Ctr Atom Res	3.320	Natl Inst Technol	4.156
Natl Phys Lab	3.308	Univ Roorkee	2.819	Def Met Res Lab	3.871	Natl Inst Technol	2.702	Indira Gandhi Ctr Atom Res	2.810
Univ Roorkee	2.484	Reg Res Lab	2.325	Csir	3.276	Def Met Res Lab	2.472	Anna Univ	2.587
Ahmedabad Text Ind Res Assoc	2.129	Natl Phys Lab	2.160	Cent Glass Ceram Res Inst	2.399	Banaras Hindu Univ	2.384	Jadavpur Univ	2.190
Sardar Patel Univ	1.987	Osmania Univ	2.160	Indian Assoc Cultivat Sci	2.258	Natl Chem Lab	2.269	Banaras Hindu Univ	2.104
Reg Engn Coll	1.916	Csir	2.078	Cent Electrochem Res Inst	2.157	Anna Univ	2.251	Shivaji Univ	2.018
% Contribution	66.193		59.407		62.605		66.407		62.003

Mathematics									
1971-80		**1981-90**		**1991-2000**		**2001-2010**		**2011-2012**	
Indian Inst Technol	14.263	Indian Inst Technol	13.066	Indian Stat Inst	15.567	Indian Inst Technol	17.771	Indian Inst Technol	17.393
Tata Inst Fundamental Res	10.678	Tata Inst Fundamental Res	9.992	Indian Inst Technol	13.617	Indian Stat Inst	12.263	Indian Stat Inst	8.012
Indian Statistical Inst	6.937	Indian Stat Inst	9.268	Tata Inst Fundamental Res	9.763	Tata Inst Fundamental Res	10.386	Tata Inst Fundamental Res	7.641
Univ Delhi	5.222	Indian Inst Sci	4.107	Indian Inst Sci	5.129	Indian Inst Sci	5.813	Aligarh Muslim Univ	5.415
Aligarh Muslim Univ	2.884	Banaras Hindu Univ	4.023	Univ Delhi	3.527	Univ Delhi	3.400	Indian Inst Sci	4.355
Banaras Hindu Univ	2.884	Univ Delhi	2.526	Inst Math Sci	2.881	Inst Math Sci	3.122	Inst Math Sci	3.116
Panjab Univ	2.728	Univ Madras	2.185	Univ Poona	2.005	Aligarh Muslim Univ	3.084	Banaras Hindu Univ	2.782
Indian Institute of Science Iisc Banglore	2.698	Univ Roorkee	2.185	Panjab Univ	1.953	Harish Chandra Res Inst	2.025	Univ Delhi	2.300
Andhra Univ	2.416	Panjab Univ	2.025	Banaras Hindu Univ	1.926	Univ Calcutta	2.025	Natl Inst Technol	1.855
Kurukshetra Univ	2.104	Univ Bombay	1.918	Univ Madras	1.900	Panjab Univ	1.375	Univ Calcutta	1.780
% Contribution	52.814		51.295		58.268		61.264		54.649

Microbiology

1971-80		1981-90		1991-2000		2001-2010		2011-2012	
Indian Inst Sci	6.760	Banaras Hindu Univ	6.294	Indian Inst Chem Biol	4.588	Council of Scientific Industrial Research Csir India	18.996	Csir	14.470
Indian Inst Exptl Med	5.361	Indian Inst Sci	5.381	Univ Delhi	4.297	Indian Institute of Technology Iit	8.347	Indian Inst Technol	7.137
Univ Delhi	4.196	Postgrad Inst Med Educ Res	4.569	Bose Inst	4.007	University of Delhi	4.230	Univ Delhi	2.933
Cent Drug Res Inst	3.730	Bose Inst	4.162	Banaras Hindu Univ	3.891	Banaras Hindu University	2.667	Banaras Hindu Univ	2.780
Banaras Hindu Univ	3.497	Indian Inst Chem Biol	3.959	Postgrad Inst Med Educ Res	3.717	Indian Institute of Technology Iit Delhi	2.475	All India Inst Med Sci	2.095
Jadavpur Univ	3.263	Cent Drug Res Inst	2.944	Jawaharlal Nehru Univ	3.484	Central Food Technological Research Institute India	2.328	Postgrad Inst Med Educ Res	2.048
Bhabha Atom Res Ctr	3.030	Maharaja Sayajirao Univ Baroda	2.843	Natl Inst Cholera Enter Dis	3.020	Indian Institute of Science Iisc Banglore	2.293	Indian Agr Res Inst	1.664
Univ Allahabad	3.030	Bhabha Atom Res Ctr	2.538	Indian Inst Sci	2.846	Institute of Microbial Technology India	2.154	Indian Inst Sci	1.357
Bose Inst	2.797	Punjab Univ	2.234	Indian Vet Res Inst	2.846	All India Institute of Medical Sciences	1.920	Panjab Univ	1.350

Cent Rice Res Inst	Univ Madras	All India Inst Med Sci	Bhabha Atomic Research Center	Anna University
2.797	2.234	2.671	1.763	1.292
% Contribution 38.461	37.158	35.367	47.173	37.126

Molecular Biology & Genetics

1971-80		1981-90		1991-2000		2001-2010		2011-2012	
Banaras Hindu Univ	7.372	Banaras Hindu Univ	9.000	Indian Inst Sci	9.899	Ctr Cellular Mol Biol	15.059	Csir	16.436
Panjab Univ	5.662	Indian Inst Sci	8.427	Banaras Hindu Univ	5.201	Indian Inst Sci	6.251	Indian Inst Technol	6.050
Univ Calcutta	4.594	Univ Calcutta	5.643	Indian Inst Chem Biol	3.915	Indian Inst Technol	4.914	Banaras Hindu Univ	3.696
Univ Delhi	4.129	Andhra Univ	3.786	All India Inst Med Sci	3.859	All India Inst Med Sci	3.839	Indian Inst Sci	3.462
Univ Madras	3.526	Univ Delhi	3.786	Univ Delhi	3.312	Univ Delhi	3.644	Univ Delhi	3.132
Andhra Univ	3.312	All India Inst Med Sci	2.714	Jawaharlal Nehru Univ	3.166	Univ Madras	3.388	All India Inst Med Sci	2.754
Punjab Agr Univ	2.991	Panjab Univ	2.500	Tata Inst Fundamental Res	3.076	Jawaharlal Nehru Univ	2.823	Panjab Univ	2.391
Karnatak Univ	2.457	Osmania Univ	1.929	Natl Inst Immunol	3.020	Banaras Hindu Univ	2.597	Ctr Cellular Mol Biol	2.028
Osmania Univ	2.350	Univ Jammu	1.929	Ctr Cellular Mol Biol	2.740	Indian Inst Chem Biol	2.575	Jawaharlal Nehru University	1.962
Univ Kerala	2.244	Univ Kerala	1.857	Univ Calcutta	2.573	Postgrad Inst Med Educ Res	1.987	Osmania Univ	1.812
% Contribution	38.637		41.571		40.761		47.077		43.723

Multi-Disciplinary

1971-80		1981-90		1991-2000		2001-2010		2011-2012	
Indian Inst Sci	5.189	Indian Inst Sci	3.894	Indian Inst Sci	9.187	Indian Inst Technol	6.961	Indian Inst Sci	8.895
Andhra Univ	4.349	Indian Agr Res Inst	3.057	Indian Inst Technol	3.360	Indian Inst Sci	5.875	Indian Inst Technol	7.895
Indian Agr Res Inst	3.083	Banaras Hindu Univ	2.882	Banaras Hindu Univ	3.007	Natl Geophys Res Inst	2.763	Csir	2.903
Univ Delhi	2.826	Univ Madras	2.566	Tata Inst Fundamental Res	2.737	Natl Inst Oceanog	2.470	Banaras Hindu Univ	2.307
Banaras Hindu Univ	2.257	Osmania Univ	2.530	Raman Res Inst	2.406	Banaras Hindu Univ	2.134	Univ Delhi	2.141
Osmania Univ	2.220	Univ Delhi	1.838	Bhabha Atom Res Ctr	2.281	Univ Delhi	1.954	Tata Inst Fundamental Res	2.060
Sri Venkateswara Univ	1.908	Sri Venkateswara Univ	1.752	Natl Geophys Res Inst	2.095	Indian Agr Res Inst	1.619	Indian Agr Res Inst	1.873
Univ Madras	1.780	Bhabha Atom Res Ctr	1.747	Univ Delhi	2.012	Tata Inst Fundamental Res	1.528	Jawaharlal Nehru Ctr Adv Sci Res	1.685
Univ Mysore	1.688	Tata Inst Fundamental Res	1.729	Natl Inst Oceanog	1.908	Jawaharlal Nehru Ctr Adv Sci Res	1.409	Bhabha Atom Res Ctr	1.592
Karnatak Univ	1.541	Univ Mysore	1.711	Indian Agr Res Inst	1.555	Birbal Sahni Inst Paleobot	1.382	Univ Pune	1.311
% Contribution	26.841		23.706		30.548		28.095		32.662

Neuroscience & Behavior

1971-80		1981-90		1991-2000		2001-2010		2011-2012	
15.250	All India Inst Med Sci	16.410	All India Inst Med Sci	16.089	Natl Inst Mental Hlth Neurosci	12.576	All India Inst Med Sci	10.547	All India Inst Med Sci
9.917	Postgrad Inst Med Educ Res	10.483	Natl Inst Mental Hlth Neuro Sci	8.492	All India Inst Med Sci	9.567	Natl Inst Mental Hlth Neurosci	8.203	Natl Inst Mental Hlth Neurosci
8.250	Banaras Hindu Univ	7.420	Postgrad Inst Med Educ Res	7.139	Postgrad Inst Med Educ Res	5.350	Sanjay Gandhi Postgrad Inst Med Sci	6.038	Sree Chitra Tirunal Inst Med Sci Technol
5.000	Christian Med Coll Hosp	6.496	Banaras Hindu Univ	5.921	Sanjay Gandhi Postgrad Inst Med Sci	5.045	Sree Chitra Tirunal Inst Med Sci Technol	5.252	Postgrad Inst Med Educ Res
3.500	Jj Grp Hosp	3.590	Indian Inst Chem Biol	5.337	Christian Med Coll Hosp	4.434	Postgrad Inst Med Educ Res	3.774	Sanjay Gandhi Postgrad Inst Med Sci
3.000	Grant Med Coll	3.590	King Georges Med Coll	3.467	Sree Chitra Tirunal Inst Med Sci Technol	3.853	Christian Med Coll Hosp	3.212	Christian Med Coll Hosp
2.893	National Institute of Mental Health Neurosciences India	3.077	Univ Hyderabad	3.427	Nizams Inst Med Sci	2.647	Panjab Univ	2.830	Panjab Univ
2.686	Madras Med Coll	2.473	Christian Med Coll Hosp	3.084	Banaras Hindu Univ	2.131	Natl Brain Res Ctr	2.264	Banaras Hindu Univ

Pharmacology & Toxicology

1971-80		1981-90		1991-2000		2001-2010		2011-2012	
Banaras Hindu Univ	8.046	Ind Toxicol Res Ctr	9.361	Ind Toxicol Res Ctr	6.529	Council of Scientific Industrial Research Csir India	10.094	Csir	10.724
Ind Toxicol Res Ctr	7.586	Banaras Hindu Univ	6.889	Cent Drug Res Inst	5.698	Panjab Univ	3.796	Jamia Hamdard	3.410
Univ Rajasthan	4.828	Postgrad Inst Med Educ Res	5.822	Banaras Hindu Univ	5.202	Jamia Hamdard	3.281	Panjab Univ	2.741
Lucknow Univ	4.372	Cent Drug Res Inst	5.137	Jadavpur Univ	4.253	Univ Madras	3.193	Jadavpur Univ	2.249
King Georges Med Coll	4.368	King Georges Med Coll	4.395	Postgrad Inst Med Educ Res	3.551	Annamalai Univ	3.173	Banaras Hindu Univ	2.214
Cent Drug Res Inst	4.023	Univ Rajasthan	3.196	Univ Madras	3.386	Jadavpur Univ	2.768	Univ Delhi	2.178
Univ Allahabad	3.678	Panjab Univ	3.139	All India Inst Med Sci	2.849	Indian Toxicology Research Institute	2.236	Indian Institute of Chemical Technology	2.003
Univ Delhi	3.563	Univ Calcutta	3.025	Def Res Dev Estab	2.436	All India Inst Med Sci	2.165	Indian Institute of Technology Iit	2.003
Kem Hosp	2.500	Ind Toxicol Res Ctr	2.393	Univ Hyderabad	3.027	Tata Inst Fundamental Res	1.916	Gb Pant Hosp	2.075
Indian Inst Sci	2.250	Maharaja Sayajirao Univ Baroda	2.393	Indian Inst Chem Biol	2.570	Indian Inst Chem Biol	1.722	Nizams Inst Med Sci	1.887
% Contribution	55.246	% Contribution	58.325	% Contribution	58.553	% Contribution	49.241	% Contribution	46.082

Panjab Univ	2.644	Univ Delhi	1.998	Panjab Univ	2.271	Cent Drug Res Inst	1.977	Annamalai Univ	1.968
Bhabha Atom Res Ctr*	2.184	Reg Res Lab	1.884	Univ Calcutta	2.225	Niper	1.690	All India Inst Med Sci	1.722
% Contribution	45.292		44.846		38.400		34.373		31.212

Physics

1971-80		1981-90		1991-2000		2001-2010		2011-2012	
Indian Inst Technol	16.417	Indian Inst Technol	14.630	Indian Inst Technol	13.079	Indian Inst Technol	16.300	Indian Inst Technol	16.331
Univ of Delhi	4.928	Bhabha Atom Res Ctr	7.542	Tata Inst Fundamental Res	8.471	Council of Scientific Industrial Research Csir India	8.557	Council of Scientific Industrial Research Csir India	7.905
Bhabha Atom Res Ctr	4.446	Tata Inst Fundamental Res	6.291	Council of Scientific Industrial Research Csir India	7.472	Tata Inst Fundamental Res	7.059	Bhabha Atom Res Ctr	7.242
Banaras Hindu Univ	4.426	Indian Inst Sci	4.908	Indian Inst Sci	6.940	Bhabha Atom Res Ctr	6.568	Tata Inst Fundamental Res	5.498
Tata Inst Fundamental Res	4.227	Council of Scientific Industrial Research Csir India	4.454	Bhabha Atom Res Ctr	6.929	Indian Inst Sci	6.007	Indian Inst Sci	5.160
Indian Inst Sci	3.369	Jadavpur Univ	3.895	Saha Inst Nucl Phys	4.038	Saha Inst Nucl Phys	4.030	Univ Delhi	4.841
Univ Allahabad	3.236	Indian Assoc Cultivat Sci	3.272	Indian Assoc Cultivat Sci	3.777	Univ Delhi	3.812	Panjab Univ	4.546

1971-80		1981-90		1991-2000		2001-2010		2011-2012	
Jadavpur Univ	2.271	Banaras Hindu Univ	3.141	Univ Delhi	3.281	Indian Assoc Cultivat Sci	3.493	Saha Inst Nucl Phys	4.197
Indian Assoc Cultivat Sci	2.093	Univ Delhi	2.831	Jadavpur Univ	2.933	Panjab Univ	3.315	Indian Association for the Cultivation of Science	2.914
Agra Coll	1.852	Saha Inst Nucl Phys	2.609	Banaras Hindu Univ	2.595	Jadavpur Univ	2.381	Banaras Hindu Univ	2.226
% Contribution	47.265	% Contribution	53.573					% Contribution	60.860

Plant & Animal Science

1971-80		1981-90		1991-2000		2001-2010		2011-2012	
Indian Vet Res Inst	8.101	Indian Vet Res Inst	7.216	Indian Vet Res Inst	7.503	Council of Scientific Industrial Research Csir India	7.570	Council of Scientific Industrial Research Csir India	8.089
Indian Agr Res Inst	6.355	Punjab Agr Univ	6.118	Council of Scientific Industrial Research Csir India	6.750	Indian Vet Res Inst	6.459	Indian Veterinary Research Institute	5.113
Punjab Agr Univ	6.333	Haryana Agr Univ	6.023	Punjab Agr Univ	5.007	Indian Agr Res Inst	3.059	Indian Agricultural Research Institute	4.407
Haryana Agr Univ	4.828	Council of Scientific Industrial Research Csir India	4.205	Haryana Agr Univ	3.950	Punjab Agr Univ	2.843	Natl Dairy Res Inst	2.975

1971-80		1981-90		1991-2000		2001-2010		2011-2012	
Univ Delhi	3.662	Coll Vet Sci Anim Husb	3.842	Natl Dairy Res Inst	2.142	Natl Dairy Res Inst	2.298	Int Crops Res Inst Semi Arid Trop	2.194
Banaras Hindu Univ	2.803	Natl Dairy Res Inst	2.506	Indian Agr Res Inst	1.825	Madras Vet Coll	2.031	Aligarh Muslim University	2.008
Council of Scientific Industrial Research Csir India	2.456	Govind Ballabh Pant Univ Agr Technol	2.204	Coll Vet Sci Anim Husb	1.787	Univ Delhi	2.026	University of Delhi	2.008
Coll Vet Sci Anim Husb	2.169	Banaras Hindu Univ	2.124	Banaras Hindu University	1.747	Govind Ballabh Pant Univ Agr Technol	1.733	National Institute of Oceanography India	1.897
Natl Dairy Res Inst	1.678	Univ Delhi	2.092	Govind Ballabh Pant Univ Agr Technol	1.704	Banaras Hindu Univ	1.589	Guru Angad Dev Vet Anim Sci Univ	1.711
Govind Ballabh Pant Univ Agr Technol	1.486	Indian Agr Res Inst	1.940	Univ Delhi	1.676	Coll Vet Sci Anim Husb	1.421	Banaras Hindu University	1.618
% Contribution	39.871		38.27		34.091		31.029		32.020

Psychiatry/Psychology

1971-80		1981-90		1991-2000		2001-2010		2011-2012	
Indian Inst Technol	9.375	Natl Inst Mental Hlth Neuro Sci	13.052	Natl Inst Mental Hlth Neurosci	10.878	Natl Inst Mental Hlth Neurosci	15.829	Natl Inst Mental Hlth Neurosci	13.755
Univ Bombay	5.114	Univ Allahabad	8.829	Univ Delhi	5.898	Indian Inst Technol	5.777	Postgrad Inst Med Educ Res	7.002
Banaras Hindu Univ	4.830	Banaras Hindu Univ	6.526	Banaras Hindu Univ	5.505	Postgrad Inst Med Educ Res	4.910	Indian Inst Technol	6.697

Rank	1971-80		1981-90		1991-2000		2001-2010		2011-2012	
1	All India Inst Med Sci	3.125	All India Inst Med Sci	4.607	Postgrad Inst Med Educ Res	4.841	All India Inst Med Sci	4.448	All India Inst Med Sci	5.576
2	Natl Inst Mental Hlth Neuro Sci *	3.125	Indian Inst Technol	4.415	Univ Allahabad	4.194	Christian Med Coll Hosp	4.438	Univ Delhi	4.110
3	Pgimer Chandigarh	3.052	Univ Bombay	4.031	Indian Inst Technol	3.810	Cent Inst Psychiat	4.304	Christian Med Coll Hosp	3.346
4	Guru Nanak Univ	2.841	Indian Inst Management	3.647	All India Inst Med Sci	3.175	Univ Delhi	3.430	Kasturba Med Coll Hosp	2.602
5	Panjab Univ	2.841	Postgrad Inst Med Educ Res	3.647	Univ Bombay	3.145	Univ Allahabad	2.892	Voluntary Hlth Serv	2.602
6	Univ Allahabad	2.841	King Georges Med Coll	2.687	Utkal Univ	2.359	Panjab Univ	1.816	University of Allahabad	2.131
7	King Georges Med Coll	2.557	Univ Delhi	2.495	Cent Inst Psychiat	1.835	Banaras Hindu Univ	1.681	University of Calcutta	2.131
8	Patna Univ	2.557		2.557						
% Contribution		42.258		53.936		45.64		49.525		49.952

Social Sciences, General

Rank	1971-80		1981-90		1991-2000		2001-2010		2011-2012	
1	Univ Delhi	6.456	Jawaharlal Nehru Univ	8.359	Univ Delhi	8.363	Univ Delhi	5.730	Univ Delhi	5.468
2	Jawaharlal Nehru Univ	5.460	Univ Delhi	6.374	Jawaharlal Nehru Univ	6.897	Indian Inst Technol	4.936	Jawaharlal Nehru Univ	5.302
3	Anthropol Survey India	4.925	Tata Inst Social Sci	6.061	Tata Inst Social Sci	3.726	Jawaharlal Nehru Univ	4.416	Indian Inst Technol	4.641
4	Tata Inst Social Sci	3.854	Anthropol Survey India	2.665	Indian Inst Management	2.537	Tata Inst Social Sci	3.209	Publ Hlth Fdn India	2.817
5	Indian Inst Technol	2.570	Indian Inst Technol	2.299	Natl Inst Sci Technol Dev Studies	2.180	Univ Hyderabad	2.116	All India Inst Med Sci	2.237

Andhra Univ	2.463	Indian Stat Inst	1.985	Indian Inst Technol	2.140	Indian Stat Inst	1.931	Tata Inst Social Sci	2.154
Indian Inst Management	2.152	Inst Econ Growth	1.724	Inst Econ Growth	1.942	All India Inst Med Sci	1.857	Indian Stat Inst	1.988
Indian Stat Inst	2.034	Indian Inst Management	1.358	Delhi Sch Econ	1.863	Indian Inst Management	1.581	Manipal Univ	1.683
Panjab Univ	1.713	Sri Venkateswara Univ	1.306	Indian Stat Inst	1.744	Panjab Univ	1.329	Indian Inst Management	1.408
Dibrugarh Univ	1.606	Banaras Hindu Univ	1.254	Anthropol Survey India	1.467	Inst Econ Growth	1.238	Univ Hyderabad	1.326
% Contribution	33.233		33.385		32.859		28.343		29.024

Space Science

1971-80		1981-90		1991-2000		2001-2010		2011-2012	
Tata Inst Fundamental Res	26.892	Tata Inst Fundamental Res	18.176	Tata Inst Fundamental Res	17.807	Tata Inst Fundamental Res	17.691	Tata Inst Fundamental Res	16.807
Indian Inst Astrophys	9.562	Indian Inst Astrophys	16.184	Indian Inst Astrophys	17.149	Indian Inst Astrophys	15.876	Indian Inst Astrophys	14.949
Univ Delhi	9.466	Uttar Pradesh State Observ	7.971	Phys Res Lab	8.254	Phys Res Lab	6.707	Indian Inst Technol	10.264
Uttar Pradesh State Observ	9.163	Banaras Hindu Univ	7.669	Interuniv Ctr Astron Astrophys	5.944	Raman Res Inst	6.356	Interuniv Ctr Astron Astrophys*	7.248
Phys Res Lab	8.964	Phys Res Lab	5.133	Raman Res Inst	5.944	Interuniv Ctr Astron Astrophys	4.775	Phys Res Lab	6.543

Indian Inst Geomagnetism	4.980	Council of Scientific Industrial Research Csir India	4.367	Banaras Hindu Univ	5.523	Iucaa	4.297	Raman Res Inst	5.776
Indian Institute of Technology	4.311	University of Delhi	3.133	Indian Inst Sci	3.893	Indian Inst Technol	4.238	Aryabhatta Res Inst Observat Sci Aries	4.922
Banaras Hindu Univ	2.988	Indian Inst Sci	2.883	Univ Delhi	3.419	Panjab Univ	3.985	Indian Inst Sci	3.842
Osmania Univ	2.988	Andhra Univ	2.657	Natl Phys Lab	2.998	Indian Inst Sci	3.820	Vikram Sarabhai Space Ctr	3.511
Panjab Univ	1.992	Raman Res Inst	2.234	Uttar Pradesh State Observ*	2.788	Natl Ctr Radio Astrophys	3.760	Indian Inst Geomagnetism	2.945
% Contribution	81.306		70.407		73.719		71.505		76.807

Table 3.18: Number of Institutions contributing Top 1% (Worldwide)
(Jan 1, 2002 to June 30, 2012)

Subjects	No. of Institutions	Proportion of Contribution/ Institution
Agricultural Sciences	546	4.08
Biology & Biochemistry	809	7.02
Chemistry	1058	11.97
Clinical Medicine	3612	6.14
Computer Science	371	7.27
Economics & Business	224	7.95
Engineering	1230	6.96
Environment/Ecology	663	4.52
Geosciences	511	5.94
Immunology	343	3.73
Materials Science	698	7.26
Mathematics	223	12.13
Microbiology	404	4.46
Molecular Biology & Genetics	520	5.72
Multi disciplinary	71	2.58
Neuroscience & Behavior	527	5.88
Pharmacology & Toxicology	479	4.25
Physics	753	13.01
Plant & Animal Science	1008	5.61
Psychiatry/Psychology	455	5.63
Social Sciences, general	879	5.86
Space Science	153	8.39

Table 3.19: Institutions (Unique) Contributing to Highly Cited Papers viz a viz
Proportion of Contribution

Subjects	Unique Institutes*	Papers**	Average Paper per Instiute	World Papers	India's Proportion Viz-a-Viz World
Agricultural Sciences	45	64	1.42	2228	2.87
Biology & Biochemistry	48	59	1.23	5676	1.04
Chemistry	111	256	2.31	12664	2.02
Clinical Medicine	151	247	1.64	22192	1.11
Computer Science	38	55	1.45	2697	2.04

Economics & Business	5	5	1.00	1781	0.28
Engineering	189	384	2.03	8559	4.49
Environment/Ecology	47	67	1.43	2997	2.24
Geosciences	35	47	1.34	3037	1.55
Immunology	2	2	1.00	1280	0.16
Materials Science	76	126	1.66	5068	2.49
Mathematics	26	39	1.50	2706	1.44
Microbiology	14	18	1.29	1803	1.00
Molecular Biology & Genetics	25	37	1.48	2975	1.24
Multi disciplinary	0	0	0.00	183	0.00
Neuroscience & Behavior	7	11	1.57	3097	0.36
Pharmacology & Toxicology	37	51	1.38	2037	2.50
Physics	68	670	9.85	9798	6.84
Plant & Animal Science	66	106	1.61	5658	1.87
Psychiatry/Psychology	4	8	2.00	2560	0.31
Social Sciences, general	33	39	1.18	5153	0.76
Space Science	12	20	1.67	1283	1.56

* Unique Institute = No. of Unique Institutions where a given institution's name is counted only once within a field, however across the fields the repeat name has been accounted for

** Papers= No. of papers contributed by these institutions during the period January 2002-June 2012

**** Total No. of highly citied papers are 1531 and total no. of unique institutions publishing these papers are 637 therefore no. of publications per institutions is = 2.4 per institute

Table 3.20: Top 5 Institutions Contributing to the HCP: Subject-wise Analysis

Agriculture Sciences	
Institute Name	**Papers**
Cent Food Technol Res Inst,	9
Banaras Hindu University	5
Indian Agr Res Inst, New Delhi	3
Bharathiar University, TN	2
Indian Inst Technol Delhi, New Delhi	2
Biology and Biochemistry	
Institute Name	**Papers**
Univ Delhi, New Delhi	4
Natl Chem Lab, Pune, Maharashtra	3
Amrita Vishwa Vidhyapeetham Univ, Cochin, Kerala	2
Bharathiar Univ, Coimbatore, Tamil Nadu	2

Inst Bioinformat, Bangalore, Karnataka, India.	2

Chimistry

Institute Name	Papers
Indian Inst Sci, Bangalore 560012, Karnataka, India.	20
Univ Hyderabad, Hyderabad 500046, Andhra Pradesh, India.	15
Natl Chem Lab, Pune 411008, Maharashtra, India.	12
Indian Inst Chem Technol, Hyderabad 500007, Andhra Pradesh, India	11
Jawaharlal Nehru Ctr Adv Sci Res, I Jakkur PO, Bangalore 560064, Karnataka, India.;	11

Clinical Medicine

Institute Name	Papers
All India Inst Med Sci, New Delhi	12
Tata Mem Hosp, Navi Mumbai, Maharashtra	9
St Johns Med Coll, Bangalore, Karnataka	8
Publ Hlth Fdn India, New Delhi	7
Banaras Hindu Univ, Varanasi, Uttar Pradesh	5

Computer Science

Institute Name	Papers
Indian Inst Technol, New Delhi	5
Indian Inst Technol,Kanpur, Uttar Pradesh	4
Inst Microbial Technol, Chandigarh	4
Indian Inst Sci, Bangalore, Karnataka	3
Indian Stat Inst, 203 BT Rd, Calcutta	3

Economics and Business

Institute Name	Papers
ATREE, Bangalore, Karnataka	1
Indian Stat Inst, Calcutta	1
ISB, Hyderabad, Andhra Pradesh	1
Management Dev Inst, Sukhrali, Gurgaon	1
SR Fatepuria Coll, Murshidabad	1

Engineering

Institute Name	Papers
Anna Univ, Coimbatore, Tamil Nadu	18
Indian Inst Technol, Kharagpur, W Bengal	14
Indian Inst Technol, Madras, Tamil Nadu	14
Indian Inst Technol, Kanpur, Uttar Pradesh	14
Indian Inst Technol Roorkee,Roorkee, Uttar Pradesh	13

Environment/Ecology	
Institute Name	**Papers**
Indian Inst Technol, Roorkee	7
Anna Univ, Coll Engn, Madras Tamil Nadu	4
Devi Ahilya Univ, Indore, Madhya Pradesh,	3
Ind Toxicol Res Ctr, Lucknow	3
Indian Inst Technol Delhi, New Delhi	3
Geoscience	
Institute Name	**Papers**
Indian Inst Sci, Bangalore, Karnataka	3
Indian Inst Technol, Kanpur, Uttar Pradesh	3
Indian Inst Trop Meteorol, Pune, Maharashtra	3
Vikram Sarabhai Space Ctr, Trivandrum, Kerala	3
Gauhati Univ,Gauhati, Assam	2
Immunology	
Institute Name	**Papers**
Indian Inst Sci Educ & Res IISER Bhopal,Bhopal 460023, India.	1
CIFA, Fish Hlth Management Div, Bhubaneswar 751002, Orissa, India	1
Material Science	
Institute Name	**Papers**
Indian Inst Sci, Dept Phys, Bangalore, Karnataka	14
Jawaharlal Nehru Ctr Adv Sci Res, Bangalore, Karnataka	10
Natl Chem Lab, Pune, Maharashtra	7
Banaras Hindu Univ, Varanasi, Uttar Pradesh	5
Indian Inst Technol, Dept Chem, Madras, Tamil Nadu	5
Mathematics	
Institute Name	**Papers**
Bharathiar Univ, Coimbatore, Tamil Nadu	5
Gandhigram Rural Univ, Gandhigram, Tamil Nadu	4
Banaras Hindu Univ, \ Varanasi, Uttar Pradesh	3
Bengal Engn & Sci Univ, Howrah, W Bengal	3
Disha Inst Management & Technol, Raipur, Chhattisgarh	3
Microbiology	
Institute Name	**Papers**
Bharathiar Univ, Coimbatore, Tamil Nadu	2
Christian Med Coll & Hosp, Vellore, Tamil Nadu	2
Indian Inst Technol, Bombay 400076, Maharashtra	2

Univ Delhi, New Delhi	2
All India Inst Med Sci, New Delhi	1
Molicular Biology and Genetics	
Institute Name	**Papers**
Univ Allahabad, Allahabad, Uttar Pradesh	8
Bharathidasan Univ, Tiruchchirappalli, Tamil Nadu	2
Cent Drug Res Inst, Lucknow, Uttar Pradesh	2
Chhatrapati Shahuji Maharaj Med Univ, Lucknow, Uttar Pradesh	2
Jiwaji Univ, Gwalior	2
18 Nuroscience and Behavior	
Institute Name	**Papers**
Tata Inst Fundamental Res, Bangalore Karnataka	4
Natl Inst Mental Hlth & Neurosci, Bangalore	2
Cent Drug Res Inst, Lucknow	1
CSIR, IITR, Lucknow	1
Karpagam Univ, Coimbatore, Tamil Nadu	1
Pharmcology and Toxicology	
Institute Name	**Papers**
NIPER, Dept Pharmaceut, Mohali, Punjab	5
Panjab Univ, Chandigarh	5
Dr Ambedkar Coll, Nagpur, Maharashtra	3
Inst Life Sci, Bhubaneswar, Orissa	2
Jadavpur Univ, Calcutta	2
Physics	
Institute Name	**Papers**
Tata Inst Fundamental Res, Mumbai Maharashtra	105
Panjab Univ, Chandigarh	84
Bhabha Atom Res Ctr, Mumbai, Maharashtra	42
Univ Delhi, Delhi	41
Bhabha Atom Res Ctr, Calcutta	31
Plant and Animal Science	
Institute Name	**Papers**
Indian Agr Res Inst, New Delhi	11
Int Crops Res Inst Semi Arid Trop, Patancheru, Andhra Pradesh	7
Annamalai Univ, Parangipettai, Tamil Nadu	4
Univ Delhi, New Delhi	4
Univ Hyderabad, Hyderabad, Andhra Pradesh	4

Psychiatry/Psychology	
Institute Name	**Papers**
All India Inst Med Sci, Delhi	5
Coll Med, Jaipur, Rajasthan	1
Natl Inst Mental Hlth & Neurosci, Bangalore 560029, Karnataka	1
Schizophrenia Res Fdn SCARF, Chennai, Tamil Nadu	1
Space Science	
Institute Name	**Papers**
Inter Univ Ctr Astron & Astrophys, Pune, Maharashtra	6
Indian Inst Astrophys, Bangalore, Karnataka	4
Natl Ctr Astrophys, Pune, Maharashtra	3
Aryabhatta Res Inst Observ Sci ARIES, Naini Tal	1
Homi Bhabha Ctr Sci Educ, Bombay, Maharashtra	1
Social Science, General	
Institute Name	**Papers**
BITS, Dean EHD,Pilani, Rajasthan	2
George Inst Global Hlth, Hyderabad	2
INCLEN Trust Int, New Delhi	2
India Inst Management, Bangalore, Karnataka	2
Madurai Kamaraj Univ, Madurai, Tamil Nadu	2

Table 3.21: Top 20 Institutions Based on the Number of High Impact Papers Published

Institutions	No. of Papers*
Tata Institute of Fundamental Research, Bombay,Maharashtra, India.	114
Panjab University, Chandigarh, India.	100
Bhabha Atomic Research Centre, Bombay, Maharashtra, India.	88
Indian Institute of Sciences, Bangalore, Karnataka, India.	73
University of Delhi, Delhi, India.	72
Banaras Hindu University, Varanasi, Uttar Pradesh, India.	60
Indian Institute of Technology, Bombay, Powai,Mumbai, Maharashtra, India.	42
Jawaharlal Nehru Centre for Advanced Scientific Research, Bangalore, Karnataka, India.	38
Anna University, Tirunelveli, TN, India	37
Inter University Centre of Astronomy & Astrophysics, Pune, Maharashtra, India.	37
University of Rajasthan, Jaipur, Rajasthan, India.	37
Indian Institute of Technology, Madras, Tamil Nadu, India.	36

Indian Institute of Technology, Kharagpur, West Bengal, India.	36
Indian Institute of Technology, Kanpur, Uttar Pradesh, India.	35
Institute of Physics, Bhubaneswar, Orissa, India.	32
University of Jammu, Jammu, India.	31
Indian Institute of Technology, Roorkee, Roorkee, Uttar Pradesh, India.	30
Indian Institute of Technology, Delhi, New Delhi, India.	29
National Chemical Laboratory, Pune, Maharashtra, India.	26
University of Hyderabad, Hyderabad, Andhra Pradesh, India.	26

No. of Papers* = The number of papers could be more here as the several institutions contributing to various collaborativee papers have been counted as many times as they appear across fields

Table 3.22: Relative Ranking of Disciplines Contributing to Publications in Top 1% viz a viz Citation Impact

	Total 1% publications				
Subjects	World	India	% (Contribution) of India to World	CPP -Highest Cited Paper*	CPP-Lowest Cited Paper**
Agricultural Sciences	2228	59	2.65	83.17	20.92
Biology & Biochemistry	5676	49	0.86	43.65	52.73
Chemistry	12664	215	1.70	12.78	12.75
Clinical Medicine	22192	160	0.72	12.62	29.84
Computer Science	2697	40	1.48	110.51	8.71
Economics & Business	1781	4	0.22	75.9	54.8
Engineering	8559	290	3.39	8.34	3.69
Environment/Ecology	2997	49	1.63	40.91	15.55
Geosciences	3037	33	1.09	27.52	15.28
Immunology	1280	2	0.16	116.7	30.46
Materials Science	5068	88	1.74	8.84	9.93
Mathematics	2706	35	1.29	304.2	7.63
Microbiology	1803	14	0.78	188.71	57.82
Molecular Biology & Genetics	2975	27	0.91	354.04	109.42
Multi disciplinary	183	0	0.00	236.88	149
Neuroscience & Behavior	3097	8	0.26	64.32	72.14
Pharmacology & Toxicology	2037	43	2.11	32.53	6.06
Physics	9798	290	2.96	12.88	11.67
Plant & Animal Science	5658	73	1.29	67.64	38.58

Psychiatry/Psychology	2560	8	0.31	86.17	26.68
Social Sciences, general	5153	26	0.50	34.36	11.44
Space Science	1283	18	1.40	125.19	52.24

Note: * & ** Indicates citation impact in terms of citations per paper for the highest in the respective subject areas
    and the lowest paper included in the top 1% of world papers
    (Data covers 10 years period from Jan 1, 2002 to June 30, 2012)

Table 3.23: Analysis of Highly Cited Papers: Proportion of Indian Contribution

Subjects	Papers			Colloboating Authors		
	Total Papers	Single Authored	Mutiauthored	Indian	Foreign	Foreign/India
Agricultural Sciences	59	4	55	145	162	1 : .9
Biology & Biochemistry	49	1	48	197	363	1 : .5
Chemistry	215	17	197	699	4363	1 : .2
Clinical Medicine	160	1	160	463	5442	1 : .1
Computer Science	40	3	37	92	34	1:03
Economics & Business	4	2	2	4	3	1 : 1
Engineering	290	17	273	811	2340	1 : .3
Environment/Ecology	49	0	49	120	448	1 : .3
Geosciences	33	0	33	88	207	1 : .4
Immunology	2	0	2	7	2	1:03
Materials Science	88	3	85	272	121	1:02
Mathematics	35	0	35	58	36	1 : 1.6
Microbiology	14	0	14	33	125	1 : .3
Molecular Biology & Genetics	27	0	27	110	812	1 : .1
Multi disciplinary	0	0	0	0	0	0
Neuroscience & Behavior	8	0	8	21	25	1 : .9
Pharmacology & Toxicology	43	0	43	139	40	1 : 3
Physics	290	31	259	3056	93655	1 : .03
Plant & Animal Science	73	1	72	206	765	1 : .3
Psychiatry/Psychology	8	0	8	11	147	1 : .1
Social Sciences, general	26	3	23	56	94	1 : .1
Space Science	18	0	18	21	1111	1 : .02

Table 3.24: Authorship Pattern: Analysis of Proportion of Indian Contribution in Collaborative (HCP) Papers

	>1%	1-9%	10-19%	20-29%	30-39%	40-49%	50-59%	60-69%	70-79%	80-89%	90-99%	100%*
Agricultural Sciences	-	4	3	2	3	3	5	2	2	1	-	34
Biology & Biochemistry	-	4	4	3	2	-	5	2	2	2	2	22
Chemistry	-	3	8	11	7	2	6	12	7	1	-	157
Clinical Medicine	4	72	37	9	6	3	2	6	4	5	-	13
Computer Science	-	-	1	1	4	1	6	3	-	1	-	23
Economics & Business	-	-	-	-	1	-	1	-	-	-	-	2
Engineering	3	6	6	17	7	3	20	14	4	3	-	207
Environment/ Ecology	-	5	4	4	-	3	5	1	1	-	25	
Geosciences	-	7	4	3	3	4	1	1	-	2	-	8
Immunology	-	-	-	-	1	-	-	-	-	-	-	1
Materials Science	-	-	6	6	4	2	9	3	1	3	-	55
Mathematics	-	-	-	3	8	-	5	3	1	-	-	15
Microbiology	-	2	1	1	3	-	1	1	-	-	5	
Molecular Biology & Genetics	2	3	1	1	-	2	1	1	-	-	16	
Multi disciplinary	-	-	-	-	-	-	-	-	-	-	-	-
Neuroscience & Behavior	-	-	2	2	1	-	-	-	-	-	-	3
Pharmacology & Toxicology	-	-	-	3	7	-	2	1	-	-	-	30
Physics	25	103	10	13	20	3	16	11	4	2	-	82
Plant & Animal Science	2	7	6	5	6	2	6	2	3	-	1	33

Psychiatry/ Psychology	-	7	-	-	-	-	-	-	-	-	1
Social Sciences, general	-	-	4	4	4	1	2	1	1	-	9
Space Science	4	7	1	-	1	-	4	-	-	-	1

*Analysis based on the number of collaborative papers: Number in the different blocks against various subject fields indicate number of papers corresponding to the proportion of Indian contribution to these papers with more than one authors

Single authored papers have not been included here

100%* = indicates all authors have Indian institutional affiliation

Table 3.27: Field-wise Total Number of Journals/Papers/Citations (Jan 1, 2002 to June 30, 2012)

Subjects	Total no. of Journals	Total Papers	Total Citations	C.P.P	% Share
Clinical Medicine	1098	1893932	27249405	13822.9	20.73
Social Sciences, general	945	399741	2338381	5066.4	7.60
Biology & Biochemistry	243	499278	9107005	4826.04	7.24
Psychiatry/Psychology	309	196603	3301041	4076.79	6.11
Chemistry	305	112781	14278491	3990	5.98
Molecular Biology & Genetics	157	255591	5740987	3923.84	5.88
Plant & Animal Science	439	471112	4282625	3909.19	5.86
Engineering	517	759513	4350259	3075.49	4.61
Neuroscience & Behavior	142	278831	5653778	3012.22	4.52
Environment/Ecology	187	264423	3280154	2524.34	3.79
Economics & Business	286	142695	1141771	2322.61	3.48
Geosciences	227	262637	2874457	2224.4	3.34
Pharmacology & Toxicology	127	176839	2377038	2051.21	3.08
Materials Science	189	444316	3712993	1970.08	2.95
Physics	185	901181	8140147	1957.35	2.93
Immunology	60	114743	2550897	1766.69	2.65
Microbiology	79	161766	2632011	1589	2.38
Agricultural Sciences	166	194837	1598206	1493.55	2.24
Computer Science	199	249758	1098889	1172.29	1.76
Mathematics	253	237291	935798	1132.76	1.70
Space Science	40	119621	1848623	610	0.91
Multi-disciplinary	24	16327	135012	174.6	0.26

Table 3.28: Relative Ranking of High Impact Making Journals

Subject Area	Total journals	Gr 1 >=100	Gr 2 20-99	Gr 3 10-19	Gr 4 5-9	Gr 5 < 5
Agricultural Sciences	166	2	5	29	50	80
Biology & Biochemistry	243	6	54	104	67	12
Chemistry	305	3	33	91	93	85
Clinical Medicine	1098	8	111	360	453	166
Computer Science	199	1	4	12	54	128
Economics & Business	286	0	20	48	101	117
Engineering	517	0	12	37	178	290
Environment/Ecology	187	3	21	44	84	35
Geosciences	227	0	13	67	102	45
Immunology	60	5	12	27	16	0
Materials Science	189	3	15	13	61	97
Mathematics	253	0	2	12	56	183
Microbiology	79	2	16	33	23	5
Molecular Biology & Genetics	157	7	42	81	25	2
Multi disciplinary	24	0	3	1	2	18
Neuroscience & Behavior	142	4	35	79	21	3
Pharmacology & Toxicology	127	3	18	45	51	10
Physics	185	3	10	25	70	77
Plant & Animal Science	439	3	18	77	213	128
Psychiatry/Psychology	309	1	41	105	137	25
Social Sciences, general	945	0	13	89	295	548
Space Science	40	1	6	10	9	14
Total	6177	55	504	1389	2161	2068
0.89	8.16	22.49	34.98	33.48		

NB: Gr 1 - referes to journals with over 100 CPP, similarly Gr 2 refers to journals with CPP between 20 to 99 and the like

Table 3.29: Top Impact Making Journal
(Journals Covered in the First Level Group with >100 CPP)

Field	Journal	Rank *	CPP**
Agiculturl Scs	Nature	166	264.2
	Science	165	131.62
Biology & Biochemistry	Physiol Rev	228	174.85
	Annu Rev Biochem	231	152.15
	Science	173	137.26
	Nature	141	135.93
	Endocrine Rev	227	113.04
	Annu Rev Biophys Biomol Struc	240	104.08
Chemistry	Nature	290	201.54
	Science	255	184.83
	Chem Rev	179	150.54
Clinical Med	Science	291	271.31
	Nature	747	248.65
	Ca-A Cancer J Clin	1089	221.08
	N Engl J Med	102	192.67
	Nat Rev Cancer	883	152.42
	Jama-J Am Med Assn	151	124.24
	Nature Med	388	118.93
	Lancet	69	107.9
Computer Science	Science	199	1,407
Economics & Business	....	....	....
Engineering	....	....	....
Environment/Ecology	Annu Rev Ecol Syst	187	8,428
	Nature	174	41,237
	Science	163	50,052
Geosciences	....	....	....
Immunology	Annu Rev Immunol	59	209.94
	Nature	58	201.55
	Science	60	177.48
	Nat Rev Immunol	54	133.12
	Nat Immunol	28	116.9
Materials Science	Nature	188	176.7

	Science	185	142.59
	Nat Mater	93	112.99
Mathematics	…….	…….	…….
Microbiology	Nature	74	146.80
	Science	71	137.48
Molecular Biology & Genetics	Nat Rev Mol Cell Biol	107	165.56
	Nature	52	146.11
	Science	71	138.57
	Cell	21	133.18
	Nat Genet	34	120.08
	Nat Rev Genet	106	110.41
	Annu Rev Cell Dev Biol	154	106.21
Multi-disciplinary	…….	…….	…….
Neuroscience & Behavior	Science	116	147.79
	Annu Rev Neurosci	140	147.24
	Nat Rev Neurosci	104	135.65
	Nature	117	129.66
Pharmacology & Toxicology	Nature	21	141.19
	Pharmacol Rev	249	115.83
	Nat Rev Drug Discov	718	106.85
Physics	Rev Mod Phys	179	168.48
	Nature	153	167.37
	Science	160	138.47
Plant & Animal Science	Nature	344	120.62
	Science	335	116.88
	Annu Rev Plant Biol	424	116.54
Psychiatry/Psychology	Nature	309	118.80
Social Sciences, General	…….	…….	…….
Space Science	Annu Rev Astron Astrophys	39	108.96

* Rank - Journal Ranking (within the field) in terms of Number of Papers Published
** CPP Arranged on the basis of high impact making journals

Table 3.30: Journal Preference Analysis – Top 1% Indian Highly Cited Papers: Citations per Paper and Impact Factor Score

(a) Agricultural Sciences				
Journal	No. of Papers	CPP*	Rank **	Impact Factor***
Science	1	131.62	2	31.201
Crit Rev Food Sci Nutr	5	21.97	4	4.789
Advan Agron	2	20.86	5	5.204
J Nutrition	1	20.46	6	3.916
Trends Food Sci Tech	2	20.12	7	3.672
Int J Food Microbiol	2	15.11	13	3.327
J Agr Food Chem	2	14.04	15	2.823
Mol Nutr Food Res	1	12.56	19	4.301
Brit J Nutr	1	12.09	21	3.724
Food Chem	9	11.15	25	3.655
Postharvest Biol Tec	1	11.03	26	2.411
Food Microbiology	1	10.72	29	3.283
Compr Rev Food Sci Food Saf	1	10.62	31	3.724
J Food Compos Anal	1	10.1	35	2.079
Food Chem Toxicol	5	9.82	38	2.999
Field Crop Res	3	9.68	39	2.474
Food Res Int	4	9.15	44	3.15
Pest Manag Sci	1	8.46	47	2.251
Crop Sci	2	8.08	52	1.641
Agr Water Manage	2	7.87	54	1.998
J Plant Nutr Soil Sci	1	7.39	61	1.596
Lwt-Food Sci Technol	1	6.23	72	2.545
Ind Crop Prod	3	6.04	75	2.469
Food Technol Biotech	1	5.88	80	1.195
Food Nutr Bull	1	5.05	86	1.922
Int J Food Sci. Technol	1	4.88	89	1.259
Am J Potato Res	1	4.14	102	1.234
Food Bioprocess Technol	1	3.81	107	3.703
Nutrition	1	3.42	115	3.025
Int J Agric Sustain	1	2.03	147	1.696

(b) Biology And Biochemistry				
Journal	**No. of Papers**	**CPP***	**Rank ****	**Impact Factor*****
Science	1	137.26	3	31.201
Nat Biotechnol	3	93.51	8	23.268
Proc Nat Acad Sci USA	1	46.15	8	9.681
Nat Chem Biol	1	40.4	20	14.69
Protein-Struct Funct Genet	1	36.88	24	3.392
Trends In Biotechnology	1	33.58	29	9.148
Nucl Acid Res	5	29.36	37	8.026
J Biol Chem	1	28.97	38	4.773
J Bone Miner Res	1	25.96	42	6.373
Biotechnol Adv	8	25.46	44	9.646
Nat Protoc	1	24.82	45	9.924
Biomacromolecules	1	20.5	59	5.479
Biosens Bioelectron	3	17.13	76	5.602
Arch Biochem Biophys	1	13.87	106	2.935
Life Sci	1	13.42	110	2.527
Process Biochem	1	13.16	115	2.627
Biochem Biophys Res Commun	1	12.91	119	2.484
Bioresource Technol	7	12.3	127	4.98
Biotechnol Progr	1	11.3	143	2.34
Colloid Surface B	4	8.69	183	3.456
J Ind Microbiol Biotechnol	1	8.67	184	2.735
J Biomol Struct Dyn	2	7.98	198	4.986
J Biosciences	1	7.43	208	1.648
J Biomed Nanotechnol	1		4.216	
(c) Chemistry				
Journal	**No. of Papers**	**CPP***	**Rank ****	**Impact Factor*****
Science	1	184.83	2	31.201
Chem Rev	13	150.54	3	40.197
Prog Polym Sci	12	91.17	4	24.1
Surf Sci Rep	1	83.21	5	11.696
Account Chem Res	9	77.57	6	21.64
Chem Soc Rev	9	58.76	8	28.76

Proc Nat Acad Sci USA	1	52.38	10	9.681
Coord Chem Rev	5	50.25	12	12.11
Nano Lett	4	44.64	13	13.198
J Photochem Photobiol C-Photo	1	43.95	14	10.36
Angew Chem Int Ed	14	38.69	15	13.455
J Am Chem Soc	9	38.51	16	9.907
Nano Today	1	30.43	22	15.355
Advan Colloid Interface Sci	6	27.06	24	8.12
Org Lett	10	20.94	34	5.862
J Phys Chem B	7	20.39	36	3.696
Acta Crystallogr B-Struct Sci	1	18.86	40	2.286
Inorg Chem	3	18.19	44	4.601
Chem Commun	8	18.13	45	6.169
J Org Chem	3	18.13	46	4.45
Langmuir	4	18.08	47	4.186
Chem-Eur J	2	18.01	48	5.925
Carbon	2	17.38	51	5.378
J Appl Cryst	1	17.01	53	5.152
Polymer	1	16.89	54	3.438
Appl Catal A-Gen	1	16.38	56	3.903
Anal Chim Acta	4	15.55	60	4.555
J Membrane Sci	2	14.93	62	3.85
Acs Nano	1	14.58	65	10.774
J Biol Inorg Chem	1	14.14	67	3.289
Cryst Growth Des	1	13.53	73	4.72
J Chem Phys	3	13.32	75	3.333
Chemphyschem	1	13.01	78	3.412
Tetrahedron	6	12.84	84	3.025
J Phys Chem A	2	12.76	85	2.946
Talanta	1	12.61	87	3.794
Pure Appl Chem	2	12.46	88	2.789
J Colloid Interface Sci	6	11.98	92	3.07
Polym Degrad Stabil	1	11.74	94	2.769
Tetrahedron-Asymmetry	1	11.74	93	2.652
Chem Phys Lett	3	11.72	95	2.337
J Cheminformatics	1	11.54	97	3.419
Bioorgan Med Chem	1	11.51	98	2.921

Journal	No. of Papers	CPP*	Rank **	Impact Factor***
Eur Polym J	2	11.49	99	2.739
Anal Bioanal Chem	1	11.34	102	3.778
Eur J Org Chem	1	11.2	104	3.329
Org Biomol Chem	2	11.15	106	3.696
Analyst	2	11.11	107	4.23
Sep Purif Technol	1	10.9	109	2.921
Dye Pigment	5	10.71	113	3.126
Dalton Trans	2	10.36	119	3.838
Synthesis-Stuttgart	1	10.25	121	2.466
Fuel	3	10.14	123	3.248
React Funct Polym	1	10.06	126	2.479
Carbohyd Polym	3	9.63	134	3.628
Colloid Surface A	1	8.94	146	2.236
Eur J Med Chem	1	8.71	150	3.346
Crystengcomm	2	8.41	152	3.842
Chem Eng J	2	7.89	161	3.461
Desalination	1	7.05	178	2.59
J Fluoresc	1	6.96	184	2.107
Biomed Chromatogr	1	6.16	196	1.966
J Therm Anal Calorim	1	5.5	208	1.604
J Phys Chem Lett	1	4.85	224	6.213
J Chem Sci	1	4.8	226	1.177
Molecules	1	4.21	241	2.386
Nanoscale	1	4.13	243	5.914
J Instrum	2		1.869	
J Nanophotonics	1		1.57	
Prog Cryst Growth Charact	1		5.75	
Prog Solid State Chem	1		4.188	
Qsar Comb Sci	1		1.55	
Tetrahedron Lett	5		2.683	
(d) Clinical Medicine				
Journal	**No. of Papers**	**CPP***	**Rank ****	**Impact Factor*** **
Nature	5	248.65	2	36.28
N Engl J Med	26	192.67	4	53.298
Jama-J Am Med Assn	5	124.24	6	30.026

Lancet	39	107.9	8	38.278
J Exp Med	1	72.92	9	13.853
Ann Intern Med	1	66.39	11	16.733
Proc Nat Acad Sci USA	2	64.65	12	9.681
Circulation	3	64.12	13	14.739
J Nat Cancer Inst	1	63.04	14	13.757
Lancet Neurol	1	57.12	15	23.462
J Clin Oncol	10	54.12	16	18.372
Lancet Infect Dis	3	51.63	18	17.391
J Amer Coll Cardiol	1	47.78	20	14.156
Lancet Oncol	2	43.93	25	22.589
Hepatology	1	40.86	27	11.665
Amer J Respir Crit Care Med	1	40.58	28	11.08
Cancer Res	3	39.69	30	7.856
Blood	1	39.34	33	9.898
Bba-Rev Cancer	1	38.31	34	9.38
Eur Heart J	1	35.52	40	10.478
Oncogene	1	33.91	42	6.373
Diabetes Care	2	30.72	50	8.087
Stem Cells	2	29.92	53	7.781
Clin Infect Dis	1	28.96	55	9.154
Plos Med	2	27.72	60	16.269
Thorax	1	26.39	69	6.84
Amer J Gastroenterol	1	26.28	70	7.282
Diabetologia	4	23.49	84	6.814
J Intern Med	1	23.29	88	5.483
Am J Transplant	2	22.37	96	6.394
Eur Resp J	1	22.32	98	5.895
Ann Rheum Dis	1	22.01	101	8.727
Brit J Cancer	1	20.79	110	5.042
Neoplasia	1	20.07	119	5.946
J Thromb Haemost	1	19.15	129	5.731
Osteoporosis Int	2	16.54	190	4.58
Nmr Biomed	1	16.16	197	3.214
Radiother Oncol	2	14.93	220	2.321
Ann N Y Acad Sci	1	14.26	242	3.155
J Amer Acad Dermatol	1	14.26	243	3.991

Neuro-Oncology	1	14.22	247	5.723
J Clin Pharmacol	1	12.36	319	2.911
Nanomedicine	1	12.1	331	5.055
Amer J Trop Med Hyg	1	11.94	340	2.592
Clin J Am Soc Nephrol	1	11.2	396	5.227
Urology	1	10.83	419	2.428
Bju Int	1	10.34	459	2.844
Amer J Infect Control	1	10.12	466	2.396
J Pharm Pharm Sci	1	9.59	504	1.646
Nanomed-Nanotechnol Biol Med	4	9.37	524	6.692
J Vasc Interven Radiol	1	9.01	557	2.075
Periton Dialysis Int	2	7.58	676	2.097
J Endourol	2	6.29	812	1.847
Int Orthop	1	4.79	951	2.025
Plos One	3	4.34	989	4.092
Hepatol Int	1		2.645	

(e) Computer Science				
Journal	**No. of Papers**	**CPP***	**Rank ****	**Impact Factor*****
Brief Bioinform	1	40.06	3	5.202
Bioinformatics	6	26.48	4	5.468
Bmc Bioinformatics	5	10.75	14	2.751
Environ Modell Softw	1	9.27	20	3.114
Evol Comput	1	8.02	26	1.061
Comput Chem Eng	2	7.99	27	2.32
Inform Sciences	2	7.02	34	2.833
Ieee Trans Commun	4	6.9	36	1.677
Comput Struct	1	6.83	37	1.874
Comput Biol Chem	1	6.72	40	1.551
Commun Acm	1	6.04	54	1.919
Comput Ind Eng	2	4.89	76	1.589
Comput Math Appl	6	4.02	99	1.747
Appl Soft Comput	3	3.31	130	2.612
Comput Sci Eng	1	3.27	132	1.422
J Netw Comput Appl	1	1.44	185	1.065
Curr Bioinform	2		0.898	

(f) Economics & Business				
Journal	**No. of Papers**	**CPP***	**Rank ****	**Impact Factor*****
Int J Manag Rev	1	11.77	50	0.604
J Econometrics	1	11.49	53	1.349
Ecol Econ	2	9.9	69	2.713
(g) Engineering				
Journal	**No. of Papers**	**CPP***	**Rank ****	**Impact Factor*****
Annu Rev Fluid Mech	2	48.15	4	12.767
Prog Energ Combust Sci	2	42.66	6	14.22
Ieee Trans Evol Computat	5	22.26	11	3.341
Ieee Trans Patt Anal Mach Int	2	18.26	13	4.908
Proc Ieee	1	16.27	16	6.81
J Power Sources	8	15.66	18	4.951
Ieee Signal Process Mag	1	14.85	21	4.066
Prog Photovoltaics	1	12.48	26	5.789
Ieee Trans Fuzzy Syst	1	12.14	28	4.26
Sensor Actuator B-Chem	16	12	30	3.898
J Hydrol	3	11.85	31	2.656
Electroanal	2	11.33	35	2.872
J Hazard Mater	97	11.28	36	4.173
Ieee J Solid-State Circuits	1	10.28	46	3.226
Ieee Trans Geosci Remot Sen	1	10.2	48	2.895
Int J Hydrogen Energ	19	9.69	53	4.054
Ieee Trans Image Processing	1	9.68	54	3.042
Ieee Trans Ind Electron	4	9.47	56	5.16
Int J Heat Mass Transfer	9	9.35	59	2.407
Chemometr Intell Lab Syst	2	9.28	60	1.92
Macromol Mater Eng	1	9.1	62	1.986
Org Electron	1	8.67	69	4.047
Waste Management	4	8.53	74	2.428
Ieee Trans Microwave Theory	1	8.53	73	1.853
Energ Fuel	3	8.52	75	2.721
Ieee Electron Dev Lett	1	8.5	77	2.849
Nucl Data Sheets	1	8.34	81	3.821

Energ Conv Manage	3	8.31	82	2.216
Ieee Trans Power Syst	3	8.04	92	2.678
Thermochim Acta	1	7.91	95	2.162
Vib Spectrosc	1	7.6	99	1.65
Prog Electromagn Res	1	7.29	105	5.298
Eur J Oper Res	4	7.28	106	1.815
Energy	4	7.04	115	3.487
Renewable Energy	8	6.93	120	2.978
Spectrochim Acta Pt A-Mol Bio	16	6.2	155	2.098
Int J Non-Linear Mech	2	6.18	157	1.209
Appl Energ	8	6.13	161	5.106
Appl Therm Eng	1	6.13	159	2.064
Int J Mech Sci	1	6.02	165	1.231
Int J Therm Sci	2	5.85	172	2.142
Image Vision Comput	1	5.66	181	1.723
Med Biol Eng Comput	1	5.62	183	1.878
Pattern Recognition Lett	1	5.61	184	2.292
J Heat Transfer	1	5.52	190	1.83
J Irrig Drain Eng-Asce	2	5.34	202	0.941
Exp Therm Fluid Sci	1	5.31	205	1.414
Cold Reg Sci Technol	1	5.08	220	1.429
Nucl Instrum Meth Phys Res B	1	5.03	224	1.207
J Global Optim	1	4.9	232	1.196
Int Commun Heat Mass Trans	1	4.75	240	1.892
Nucl Instrum Meth Phys Res A	8	4.68	246	1.207
Mech Mach Theor	1	4.45	268	1.366
Ieee Trans Dielect Electr In	1	4.44	270	1.094
Appl Math Comput	4	4.39	274	1.317
Nav Res Log	1	4.29	279	1.038
Neurocomputing	1	4.07	299	1.58
Signal Process	1	3.84	325	1.503
Expert Syst Appl	1	3.82	328	2.203
Ieee Sens J	1	3.75	334	1.52
Ind Eng Chem Res	2	3.7	340	2.237
Elec Power Syst Res	1	3.63	344	1.478
Microelectron Rel	1	3.5	351	1.167

Journal	No. of Papers	CPP*	Rank**	Impact Factor***
Math Comput Modelling	1	3.42	360	1.346
Heat Mass Transfer	2	3.24	378	0.896
Int J Adv Manuf Technol	2	3.17	380	1.103
J Franklin Inst-Eng Appl Math	1	3.15	382	2.724
Nucl Eng Des	1	2.83	408	0.765
Algorithmica	1	2.62	426	0.604
Int Rev Electr Eng-Iree	1	1.74	476	
Acta Astronaut	1	1.48	493	0.569
Desalin Water Treat	1		0.614	

(h) Environment/Ecology

Journal	No. of Papers	CPP*	Rank**	Impact Factor***
Nature	2	127.67	2	36.28
Science	4	124.82	3	31.201
Proc Nat Acad Sci USA	1	51.02	6	9.681
Adv Environ Res	2	24.88	16	
Environ Sci Technol	2	21.31	23	5.228
Water Res	5	19.15	27	4.865
Environ Int	2	19.13	28	5.297
Global Environ Change	1	17.4	34	6.868
Crit Rev Environ Sci Technol	2	16.89	37	4.841
Biol Conserv	1	16.7	38	4.115
Chemosphere	1	14.7	42	3.206
Mar Pollut Bull	1	11.05	61	2.503
Biomass Bioenerg	4	10.8	62	3.646
Mol Ecol Resour	1	10.75	64	3.062
Energy Environ Sci	1	10.29	67	9.61
Renew Sustain Energy Rev	15	9.64	72	6.018
Ecotoxicol Environ Safety	1	9.21	76	2.294
Hydrol Process	1	9.14	80	2.488
J Environ Manage	1	8.29	90	3.245
Int Biodeterior Biodegrad	1	7.22	115	2.074

(i) Geosciences

Journal	No. of Papers	CPP*	Rank**	Impact Factor***
Science	3	76.58	1	31.201

Nature	2	67	2	36.28
Bull Amer Meteorol Soc	1	22.87	8	6.026
Global Biogeochem Cycle	1	20.12	13	4.785
Earth Planet Sci Lett	1	19.77	14	4.18
Nat Geosci	2	18.69	17	11.754
J Geophys Res-Atmos	5	15.79	29	3.021
Clim Dynam	1	15.62	31	4.602
Atmos Chem Phys	1	15.44	33	5.52
Atmos Environ	1	14.86	34	3.465
Quart J Roy Meteorol Soc	1	13.23	46	2.907
Geophys Res Lett	2	12.9	51	3.792
Palaeogeogr Palaeoclimatol	1	12.41	55	2.392
Gondwana Res	3	11.27	62	6.659
Appl Geochem	1	10.7	69	2.176
Appl Clay Sci	1	8.14	103	2.474
Quatern Int	3	6.78	134	1.874
Int J Remote Sens	1	6.17	154	1.117
Geochem J	1	4.84	185	0.711
Curr Sci	1	4.73	187	0.935

(j) Immunology

Journal	No. of Papers	CPP*	Rank **	Impact Factor***
Int Rev Immunol	1	17.7	21	3.426
Develop Comp Immunol	1	12.39	33	3.268

(k) Materials Science

Journal	No. of Papers	CPP*	Rank **	Impact Factor***
Science	1	142.59	2	31.201
Nat Mater	6	112.99	3	32.841
Prog Mater Sci	3	63.44	7	18.216
Advan Mater	11	42.36	8	13.877
Int Mater Rev	2	30.68	11	6.962
Adv Funct Mater	5	29.19	12	10.179
Biomaterials	7	28.91	13	7.404
Chem Mater	11	28.54	14	7.286
Curr Opin Solid State Mat Sci	2	26.96	15	4.233

Nat Nanotechnol	1	26.27	16	27.27
Small	2	20.07	18	8.349
Acta Mater	4	16.82	21	3.755
J Mater Chem	9	14.25	23	5.968
Composites Sci Technol	1	14.21	24	3.144
Scripta Mater	1	11.65	27	2.699
Solar Energ Mater Solar Cells	3	11.37	28	4.542
Corros Sci	2	10.98	29	0.692
Nanotechnol	1	9.37	33	3.979
Surf Coat Tech	1	8.9	36	1.867
Mater Sci Eng A-Struct Mater	1	8.74	39	2.003
J Mater Sci-Mater Med	1	8.25	45	2.316
Prog Org Coating	1	8.05	47	1.977
Thin Solid Films	2	7.76	49	1.89
Mater Lett	1	7.58	52	2.307
Mater Chem Phys	1	7.35	54	2.234
Sci Technol Adv Mater	1	7.3	55	3.513
J Nanopart Res	1	6.9	61	3.287
J Alloys Compounds	1	6.55	69	2.289
Polym Eng Sci	1	6.28	72	1.302
J Mater Sci	2	5.73	80	2.015
Mater Manuf Process	1	2.62	142	1.058
Mater Sci Eng C	1	2.18	154	0.229

(I) Mathematics				
Journal	**No. of Papers**	**CPP***	**Rank ****	**Impact Factor*****
J Amer Statist Assn	1	16.15	4	1.992
Ann Probab	1	7	35	1.789
Stat Sinica	1	6.49	40	1.017
Eur J Appl Math	2	5.31	62	0.656
J Math Anal Appl	1	4.75	75	1.001
Nonlinear Anal-Real World App	9	4.49	89	2.043
J Comput Appl Math	2	4.34	95	1.112
Z Angew Math Phys	1	4.11	103	0.951
Commun Nonlinear Sci Numer Si	8	4.03	104	2.806
Numer Func Anal Optimiz	1	2.81	164	0.711

Journal	No. of Papers	CPP*	Rank **	Impact Factor***
Electron J Probab	1	2.63	179	0.713
Fixed Point Theory Appl	5	2.47	190	1.634
Int J Biomath	2		0.364	

(m) Microbiology				
Journal	**No. of Papers**	**CPP***	**Rank ****	**Impact Factor*****
Nature	1	146.8	1	36.28
Microbiol Mol Biol Rev	1	92.14	3	13.018
Clin Microbiol Rev	1	82.92	4	16.129
Mol Microbiol	1	27.39	12	5.01
Antimicrob Agents Chemother	1	19.43	20	4.841
Plos Pathog	1	16.07	25	9.127
Appl Microbiol Biotechnol	2	12.47	34	3.425
Int J Antimicrobial Agents	1	10.83	47	4.128
Acta Biomater	2	9.61	53	4.865
Bmc Microbiol	1	7.82	61	3.044
Parasitol Res	2	6.27	69	2.149

(n) Molecular Biology & Genetics				
Journal	**No. of Papers**	**CPP***	**Rank ****	**Impact Factor*****
Nat Rev Mol Cell Biol	1	165.56	1	39.123
Science	1	138.57	3	31.201
Cell	2	133.18	4	32.403
Nat Genet	3	120.08	5	35.532
Genome Res	1	50.48	18	13.608
Proc Nat Acad Sci USA	1	38.2	24	9.681
Dna Res	1	20.09	49	5.164
J Theor Biol	1	9.45	136	2.208
Mol Cell Biochem	1	8.74	139	2.057
Cell Mol Biol	14	7.38	146	0.975
Sci Transl Med 3 (111):	1		7.804	

(o) Multi Disciplinary				
Journal	**No. of Papers**	**CPP***	**Rank ****	**Impact Factor*****
-	-	-	-	-

(p) Neuroscience & Behavior			
Journal	**No. of Papers**	**CPP***	**Rank ****	**Impact Factor***
Nat Rev Neurosci	1	135.65	3	30.445
Neuron	2	71.25	5	14.736
Prog Neurobiol	1	48.06	9	8.874
J Neurosci	1	35.1	18	7.115
Neurology	1	31.34	19	8.312
Neurobiol Aging	1	18.83	46	6.189
Brain Res Rev	1		10.342	

(q) Pharmacology & Toxicology			
Journal	**No. of Papers**	**CPP***	**Rank ****	**Impact Factor***
Advan Drug Delivery Rev	3	52.7	6	11.502
Drug Discov Today	2	26.09	13	6.828
J Control Release	8	25.97	14	5.732
Crit Rev Ther Drug Carr Syst	1	22.5	20	2.609
Biochem Pharmacol	5	17.77	26	4.705
Brit J Pharmacol	2	17.28	30	4.409
Int J Pharm	6	13.5	41	3.35
Toxicol Lett	2	13.49	42	3.23
Eur J Pharm Sci	2	13.06	44	3.212
J Pharm Sci	1	11.72	51	3.055
Eur J Pharmacol	1	11.7	52	2.516
J Ethnopharmacol	1	11.35	53	3.014
J Pharmaceut Biomed Anal	1	10.78	58	2.967
Toxicol Vitro	1	9.28	71	2.775
Phytother Res	1	8.2	85	2.086
J Appl Toxicol	1	7.89	87	2.478
Pharmacol Rep	2	6.68	100	2.445
Mar Drugs	2	4.7	118	3.854
Curr Neuropharmacol	1		2.847	

(r) Physics			
Journal	**No. of Papers**	**CPP***	**Rank ****	**Impact Factor***
Rev Mod Phys	3	168.48	1	43.933

Nature	1	167.37	2	36.28
Science	3	138.47	3	31.201
Phys Rep-Rev Sect Phys Lett	9	82.51	4	20.394
Advan Phys	1	78.74	5	37
Rep Progr Phys	5	46.69	7	14.72
Phys Rev Lett	78	27.56	11	7.37
Nucl Phys B	3	20.63	13	4.661
Phys Lett B	29	16.64	16	3.955
Phys Rev D	41	15.83	17	4.558
J High Energy Phys	25	14.1	20	5.831
Appl Phys Lett	4	13.98	21	3.844
Phys Rev B	16	12.36	27	3.691
Eur Phys J C	7	11.27	30	3.631
Ann Phys N Y	1	11.21	31	2.857
Nucl Fusion	1	10.9	32	4.09
Phys Rev C	18	10.73	35	3.308
Class Quantum Gravity	5	10.22	37	3.32
J Phys Chem C	10	10.2	38	4.805
Phys Rev A	2	9.66	40	2.878
Commun Math Phys	1	9.19	48	1.941
Comput Phys Commun	2	8.2	55	3.268
J Appl Phys	2	8.18	56	2.168
Phys Plasmas	2	8.09	59	2.147
J Phys-Condens Matter	2	8.07	60	2.546
Nucl Phys A	3	7.72	63	1.54
J Phys G-Nucl Particle Phys	3	7.27	71	4.178
J Phys-B-At Mol Opt Phys	1	6.96	80	1.875
Physica A	1	6.5	89	1.373
Gen Relativ Gravit	2	5.96	97	2.069
Mod Phys Lett A	2	4.73	115	1.083
Jetp Lett-Engl Tr	1	4.42	119	1.352
Physica E	1	4.42	118	1.532
Superlattice Microstruct	1	4.2	122	1.487
Curr Appl Phys	1	4.12	124	1.9
Int J Mod Phys A	1	3.66	131	1.053
Annu Rev Condens Matter Phys	1		12.389	
Crit Rev Solid State Mat Sci	1		9.467	

(s) Plant & Animal Science			
Journal	No. of Papers	CPP*	Rank **	Impact Factor***
Nature	4	120.62	1	36.28
Science	3	116.88	2	31.201
Plant Cell	1	49.01	6	8.987
Proc Nat Acad Sci USA	2	48.86	7	9.681
Trends Plant Sci	7	43.72	8	11.047
Curr Opin Plant Biol	2	41.55	9	9.272
Plant Physiol	1	31.27	12	6.535
Plant Mol Biol	1	20.48	20	4.15
J Exp Bot	3	19.64	22	5.364
Theor Appl Genet	3	18.12	26	3.297
Plant Biotechnol J	1	15.7	30	5.442
Ann Bot	1	15.35	32	4.03
Aquat Toxicol	1	15.04	33	3.761
Photosynth Res	1	12.67	56	3.243
Plant Sci	3	11.56	66	2.945
Aquaculture	1	10.98	77	2.041
Environ Exp Bot	2	10.52	85	2.985
Estuar Coast Shelf Sci	1	10.24	89	2.247
Fish Shellfish Immunol	1	10.24	90	3.322
Plant Physiol Biochem	4	10.2	92	2.838
Plant Cell Rep	3	9.73	103	2.274
Mol Breeding	1	9.15	120	2.852
Vet Parasitol	1	9.06	126	2.579
J Plant Physiol	1	8.99	131	2.791
Aquat Bot	1	8.97	132	1.516
J Insect Physiol	1	8.73	139	2.236
Taxon	1	8.59	145	2.703
J Mar Res	1	8.33	152	0.766
Biol Control	1	8.21	158	2.003
Phycologia	1	8.07	165	2
Bmc Plant Biol	1	7.41	189	3.447
J Stored Prod Res	1	6.97	210	1.414
Euphytica	2	6.69	220	1.554

Plant Cell Tissue Organ Cult	1	6.63	223	3.09
Curr Sci	4	6.13	244	0.935
Wood Sci Technol	1	6.06	252	1.727
Aquac Res	1	5.7	276	1.203
J Integr Plant Biol	1	5.1	304	2.534
Pl Mol Biol Rep	1	5.07	308	2.453
S Afr J Bot	2	3.82	362	1.659
Acta Physiol Plant	2	2.78	406	1.639
Phytotaxa	1		1.797	

(t) Psychiatry/Psychology				
Journal	**No. of Papers**	**CPP***	**Rank ****	**Impact Factor*****
Arch Gen Psychiat	2	69.15	5	12.016
J Clin Psychiat	2	25.93	22	5.799
Brit J Psychiat	1	25.79	23	6.619
Int J Meth Psychiatr Res	1	17.13	56	2.462
World Psychiatry	2		6.233	

(u) Social Sciences, General				
Journal	**No. of Papers**	**CPP***	**Rank ****	**Impact Factor*****
Science	1	35.17	4	31.201
Ageing Res Rev	1	23.59	11	6.174
Soc Sci Med	1	15.17	32	2.699
Energy Educ Sci Technol-Pt A	4	11.91	61	31.677
Aids Educ Prev	1	11.35	76	1.59
Public Health Nutr	2	11.14	81	2.169
World Develop	1	9.07	125	1.537
Scientometrics	2	7.33	193	1.966
Ann N Y Acad Sci	1	7.15	205	3.155
Energ Policy	2	7.1	206	2.723
Res Develop Disabil	1	6.36	269	3.405
Inform Soc	1	6.28	273	
Neurorehabilitation	1	6.11	288	1.635
Health Policy	1	5.93	303	1.506
J Archaeol Sci	1	5.84	308	1.914
Bmc Public Health	1	5.17	372	1.997

Journal	No. of Papers	CPP*	Rank **	Impact Factor***
J Health Popul Nutr	1	4.88	412	0.954
J Peasant Stud	1	3.4	607	2.55
Aaohn J	1		1.468	
Popul Health Metr	1		1.024	
(v) Space Science				
Journal	**No. of Papers**	**CPP***	**Rank ****	**Impact Factor***
Nature	2	63.34	2	36.28
Science	3	59.12	3	31.201
Astrophys J Suppl Ser	4	50.64	4	13.456
Astrophys J	1	23.51	6	6.024
Astron J	2	22.89	7	4.035
Mon Notic Roy Astron Soc 3	3	17.25	8	4.9
Astrophys J Lett	1	9.62	19	5.526
Int J Mod Phys D	2	6.13	24	1.183

* CPP= Citations Per paper published in the corresponding journal - CPP based on 10 year period (2002-June 2012)

**Rank out of total number of journals in a given discipline based on Citations per paper (CPP)

*** Impact Factor of the corresponding journal for the year 2012

Table 3.31: Journal Analysis of Top 1% Indian Highly Cited Papers: Total Number of Journals vs Number of Unique Journals Publishing Indian HCPs

Subject	Papers	Total Jls*	Unique Jls**
Agiculture Scs	59	166	29
Biology and Biochemistry	49	243	24
Chemistry	214	305	72
Clinial Medicine	161	1098	57
Computer Science	40	199	17
Economics & Business	4	286	3
Engineering	290	518	72
Environment/Ecology	49	187	20
Geosciences	33	227	20
Immunology	2	60	2
Materials Science	88	189	33
Mathematics	35	23	13
Microbiology	14	79	11
Molecular Biology & Genetics	27	157	11
Multi disciplinary	-	-	-

Neuroscience & Behavior	8	142	7
Pharmacology & Toxicology	43	127	19
Physics	290	185	38
Plant & Animal Science	73	439	42
Psychiatry/Psychology	8	309	5
Social Sciences, general	26	945	20
Space Science	18	40	8

Papers: Disciplibe-wise No, of Indian Highly Cited papers (HCP) (Out of 1531)
*Total number of Journals in Various subject disciplines in ESI database
  No. of Unique journals in a discipline in which Indian HCP's have been published

Table 3.32: Proportion of Papers Across All the Four Quartiles by Journal Impact Based on Average CPP

Field	Quartiles				
	Ist	2nd	3rd	4th	NA
Agricultural Sciences	16	8	5	1	-
Biology and Biochemistry	12	5	2	4	1
Chemistry	32	26	7	2	6
Clinical Medicine	41	9	3	2	1
Computer Science	10	3	2	1	1
Economics & Business	3	-	-	-	-
Engineering	35	17	15	4	1
Environment/Ecology	11	9	-	-	-
Geosciences	13	3	2	2	-
Immunology	-	1	1	-	-
Materials Science	22	8	2	-	-
Mathematics	4	5	3	-	1
Microbiology	5	2	3	2	-
Molecular Biology & Genetics	6	1	-	3	-
Multi disciplinary	-	-	-	-	-
Neuroscience & Behavior	5	1	-	-	1
Pharmacology & Toxicology	6	7	3	2	1
Physics	20	9	7	-	2
Plant & Animal Science	21	12	6	2	1
Psychiatry/Psychology	4	-	-	-	1
Social Sciences, general	10	7	1	-	2
Space Science	6	1	1	-	-

Collaboration

Table 3.33: India's International Collaboration Links
(Decadewise/Subjecwise/Countrywise contribution of Collaboration in Various Subject Disciplines)

Subjects	Decadewise Analysis										Overall Scenario for ~4 decades period		Countries with Maximum Share	
	1971-80		1981-90		1991-2000		2001-2010		2011-2012		Total Unique Countries	Total Share (%)	No. of Countries	Share (%)
	NOCC	Share (%)	NOCC	Share (%)	NOCC	Share (%)	NOCC	Share (%)	NOCC	Share (%)				
Agricultural Sciences	27	1.599	43	3.391	77	8.176	107	19.152	84	22.961	132	11.314	13	64.5
Biology & Biochemistry	34	3.52	50	7.328	79	17.371	117	25.931	98	11.071	145	18.5	26	86
Chemistry	42	3.638	65	6.977	96	12.848	114	21.533	98	24.863	145	16.164	28	89.16
Clinical Medicine	65	5.521	80	8.577	120	18.747	159	28.56	148	40.134	186	24.909	42	88.92
Computer Science	19	10.744	25	13.108	50	18.578	76	29.859	62	35.201	94	25.894	16	83.32
Economics & Business	19	13.552	29	31.375	44	39.917	65	51.49	49	62.824	84	41.197	14	83.44
Engineering	55	6.366	75	8.971	100	15.283	125	27.054	108	29.557	154	20.741	29	86
Environment/Ecology	19	6.13	47	5.462	65	14.129	126	29.492	102	37.874	146	23.431	28	83.2
Geosciences	43	7.558	59	8.896	58	14.637	93	28.23	73	34.752	120	21.573	19	82.07
Immunology	16	2.269	40	6.553	66	26.109	115	32.532	112	50.809	145	29.626	27	81.17
Materials Science	26	6.356	47	7.087	62	13.744	85	22.269	78	24.772	103	19.809	18	83.05
Mathematics	33	17.087	59	17.622	74	26.384	103	32.461	75	35.982	124	30.154	21	83.01
Microbiology	14	4.412	29	7.938	67	27.461	94	34.779	88	34.194	129	36.026	26	83.05
Molecular Biology & Genetics	34	4.087	47	10.083	81	20.853	116	30.11	99	37.658	118	22.091	24	84.4
Multi disciplinary	34	1.985	40	3.209	59	8.827	115	21.453	109	38.567	148	12.784	26	81.67
Neuroscience & Behavior	19	5.791	23	11.192	51	14.068	85	23.795	71	32.024	100	22.34	20	80.64

Pharmacology & Toxicology	22	9.232	40	9.369	64	12.982	107	17.003	77	21.983	124	16.471	23	82.09
Physics	49	5.791	74	14.636	94	42.016	112	67.751	93	118.11	141	52.901	31	85.72
Plant & Animal Science	47	3.39	64	4.384	76	8.623	118	18.187	98	25.052	145	18.5	32	85.74
Psychiatry/Psychology	19	7.095	33	15.404	56	30.232	106	84.871	69	101.364	119	48.528	27	78.11
Social Sciences, general	31	4.355	44	7.38	58	11.768	120	36.491	86	56.167	136	21.015	23	80.69
Space Science	23	11.905	44	17.599	61	57.087	83	152.052	72	244.475	102	103.968	26	87.66

NOCC: Number of Collaborating Countries with India

Share(%): Share (%) of India Total

Total Countries Collaborated with India during last ~four decades

*No. of Countries contributing Above total Average in each discipline

**Proportion of Contribution(based on cut off Avg. in each discipline)

Table 3.34: India's Leading International Research Partners in the last Over Four Decades (Share (%) of India's Total Collaborative Papers)

Agricultural Sciences

1971-80		1981-90		1991-2000		2001-10		2011-12	
USA	0.664	USA	1.038	USA	1.757	USA	4	USA	4.886
England	0.157	England	0.343	England	0.839	England	1.278	Australia	1.595
Fed Rep Ger	0.14	Fed Rep Ger	0.333	Japan	0.498	Germany	1.015	Germany	1.139
Canada	0.087	Canada	0.238	Germany	0.488	Australia	0.946	Canada	1.063
Japan	0.07	Japan	0.2	Philippines	0.488	Canada	0.794	England	0.937
Denmark	0.07	Philippines	0.143	Canada	0.469	Japan	0.774	Japan	0.835
Switzerland	0.035	Australia	0.143	Australia	0.42	Philippines	0.601	Mexico	0.81
Hungary	0.035	Niger	0.095	Netherlands	0.293	South Korea	0.525	Philippines	0.785
Libya	0.017	Germany	0.067	Syria	0.254	Peoples R China	0.525	South Korea	0.684
USSR	0.017	Nigeria	0.067	Niger	0.195	Mexico	0.477	Saudi Arabia	0.658
Other Countries With (0.017 Share (%))West Germany, Yugoslavia, Turkey, Spain, Brazil, Iran, Phillipines, Australia, Nigeria, France, Sweden, Thailand, Sudan, Iraq, Poland And Italy	0.017	Other Countries With (0.048 Share (%)) France, Denmark, Kenya, Switzerland, Sweden; (0.38% With Counries Such As) Netherlands, Scotland, Thailand		Other Countries With (0.185 Share (%)) Scotland, Wales; (.146%) France, (.137) Nigeria, Malawi ; (.127%) China		Other Countries (With Proportionate Contribution) France (.463), Netherlands (.456), Iran (.359), Italy (.325)		Other Countries (With Proportionate Contribution) Peoples R China (0.608), Taiwan (0.557), Italy (0.532)	

Biology & Biochemistry

1971-80		1981-90		1991-2000		2001-10		2011-12	
USA	1.434	USA	2.962	USA	5.959	USA	7.655	USA	2.854

1971-80		1981-90		1991-2000		2001-10		2011-12	
England	0.363	Fed Rep Ger	0.899	Germany	1.837	Germany	2.043	Germany	0.671
Fed Rep Ger	0.277	England	0.708	England	1.366	Japan	1.652	England	0.587
Japan	0.22	France	0.442	Japan	1.238	England	1.497	South Korea	0.462
France	0.21	Japan	0.411	France	0.743	France	1.103	Japan	0.455
Canada	0.182	Canada	0.388	Canada	0.655	South Korea	0.985	France	0.447
Australia	0.076	Scotland	0.175	Sweden	0.383	Canada	0.762	Saudi Arabia	0.424
Sweden	0.076	Italy	0.114	Italy	0.351	Peoples R China	0.625	Australia	0.387
Denmark	0.076	Germany	0.099	Denmark	0.335	Italy	0.557	Canada	0.356
Scotland	0.076	Denmark	0.099	Scotland	0.24	Australia	0.553	Peoples R China	0.322
				North Ireland	0.24	Sweden	0.534		
Chemistry									
USA	1.383	USA	2.478	USA	3.562	USA	4.338	USA	4.38
Fed Rep Ger	0.405	Fed Rep Ger	0.713	Germany	1.564	Germany	2.708	Germany	2.123
England	0.287	England	0.589	Japan	0.928	Japan	1.608	South Korea	1.997
Japan	0.232	Japan	0.546	England	0.876	England	1.233	Japan	1.306
Canada	0.159	Canada	0.526	France	0.788	South Korea	1.173	Malaysia	1.274
France	0.146	France	0.291	Canada	0.588	France	1.083	France	1.169
Italy	0.146	Italy	0.165	Italy	0.55	Malaysia	0.871	England	1.053
Australia	0.141	Australia	0.162	Australia	0.346	Italy	0.732	Saudi Arabia	1.042
Switzerland	0.1	Netherlands	0.115	Malaysia	0.31	Taiwan	0.669	Spain	0.734
Czechoslovakia	0.082	Germany	0.101	Spain	0.244	Canada	0.652	Italy	0.716
Scotland	0.082	Wales	0.096	Switzerland	0.167				
Clinical Medicine									
USA	2.55	USA	2.489	USA	5.827	USA	7.779	USA	8.395

	1971-80		1981-90		1991-2000		2001-10		2011-12
England	0.741	England	1.563	England	2.235	England	2.631	England	3.103
Canada	0.181	Fed Rep Ger	0.63	Canada	1.033	Canada	1.105	Australia	1.852
Fed Rep Ger	0.181	Japan	0.322	Australia	0.916	Australia	1.069	Canada	1.697
Sweden	0.181	France	0.309	Germany	0.747	Germany	1.069	Germany	1.296
Scotland	0.154	Australia	0.309	France	0.606	France	0.99	France	1.218
Denmark	0.136	Canada	0.27	Japan	0.517	Japan	0.904	Switzerland	1.122
Australia	0.127	Italy	0.225	Switzerland	0.474	Switzerland	0.754	Peoples R China	1.118
Switzerland	0.127	Scotland	0.199	Peoples R China	0.38	Peoples R China	0.717	Japan	1.013
France	0.108	Sweden	0.199	Italy	0.362	Netherlands	0.608	Netherlands	0.89

Computer Science

	1971-80		1981-90		1991-2000		2001-10		2011-12
USA	3.643	USA	6.485	USA	8.299	USA	11.01	USA	9.176
Canada	1.639	Canada	1.98	Canada	1.876	Canada	2.057	Canada	2.814
England	1.639	England	1.16	Germany	1.567	Germany	1.688	France	1.958
Japan	0.729	Japan	0.478	Japan	0.771	Singapore	1.516	England	1.876
Fed Rep Ger	0.546	Fed Rep Ger	0.41	England	0.694	England	1.292	Germany	1.713
France	0.364	France	0.341	France	0.617	France	1.279	Peoples R China	1.713
Australia	0.182	Australia	0.273	Singapore	0.617	Peoples R China	1.134	Singapore	1.468
Belgium	0.182	Belgium	0.273	Australia	0.54	Japan	0.989	South Korea	1.346
Iraq	0.182	Iraq	0.205	Italy	0.308	Australia	0.831	Japan	0.979
Bangladesh	0.182	Bangladesh	0.137	Netherlands	0.283	South Korea	0.804	Australia	0.938
Other Countries With (0.182 Share (%)) Iraq, Italy, Japan, Malaysia, Netherlands, Scotland, Taiwan	0.182	Other Countries With (0.137 Share (%)) Germany, Hungary, Italy, Libya, Netherlands, Singapore	0.137						

Economics & Business

1971-80		1981-90		1991-2000		2001-10		2011-12	
USA	7.794	USA	16.375	USA	19.076	USA	18.921	USA	21.417
Canada	1.319	England	3.625	Canada	2.844	England	4.912	England	7.874
England	1.319	Canada	3.5	England	2.725	Canada	2.608	Canada	3.15
Australia	0.6	Australia	1	Australia	1.54	Peoples R China	2.486	Germany	2.992
Japan	0.36	Japan	0.875	Netherlands	1.422	Australia	2.183	Australia	2.835
Israel	0.24	Fed Rep Ger	0.75	Belgium	1.303	Japan	2.001	Singapore	2.362
Nigeria	0.24	Belgium	0.625	Japan	1.066	Netherlands	1.88	France	2.047
Scotland	0.24	Austria	0.5	Spain	0.948	Germany	1.577	Netherlands	2.047
Switzerland	0.24	Italy	0.5	France	0.829	France	1.092	Peoples R China	1.732
Austria	0.12	Switzerland	0.5	Austria	0.711	Singapore	1.092	Italy	1.575
Other Countries With (0.12 Share (%))Belgium, Cyprus, Fed Reb Ger, France, Netherlands, New Zealand, Norway, Singapore, Thaiand	0.12			Germany	0.711			Japan	1.575

Engineering

1971-80		1981-90		1991-2000		2001-10		2011-12	
USA	2.304	USA	3.223	USA	4.782	USA	5.969	USA	5.656
England	1.072	Canada	1.299	Germany	1.604	Germany	2.474	South Korea	2.16
Canada	1.063	England	0.76	Canada	1.236	South Korea	1.448	Germany	1.818
Fed Rep Ger	0.313	Fed Rep Ger	0.683	England	0.978	Japan	1.444	France	1.518
Australia	0.152	Japan	0.293	Japan	0.943	England	1.34	England	1.253

France	0.152	Australia	0.282	France	0.679	France	1.28	Peoples R China	1.15
Scotland	0.107	France	0.282	Italy	0.477	Canada	1.138	Canada	1.119
Italy	0.098	Italy	0.149	Singapore	0.331	Peoples R China	0.849	Japan	1.088
Switzerland	0.089	Netherlands	0.128	Australia	0.289	Australia	0.698	Malaysia	0.937
West Germany	0.089	Germany	0.103	Netherlands	0.247	Italy	0.64	Australia	0.875
		Switzerland	0.103						

Environment/Ecology

1971-80		1981-90		1991-2000		2001-10		2011-12	
USA	2.167	USA	2.005	USA	4.333	USA	5.935	USA	6.857
Canada	0.897	England	0.507	England	1.199	Germany	1.923	England	2.464
England	0.897	Canada	0.485	Germany	1.157	England	1.79	Germany	2.014
Switzerland	0.448	Japan	0.419	Japan	1.052	Japan	1.517	Australia	1.907
Fed Rep Ger	0.374	Fed Rep Ger	0.286	France	0.778	France	1.509	South Korea	1.628
Sweden	0.224	Thailand	0.176	Canada	0.715	South Korea	1.351	Canada	1.586
Wales	0.224	Australia	0.11	Netherlands	0.442	Canada	1.152	France	1.393
Kenya	0.149	Belgium	0.11	Scotland	0.337	Netherlands	1.061	Japan	1.371
Belgium	0.075	France	0.088	Philippines	0.231	Peoples R China	0.92	Malaysia	1.286
Bundes Republik	0.075	Philippines	0.088	Sweden	0.231	Australia	0.904	Peoples R China	1.264
Other Countries With (0.075 Share (%))France, Ireland, Italy, Kuwait, Netherlands, Philippines, Tanzania, Thailand	0.075								

Geosciences

1971-80		1981-90		1991-2000		2001-10		2011-12	
USA	2.923	USA	2.798	USA	3.865	USA	5.974	USA	6.781
England	0.964	Canada	0.945	Germany	1.855	Germany	3.297	Germany	2.793
Canada	0.844	England	0.908	Japan	1.385	Japan	2.677	Japan	2.777
Fed Rep Ger	0.422	Fed Rep Ger	0.87	England	1.188	England	1.771	South Korea	1.986
Japan	0.301	Japan	0.41	Canada	0.838	France	1.674	France	1.937
France	0.211	Australia	0.385	France	0.778	South Korea	1.428	England	1.566
Ussr	0.181	France	0.311	Russia	0.462	Canada	1.142	Canada	1.388
Italy	0.121	Ussr	0.211	Italy	0.436	Australia	0.901	Australia	1.34
Wales	0.121	Netherlands	0.174	Australia	0.393	Peoples R China	0.66	Peoples R China	1.162
Australia	0.09	Brazil	0.162	Sweden	0.333	Brazil	0.63	Italy	0.807
Other Countries With (0.09 Share (%))Norway, Scotland, Sweden	0.09								

Immunology

1971-80		1981-90		1991-2000		2001-10		2011-12	
USA	0.768	USA	1.792	USA	8.185	USA	10.882	USA	12.713
England	0.402	England	0.941	England	3.115	England	3.062	England	4.818
Canada	0.219	Fed Rep Ger	0.516	Japan	2.173	France	1.669	Canada	3.198
Fed Rep Ger	0.146	France	0.456	Germany	1.775	Canada	1.64	Switzerland	2.996
Japan	0.11	Australia	0.425	France	1.014	Japan	1.524	France	2.308
Netherlands	0.11	Switzerland	0.364	Netherlands	1.014	Australia	1.306	Australia	2.186
Australia	0.073	Sweden	0.334	Sweden	0.688	Switzerland	1.146	Belgium	1.822
Italy	0.073	Canada	0.213	Australia	0.58	Germany	1.045	Japan	1.781
Sweden	0.073	Japan	0.182	Scotland	0.58	Peoples R China	0.972	Germany	1.741

Tanzania	Bangladesh	Switzerland	Netherlands		Thailand
0.073	0.037	0.091	0.58	0.885	1.538
	Other Countries (With Proportionate Contribution-0.037% (Each Country))France, New Zealand, Sri Lanka, USSR, West Germany	Other Countries With (0.091 Share (%))Nepal, Netherlands			

Materials Science

1971-80		1981-90		1991-2000		2001-10		2011-12	
USA	2.105	USA	2.238	USA	3.619	USA	4.455	USA	4.626
England	1.053	England	0.935	Germany	1.731	Germany	3.02	South Korea	3.621
Canada	0.902	Fed Rep Ger	0.764	France	1.363	Japan	2.797	Germany	2.104
Fed Rep Ger	0.451	Canada	0.645	Japan	1.308	South Korea	2.559	Japan	2.04
France	0.226	France	0.487	England	1.097	France	1.424	France	1.592
Japan	0.188	Japan	0.395	Canada	0.595	England	1.175	Singapore	0.983
Wales	0.15	Belgium	0.105	Italy	0.587	Taiwan	0.751	England	0.961
Belgium	0.113	Germany	0.105	Singapore	0.423	Italy	0.71	Saudi Arabia	0.897
Denmark	0.113	Italy	0.105	Sweden	0.251	Peoples R China	0.516	Italy	0.769
Norway	0.113	Trinid Tobago	0.105	Australia	0.196	Canada	0.505	Taiwan	0.758
Sweden	0.113	Spain	0.196						

Mathematics

1971-80		1981-90		1991-2000		2001-10		2011-12	
USA	8.095	USA	6.942	USA	8.881	USA	9.069	USA	7.522
Canada	3.246	Canada	2.962	Canada	3.471	Germany	3.528	Germany	2.779
England	1.212	England	1.181	Germany	2.22	Canada	2.337	Peoples R China	2.666

1971-80		1981-90		1991-2000		2001-10		2011-12	
Fed Rep Ger	0.626	Japan	1.447	England	0.89	France	2.311	Canada	2.414
France	0.508	France	1.447	France	0.763	England	1.751	Saudi Arabia	2.189
Australia	0.391	Fed Rep Ger	1.068	Italy	0.672	Peoples R China	1.576	France	1.909
Japan	0.352	Australia	0.984	Japan	0.654	Japan	1.471	South Korea	1.684
Italy	0.313	Italy	0.618	Netherlands	0.509	South Korea	1.366	Turkey	1.375
Netherlands	0.274	Netherlands	0.408	Belgium	0.254	Italy	1.278	Italy	1.319
Kuwait	0.235	Switzerland	0.351	Australia	0.254	Spain	0.858	Malaysia	1.319
Wales	0.235								

Microbiology

1971-80		1981-90		1991-2000		2001-10		2011-12	
USA	1.009	USA	1.944	USA	6.404	USA	9.413	USA	8.471
Japan	0.631	England	1.62	Japan	3.573	Germany	3.2	Germany	2.686
Czechoslovakia	0.504	Fed Rep Ger	0.81	Germany	2.691	Japan	2.968	England	2.428
England	0.504	Japan	0.54	England	2.181	England	2.295	Japan	1.911
Belgium	0.252	Norway	0.324	France	1.067	South Korea	2.133	Switzerland	1.498
Italy	0.252	France	0.27	Canada	0.928	France	1.901	Australia	1.291
Netherlands	0.252	Germany	0.27	Netherlands	0.835	Bangladesh	1.438	France	1.24
Portugal	0.252	Sweden	0.27	Sweden	0.742	Australia	1.02	South Korea	1.24
Australia	0.126	Scotland	0.216	Bangladesh	0.65	Peoples R China	0.974	Taiwan	1.085
Brazil	0.126	Australia	0.162	Scotland	0.51	Switzerland	0.881	Canada	0.981
Other Countries With (0.126 Share (%))Canada, Iran, West Germany	0.126	Other Countries With (0.162 Share (%))Belgium, Switzerland	0.162			Other Countries (With Proportionate Contribution-0.835% (Each Country)) Netherlands, Taiwan	0.835	Other Countries With (0.98 Share (%))Peoples R China, Saudi Arabia	0.981

Molecular Biology & Genetics

1971-80		1981-90		1991-2000		2001-10		2011-12	
USA	1.473	USA	3.668	USA	8.015	USA	9.596	USA	10.237
England	0.605	Fed Rep Ger	1.235	Germany	2.184	Germany	2.601	Germany	2.39
Fed Rep Ger	0.434	England	0.803	Japan	1.473	Japan	2.047	England	1.83
Japan	0.224	Canada	0.659	England	1.327	England	1.575	Japan	1.55
Canada	0.171	Japan	0.611	France	0.951	France	1.409	France	1.401
France	0.171	France	0.575	Canada	0.606	Canada	0.793	Australia	1.203
Australia	0.158	Scotland	0.276	Denmark	0.387	Italy	0.738	Saudi Arabia	1.137
West Germany	0.131	Italy	0.24	Italy	0.366	Australia	0.732	Canada	1.072
Scotland	0.092	Netherlands	0.192	Switzerland	0.345	South Korea	0.716	Peoples R China	1.022
Denmark	0.079	Germany	0.144	Sweden	0.334	Peoples R China	0.599	South Korea	0.989
		Sweden	0.144						

Multi Disciplinary

1971-80		1981-90		1991-2000		2001-10		2011-12	
USA	0.63	USA	1.155	USA	2.639	USA	4.963	USA	8.785
England	0.264	England	0.247	Germany	0.929	England	1.492	England	3.086
Canada	0.183	Canada	0.195	England	0.737	Germany	1.401	Germany	2.029
France	0.115	Japan	0.182	Japan	0.619	Japan	1.058	France	1.89
Japan	0.115	Fed Rep Ger	0.169	France	0.457	France	1.049	Peoples R China	1.362
Australia	0.092	France	0.156	Australia	0.31	Canada	0.778	Australia	1.279
Fed Rep Ger	0.092	Australia	0.117	Canada	0.295	Australia	0.615	Canada	1.279
USSR	0.069	Italy	0.117	Italy	0.221	Italy	0.533	Japan	1.195
Bulgaria	0.034	USSR	0.091	Belgium	0.147	Peoples R China	0.533	Italy	1.112
Italy	0.034	Sweden	0.065	Israel	0.147	Sweden	0.515	Netherlands	0.973

Neuroscience & Behavior

1971-80		1981-90		1991-2000		2001-10		2011-12	
USA	1.033	USA	4.122	USA	6.077	USA	8.782	USA	7.552
England	0.62	Fed Rep Ger	1.413	England	1.169	England	2.239	England	2.127
Scotland	0.62	Sweden	0.824	Canada	1.13	Germany	1.25	Germany	1.606
Switzerland	0.413	England	0.707	Japan	0.974	Canada	1.207	Australia	1.519
Canada	0.207	Scotland	0.589	Germany	0.896	Australia	0.771	Canada	1.302
Czechoslovakia	0.207	France	0.471	Italy	0.312	Japan	0.741	Italy	1.215
Denmark	0.207	Hungary	0.353	Sweden	0.312	Italy	0.683	Saudi Arabia	1.128
Greece	0.207	Japan	0.353	Australia	0.273	Sweden	0.407	Japan	0.825
Indonesia	0.207	Canada	0.236	France	0.273	Switzerland	0.407	Peoples R China	0.825
Japan	0.207	Libya	0.236	Austria	0.195	Peoples R China	0.378	Spain	0.825
Other Countries With (0.207 Share (%)) Libya, Netherlands, Nigeria, North Ireland, Norway, S Korea, Taiwan, Thailand	0.207	Other Countries With (0.236Share (%)) Nigeria, S Korea, Switzerland	0.236						

Pharmacy

1971-80		1981-90		1991-2000		2001-10		2011-12	
USA	6.072	USA	3.473	USA	4.616	USA	5.449	USA	6.08
Fed Rep Ger	0.486	Fed Rep Ger	1.242	Germany	1.705	Germany	0.961	Saudi Arabia	1.553
Other Countries With (0.034 Share (%)) Scotland, Sweden, Switzerland	0.034	Switzerland	0.065						

1971-80	1981-90	1991-2000	2001-10	2011-12
Scotland 0.486	Japan 0.875	England 1.081	England 0.95	Malaysia 1.393
England 0.364	England 0.593	Japan 0.749	Japan 0.782	South Korea 0.842
Sweden 0.304	Canada 0.339	France 0.395	France 0.553	England 0.827
France 0.243	Denmark 0.254	Italy 0.354	Italy 0.548	Germany 0.784
Finland 0.182	France 0.254	Canada 0.312	Canada 0.537	Japan 0.624
Canada 0.121	Belgium 0.226	South Korea 0.291	South Korea 0.458	Australia 0.609
Czechoslovakia 0.121	Spain 0.198	Belgium 0.25	Switzerland 0.397	Switzerland 0.595
Japan 0.121	Italy 0.169	Peoples R China 0.229	Belgium 0.38	Belgium 0.566
Germany	France 0.169			

Physics

1971-80	1981-90	1991-2000	2001-10	2011-12
USA 0.188	USA 3.832	USA 7.984	USA 9.323	USA 9.777
England 0.064	Fed Rep Ger 1.404	Germany 4.563	Germany 6.8	Germany 7.267
Fed Rep Ger 0.059	Italy 1.145	France 3.011	Japan 4.546	France 5.945
France 0.047	France 0.992	Italy 2.432	France 3.914	South Korea 5.38
England 0.046	England 0.982	England 1.947	South Korea 3.585	England 4.22
Canada 0.042	Canada 0.871	Japan 1.947	Russia 2.878	Russia 4.03
Japan 0.026	Japan 0.649	Russia 1.548	Peoples R China 2.774	Italy 3.927
Switzerland 0.019	Switzerland 0.554	Canada 1.419	England 2.763	Peoples R China 3.736
Sweden 0.016	Ussr 0.396	Switzerland 1.326	Italy 2.327	Spain 3.494
Australia 0.014	Sweden 0.38	Peoples R China 1.157	Taiwan 1.996	Japan 3.164

Plant & Animal Science

1971-80	1981-90	1991-2000	2001-10	2011-12
USA 1.153	USA 1.267	USA 2.189	USA 3.808	USA 4.83
Fed Rep Ger 0.312	Fed Rep Ger 0.473	Germany 1.007	Germany 1.493	South Korea 1.43
England 0.302	England 0.4	England 0.758	Japan 1.125	Australia 1.282

Canada	0.248	Canada	0.331	Japan	0.752	England	1.102	Japan	1.263
Japan	0.215	Japan	0.326	Canada	0.467	South Korea	0.923	Germany	1.245
France	0.151	Scotland	0.158	France	0.315	France	0.694	England	1.115
Scotland	0.097	France	0.137	Australia	0.255	Australia	0.634	Peoples R China	1.096
Australia	0.065	Sweden	0.105	Belgium	0.224	Canada	0.634	France	0.817
Denmark	0.054	Nigeria	0.089	Denmark	0.188	Peoples R China	0.593	Canada	0.78
Finland	0.054	Denmark	0.084	Philippines	0.176	Scotland	0.413	Mexico	0.669
Other Countries With (0.054 Share (%))Philippines, Sweden	0.054								

Psychiatry/Psychology

1971-80		1981-90		1991-2000		2001-10		2011-12	
USA	4.219	USA	6.183	USA	9.762	USA	15.945	USA	18.569
Canada	0.539	England	1.57	England	4.286	England	9.879	England	8.067
Australia	0.359	Canada	1.276	Canada	2.381	Australia	5.026	Peoples R China	4.718
England	0.359	Australia	0.785	Australia	1.349	Canada	3.235	Australia	3.805
Switzerland	0.269	Netherlands	0.491	Switzerland	1.27	Germany	2.657	Canada	3.196
Nigeria	0.18	Switzerland	0.491	Germany	1.032	Peoples R China	2.253	Japan	3.196
Austria	0.09	Egypt	0.393	Japan	0.794	Switzerland	2.253	Germany	3.044
Colombia	0.09	Hong Kong	0.393	Malaysia	0.635	Brazil	2.08	Mexico	2.892
Cyprus	0.09	Fed Rep Ger	0.393	Peoples R China	0.556	Japan	1.906	Netherlands	2.74
Denmark	0.09	Malaysia	0.294	Netherlands	0.476	Netherlands	1.675	Switzerland	2.74

	1971-80		1981-90		1991-2000		2001-10		2011-12	
Other Countries With (0.09 Share (%))Iran, Israel, Nepal, Netherlands, Norway, Philippines, Sudan, Turkey, West Germany	0.09	Other Countries With (0.294 Share (%))Philippines, Sudan, USSR	0.294	Zimbabwe	0.476					
			0.294							

Social Science

1971-80		1981-90		1991-2000		2001-10		2011-12	
USA	2.391	USA		USA	5.059	USA	11.501	USA	15.514
England	0.43	England	3.231	England	1.449	England	4.533	England	7.279
Canada	0.287	Canada	1.084	Canada	0.499	Australia	1.444	Canada	3.549
Switzerland	0.143	Australia	0.482	Australia	0.451	Canada	1.444	Australia	2.548
Scotland	0.096	Fed Rep Ger	0.361	Netherlands	0.451	Switzerland	1.007	Netherlands	1.729
Australia	0.072	Switzerland	0.241	Switzerland	0.261	Netherlands	0.94	Germany	1.228
France	0.072	Japan	0.201	Germany	0.238	France	0.705	Switzerland	1.228
Iran	0.072	France	0.12	Japan	0.214	Japan	0.705	Peoples R China	1.137
Nigeria	0.072	Netherlands	0.1	Thailand	0.214	South Africa	0.672	Sweden	1.137
North Ireland	0.072	Austria	0.1	Belgium	0.166	Belgium	0.621	Japan	0.955
Norway	0.072	Other Countries With (0.08 Share (%))Belgium, Denmark, Germany, Italy, Thailand	0.08	France	0.166				
			0.08							

Space Science

1971-80		1981-90		1991-2000		2001-10		2011-12	
USA	5.155	USA	5.967	USA	15.382	USA	24.495	USA	28.211
Fed Rep Ger	1.218	England	1.9	Germany	4.704	Germany	10.517	Germany	18.127
Italy	0.937	Fed Rep Ger	1.2	France	4.098	France	9.195	France	13.745
Sweden	0.469	Italy	1.067	England	3.838	Japan	8.612	England	12.485
Wales	0.469	Canada	0.867	Italy	3.29	England	7.834	Russia	12.005
Australia	0.375	Netherlands	0.867	Japan	1.76	Russia	7.484	Spain	11.104
Canada	0.375	Ussr	0.633	Australia	1.732	Australia	6.026	Italy	9.964
England	0.375	Australia	0.467	Russia	1.674	Italy	6.026	South Korea	8.944
Japan	0.375	Japan	0.433	Brazil	1.645	Peoples R China	5.832	Netherlands	8.824
Netherlands	0.375	France	0.4	Canada	1.558	South Korea	5.054	Canada	8.643

**Table 3.35: Overall Collaboration Pattern (1971-2012):
Top 20 Countries Collaborating with India**

Collaborative Countries	Share(%) of India Total
USA	2.48
England	0.786
Australia	0.583
Germany	0.557
Canada	0.539
Japan	0.495
Philippines	0.409
Netherlands	0.267
Peoples R China	0.256
Mexico	0.254
France	0.239
South Korea	0.239
Italy	0.194
Iran	0.148
Kenya	0.139
Spain	0.139
Taiwan	0.139
Scotland	0.137
Brazil	0.133
Switzerland	0.13

**Table 3.36: International Organisations Collaborating with India
(Decade-wise Analysis)**

Organisation	Number of Co-Authored Papers				
	1971-80	1981-90	1991-2000	2001-10	Total
University of California System	107	232	853	2142	3334
Max Planck Society	37	105	379	1290	1811
Chinese Academy of Sciences	-	-	283	949	1232
State University of New York Suny System	37	57	262	732	1088
University of Michigan	22	64	360	632	1078
Purdue University	28	59	340	640	1067
University of Maryland College Park	24	60	286	684	1054
Johns Hopkins University	41	269	725	1035	

University of Illinois System	64	114	-	807	985
University of California Berkeley	-	55	207	681	943
Cnrs	-	69	238	634	941
Massachusetts Institute of Technology Mit	42	54	284	561	941
Princeton University	-	-	254	687	941
University of Science Technology China	-	-	176	704	880
University of Tokyo	-	-	-	872	872
Harvard University	-	58	218	501	777
Korea University	-	-	-	748	748
University of Oxford	40	-	242	457	739
Pres University of Lyon	-	-	234	501	735
California Institute of Technology	-	-	264	463	727
University of Cambridge	36	78	167	440	721
Tohoku University	-	-	-	713	713
Indiana University	-	-	170	536	706
University of Claude Bernard Lyon 1	-	-	232	473	705
Hungarian Academy of Sciences	-	-	257	447	704
University of Sains Malaysia	-	-	-	699	699
Michigan State University	-	-	166	530	696
Imperial College London	39	-	-	655	694
University of Manchester	32	108	-	554	694
Istituto Nazionale Di Fisica Nucleare	-	-	267	423	690
University of Illinois Chicago	-	-	176	509	685
Columbia University	-	-	-	681	681
University of Notre Dame	-	65	191	423	679
Helmholtz Association	-	-	203	472	675
National Taiwan University	-	-	-	669	669
Sungkyunkwan University	-	-	-	661	661
Rwth Aachen University	-	-	228	431	659
Seoul National University	-	-	-	659	659
Boston University	-	-	237	419	656
Inst Theoret Expt Phys	-	-	-	643	643
Yonsei University	-	-	-	641	641
Texas A M University System	-	-	198	440	638
University of Rochester	-	-	169	468	637
University of Amsterdam	-	-	174	456	630

Tokyo Institute of Technology	-	-	-	624	624
Pierre Marie Curie University Paris 6	-	54	-	566	620
University of Arizona	-	-	189	426	615
University of California Riverside	-	-	-	608	608
University of Zurich	-	-	190	417	607
King Edward Mem Hosp	-	88	-	517	605
Iowa State University	26	-	-	577	603
Consiglio Nazionale Delle Ricerche Cnr	-	-	180	421	601
Int Ctr Genet Engn Biotechnol	-	-	-	600	600
University of Washington	-	-	-	596	596
University of Washington Seattle	-	-	-	596	596
Radboud University Nijmegen	-	-	171	418	589
Kyungpook National University	-	-	-	586	586
Brookhaven National Laboratory	-	-	-	585	585
National University of Singapore	-	-	-	585	585
University of California Davis	-	-	176	409	585
University of Melbourne	-	-	-	585	585
Czech Academy of Sciences	-	-	-	558	558
University College London	30	72	-	447	549
Nagoya University	-	-	-	544	544
University of Paris Sud Paris Xi	-	-	-	543	543
University of Sydney	-	-	-	530	530
University of Bonn	21	65	-	418	504
Osaka University	-	-	-	496	496
Penn State University	-	-	-	496	496
Mcgill University	-	-	-	490	490
Ohio State University	-	-	-	484	484
Lomonosov Moscow State University	-	-	-	482	482
Florida State University	-	-	-	480	480
Johannes Gutenberg University of Mainz	-	-	-	463	463
University of Cincinnati	-	-	-	461	461
Lund University	-	-	-	460	460
Howard University	-	-	-	458	458
Lawrence Berkeley National Laboratory	-	-	-	452	452
Kyoto University	-	-	-	451	451
Goethe University Frankfurt	-	-	-	449	449

University of Hawaii System	-	-	-	449	449
Virginia Polytechnic Institute	-	-	-	432	432
Rice University	-	-	-	428	428
Brown University	-	-	-	427	427
University of Toronto	28	-	-	395	423
University of Virginia	-	-	-	415	415
Osaka City University	-	-	-	414	414
University of California Los Angeles	24	-	-	390	414
University of Paris Diderot Paris Vii	-	-	-	413	413
University of Southampton	-	-	-	413	413
University of Munich	-	-	-	407	407
Swiss Federal Institute of Technology Lausanne	-	-	-	405	405
University of Sao Paulo	-	-	-	404	404
University of Nebraska System	-	-	-	398	398
University of Tsukuba	-	-	-	392	392
Charles University Prague	-	-	-	389	389
University of California San Diego	-	56	262	-	318
National Aeronautics And Space Administration Nasa	36	77	197	-	310
United States Department of Defense	-	75	173	-	248
European Organization for Nuclear Research Cern	-	-	237	-	237
University of Geneva	-	-	233	-	233
Sapienza University Rome	-	-	232	-	232
University of Alabama System	-	-	225	-	225
University of London	-	225	-	-	225
Carnegie Mellon University	-	-	220	-	220
University of Hamburg	-	-	218	-	218
University of Florence	-	-	207	-	207
Bulgarian Academy of Sciences	-	-	205	-	205
Los Alamos National Laboratory	-	-	205	-	205
Desy	-	-	201	-	201
University of Perugia	-	-	194	-	194
University of Lausanne	-	-	191	-	191
Texas A M University College Station	-	-	182	-	182
Spic Sci Fdn	-	-	173	-	173
Univ Cyprus	-	-	166	-	166

Univ Santiago	-	-	164	-	164
University of Alberta	30	87	-	-	117
University of Wisconsin System	40	68	-	-	108
University of Calgary	31	60	-	-	91
University of Wisconsin Madison	34	56	-	-	90
University of Waterloo	34	55	-	-	89
Fdn Med Res	-	87	-	-	87
Univ Coll Sci Technol	-	87	-	-	87
Haffkine Inst	-	86	-	-	86
University of Minnesota Twin Cities	21	64	-	-	85
University of Texas Austin	30	54	-	-	84
Natl Inst Commun Dis	-	77	-	-	77
Bareilly Coll	-	76	-	-	76
Rutgers State University	-	74	-	-	74
University of Illinois Urbana Champaign	-	74	-	-	74
University of Pittsburgh	-	74	-	-	74
University of North Dakota Grand Forks	72	-	-	-	72
Technical University of Berlin	-	66	-	-	66
University of Edinburgh	-	66	-	-	66
Hiroshima University	-	63	-	-	63
Alchem Res Ctr	-	58	-	-	58
RBS Coll	-	58	-	-	58
University of Florida	-	58	-	-	58
United States Department of Agriculture Usda	-	56	-	-	56
Ruhr University Bochum	-	54	-	-	54
University of Fukui	-	53	-	-	53
East Carolina University	37	-	-	-	37
University of Sheffield	33	-	-	-	33
University of Victoria	31	-	-	-	31
University of Chicago	28	-	-	-	28
Birkbeck University London	27	-	-	-	27
University of Glasgow	27	-	-	-	27
University of North Carolina Chapel Hill	27	-	-	-	27
Memorial University Newfoundland	26	-	-	-	26
Swansea University	25	-	-	-	25
Cornell University	23	-	-	-	23
University of California San Francisco	23	-	-	-	23

University of Western Ontario	23	-	-	-	23
Ravenshaw Coll	22	-	-	-	22
	1358	3501	12775	53301	70935

Table 3.37: International Organisations Collaborating Frequently with India (1971-2012)

Organisation	Number of Co-Authored Papers	Share (%) of Total International Collaboration
University of California System	3334	4.70
Max Planck Society	1811	2.55
Chinese Academy of Sciences	1232	1.74
State University of New York Suny System	1088	1.53
University of Michigan	1078	1.52
Purdue University	1078	1.52
University of Maryland College Park	1067	1.50
Johns Hopkins University	1054	1.49
University of Illinois System	1035	1.46
University of California Berkeley	985	1.39
Cnrs	943	1.33
Massachusetts Institute of Technology Mit	941	1.33
Princeton University	941	1.33
University of Science Technology China	941	1.33
University of Tokyo	880	1.24
Harvard University	872	1.23
Korea University	777	1.10
University of Oxford	748	1.05
Pres University of Lyon	739	1.04
California Institute of Technology	735	1.04

Table 3.40: Top Twenty Collaborating Indian Institutions and the Proportion of their National/International Collaboration

Organization	Total Publications	Colloborative Publications			Propostion (%) of Colloboration		Proportion of Papers Contributed (1971-2012)
		Total	International	National	National	International	
Indian Institute of Science Iisc Banglore	31298	28858	7609	21249	73.63	26.37	11.15
Bhabha Atomic Research Center	22556	69804	15716	54088	77.49	22.51	8.04
Banaras Hindu University	19283	17430	5129	12301	70.57	29.43	6.87
University of Delhi	18197	13981	3822	10159	72.66	27.34	6.49
Indian Institute of Technology Iit Kharagpur	17608	10901	3542	7359	67.51	32.49	6.27
All India Institute of Medical Sciences	16263	12781	4737	8044	62.94	37.06	5.79
Indian Institute of Technology Iit Delhi	15613	12668	3213	9455	74.64	25.36	5.56
Indian Institute of Technology Iit Madras	15551	11514	3285	8229	71.47	28.53	5.54
Indian Institute of Technology Iit Kanpur	14553	9287	3485	5802	62.47	37.53	5.19
Tata Institute of Fundamental Research	14067	114245	28676	85569	74.90	25.10	5.02
Indian Institute of Technology Iit Bombay	13766	21311	6773	14538	68.22	31.78	4.91
Jadavpur University	11609	10095	2395	7700	76.28	23.72	4.13
Pgimer Chandigarh	10716	5257	1422	3835	72.95	27.05	3.82
National Chemistry Laboratory Pune	9891	10183	1335	8848	86.89	13.11	3.53
University of Calcutta	9400	7101	1746	5355	75.41	24.59	3.35
Aligarh Muslim University	8899	9965	3008	6957	69.81	30.19	3.17
Indian Association for the Cultivation of Science	8566	6082	2075	4007	65.88	34.12	3.05
University of Madras	7815	5910	1712	4198	71.03	28.97	2.79
Indian Institute of Chemical Technology	7765	3599	897	2702	75.08	24.92	2.77
Panjab University	8703	123334	28309	95025	77.05	22.95	2.56

Table 3.41: Collaboration Pattern of India's High Impact Co-authored Papers

Subject	Total HCP per Subject	Proportion of Indian Contribution					Ratio 20%	Ratio 50%	Ratio 90%	90-99%	Ratio of Papers with 100% Indian Contribution
		<=20	<=50	<90	>=90-99	100					
Agricultural Sciences	59	9	11	5	0	34	15.25	18.64	8.47	0.00	57.63
Biology & Biochemistry	49	9	10	6	2	22	18.37	20.41	12.24	4.08	44.90
Chemistry	214	14	23	20	157	6.54	10.75	9.35	73.36	0.00	
Clinical Medicine	161	116	16	16	0	13	72.05	9.94	9.94	0.00	8.07
Computer Science	40	1	12	4	0	23	2.50	30.00	10.00	0.00	57.50
Economics & Business	4	0	2	0	0	2	0.00	50.00	0.00	0.00	50.00
Engineering	290	20	42	21	0	207	6.90	14.48	7.24	0.00	71.38
Environment/Ecology	49	10	7	7	0	25	20.41	14.29	14.29	0.00	51.02
Geosciences	33	12	10	3	0	8	36.36	30.30	9.09	0.00	24.24
Immunology	2	0	1	0	0	1	0.00	50.00	0.00	0.00	50.00
Materials Science	88	11	15	7	0	55	12.50	17.05	7.95	0.00	62.50
Mathematics	35	1	15	4	0	15	2.86	42.86	11.43	0.00	42.86
Microbiology	14	3	5	1	0	5	21.43	35.71	7.14	0.00	35.71
Molecular Biology & Genetics	27	7	1	3	0	16	25.93	3.70	11.11	0.00	59.26
Multi disciplinary	0	0	0	0	0	0	0.00	0.00	0.00	0.00	0.00
Neuroscience & Behavior	8	2	3	0	0	3	25.00	37.50	0.00	0.00	37.50
Pharmacology & Toxicology	43	1	11	1	0	30	2.33	25.58	2.33	0.00	69.77
Physics	290	141	47	17	0	82	48.62	16.21	5.86	0.00	28.28
Plant & Animal Science	73	17	15	6	1	33	23.29	20.55	8.22	1.37	45.21
Psychiatry/Psychology	8	7	0	0	0	1	87.50	0.00	0.00	0.00	12.50
Social Sciences, general	26	7	8	2	0	9	26.92	30.77	7.69	0.00	34.62
Space Science	18	13	5	0	0	0	72.22	27.78	0.00	0.00	0.00

APPENDIX - I: Discipline-wise Highly Cited Papers (A-V)

(a) Agricultural Sciences

Title	Indian Authors	No. of Indian Authors	Total Author	Ratio I/F*	Institue Name	Source
Studies on the Antioxidant Activity of Pomegranate (Punica Granatum) Peel and Seed Extracts Using in Vitro Models	Singh RP, Murthy KNC, Jayaprakasha GK	3	3	100.00	Cent Food Technol Res Inst;, Govt Coll Pharm, Bangalore	J Agr Food Chem 50 (1): 81-86 Jan 2 2002
Morphological, Thermal and Rheological Properties of Starches From Different Botanical Sources	Narpinder Singh, Jaspreet Singh, Lovedeep Kaur, Navdeep Singh Sodhi, Balmeet Singh Gill	5	5	100.00	Department of Food Science and Technology, Guru Nanak Dev University, Amritsar-143005, India	Food Chem - Volume 81, Issue 2, May 2003, Pages 219–231
Anti-Oxidant Activity and Total Phenolic Content of Some Asian Vegetables	Charanjit Kaur, Harish C Kapoor	2	2	100.00	International Journal of Food Science and Technology	Int J Food Sci Technol, Vol. 37, No. 2, Pp. 153-161, 2002
Biological Properties of Curcumin-Cellular and Molecular Mechanisms of Action	Lokesh BR	1	3	33.33	Department of Physiology and Molecular Medicine, Medical College of Ohio, Block Health Science Building, 3035 Arlington Avenue, Toledo, Oh 43614-5804, USA	Crit Rev Food Sci Nutr. 2004; 44(2): 97-111.
Biodegradable Films and Composite Coatings: Past, Present and Future	Tharanathan RN	1	1	100.00	Central Food Technological Research Institute, Mysore	Trends in Food Science & Technology
Chitin - the Undisputed Biomolecule of Great Potential	Tharanathan RN, Kittur FS	2	2	100.00	Department of Biochemistry and Nutrition, Central Food Technological Research Institute, Mysore-570 013, India.	Crit Rev Food Sci Nutr. 2003; 43(1): 61-87.

Title	Authors				Institution	Source
Analysis of Genetic Diversity in Crop Plants - Salient Statistical Tools and Considerations	Mohammadi SA, Prasanna BM	1	2	50.00	Indian Agr Res Inst.	Crop Sci 43 (4): 1235-1248 Jul-Aug 2003
Resistant Starch - A Review	Sajilata MG, Singhal RS, Kulkarni PR	3	3	100.00	Inst Chem Technol, Maharashtra	Compr Rev Food Sci Food Saf 5 (1): 1-17 Jan 2006
Antibacterial and Antioxidant Activities of Grape (Vitis Vinifera) Seed Extracts	Jayaprakasha GK, Tamil Selvi, Sakariah KK	3	3	100.00	Cent Food Technol Res Inst,	Food Res Int 36 (2): 117-122 2003
Opportunities and Challenges in High Pressure Processing of Foods	Rastogi NK, Raghavarao KSMS	2	5	40.00	Cent Food Tech Res Inst.	Crit Rev Food Sci Nutr 47 (1): 69-112 2007
Early Life Origins of Insulin Resistance and Type 2 Diabetes in India and Other Asian Countries	Yajnik CS	1	1	100.00	King Edward Mem Hosp & Res Ctr, Maharashtra	J Nutr 134 (1): 205-210 Jan 2004
Chitin/Chitosan: Modifications and Their Unlimited Application Potential - An Overview	Harish Prashanth KV, Tharanathan RN	2	2	100.00	Central Food Technological Research Institute, Mysore	Trends in Food Science & Technology Volume 18, Issue 3, March 2007, Pages 117–131
Exploitation of Natural Products As An Alternative Strategy to Control Postharvest Fungal Rotting of Fruit and Vegetables	Tripathi Pramila, Dubey NK	2	2	100.00	Bhu	Postharvest Biol Technol 32 (3): 235-245 Jun 2004
Studies on the Antioxidant Activity of Indian Laburnum (Cassia Fistula L.): A Preliminary Assessment of Crude Extracts From Stem Bark, Leaves, Flowers and Fruit Pulp	Mohan PS	1	3	33.33	Bharathiar Uni, Tamil Nadu	Food Chem 79 (1): 61-67 Oct 2002

Title	Authors				Affiliation	Citation
Free Radical Scavenging Activity of An Aqueous Extract of Potato Peel	Nandita Singh, PS Rajini	2	2	100.00	Cent Food Technol Res Inst, Food Protectants & Infestat Control Dept, Mysore 570013, Karnataka, India	Food Chem 85 (4): 611-616 May 2004
Antibiotic Resistance in Food Lactic Acid Bacteria - A Review	Mathur S, Singh R	2	2	100.00	Natl Dairy Res Inst, Natl Collect Dairy Cultures, Dairy Microbiol Div, Karnal 243122, Punjab, India.	Int J Food Microbiol 105 (3): 281-295 Dec 15 2005
Perspectives For Chitosan Based Antimicrobial Films in Food Applications	Dutta PK, Shipra Tripathia, Mehrotraa GK, Joydeep Duttab	4	4	100.00	Department of Chemistry, Motilal Nehru National Institute of Technology, Allahabad 211004, India; Regenerative Medicine, Reliance Life Sciences Pvt. Ltd., R-282, Ttc Area of Midc, Thane Belapur Road, Rabale, Navi Mumbai-400701, India Reliance Life Sci Pvt Ltd, Navi Mumbai	Food Chem 114 (4): 1173-1182 Jun 15 2009
In Vitro Antioxidant Activities of Methanol Extracts of Five Phyllanthus Species From India	Kumaran A, Karunakaran RJ	2	2	100.00	Madras Christian Coll, Dept Chem, Madras 600059, Tamil Nadu, India.	Lwt-Food Sci Technol 40 (2): 344-352 2007
An Evaluation of Multipurpose Oil Seed Crop For Industrial Uses (Jatropha Curcas L.): A Review	Kumar A, Sharma S	2	2	100.00	Indian Inst Technol, Ctr Rural Dev & Technol, Delhi 110016, India.	Ind Crops Products 28 (1): 1-10 Jul 2008
The Antioxidant and Free Radical Scavenging Activities of Processed Cowpea (Vigna Unguiculata (L.) Walp.) Seed Extracts	Siddhuraju P	1	2	50.00	Karpagam Arts & Sci Coll, Dept Biotechnol, Coimbatore 641021, Tamil Nadu, India.	Food Chem 101 (1): 10-19 2007

Title	Authors				Institution	Journal Reference
Agroforestry As A Strategy For Carbon Sequestration	Mohan Kumar B	1	3	33.33	Kerala Agricultural University, Thrissur 680656, Kerala	Journal of Plant Nutrition and Soil Science Volume 172, Issue 1, Pages 10–23, February, 2009
Efficiency of Fertilizer Nitrogen in Cereal Production: Retrospects and Prospects	Himanshu Pathak	1	5	20.00	Indian Agr Res Inst, Unit Stimulat & Informat, New Delhi 110012, India	Advan Agron 87: 85-156 2005
Alpha-Amylases From Microbial Sources - An Overview on Recent Developments	Swetha Sivaramakrishnan, Dhanya Gangadharan, Kesavan Madhavan Nampoothiri, Ashok Pandey	4	5	80.00	Regional Research Laboratory, Csir, Trivandrum	Food Technol. Biotechnol. 44 (2) 173–184
Antioxidant Capacity and Other Bioactivities of the Freeze-Dried Amazonian Palm Berry, Euterpe Oleraceae Mart. (Acai)	Agarwal A	1	10	10.00	Nat Remedies, Bangalore 560100, Karnataka, India.	J Agr Food Chem 54 (22): 8604-8610 Nov 1 2006
Fruit Ripening Phenomena - An Overview	Prasannaa V, Prabhaa TN, Tharanathana RN	3	3	100.00	Central Food Technological Research Institute, Mysore, 570020, Karnataka	Critical Reviews in Food Science and Nutrition Volume 47, Issue 1, 2007
Cancer Preventive Properties of Ginger: A Brief Review	Yogeshwer Shukla, Madhulika Singh	2	2	100.00	Industrial Toxicology Research Centre, P.O. Box 80, MG Marg, Lucknow	Food and Chemical Toxicology Volume 45, Issue 5, May 2007, Pages 683–690
Weed Management in Direct-Seeded Rice	Rao AN, Sivaprasad B, Ladha JK	3	5	60.00	International Rice Research Institute (Irri), Irri-India Office, National Agriculture Science Center (Nasc) Complex, New Delhi	Advances in Agronomy Volume 93, 2007, Pages 153–255

Title	Authors				Affiliation	Citation
Black Pepper and Its Pungent Principle-Piperine: A Review of Diverse Physiological Effects	Srinivasan K	1	1	100.00	Cent Food Technol Res Inst, Dept Biochem & Nutr, Mysore 570020, Karnataka, India.	Crit Rev Food Sci Nutr 47 (8): 735-748 2007
A Comparison of Chemical, Antioxidant and Antimicrobial Studies of Cinnamon Leaf and Bark Volatile Oils, Oleoresins and Their Constituents	Singh G, Maurya S	2	4	50.00	Ddu Gorakhpur Univ, Chem Dept, Gorakhpur 273009, Uttar Pradesh, India.	Food Chem Toxicol 45 (9): 1650-1661 Sep 2007
Dpph Antioxidant Assay Revisited	Sharma OP, Bhat TK	2	2	100.00	Indian Vet Res Inst, Biochem Lab, Reg Stn, Palampur 176061, Himachal Prades, India.	Food Chem 113 (4): 1202-1205 Apr 15 2009
Antioxidant Properties of Various Solvent Extracts of Mulberry (Morus Indica L.) Leaves	Arabshahi-Delouee S, Urooj A	2	2	100.00	Univ Mysore, Dept Food Sci & Nutr, Mysore 570006, Karnataka, India.	Food Chem 102 (4): 1233-1240 2007
The Antibacterial Properties of A Novel Chitosan-Ag-Nanoparticle Composite	Sanpui P, Murugadoss A, Prasad PVD, Ghosh SS, Chattopadhyay A	5	5	100.00	Indian Inst Technol, Dept Biotechnol, Gauhati 781039, India.	Int J Food Microbiol 124 (2): 142-146 May 31 2008
Indian Medicinal Herbs As Sources of Antioxidants	Ali SS, Kasoju N, Luthra A, Singh A, Sharanabasava H, Sahu A, Bora U	7	7	100.00	Indian Inst Technol, Dept Biotechnol, Gauhati 781039, Assam, India.	Food Res Int 41 (1): 1-15 2008
Phenotypic and Genotypic Analysis of Drought-Resistance Traits For Development of Rice Cultivars Adapted to Rainfed Environments	Chandra Babu R, Manikanda Boopathi N	2	4	50.00	Center For Plant Molecular Biology, Tamil Nadu Agricultural University, Coimbatore	Field Crops Research Volume 109, Issues 1-3, October-December 2008, Pages 1-23
In Vitro Antioxidant Activities of Three Selected Brown Seaweeds of India	Kumar Chandini S, Ganesan P, Bhaskar N	3	3	100.00	Central Food Technological Research Institute, Mysore	Food Chemistry Volume 107, Issue 2, 15 March 2008, Pages 707-713

Title	Authors				Institution	Reference
Smart Investments in Sustainable Food Production: Revisiting Mixed Crop-Livestock Systems	Parthasarathy Rao P	1	17	5.88	International Crops Research Institute For the Semi-Arid Tropics (Icrisat), Hyderabad	Science 12 February 2010: Vol. 327 No. 5967 Pp. 822-825
Detection of Aspergillus Spp. and Aflatoxin B-1 in Rice in India	Reddy KRN, Reddy CS, Muralidharan K	3	3	100.00	Directorate of Rice Research, Rajendranagar, Hyderabad	Food Microbiology Volume 26, Issue 1, February 2009, Pages 27–31
Antioxidant Activity of Fresh and Dry Fruits Commonly Consumed in India	C Vijaya Kumar Reddy, Sreeramulu D, Raghunath M	3	3	100.00	National Institute of Nutrition, Hyderabad	Food Research International Volume 43, Issue 1, January 2010, Pages 285–288
In Vitro Antioxidant and in Vivo Anti-Inflammatory Potential of Crude Polysaccharide From Turbinaria Ornata (Marine Brown Alga)	Subash Ananthi, Hanumantha Rao Balaji Raghavendran, Adoor Gopalan Sunil, Veeraraghavan Gayathri, Ganapathy Ramakrishnan, Hannah R Vasanthi	6	6	100.00	Sri Ramachandra University, Porur, Chennai	Food and Chemical Toxicology Volume 48, Issue 1, January 2010, Pages 187–192
The Challenges of Wastewater Irrigation in Developing Countries	Minhas PS	1	7	14.29	Indian Council of Agricultural Research (Icar), Kab-Ii, Pusa, New Delhi-110 012, India	Agricultural Water Management Volume 97, Issue 4, April 2010, Pages 561–568 Comprehensive Assessment of Water Management in Agriculture
Health Risk Assessment of Heavy Metals Via Dietary Intake of Foodstuffs From the Wastewater Irrigated Site of A Dry Tropical Area of India	Anita Singh, Rajesh Kumar Sharma, Madhoolika Agrawal	3	4	75.00	Banaras Hindu University, Varanasi	Food and Chemical Toxicology Volume 48, Issue 2, February 2010, Pages 611–619

Title	Author				Institution	Publication
Obesity: Genes, Brain, Gut, and Environment	Undurti N Das	1	1	100.00	Jawaharlal Nehru Technological University, Kakinada, Andhra Pradesh, India	Nutrition Volume 26, Issue 5, May 2010, Pages 459–473
Managing Water in Rainfed Agriculture-The Need For A Paradigm Shift	Suhas P Wani	1	9	11.11	International Crops Research Institute For Semi-Arid Tropics (Icrisat), Patancheru 502324, India	Agricultural Water Management Volume 97, Issue 4, April 2010, Pages 543–550 Comprehensive Assessment of Water Management in Agriculture
Sequencing the Potato Genome: Outline and First Results to Come From the Elucidation of the Sequence of the World's Third Most Important Food Crop	Swarup K Chakrabati	1	16	6.25	Central Potato Research Institute, Shimla	American Journal of Potato Research November 2009, Volume 86, Issue 6, Pp 417-429
The Top 100 Questions of Importance to the Future of Global Agriculture	Ravindranath NH	1	55	1.82	Indian Institute of Science, Bangalore	International Journal of Agricultural Sustainability Volume 8, Issue 4, 2010
Dietary Carbohydrates Glycaemic Load, Food Groups and Newly Detected Type 2 Diabetes Among Urban Asian Indian Population in Chennai, India (Chennai Urban Rural Epidemiology Study 59)	Mohan Viswanathan, Radhika Ganesan, Mohan Sathya, Rangaswamy, Ramjothi Tamil, Selvi Ganesan, Anbazhagan, Sudha Vasudevan	6	6	100.00	Madras Diabetes Research Foundation and Dr Mohan's Diabetes Specialities Centre, Who Collaborating Centre For Non-Communicable Diseases, Gopalapuram, Chennai	British Journal of Nutrition / Volume 102 / Issue 10 / November 2009, Pp 1498-1506

Title	Authors				Affiliation	Journal
Population Structure and Linkage Disequilibrium in Us Barley Germplasm: Implications For Association Mapping	Prasanna R Bhat	1	18	5.56	Monsanto Research Centre, Bangalore	Crop Science (2010)
Chemical Profile, Antifungal, Antiaflatoxigenic and Antioxidant Activity of Citrus Maxima Burm. and Citrus Sinensis (L.) Osbeck Essential Oils and Their Cyclic Monoterpene, Dl-Limonene	Priyanka Singh, Ravindra Shukla, Bhanu Prakash, Ashok Kumar, Prashant Kumar Mishra, Nawal Kishore Dubey, Shubhra Singh	7	7	100.00	Banaras Hindu University, Varanasi; Allahabad Agricultural Institute	Food and Chemical Toxicology Volume 48, Issue 6, June 2010, Pages 1734–1740
Oil Palm Fiber (Opf) and Its Composites: A Review	Shinoj S, Visvanathan R, Kochubabu M	3	4	75.00	A Directorate of Oil Palm Research, Indian Council of Agricultural Research, Pedavegi, Eluru, Andhra Pradesh 534 450, India B Department of Food & Agricultural Process Engineering, Tamil Nadu Agricultural University, Coimbatore, Tamil Nadu 641 003, India	Industrial Crops and Products Volume 33, Issue 1, January 2011, Pages 7–22
Exploring Solid Lipid Nanoparticles to Enhance the Oral Bioavailability of Curcumin	Vandita Kakkar, Sukhjit Singh, Dinesh Singla, Indu Pal Kaur	4	4	100.00	1Department of Pharmaceutical Sciences, University Institute of Pharmaceutical Sciences, Panjab University, Chandigarh, India 2Formulation Development, Panacea Biotec Ltd, Lalru, Punjab, India	Molecular Nutrition & Food Research Volume 55, Issue 3, Pages 495–503, March 2011

Title	Authors				Institution	Journal
Efficacy of Chemically Characterized Ocimum Gratissimum L. Essential Oil As An Antioxidant and A Safe Plant Based Antimicrobial Against Fungal and Aflatoxin B-1 Contamination of Spices	Bhanu Prakash, Ravindra Shukla, Priyanka Singh, Prashant Kumar Mishra, Nawal Kishore Dubey, Ravindra Nath Kharwar	6	6	100.00	Banaras Hindu University, Varanasi	Food Research International Volume 44, Issue 1, January 2011, Pages 385–390
Biofortification: A New Tool to Reduce Micronutrient Malnutrition	Meenakshi JV	1	5	20.00	Delhi Sch Econ, Delhi, India.	Food Nutr Bull 32 (1): S31-S40 Suppl. S Mar 2011
Root Biology and Genetic Improvement For Drought Avoidance in Rice	Gowda VRP, Shashidhar HE	2	5	40.00	Univ Agr Sci Bangalore, Dept Biotechnol, Coll Agr, Gkvk, Bangalore 560065, Karnataka, India.	Field Crop Res 122 (1): 1-13 Apr 28 2011
Value-Addition of Agricultural Wastes For Augmented Cellulase and Xylanase Production Through Solid-State Tray Fermentation Employing Mixed-Culture of Fungi	Kaur S, Oberoi HS	2	5	40.00	Banaras Hindu Univ, Inst Agr Sci, Dept Mycol & Plant Pathol, Varanasi 221005, Uttar Pradesh, India.	Ind Crops Products 34 (1): 1160-1167 Jul 2011
Relations of Rice Seeding Rates to Crop and Weed Growth in Aerobic Rice	Singh VP, Kumar A	2	4	50.00	Govind Ballabh Pant Univ Agr & Technol, Up	Field Crop Res 121 (1): 105-115 Feb 28 2011
Antimicrobial Potential and Chemical Composition of Eucalyptus Globulus Oil in Liquid and Vapour Phase Against Food Spoilage Microorganisms	Tyagi AK, Malik A	2	2	100.00	Indian Inst Technol Delhi, Appl Microbiol Lab, Ctr Rural Dev & Technol, New Delhi 110016, India	Food Chem 126 (1): 228-235 May 1 2011
Antioxidant and Free Radical Scavenging Capacity of the Underutilized Legume, Vigna Vexillata (L.) A. Rich	Sowndhararajan K, Siddhuraju P, Manian S	3	3	100.00	Bharathiar Univ, Tamil Nadu-2	J Food Compos Anal 24 (2): 160-165 Mar 2011

Title	Indian Authors	No. of Indian Authors	Total Author	Ratio I/F*	Institue Name	Source
Field-Evolved Resistance to Bt Toxin Cry1ac in the Pink Bollworm, Pectinophora Gossypiella (Saunders) (Lepidoptera: Gelechiidae), From India	Dhurua S, Gujar GT	2	2	100.00	Indian Agr Res Inst, Div Entomol, New Delhi 110012, India	Pest Manag Sci 67 (8): 898-903 Aug 2011
Shelf-Life Extension of Fresh Ready-To-Bake Pizza by the Application of Modified Atmosphere Packaging	Goyal GK, Wani AA	2	3	66.67	Islamic Univ Sci & Technol, Dept Food Technol, Awantipora, Jammu & Kashmir, India. Natl Dairy Res Inst, Dairy Technol Div, Karnal 132001, Haryana, India.	Food Bioprocess Technol 5 (3): 1028-1037 Apr 2012
(b) Biology and Biochemistry						
Recent Developments in Ring Opening Polymerization of Lactones For Biomedical Applications	Varma IK	1	2	50.00	Indian Inst Technol, Ctr Polymer Sci & Engn, New Delhi 110016, India.	Biomacromolecules 4 (6): 1466-1486 Nov-Dec 2003
Application of Conducting Polymers to Biosensors	Gerard M, Chaubey A, Malhotra BD	3	3	100.00	Natl Phys Lab, Dr Krishnan Marg, New Delhi 110012, India.Deemed Univ, Allahabad Agr Inst, Dept Chem, Allahabad, Uttar Pradesh, India.	Biosens Bioelectron 17 (5): 345-359 May 2002
Production, Purification, Characterization, and Applications of Lipases	Sharma R, Banerjee UC	2	3	66.67	Natl Inst Pharmaceut Educ & Res, Mohali 160062, Punjab, India.	Biotechnol Adv 19 (8): 627-662 Dec 2001

Title	Authors				Affiliation	Journal
Map Kinase Phosphatase As A Locus of Flexibility in A Mitogen-Activated Protein Kinase Signaling Network	Bhalla US,	1	3	33.33	Natl Ctr Biol Sci, Bangalore 560065, Karnataka, India.	Science 297 (5583): 1018-1023 Aug 9 2002
Silver Nanoparticles As A New Generation of Antimicrobials	Rai M, Yadav A, Gade A	3	3	100.00	Sgb Amravati Univ, Dept Biotechnol, Amravati 444602, Maharashtra, India.	Biotechnol Adv 27 (1): 76-83 Jan-Feb 2009
Human Protein Reference Database-2009 Update	Prasad TSK, Goel R, Kandasamy K, Keerthikumar S, Kumar S, Mathivanan S, Telikicherla D, Raju R, Shafreen B, Venugopal A, Balakrishnan L, Marimuthu A, Banerjee S, Somanathan DS, Sebastian A, Rani S, Ray S, Kishore CJH, Kanth S, Ahmed M, Kashyap MK, Mohmood R, Ramachandra YL, Krishna V, Rahiman BA, Mohan S, Ranganathan P, Ramabadran S	29	30	96.67	Int Tech Pk, Inst Bioinformat, Bangalore 560066, Karnataka, India. Kuvempu Univ, Dept Biotechnol, Shankaraghatta, Karnataka, India.	Nucl Acid Res 37: D767-D772 Sp. Iss. Si Jan 2009

Title	Authors	41	42	%	Affiliation	Reference
Human Protein Reference Database - 2006 Update	Mishra GR, Suresh M, Kumaran K, Kannabiran N, Suresh S, Bala P, Shivakumar K, Anuradha N, Reddy R, Raghavan TM, Menon S, Hanumanthu G, Gupta M, Upendran S, Gupta S, Mahesh M, Jacob B, Mathew P, Chatterjee P, Arun KS, Sharma S, Chandrika KN, Deshpande N, Palvankar K, Raghavnath R, Krishnakanth R, Karathia H, Rekha B, Nayak R, Vishnupriya G, Kumar HGM, Nagini M, Kumar GSS, Jose R, Deepthi P, Mohan SS, Gandhi TKB, Harsha HC, Deshpande KS, Sarker M, Prasad TSK	41	42	97.62	Inst Bioinformat, Bangalore 560066, Karnataka, India.	Nucl Acid Res 34: D411-D414 Sp. Iss. Si Jan 1 2006
Multiple Biological Activities of Curcumin: A Short Review	Srimal RC	1	4	25.00	Ind Toxicol Res Ctr, Lucknow, Uttar Pradesh, India.	Life Sci 78 (18): 2081-2087 Mar 27 2006
Dissecting Protein-Protein Recognition Sites	Chakrabarti P	1	2	50.00	Bose Inst, Dept Biochem, Calcutta 700054, W Bengal, India.	Protein-Struct Funct Genet 47 (3): 334-343 May 15 2002
Cold, Salinity and Drought Stresses: An Overview	Mahajan S, Tuteja N	2	2	100.00	Int Ctr Genet Engn & Biotechnol, Aruna Asaf Ali Marg, New Delhi 110067, India.	Arch Biochem Biophys 444 (2): 139-158 Dec 15 2005
Self-Assembled Monolayers As A Tunable Platform For Biosensor Applications	Chaki NK, Vijayamohanan K	2	2	100.00	Natl Chem Lab, Phys & Mat Chem Div, Dr Homi Bhabha Rd, Pune 411008, Maharashtra, India.	Biosens Bioelectron 17 (1-2): 1-12 Jan 2002

Title	Authors			%	Affiliation	Citation
Ever-Fluctuating Single Enzyme Molecules: Michaelis-Menten Equation Revisited	Cherayil BJ	1	9	11.11	Indian Inst Sci, Dept Inorgan & Phys Chem, Bangalore 560012, Karnataka, India.	Nat Chem Biol 2 (2): 87-94 Feb 2006
Extracellular Biosynthesis of Silver Nanoparticles Using the Fungus Fusarium Oxysporum	Ahmad A, Mukherjee P, Senapati S, Mandal D, Khan MI, Kumar R, Sastry M	7	7	100.00	Natl Chem Lab, Mat Chem Div, Pune 411008, Maharashtra, India.	Colloid Surface B 28 (4): 313-318 May 1 2003
The Hupopsi's Molecular Interaction Format - A Community Standard For the Representation of Protein Interaction Data	K Shanker		39	0.00	Inst Bioinformat, Bangalore 560066, Karnataka, India.	Nat Biotechnol 22 (2): 177-183 Feb 2004
Curcumin, A Novel P300/Creb-Binding Protein-Specific Inhibitor of Acetyltransferase, Represses the Acetylation of Histone/Nonhistone Proteins and Histone Acetyltransferase-Dependent Chromatin Transcription	Balasubramanyam K, Varier RA, Altaf M, Swaminathan V, Siddappa NB, Ranga U, Kundu TK	7	7	100.00	Jawaharlal Nehru Ctr Adv Sci Res, Mol Virol Lab, Mol Biol & Genet Unit, Bangalore 560064, Karnataka, India.	J Biol Chem 279 (49): 51163-51171 Dec 3 2004
Synthesis of Gold Nanotriangles and Silver Nanoparticles Using Aloe Vera Plant Extract	Chandran SP, Chaudhary M, Pasricha R, Ahmad A, Sastry M	5	5	100.00	Natl Chem Lab, Biochem Sci Div, Mat Chem Div, Nanosci Grp, Pune 411008, Maharashtra, India.	Biotechnol Progr 22 (2): 577-583 Mar-Apr 2006
Global Patterns of Diversification in the History of Modern Amphibians	Biju SD	1	8	12.50	Univ Delhi, Sch Environm Studies, Ctr Environm Management Degraded Ecosyst, Delhi 110007, India.	Proc Nat Acad Sci USA 104 (3): 887-892 Jan 16 2007
Microbial and Plant Derived Biomass For Removal of Heavy Metals From Wastewater	Ahluwalia SS, Goyal D	2	2	100.00	Thapar Inst Engn & Technol, Dept Biotechnol & Environm Sci, Patiala 147004, Punjab, India.	Bioresource Technol 98 (12): 2243-2257 Sep 2007

Title	Authors				Affiliation	Source
Experimental and Kinetic Studies on Methylene Blue Adsorption by Coir Pith Carbon	Namasivayam C	1	2	50.00	Bharathiar Univ, Dept Environm Sci, Environm Chem Div, Coimbatore 641046, Tamil Nadu, India.	Bioresource Technol 98 (1): 14-21 Jan 2007
Agricultural Waste Material As Potential Adsorbent For Sequestering Heavy Metal Ions From Aqueous Solutions - A Review	Sud D, Mahajan G, Kaur MP	3	3	100.00	St Longowal Inst Engn & Technol, Dept Chem, Longowal 148106, India.	Bioresource Technol 99 (14): 6017-6027 Sep 2008
Validation of Housekeeping Genes As Internal Control For Studying Gene Expression in Rice by Quantitative Real-Time Pcr	Jain M, Nijhawan A, Tyagi AK, Khurana JP	4	4	100.00	Univ Delhi, Interdisciplinary Ctr Plant Genom, South Campus, New Delhi 110021, India.	Biochem Biophys Res Commun 345 (2): 646-651 Jun 30 2006
Heat Shock Genes - Integrating Cell Survival and Death	Kachru S, Schulz M, Trivedi SP	3	3	100.00	Tata Inst Fundamental Res, Bombay 400005, Maharashtra, India.	J High Energy Phys (10): Art. No.-007 Oct 2003
Assay For Quantitative Determination of Glutathione and Glutathione Disulfide Levels Using Enzymatic Recycling Method	Saibal K Biswas	1	3	33.33	Dr Ambedkar Coll, Dept Biochem & Biotechnol, Nagpur, Maharashtra, India.	Nat Protoc 1 (6): 3159-3165 2006
Bioconversion of Lignocellulosic Biomass: Biochemical and Molecular Perspectives	Kumar, R, Singh, S	2	3	66.67	Defence Research and Development Organisation India, Biotechnology Division, Dehradun, India; Hindu Rao Hospital, Department of Pathology, New Delhi, India	J Ind Microbiol Biotechnol 35 (5): 377-391 May 2008
An Overview of Enzymatic Production of Biodiesel	Ranganathan SV, Narasimhan SL, Muthukumar K	3	3	100.00	Anna Univ, Ac Coll Technol, Dept Chem Engn, Madras 600025, Tamil Nadu, India	Bioresource Technol 99 (10): 3975-3981 Jul 2008

Title	Authors				Affiliation	Reference
Enzyme Stability and Stabilization - Aqueous and Non-Aqueous Environment	Iyer PV, Ananthanarayan L	2	2	100.00	Inst Chem Technol, Food Engn & Technol Dept, Bombay 400019, Maharashtra, India.	Process Biochem 43 (10): 1019-1032 Oct 2008
Biodegradable Polymeric Nanoparticles Based Drug Delivery Systems	Kumari A, Yadav SK, Yadav SC	3	3	100.00	Csir, Inst Himalayan Bioresource Technol, Div Biotechnol, Palampur 176061, Himachal Prades, India	Colloid Surface B 75 (1): 1-18 Jan 1 2010
The Rice Annotation Project Database (Rap-Db): 2008 Update	Jitendra P Khurana, Saurabh Raghuvanshi, Akhilesh K Tyagi, Nagendra K Singh	4	91	4.40	Univ Delhi, Dept Plant Mol Biol, New Delhi 110021, India. Indian Agr Res Inst, Natl Res Ctr Plant Biotechnol, New Delhi 110012, India.	Nucl Acid Res 36: D1028-D1033 Sp. Iss. Si Jan 2008
Electrospinning: A Fascinating Fiber Fabrication Technique	Bhardwaj N, Kundu SC	2	2	100.00	Indian Inst Technol, Dept Biotechnol, Kharagpur 721302, W Bengal, India	Biotechnol Adv 28 (3): 325-347 May-Jun 2010
Novel Chitin and Chitosan Nanofibers in Biomedical Applications	Jayakumar R, Prabaharan M, Nair SV	3	4	75.00	Amrita Vishwa Vidhyapeetham Univ, Res Ctr, Cochin 682041, Kerala, India. Srm Univ, Dept Chem, Fac Engn & Technol, Kattankulathur 603203, India.	Biotechnol Adv 28 (1): 142-150 Jan-Feb 2010
Next-Generation Sequencing Technologies and Their Implications For Crop Genetics and Breeding	Rajeev K Varshney, Spurthi N Nayak	2	4	50.00	Centre of Excellence in Genomics (Ceg), International Crops Research Institute For the Semi-Arid Tropics (Icrisat), Patancheru, 502324, A.P., India	Trends in Biotechnology 27 (9), Pp. 522-530, 2009

Title	Authors			%	Affiliation	Journal
Production of Recombinant Proteins by Microbes and Higher Organisms	Vaishnav P	1	2	50.00	Gidc, Ankleshwar 393002, Gujarat, India.	Biotechnol Adv 27 (3): 297-306 May-Jun 2009
A Systematic Molecular Dynamics Study of Nearest-Neighbor Effects on Base Pair and Base Pair Step Conformations and Fluctuations in B-Dna	Jayaram B, Singh T	2	18	11.11	Indian Inst Technol, Dept Chem, New Delhi 110016, India.; Assam University, Department of Chemistry, Silchar, India	Nucl Acid Res 38 (1): 299-313 Jan 2010
100 Ns Molecular Dynamics Simulations to Study Intramolecular Conformational Changes in Bax	Koshy C, Parthiban M, Sowdhamini R	3	3	100.00	Natl Ctr Biol Sci, Tata Inst Fundamental Res, Bangalore 560065, Karnataka, India. Bharathiar Univ, Bioinformat Ctr, Coimbatore 641046, Tamil Nadu, India.	J Biomol Struct Dyn 28 (1): 71-83 Aug 2010
A Stoichiometry Driven Universal Spatial Organization of Backbones of Folded Proteins: Are There Chargaff's Rules For Protein Folding?	Mittal A, Jayaram B, Shenoy S, Bawa TS	4	4	100.00	Indian Inst Technol, Sch Biol Sci, New Delhi 110016, India	J Biomol Struct Dyn 28 (2): 133-142 Oct 2010
An Overview of the Recent Developments in Polylactide (Pla) Research	Nampoothiri KM, Nair NR, John RP	3	3	100.00	Csir, Div Biotechnol, Niist, Thiruvananthapuram 695019, Kerala, India	Bioresource Technol 101 (22): 8493-8501 Nov 2010
Synthesis of Silver Nanoparticles Using Acalypha Indica Leaf Extracts and Its Antibacterial Activity Against Water Borne Pathogens	Krishnaraj C, Jagan EG, Rajasekar S, Selvakumar P, Kalaichelvan PT, Mohan N	6	6	100.00	Univ Madras, Ctr Adv Studies Bot, Madras 600025, Tamil Nadu, India.	Colloid Surface B 76 (1): 50-56 Mar 1 2010

Title	Authors				Affiliation	Journal
Bioconjugated Quantum Dots For Cancer Research: Present Status, Prospects and Remaining Issues	Mundayoor S, Omkumar RV, Anas A, Biju V	4	5	80.00	Rajiv Gandhi Center For Biotechnology (Rgcb), Trivandrum, 695014, India, National Centre For Aquatic Animal Health, Cochin University of Science and Technology, Kochi, 682 016, India; Center For Arthropod Bioresources and Biotechnology, Kerala University, Trivandrum, India	Biotechnol Adv 28 (2): 199-213 Mar-Apr 2010
: Micro and Macroalgal Biomass: A Renewable Source For Bioethanol	Anisha GS, Nampoothiri KM, Pandey A	3	4	75.00	Govt Coll, Dept Zool, Palakkad, Kerala, India. Natl Inst Interdisciplinary Sci Aznd Technol, Div Biotechnol, Trivandrum 695019, Kerala, India.	Bioresource Technol 102 (1): 186-193 Sp. Iss. Si Jan 2011
Biomaterials Based on Chitin and Chitosan in Wound Dressing Applications	Jayakumara R, Prabaharanb M, Sudheesh 1Kumara PT, Naira SV	4	5	80.00	Amrita Vishwa Vidhyapeetham Univ, Amrita Inst Med Sci & Res Ctr, Amrita Ctr Nanosci & Mol Med, Cochin 682041, Kerala, India.	Biotechnol Adv 29 (3): 322-337 May-Jun 2011
Bioprospecting For Hyper-Lipid Producing Microalgal Strains For Sustainable Biofuel Production	Karthikeyan S	1	6	16.67	Tamil Nadu Agr Univ, Coimbatore 641003, Tamil Nadu, India.	Bioresource Technol 102 (1): 57-70 Sp. Iss. Si Jan 2011
Multifunctional Magnetic Quantum Dots For Cancer Theranostics	Singh SP	1	1	100.00	Natl Phys Lab, New Delhi 110012, India	J Biomed Nanotechnol 7 (1): 95-97 Sp. Iss. Si Feb 2011

Title	Authors				Affiliation	Journal
Effects of Intravenous Zoledronic Acid Plus Subcutaneous Teriparatide [Rhpth(1-34)] in Postmenopausal Osteoporosis	Hanumantha Rao	1	12	8.33	Novartis Healthcare Private Ltd, Hyderabad, Andhra Pradesh, India.	J Bone Miner Res 26 (3): 503-511 Mar 2011
Recent Advances in Polyaniline Based Biosensors	Dhand C, Das M, Datta M, Malhotra BD	4	4	100.00	Csir, Natl Phys Lab, Biomed Instrumentat Sect, Mat Phys & Engn Div, Ctr Biomol Elect, Dept Sci & T, New Delhi 110012, India. Univ Delhi, Dept Chem, New Delhi 110007, India.	Biosens Bioelectron 26 (6): 2811-2821 Feb 15 2011
Synthesis of Silver Nanoparticles in An Aqueous Suspension of Graphene Oxide Sheets and Its Antimicrobial Activity	Das MR, Sarma RK, Saikia R, Kale VS, Shelke MV, Sengupta P	6	6	100.00	Csir, Ne Inst Sci & Technol, Div Mat Sci, Jorhat 785006, Assam, India.	Colloid Surface B 83 (1): 16-22 Mar 1 2011
Plant Growth Promoting Rhizobacteria and Endophytes Accelerate Phytoremediation of Metalliferous Soils	Prasad MNV	1	4	25.00	Univ Hyderabad, Dept Plant Sci, Hyderabad 500046, Andhra Pradesh, India.	Biotechnol Adv 29 (2): 248-258 Mar-Apr 2011
Screening Ethnically Diverse Human Embryonic Stem Cells Identifies A Chromosome 20 Minimal Amplicon Conferring Growth Advantage	Maneesha S Inamdar, Parvathy Venu	2	124	1.61	Jawaharlal Nehru Ctr Adv Sci Res, Indian Inst Sci,, Bangalore 560012, Karnataka, India.	Nat Biotechnol 29 (12): 1132-U113 Dec 2011
Draft Genome Sequence of Pigeonpea (Cajanus Cajan), An Orphan Legume Crop of Resource-Poor Farmers	Rajeev K Varshney, Rachit K Saxena, Sarwar Azam, Reetu Tuteja, Hari D Upadhyaya, Trushar Shah & K B Saxena	7	31	22.58	Int Crops Res Inst Semi Arid Trop, Ap	Nat Biotechnol 30 (1): 83-U128 Jan 2012

Title	Indian Authors	No. of Indian Authors	Total Author	Ratio I/F*	Institue Name	Source
The Intact Molecular Interaction Database in 2012	Mahadevan U, Raghunath A,	2	22	9.09	Mol Connect Private Ltd, Bangalore 560004, Karnataka, India.	Nucl Acid Res 40 (D1): D841-D846 Jan 2012

(c) Chemistry

Title	Indian Authors	No. of Indian Authors	Total Author	Ratio I/F*	Institue Name	Source
Metal Carboxylates With Open Architectures	Rao CNR, Natarajan S, Vaidhyanathan R	3	3	100.00	Jawaharlal Nehru Ctr Adv Sci Res, Bangalore	Angew Chem Int Ed 43 (12): 1466-1496 2004
Recent Advances in the Baylis-Hillman Reaction and Applications	Basavaiah D, Rao AJ, Satyanarayana T	3	3	100.00	Univ Hyderabad, Sch Chem, Hyderabad 500046, Andhra Pradesh, India.	Chem Rev 103 (3): 811-891 Mar 2003
Hydrogen Bridges in Crystal Engineering: Interactions Without Borders	Desiraju GR	1	1	100.00	Univ Hyderabad, Sch Chem, Hyderabad 500046, Andhra Pradesh, India.	Account Chem Res 35 (7): 565-573 Jul 2002
Recent Applications of the Suzuki-Miyaura Cross-Coupling Reaction in Organic Synthesis	Kotha S, Lahiri K, Kashinath D	3	3	100.00	Indian Inst Technol, Dept Chem, Bombay 400076, Maharashtra, India.	Tetrahedron 58 (48): 9633-9695 Nov 25 2002
Chitosan Chemistry and Pharmaceutical Perspectives	Ravi Kumar MNV	1	5	20.00	Niper, Dept Pharmaceut, Mohali 160062, Punjab, India.	Chem Rev 104 (12): 6017-6084 Dec 2004
Supramolecular Gels: Functions and Uses	Sangeetha NM, Maitra U	2	2	100.00	Indian Inst Sci, Dept Organ Chem, Bangalore 560012, Karnataka, India.	Chem Soc Rev 34 (10): 821-836 2005
Graphene: The New Two-Dimensional Nanomaterial	Rao CNR, Sood AK, Subrahmanyam KS, Govindaraj A	4	4	100.00	Indian Inst Sci, Jawaharlal Nehru Ctr Adv Sci Res, Bangalore 560012, Karnataka, India.	Angew Chem Int Ed 48 (42): 7752-7777 2009

Title	Authors				Affiliation	Source
Inorganic Nanowires	Rao CNR, Deepak FL, Gundiah G, Govindaraj A	4	4	100.00	Indian Inst Sci, Jawaharlal Nehru Ctr Adv Sci Res, Chem & Phys Mat Unit, Jakkur Po, Bangalore 560064, Karnataka, India. Indian Inst Sci, Jawaharlal Nehru Ctr Adv Sci Res, Chem & Phys Mat Unit, Bangalore 560064, Karnataka, India.	Prog Solid State Chem 31 (1-2): 5-147 2003
Chemical and Biochemical Transformations in Ionic Liquids	Jain N, Kumar A, Chauhan S, Chauhan SMS	4	4	100.00	Univ Delhi, Dept Chem, Bioorgan Lab, Delhi 110007, India.	Tetrahedron 61 (5): 1015-1060 Jan 31 2005
Structural Diversity and Chemical Trends in Hybrid Inorganic-Organic Framework Materials	Cheetham AK, Rao CNR	2	3	66.67	Jawaharlal Nehru Ctr Adv Sci Res, Bangalore 560064, Karnataka, India.	Chem Commun (46): 4780-4795 2006
Layered Double Hydroxide Supported Nanopalladium Catalyst For Heck-, Suzuki-, Sonogashira-, and Stille-Type Coupling Reactions of Chloroarenes	Choudary BM, Madhi S, Chowdari NS, Kantam ML, Sreedhar B	5	5	100.00	Indian Inst Chem Technol, Hyderabad 500007, Andhra Pradesh, India.	J Am Chem Soc 124 (47): 14127-14136 Nov 27 2002
Strategies For Heterocyclic Construction Via Novel Multicomponent Reactions Based on Isocyanides and Nucleophilic Carbenes	Nair V, Rajesh C, Vinod Au, Bindu S, Sreekanth AR, Mathen JS, Balagopal L	7	7	100.00	Csir, Organ Chem Div, Trivandrum 695019, Kerala, India.	Account Chem Res 36 (12): 899-907 Dec 2003
Chromophore-Functionalized Gold Nanoparticles	Thomas KG, Kamat PV	2	2	100.00	Csir, Reg Res Lab, Photosci & Photon Div, Trivandrum 695019, Kerala, India.	Account Chem Res 36 (12): 888-898 Dec 2003

Title	Authors			%	Affiliation	Citation
Interparticle Coupling Effect on the Surface Plasmon Resonance of Gold Nanoparticles: From Theory to Applications	Ghosh SK, Pal T	2	2	100.00	Raidigi Coll, Dept Chem, Raidigi 743383, India. Indian Inst Technol, Dept Chem, Kharagpur 721302, W Bengal, India.	Chem Rev 107 (11): 4797–4862 Nov 2007
Crystal Engineering: A Holistic View	Desiraju GR	1	1	100.00	Univ Hyderabad, Sch Chem, Hyderabad 500046, Andhra Pradesh, India.	Angew Chem Int Ed 46 (44): 8342–8356 2007
Removal of Congo Red From Water by Adsorption Onto Activated Carbon Prepared From Coir Pith, An Agricultural Solid Waste	Namasivayam C, Kavitha D	2	2	100.00	Bharathiar Univ, Dept Environm Sci, Environm Chem Lab, Coimbatore 641046, Tamil Nadu, India.	Dye Pigment 54 (1): 47–58 Jul 2002
Recent Advances in Transition Metal Catalyzed Oxidation of Organic Substrates With Molecular Oxygen	Punniyamurthy T, Velusamy S, Iqbal J	3	3	100.00	Indian Inst Technol, Dept Chem, Gauhati 781039, Guwahati, India. Dr Reddys Res Fdn, Hyderabad 500050, Andhra Pradesh, India.	Chem Rev 105 (6): 2329–2363 Jun 2005
Pi-Organogels of Self-Assembled P-Phenylenevinylenes: Soft Materials With Distinct Size, Shape, and Functions	Ajayaghosh A, Praveen VK	2	2	100.00	Csir, Natl Inst Interdisciplinary Sci & Technol, Chem Sci & Technol Div, Photosci & Photon Grp, Trivandrum 695019, Kerala, India.	Account Chem Res 40 (8): 644–656 Aug 2007
Polymers in Sensor Applications	Adhikari B, Majumdar S	2	2	100.00	Indian Inst Technol, Ctr Mat Sci, Kharagpur 721302, W Bengal, India.	Prog Polym Sci 29 (7): 699–766 Jul 2004
Chemical Modification of Silica Surface by Immobilization of Functional Groups For Extractive Concentration of Metal Ions	Jal PK, Patel S, Mishra B	3	3	100.00	Sambalpur Univ, Dept Chem, Ctr Studies Surface Sci & Technol, Jyoti Vihar 768019, India.	Talanta 62 (5): 1005–1028 Apr 19 2004

Title	Authors			%	Affiliation	Source
Biocompatibility of Gold Nanoparticles and Their Endocytotic Fate Inside the Cellular Compartment: A Microscopic Overview	Shukla R, Bansal V, Chaudhary M, Basu A, Bhonde RR, Sastry M	6	6	100.00	Natl Chem Lab, Mat Chem Div, Pune 411008, Maharashtra, India. Natl Inst Virol, Pune 411001, Maharashtra, India.	Langmuir 21 (23): 10644-10654 Nov 8 2005
Solid Polymer Electrolyte Membranes For Fuel Cell Applications - A Review	Smitha B, Sridhar S, Khan AA	3	3	100.00	Indian Inst Chem Technol, Div Chem Engn, Membrane Separat Grp, Hyderabad 500007, Andhra Pradesh, India.	J Membrane Sci 259 (1-2): 10-26 Aug 15 2005
Equilibrium, Kinetics, Mechanism, and Process Design For the Sorption of Methylene Blue Onto Rice Husk	Vadivelan V, Kumar KV	2	2	100.00	Anna Univ, Ac Tech, Dept Chem Engn, Madras 600025, Tamil Nadu, India.	J Colloid Interface Sci 286 (1): 90-100 Jun 1 2005
Analytical Applications of Room-Temperature Ionic Liquids: A Review of Recent Efforts	Pandey S	1	1	100.00	Indian Inst Technol, Dept Chem, New Delhi 110016, India.	Anal Chim Acta 556 (1): 38-45 Jan 18 2006
Metal Complexation by Chitosan and Its Derivatives: A Review	Varma AJ, Deshpande SV,	2	3	66.67	Natl Chem Lab, Div Chem Engn, Polymer Sci & Engn Unit, Pune 411008, Maharashtra, India.	Carbohyd Polym 55 (1): 77-93 Jan 1 2004
Use of Activated Carbons Prepared From Sawdust and Rice-Husk For Adsorption of Acid Dyes: A Case Study of Acid Yellow 36	Malik PK	1	1	100.00	Jadavpur Univ, Dept Chem, Ctr Surface Sci, Calcutta, W Bengal, India.	Dye Pigment 56 (3): 239-249 Mar 2003
Crystal Engineering of Coordination Polymers Using 4, 4 '-Bipyridine As A Bond Between Transition Metal Atoms	Biradha K, Sarkar M, Rajput L	3	3	100.00	Indian Inst Technol, Dept Chem, Kharagpur 721302, W Bengal, India.	Chem Commun (40): 4169-4179 Oct 28 2006

Title	Authors				Affiliation	Reference
Electrophilicity Index	Chattaraj PK, Sarkar U, Roy DR	3	3	100.00	Indian Inst Technol, Dept Chem, Kharagpur 721302, W Bengal, India.	Chem Rev 106 (6): 2065-2091 Jun 2006
Organotin Assemblies Containing Sn–O Bonds	Chandrasekhar V, Nagendran S, Baskar V	3	3	100.00	Indian Inst Technol, Dept Chem, Kanpur 208016, Uttar Pradesh, India.	Coord Chem Rev 235 (1-2): 1-52 Dec 2002
Advances in the Baylis-Hillman Reaction-Assisted Synthesis of Cyclic Frameworks	Singh V, Batra S	2	2	100.00	Cent Drug Res Inst, Med & Proc Chem Div, Pob 173, Lucknow 226001, Uttar Pradesh, India.	Tetrahedron 64 (20): 4511-4574 May 12 2008
N-Heterocyclic Carbenes: Reagents, Not Just Ligands!	Nair V, Bindu S, Sreekumar V	3	3	100.00	Csir, Reg Res Lab, Div Organ Chem, Trivandrum 695019, Kerala, India.	Angew Chem Int Ed 43 (39): 5130-5135 2004
C-H Center Dot Center Dot Center Dot O and Other Weak Hydrogen Bonds. From Crystal Engineering to Virtual Screening	Desiraju GR	1	1	100.00	Univ Hyderabad, Sch Chem, Hyderabad 500046, Andhra Pradesh, India.	Chem Commun (24): 2995-3001 2005
Construction of Enantiopure Pyrrolidine Ring System Via Asymmetric [3+2]-Cycloaddition of Azomethine Ylides	Pandey G, Banerjee P, Gadre SR	3	3	100.00	Natl Chem Lab, Div Organ Chem Synth, Pune 411008, Maharashtra, India.	Chem Rev 106 (11): 4484-4517 Nov 2006
Single-Walled Carbon Nanotube Induces Oxidative Stress and Activates Nuclear Transcription Factor-Kappa B in Human Keratinocytes	Barrera Enrique V	1	8	12.50	Ctr Dna Fingerprinting & Diagnost, Immunol Lab, Hyderabad 500076, Andhra Pradesh, India.	Nano Lett 5 (9): 1676-1684 Sep 2005
Catalysts For Combustion of Methane and Lower Alkanes	Banerjee S, Choudhary VR	2	3	66.67	Natl Chem Lab, Div Chem Engn, Pune 411008, Maharashtra, India.	Appl Catal A-Gen 234 (1-2): 1-23 Aug 8 2002

Title	Authors			%	Institution	Citation
Transition Metal Catalyzed [2+2+2] Cycloaddition and Application in Organic Synthesis	Kotha S, Brahmachary E, Lahiri K	3	3	100.00	Indian Inst Technol, Dept Chem, Bombay 400076, Maharashtra, India.	Eur J Org Chem (22): 4741-4767 Nov 11 2005
Visible Light Induced Photocatalytic Degradation of Organic Pollutants	Chatterjee D, Dasgupta S	2	2	100.00	Cent Mech Engn Res Inst, Chem Sect, Durgapur 713209, India.	J Photochem Photobiol C-Photo 6 (2-3): 186-205 Oct 2005
The Baylis-Hillman Reaction: A Novel Source of Attraction, Opportunities, and Challenges in Synthetic Chemistry	Basavaiah D, Rao KV, Reddy RJ	3	3	100.00	Univ Hyderabad, Sch Chem, Hyderabad 500046, Andhra Pradesh, India.	Chem Soc Rev 36 (10): 1581-1588 2007
Biological Water: Femtosecond Dynamics of Macromolecular Hydration	Bagchi B	1	4	25.00	Indian Inst Sci, Solid State & Struct Chem Unit, Bangalore 560012, Karnataka, India.	J Phys Chem B 106 (48): 12376-12395 Dec 5 2002
Solvation Dynamics and Proton Transfer in Supramolecular Assemblies	Bhattacharyya K	1	1	100.00	Indian Assoc Cultivat Sci, Dept Phys Chem, Calcutta 700032, W Bengal, India.	Account Chem Res 36 (2): 95-101 Feb 2003
Zirconium(Iv) Chloride Catalyzed One-Pot Synthesis of 3, 4-Dihydropyrimidin-2(1H)-Ones	Reddy CV, Mahesh M, Raju PVK, Babu TR, Reddy VVN	5	5	100.00	Indian Inst Chem Technol, Div Organ Chem 2, Hyderabad 500007, Andhra Pradesh, India.	Tetrahedron Lett 43 (14): 2657-2659 Apr 1 2002
Selective Detection of Cysteine and Glutathione Using Gold Nanorods	Sudeep PK, Joseph STS, Thomas KG	3	3	100.00	Csir, Reg Res Lab, Trivandrum 695019, Kerala, India	J Am Chem Soc 127 (18): 6516-6517 May 11 2005
Water Dynamics in the Hydration Layer Around Proteins and Micelles	Bagchi B	1	1	100.00	Indian Inst Sci, Solid State & Struct Chem Unit, Bangalore 560012, Karnataka, India.	Chem Rev 105 (9): 3197-3219 Sep 2005

Title	Authors				Affiliation	Citation
Characterization of A Planar Cyclic Form of Water Hexamer in An Organic Supramolecular Complex: An Unusual Self-Assembly of Bimesityl-3,3'-Dicarboxylic Acid	Moorthy JN, Natarajan R, Venugopalan P	3	3	100.00	Indian Inst Technol, Dept Chem, Kanpur 208016, Uttar Pradesh, India.Panjab Univ, Dept Chem, Chandigarh 160014, India.	Angew Chem Int Ed 41 (18): 3417-3420 2002
Inorganic Nanotubes	Rao CNR, Nath M	2	2	100.00	Jawaharlal Nehru Ctr Adv Sci Res, Chem & Phys Mat Unit, Bangalore 560064, Karnataka, India.Indian Inst Sci, Solid State & Struct Chem Unit, Bangalore 560012, Karnataka, India.	Dalton Trans (1): 1-24 2003
Biodiesel Production From High Ffa Rubber Seed Oil	Ramadhas AS, Jayaraj S, Muraleedharan C	3	3	100.00	Natl Inst Technol, Dept Engn Mech, Calicut 673601, Kerala, India.	Fuel 84 (4): 335-340 Mar 2005
Efficient and Simple Colorimetric Fluoride Ion Sensor Based on Receptors Having Urea and Thiourea Binding Sites	Jose DA, Kumar DK, Ganguly B, Das A	4	4	100.00	Cent Salt & Marine Chem Res Inst, Bhavnagar 364002, Gujarat, India.	Org Lett 6 (20): 3445-3448 Sep 30 2004
Molecular Iodine-Catalyzed Efficient and Highly Rapid Synthesis of Bis(Indolyl) Methanes Under Mild Conditions	Bandgar BP, Shaikh KA	2	2	100.00	Swami Ramanand Teerth Marathwada Univ, Sch Chem Sci, Organ Chem Res Lab, Vishnupuri 431606, Nanded, India	Tetrahedron Lett 44 (9): 1959-1961 Feb 24 2003
Organogels As Scaffolds For Excitation Energy Transfer and Light Harvesting	Ajayaghosh A, Praveen VK, Vijayakumar C	3	3	100.00	Csir, Nist, Chem Sci & Technol Div, Photosci & Photon Grp, Trivandrum 695019, Kerala, India.	Chem Soc Rev 37 (1): 109-122 2008

Title	Authors				Affiliation	Citation
Coexistence of Water Dimer and Hexamer Clusters in 3D Metal-Organic Framework Structures of Ce(iii) and Pr(iii) With Pyridine-2, 6-Dicarboxylic Acid	Ghosh SK, Bharadwaj PK	2	2	100.00	Indian Inst Technol, Dept Chem, Kanpur 208016, Uttar Pradesh, India.	Inorg Chem 42 (25): 8250-8254 Dec 15 2003
Rapid Synthesis of Au, Ag, and Bimetallic Au Core-Ag Shell Nanoparticles Using Neem (Azadirachta Indica) Leaf Broth	Shankar SS, Rai A, Ahmad A, Sastry M	4	4	100.00	Natl Chem Lab, Mat Chem Div, Pune 411008, Maharashtra, India.	J Colloid Interface Sci 275 (2): 496-502 Jul 15 2004
Fungus-Mediated Synthesis of Silver Nanoparticles and Their Immobilization in the Mycelial Matrix: A Novel Biological Approach to Nanoparticle Synthesis	Mukherjee P, Ahmad A, Mandal D, Senapati S, Sainkar SR, Khan MI, Parishcha R, Ajaykumar PV, Alam M, Kumar R, Sastry M	11	11	100.00	Natl Chem Lab, Div Mat Chem, Pune 411008, Maharashtra, India.	Nano Lett 1 (10): 515-519 Oct 2001
Stable Polymeric Materials For Nonlinear Optics: A Review Based on Azobenzene Systems	Yesodha SK, Pillai CKS	2	3	66.67	Csir, Reg Res Lab, Trivandrum 695019, Kerala, India.	Prog Polym Sci 29 (1): 45-74 Jan 2004
Crystal Engineering: From Weak Hydrogen Bonds to Co-Ordination Bonds	Biradha K	1	1	100.00	Indian Inst Technol, Dept Chem, Kharagpur 721302, W Bengal, India.	Crystengcomm 5: 374-384 Oct 13 2003
Responsive Polymers in Controlled Drug Delivery	Bajpai AK, Shukla SK, Bhanu S, Kankane S	4	4	100.00	Govt Autonomous Sci Coll, Dept Chem, Bose Mem Res Lab, Jabalpur 482001, Mp, India.	Prog Polym Sci 33 (11): 1088-1118 Nov 2008
Ionic Liquid As Catalyst and Reaction Medium. the Dramatic Influence of A Task-Specific Ionic Liquid, [Bmlm]Oh, in Michael Addition of Active Methylene Compounds to Conjugated Ketones, Carboxylic Esters, and Nitriles	Ranu BC, Banerjee S	2	2	100.00	Indian Assoc Cultivat Sci, Dept Organ Chem, Calcutta 700032, W Bengal, India.	Org Lett 7 (14): 3049-3052 Jul 7 2005

Title	Authors				Affiliation	Citation
Solvation Dynamics of Coumarin-153 in A Room-Temperature Ionic Liquid	Karmakar R, Samanta A	2	2	100.00	Univ Hyderabad, Sch Chem, Hyderabad 500046, Andhra Pradesh, India.	J Phys Chem A 106 (18): 4447-4452 May 9 2002
Investigation Into the Interaction Between Surface-Bound Alkylamines and Gold Nanoparticles	Kumar A, Mandal S, Selvakannan PR, Pasricha R, Mandale AB, Sastry M	6	6	100.00	Natl Chem Lab, Mat Chem Div, Pune 411008, Maharashtra, India.	Langmuir 19 (15): 6277-6282 Jul 22 2003
Grafting: A Versatile Means to Modify Polymers – Techniques, Factors and Applications	Bhattacharya A, Misra BN	2	2	100.00	Cent Salt & Marine Chem Res Inst, Bhavnagar 364002, Gujarat, India. Himachal Pradesh Univ, Dept Chem, Shimla 171005, Himachal Prades, India.	Prog Polym Sci 29 (8): 767-814 Aug 2004
Self-Organization of Disc-Like Molecules: Chemical Aspects	Kumar S	1	1	100.00	Raman Res Inst, Cv Raman Ave, Bangalore 560080, Karnataka, India.	Chem Soc Rev 35 (1): 83-109 2006
Cu(Otf)(2): A Reusable Catalyst For High-Yield Synthesis of 3, 4-Dihydropyrimidin-2(1H)-Ones	Paraskar AS, Dewkar GK, Sudalai A	3	3	100.00	Natl Chem Lab, Proc Dev Div, Pashan Rd, Pune 411008, Maharashtra, India.	Tetrahedron Lett 44 (16): 3305-3308 Apr 14 2003
Structure of A Self-Assembled Chain of Water Molecules in A Crystal Host	Pal S, Sankaran NB, Samanta A	3	3	100.00	Univ Hyderabad, Sch Chem, Hyderabad 500046, Andhra Pradesh, India.	Angew Chem Int Ed 42 (15): 1741-1743 2003
Advancements in Development and Characterization of Biodiesel: A Review	Sharma YC, Singh B, Upadhyay SN	3	3	100.00	Banaras Hindu Univ, Inst Technol, Dept Appl Chem, Environm Engn & Res Labs, Varanasi 221005, Uttar Pradesh, India.	Fuel 87 (12): 2355-2373 Sep 2008
Catalytic Activities of Schiff Base Transition Metal Complexes	Gupta KC, Sutar AK	2	2	100.00	Indian Inst Technol, Dept Chem, Polymer Res Lab, Roorkee 247667, Uttar Pradesh, India.	

Title	Authors				Affiliation	Journal
Removal of Lead and Chromium From Wastewater Using Bagasse Fly Ash - A Sugar Industry Waste	Gupta VK, Ali I	2	2	100.00	Indian Inst Technol, Dept Chem, Roorkee 247667, Uttar Pradesh, India. Natl Inst Hydrol, Roorkee 247667, Uttar Pradesh, India.	J Colloid Interface Sci 271 (2): 321-328 Mar 15 2004
Enantioselective Palladium-Catalyzed Transformations	Bell HP	1	3	33.33	Indian Inst Technol, Dept Chem, Kanpur 208016, Uttar Pradesh, India.	Chem Rev 104 (7): 3453-3516 Jul 2004
Recent Advances in Carbon-Carbon Bond-Forming Reactions Involving Homoenolates Generated by Nhc Catalysis	Nair V, Vellalath S, Babu BP	3	3	100.00	Csir, Organ Chem Sect, Niist, Trivandrum 695019, Kerala, India.	Chem Soc Rev 37 (12): 2691-2698 2008
Nanoparticles For Bioimaging	Sharrna P	1	5	20.00	Univ Delhi, Dept Chem, St Stephens Coll, Delhi 110007, India.	Advan Colloid Interface Sci 123: 471-485 Nov 16 2006
Adsorption of Lysozyme Over Mesoporous Molecular Sieves Mcm-41 and Sba-15: Influence of Ph and Aluminum Incorporation	A Vinu, V Murugesan	2	3	66.67	Anna Univ, Dept Chem, Madras 600025, Tamil Nadu, India.	J Phys Chem B 108 (22): 7323-7330 Jun 3 2004
On Some Aspects of Variable Selection For Partial Least Squares Regression Models	Roy PP, Roy K	2	2	100.00	Jadavpur Univ, Drug Theoret & Cheminformat Lab, Div Med & Pharmaceut Chem, Dept Pharmaceut Technol, Fac Engn & Technol, Calcutta 700032, India.	Qsar Comb Sci 27 (3): 302-313 Mar 2008
Separation of Organic-Organic Mixtures by Pervaporation - A Review	Smitha B, Suhanya D, Sridhar S, Ramakrishna M	4	4	100.00	Indian Inst Chem Technol, Div Chem Engn, Membrane Seperat Grp, Hyderabad 500007, Andhra Pradesh, India.	J Membrane Sci 241 (1): 1-21 Sep 15 2004

Title	Authors				Affiliation	Citation
Equilibrium, Kinetics and Thermodynamic Studies For Adsorption of As(iii) on Activated Alumina	Singh TS, Pant KK	2	2	100.00	Indian Inst Technol, Dept Chem Engn, New Delhi 110016, India.	Sep Purif Technol 36 (2): 139-147 Apr 15 2004
Smart Polymers: Physical Forms and Bioengineering Applications	Kumar A, Galaev IY, Mattiasson B	3	4	75.00	Indian Inst Technol, Dept Biol Sci & Bioengn, Kanpur 208016, Uttar Pradesh, India.	Prog Polym Sci 32 (10): 1205-1237 Oct 2007
Adsorptive Removal of Phenol by Bagasse Fly Ash and Activated Carbon: Equilibrium, Kinetics and Thermodynamics	Srivastava VC, Swamy MM, Mall ID, Prasad B, Mishra IM	5	5	100.00	Indian Inst Technol, Dept Chem Engn, Roorkee 247667, Uttar Pradesh, India.	Colloid Surface A 272 (1-2): 89-104 Jan 5 2006
Removal of Orange-G and Methyl Violet Dyes by Adsorption Onto Bagasse Fly Ash - Kinetic Study and Equilibrium Isotherm Analyses	Mall ID, Srivastava VC, Agarwal NK	3	3	100.00	Ind Technol Inst, Dept Chem Engn, Roorkee 247667, Uttar Pradesh, India.	Dye Pigment 69 (3): 210-223 2006
Structure and Photocatalytic Activity of Ti1-Xmxo2 +/-Delta (M =W, V, Ce, Zr, Fe, and Cu) Synthesized by Solution Combustion Method	Nagaveni K, Hegde MS, Madras G	3	3	100.00	Indian Inst Sci, Dept Chem Engn, Bangalore 560012, Karnataka, India.	J Phys Chem B 108 (52): 20204-20212 Dec 30 2004
Mitsunobu and Related Reactions: Advances and Applications	Swamy KCK, Kumar NNB, Balaraman E, Kumar KVPP	4	4	100.00	Univ Hyderabad, Sch Chem, Hyderabad 500046, Andhra Pradesh, India.	Chem Rev 109 (6): 2551-2651 Jun 2009
Basic Dye (Methylene Blue) Removal From Simulated Wastewater by Adsorption Sawdust: A Timber Using Indian Rosewood Industry Waste	Garg VK, Amita M, Kumar R, Gupta R	4	4	100.00	Guru Jambheshwar Univ, Dept Environm Sci & Engn, Hisar 125001, Haryana, India.	Dye Pigment 63 (3): 243-250 Dec 2004

Title	Authors				Affiliation	Reference
A Review on Experimental Studies of Surfactant Adsorption At the Hydrophilic Solid-Water Interface	Paria S, Khilar KC	2	2	100.00	Indian Inst Technol, Dept Chem Engn, Bombay 400076, Maharashtra, India.	Advan Colloid Interface Sci 110 (3): 75-95 Aug 31 2004
Visible-Light-Driven Photocatalysis on Fluorine-Doped Tio2 Powders by the Creation of Surface Oxygen Vacancies	Labhsetwar NK	1	5	20.00	Natl Environm Engn Res Inst, Environm Mat Unit, Nagpur 440020, Maharashtra, India.	Chem Phys Lett 401 (4-6): 579-584 Jan 11 2005
Green Chemistry Approaches to the Synthesis of 5-Alkoxycarbonyl-4-Aryl-3,4-Dihydropyrimidin-2(1H)-Ones by A Three-Component Coupling of One-Pot Condensation Reaction: Comparison of Ethanol, Water, and Solvent-Free Conditions	Bose DS, Fatima L, Mereyala HB	3	3	100.00	Indian Inst Chem Technol, Div Organ Chem 3, Hyderabad 500007, Andhra Pradesh, India.	J Org Chem 68 (2): 587-590 Jan 24 2003
Perchloric Acid Adsorbed on Silica Gel As A New, Highly Efficient, and Versatile Catalyst For Acetylation of Phenols, Thiols, Alcohols, and Amines	Chakraborti AK, Gulhane R	2	2	100.00	Niper, Dept Med Chem, Sector 67, Sas Nagar 160062, Punjab, India.	Chem Commun (15): 1896-1897 2003
Poly(Ethylene Glycol) (Peg) As A Reusable Solvent Medium For Organic Synthesis. Application in the Heck Reaction	Chandrasekhar S, Narsihmulu C, Sultana SS, Reddy NR	4	4	100.00	Indian Inst Chem Technol, Organ Chem Div 1, Hyderabad 500007, Andhra Pradesh, India.	Org Lett 4 (25): 4399-4401 Dec 12 2002
Self Assembled Monolayers on Silicon For Molecular Electronics	Aswal DK, Yakhmi JV,	2	5	40.00	Bhabha Atom Res Ctr, Tech Phys & Prototype Engn Div, Bombay 400085, Maharashtra, India.	Anal Chim Acta 568 (1-2): 84-108 May 24 2006

Title	Authors				Affiliation	Reference
Puckered-Boat Conformation Hexameric Water Clusters Stabilized in A 2D Metal-Organic Framework Structure Built From Cu(II) and 1, 2, 4, 5-Benzenetetracarboxylic Acid	Ghosh SK, Bharadwaj PK	2	2	100.00	Indian Inst Technol, Dept Chem, Kanpur 208016, Uttar Pradesh, India.	Inorg Chem 43 (17): 5180-5182 Aug 23 2004
Guest-Induced Asymmetry in A Metal-Organic Porous Solid With Reversible Single-Crystal-To-Single-Crystal Structural Transformation	Mostafa G	1	4	25.00	Jadavpur Univ, Dept Phys, Calcutta 700032, W Bengal, India.	J Am Chem Soc 127 (49): 17152-17153 Dec 14 2005
An Overview on the Degradability of Polymer Nanocomposites	Pandey JK, Reddy KR, Kumar AP, Singh RP	4	4	100.00	Natl Chem Lab, Div Polymer Sci & Engn, Dr Homi Bhabha Rd, Pune 411008, Maharashtra, India.	Polym Degrad Stabil 88 (2): 234-250 May 2005
Graphene-Based Electrochemical Supercapacitors	Vivekchand SRC, Rout CS, Subrahmanyam KS, Govindaraj A, Rao CNR	5	5	100.00	Csir, Chem & Phys Mat Unit, Ctr Excellence Chem, Bangalore 560064, Karnataka, India. Jawaharlal Nehru Ctr Adv Sci Res, Dst Unit Nanosci, Bangalore 560064, Karnataka, India.	J Chem Sci 120 (1): 9-13 Jan 2008
Intramolecular 1, 3-Dipolar Cycloaddition Reactions in Targeted Syntheses	Nair V, Suja TD	2	2	100.00	Csir, Chem & Phys Mat Unit, Ctr Excellence Chem, Bangalore 560064, Karnataka, India. Jawaharlal Nehru Ctr Adv Sci Res, Dst Unit Nanosci, Bangalore 560064, Karnataka, India.	Tetrahedron 63 (50): 12247-12275 Dec 10 2007
Dynamic Stokes Shift and Excitation Wavelength Dependent Fluorescence of Dipolar Molecules in Room Temperature Ionic Liquids	Samanta A	1	1	100.00	Univ Hyderabad, Sch Chem, Hyderabad 500056, Andhra Pradesh, India.	J Phys Chem B 110 (28): 13704-13716 Jul 20 2006

Title	Authors				Affiliation	Citation
Kinetics and Thermodynamics of Methylene Blue Adsorption on Neem (Azadirachta Indica) Leaf Powder	Bhattacharyya KG, Sharma A	2	2	100.00	Gauhati Univ, Dept Chem, Assam, India.	Dye Pigment 65 (1): 51-59 Apr 2005
Efficient Sensitization of Nanocrystalline Tio2 Films by A Near-Ir-Absorbing Unsymmetrical Zinc Phthalocyanine	Vijaykumar C, Chandrasekharam M, Lakshmikantam M	4	11	36.36	Indian Inst Chem Technol, Nanomat Lab, Inorgan & Phys Chem Div, Hyderabad 500007, Andhra Pradesh, India.	Angew Chem Int Ed 46 (3): 373-376 2007
Progress in Preparation, Processing and Applications of Polyaniline	Bhadra S, Khastgir D, Singha NK	3	4	75.00	Indian Inst Technol, Ctr Rubber Technol, Kharagpur 721302, W Bengal, India.	Prog Polym Sci 34 (8): 783-810 Aug 2009
Polybenzoxazines - New High Performance Thermosetting Resins: Synthesis and Properties	Ghosh NN	1	3	33.33	Birla Inst Technol & Sci Pilani, Dept Chem, Zuarinagar 403726, Goa, India.	Prog Polym Sci 32 (11): 1344-1391 Nov 2007
On the Optical Properties of the Imidazolium Ionic Liquids	Paul A, Mandal PK, Samanta A	3	3	100.00	Univ Hyderabad, Sch Chem, Hyderabad 500046, Andhra Pradesh, India.	J Phys Chem B 109 (18): 9148-9153 May 12 2005
New Emerging Trends in Synthetic Biodegradable Polymers - Polylactide: A Critique	Gupta AP, Kumar V	2	2	100.00	Univ Delhi, Fac Technol, Delhi Coll Engn, Dept Appl Chem & Polymer Technol, New Delhi 110042, India.	Eur Polym J 43 (10): 4053-4074 Oct 2007
Engaging Zwitterions in Carbon-Carbon and Carbon-Nitrogen Bond-Forming Reactions: A Promising Synthetic Strategy	Nair V, Menon RS, Sreekanth AR, Abhilash N, Biju AT	5	5	100.00	Csir, Reg Res Lab, Organ Chem Sect, Trivandrum 695019, Kerala, India.	Account Chem Res 39 (8): 520-530 Aug 15 2006
Catalytic Asymmetric Henry Reaction	Boruwa J, Gogoi N, Saikia PP, Barua NC	4	4	100.00	Csir, Nat Prod Chem Div, Reg Res Lab, Jorhat 785006, Assam, India.	Tetrahedron-Asymmetry 17 (24): 3315-3326 Dec 27 2006

Title	Authors				Affiliation	Citation
Anticancer and Antimicrobial Metallopharmaceutical Agents Based on Palladium, Gold, and Silver N-Heterocyclic Carbene Complexes	Ray S, Mohan R, Singh JK, Samantaray MK, Shaikh MM, Panda D, Ghosh P	7	7	100.00	Indian Inst Technol, Natl Single Crystal X Ray Diffract Facil, Sch Biosci & Bioengn, Dept Chem, Bombay 400076, Maharashtra, India.	J Am Chem Soc 129 (48): 15042-15053 Dec 5 2007
Metal-Salen Complexes As Efficient Catalysts For the Oxygenation of Heteroatom Containing Organic Compounds - Synthetic and Mechanistic Aspects	Venkataramanan NS, Kuppuraj G, Rajagopal S	3	3	100.00	Madurai Kamaraj Univ, Sch Chem, Madurai 625021, Tamil Nadu, India	Coord Chem Rev 249 (11-12): 1249-1268 Jun 2005
Adsorption of As(lii) From Aqueous Solutions by Iron Oxide-Coated Sand	Gupta VK, Saini VK, Jain N	3	3	100.00	Ind Technol Inst, Dept Chem, Roorkee 247667, Uttar Pradesh, India. Cent Bldg Res Inst, Roorkee 247667, Uttar Pradesh, India.	J Colloid Interface Sci 288 (1): 55-60 Aug 1 2005
A Study of the Effect of Annealing on Fe-Doped Linbo3 by Hrxrd, Xrt and Ft-Ir	Bhagavannarayana G, Ananthamurthy RV, Budakoti GC, Kumar B, Bartwal KS	5	5	100.00	Natl Phys Lab, Mat Characterizat Div, Dr Ks Krishnan Rd, New Delhi 110012, India. Univ Delhi, Dept Phys & Astrophys, Delhi 110007, India. Ctr Adv Technol, Laser Mat Div, Indore 452013, India.	J Appl Cryst 38: 768-771 Part 5 Oct 2005
Supramolecular Gelling Agents: Can They Be Designed?	Dastidar P	1	1	100.00	Indian Assoc Cultivat Sci, Dept Organ Chem, 2A & 2B, Raja Sc Mullick Rd, Calcutta 700032, W Bengal, India	Chem Soc Rev 37 (12): 2699-2715 2008
Review on Gel Polymer Electrolytes For Lithium Batteries	Stephan AM	1	1	100.00	Cent Electrochem Res Inst, Electrochem Power Syst Div, Karaikkudi 630006, Tamil Nadu, India	Eur Polym J 42 (1): 21-42 Jan 2006

Title	Authors				Affiliation	Citation
Structural Engineering of Polyurethane Coatings For High Performance Applications	Chattopadhyay DK, Raju KVSN	2	2	100.00	Indian Inst Chem Technol, Organ Coatings & Polymers Div, Hyderabad 500007, Andhra Pradesh, India	Prog Polym Sci 32 (3): 352-418 Mar 2007
Adsorption of A Few Heavy Metals on Natural and Modified Kaolinite and Montmorillonite: A Review	Bhattacharyya KG, Sen Gupta S	2	2	100.00	Gauhati Univ, Dept Chem, Gauhati 781014, Assam, India. Bn Coll, Dept Chem, Dhubri 783324, Assam, India	Advan Colloid Interface Sci 140 (2): 114-131 Aug 5 2008
Recent Developments on Ion-Exchange Membranes and Electro-Membrane Processes	Nagarale RK, Gohil GS, Shahi VK	3	3	100.00	Cent Salt & Marine Chem Res Inst, Bhavnagar 364002, Gujarat, India	Advan Colloid Interface Sci 119 (2-3): 97-130 Feb 28 2006
Dendrimers in Oncology: An Expanding Horizon	Tekade RK, Kumar PV, Jain NK	3	3	100.00	Dr Hari Singh Gour Vishwavidyalaya, Dept Pharmaceut Sci, Pharmaceut Res Lab, Sagar 470003, India	Chem Rev 109 (1): 49-87 Jan 2009
N-Heterocyclic Carbene-Catalyzed Reaction of Chalcones and Enals Via Homoenolate: An Efficient Synthesis of 1, 3, 4-Trisubstituted Cyclopentenes	Nair V, Vellalath S, Poonoth M, Suresh E	4	4	100.00	Csir, Reg Res Lab, Organ Chem Sect, Trivandrum 695019, Kerala, India	J Am Chem Soc 128 (27): Art. No.- Ja0625677 Jul 12 2006
Synthesis and Biological Activity of Schiff and Mannich Bases Bearing 2, 4-Dichloro-5-Fluorophenyl Moiety	Karthikeyan MS, Prasad DJ, Poojary B, Bhat KS, Holla BS, Kumari NS	6	6	100.00	Mangalore Univ, Dept Chem, Mangalagangothri 574199, India. Justice Ks Hegde Acad, Dept Biochem, Mangalore, India	Bioorgan Med Chem 14 (22): 7482-7489 Nov 15 2006
Recent Contributions From the Baylis-Hillman Reaction to Organic Chemistry	Basavaiah D, Reddy BS, Badsara SS	3	3	100.00	Univ Hyderabad, Sch Chem, Hyderabad 500046, Andhra Pradesh, India	Chem Rev 110 (9): 5447-5674 Sep 2010

Title	Authors				Affiliation	Reference
Equilibrium and Kinetic Studies on Basic Dye Adsorption by Oil Palm Fibre Activated Carbon	Tan IAW, Hameed BH, Ahmad AL	3	3	100.00	Univ Sci Malaysia, Sch Chem Engn, Engn Campus, Penang 14300, India	Chem Eng J 127 (1-3): 111-119 Mar 1 2007
Semiconductor Quantum Dots and Metal Nanoparticles: Syntheses, Optical Properties, and Biological Applications	Biju, V, Ishikawa, M	2	5	40.00	Univ Kerala, Ctr Arthropod Bioresources & Biotechnol, Trivandrum 695001, Kerala, India.	Anal Bioanal Chem 391 (7): 2469-2495 Aug 2008
Prospects of Conducting Polymers in Biosensors	Malhotra BD, Chaubey A, Singh SP	3	3	100.00	Natl Phys Lab, Biomol Elect & Conduct Polymer Res Grp, Dr Ks Krishnan Marg, New Delhi 110012, India. Reg Res Lab, Jammu 180001, India.	Anal Chim Acta 578 (1): 59-74 Sep 18 2006
Chemically Modified Graphene Sheets Produced by the Solvothermal Reduction of Colloidal Dispersions of Graphite Oxide	Nethravathi C, Rajamathi M	2	2	100.00	St Josephs Coll, Dept Chem, Mat Res Grp, 36 Lalbagh Rd, Bangalore 560027, Karnataka, India	Carbon 46 (14): 1994-1998 Nov 2008
An Efficient and One-Pot Synthesis of Imidazolines and Benzimidazoles Via Anaerobic Oxidation of Carbon-Nitrogen Bonds in Water	Gogol P, Konwar D	2	2	100.00	Reg Res Lab, Synthet Organ Chem Div, Jorhat 785006, Assam, India.	Tetrahedron Lett 47 (1): 79-82 Jan 2 2006
Biofibres and Biocomposites	John MJ, Thomas S	2	2	100.00	Mahatma Gandhi Univ, Sch Chem Sci, Priyadarshini Hills Po, Kottayam 686560, Kerala, India	Carbohyd Polym 71 (3): 343-364 Feb 8 2008
Jatropha-Palm Biodiesel Blends: An Optimum Mix For Asia	Sarin R, Sharma M, Sinharay S, Malhotra RK	4	4	100.00	Indian Oil Corp Ltd, R&D Ctr, Sector 13, Faridabad 121007, Haryana, India	Fuel 86 (10-11): 1365-1371 Jul-Aug 2007

Title	Authors				Affiliation	Citation
Chitin and Chitosan Polymers: Chemistry, Solubility and Fiber Formation	Pillai CKS, Paul W, Sharma CP	3	3	100.00	Sree Chitra Tirunal Inst Med Sci & Technol, Biomed Technol Wing, Div Biosurface Technol, Thiruvananthapuram 695012, Kerala, India	Prog Polym Sci 34 (7): 641-678 Jul 2009
Red-, Blue-, or No-Shift in Hydrogen Bonds: A Unified Explanation	Joseph J, Jemmis ED	2	2	100.00	Univ Hyderabad, Sch Chem, Cent Univ Po, Hyderabad 500046, Andhra Pradesh, India. Indian Inst Sci, Dept Inorgan & Phys Chem, Bangalore 560012, Karnataka, India	J Am Chem Soc 129 (15): 4620-4632 Apr 18 2007
Host-Guest Complexation of Neutral Red With Macrocyclic Host Molecules: Contrasting Pk(A) Shifts and Binding Affinities For Cucurbit[7]Uril and Beta-Cyclodextrin	Mohanty J, Bhasikuttan AC, Pal H	3	4	75.00	Bhabha Atom Res Ctr, Radiat & Photochem Div, Bombay 400085, Maharashtra, India.	J Phys Chem B 110 (10): 5132-5138 Mar 16 2006
Hydrothermal Technology For Nanotechnology	Byrappa K	1	2	50.00	Univ Mysore, Dos Geol, Pb 21, Manasagangotri Po, Mysore 570006, Karnataka, India	Prog Cryst Growth Charact 53 (2): 117-166 2007

The Alice Experiment At the Cern Lhc	Aggarwal M, Bhati A, Kumar L, Kumar N, Ahammed Z, Baskar P, Chattopadhyay S, Das T, Dubey AE, Dutta Majumdar M, Ganti M, Ghosh P, Mohanty B, Mondal M, Mondal N, Nayak T, Pal S, Prasad S, Saini J, Singaraju R, Singhal V, Sinha B, Ahmad A, Ahmad N, Ahmad S, Danish M Azmi, Irfan M, Kamal A, Mohsin Khan M, Badyal SK, Bala R, Gupta A, Gupta V, Mahajan A, Majahan S, Mangotra K, Potukuchi B, Sambyal S, Sharma S, Sinha T, Roy P, Pal S, Dutta Majumdar A, Das I, Chattopadhyay S, Bose S, Dash A, Jena C, Mahapatra D, Sahoo R, Viyogi Y, Nandi B, Pujahari P, Varma R, Raniwala R, Raniwala S, Sushil Kumar K	57	1153	4.94	Bhabha Atom Res Ctr, Bombay, Maharashtra, India. Univ Rajasthan, Dept Phys, Jaipur 302004, Rajasthan, India. Indian Inst Technol, Bombay 400076, Maharashtra, India. Inst Phys, Bhubaneswar 751007, Orissa, India. Saha Inst Nucl Phys, Calcutta, India. Univ Jammu, Dept Phys, Jammu 180004, India. Aligarh Muslim Univ, Dept Phys, Aligarh 202002, Uttar Pradesh, India. Panjab Univ, Dept Phys, Chandigarh 160014, India. Ctr Variable Energy Cyclotron, Calcutta, India.	J Instrum 3: Art. No.-S08002 Aug 2008
Chromium(Iii) Selective Membrane Sensors Based on Schiff Bases As Chelating Ionophores	Singh AK, Gupta VK, Gupta B	3	3	100.00	Indian Inst Technol, Dept Chem, Roorkee 247667, Uttar Pradesh, India.	Anal Chim Acta. 2007 Feb 28, 585(1): 171-8. Epub 2006 Dec 6.
Poly(Lactic Acid) Fiber: An Overview	Gupta B, Revagade N	2	3	66.67	Indian Inst Technol, Dept Text Technol, New Delhi 110016, India	Prog Polym Sci 32 (4): 455-482 Apr 2007

Title	Authors			%	Affiliation	Reference
An Extremely Efficient Three-Component Reaction of Aldehydes/Ketones, Amines, and Phosphites (Kabachnik-Fields Reaction) For the Synthesis of Alpha-Aminophosphonates Catalyzed by Magnesium Perchlorate	Bhagat S, Chakraborti AK	2	2	100.00	Natl Inst Pharmaceut Educ & Res, Dept Med Chem, Sector 67, Sas Nagar 160062, Punjab, India	J Org Chem 72 (4): 1263-1270 Feb 16 2007
Chitosan-Modifications and Applications: Opportunities Galore	Mourya VK, Inamdar NN	2	2	100.00	Govt Coll Pharm, Dept Pharmaceut, Aurangabad 431001, Maharashtra, India.	React Funct Polym 68 (6): 1013-1051 Jun 2008
Highly Enantioselective Organocatalytic Direct Aldol Reaction in An Aqueous Medium	Maya V, Raj M, Singh VK	3	3	100.00	Indian Inst Technol, Dept Chem, Kanpur 208016, Uttar Pradesh, India.	Org Lett 9 (13): 2593-2595 Jun 21 2007
Efficient Cuo-Nanoparticle-Catalyzed C-S Cross-Coupling of Thiols With Iodobenzene	Rout L, Sen TK, Punniyamurthy T	3	3	100.00	Indian Inst Technol, Dept Chem, Gauhati 781039, India	Angew Chem Int Ed 46 (29): 5583-5586 2007
Significant Progress in Predicting the Crystal Structures of Small Organic Molecules - A Report on the Fourth Blind Test	Jose J, Gadre SR, Desiraju GR, Thakur TS	4	31	12.90	Univ Poona, Pune 411007, Maharashtra, India. Univ Hyderabad, Sch Chem, Hyderabad 500046, Andhra Pradesh, India.	Acta Crystallogr B-Struct Sci 65: 107-125 Part 2 Apr 2009
Unconventional Mixed-Valent Complexes of Ruthenium and Osmium	Gautam Kumar Lahiri	1	2	50.00	Indian Inst Technol, Dept Chem, Bombay 400076, Maharashtra, India.	Angew Chem Int Ed 46 (11): 1778-1796 2007
Conformational Polymorphism in Organic Crystals	Nangia A	1	1	100.00	Univ Hyderabad, Sch Chem, Hyderabad 500046, Andhra Pradesh, India	Account Chem Res 41 (5): 595-604 May 2008
Aggregation Behavior of Ionic Liquids in Aqueous Solutions: Effect of Alkyl Chain Length, Cations, and Anions	Singh T, Kumar A	2	2	100.00	Cent Salt & Marine Chem Res Inst, Bhavnagar 364002, Gujarat, India	J Phys Chem B 111 (27): 7843-7851 Jul 12 2007

Title	Authors				Affiliation	Citation
Open-Framework Structures of Transition-Metal Compounds	Natarajan S, Mandal S	2	2	100.00	Indian Inst Sci, Solid State & Struct Chem Unit, Framework Solids Lab, Bangalore 560012, Karnataka, India	Angew Chem Int Ed 47 (26): 4798-4828 2008
Alkali-Metal-Induced Enhancement of Hydrogen Adsorption in C-60 Fullerene: An Ab Initio Study	Chandrakumar KRS, Ghosh SK	2	2	100.00	Bhabha Atom Res Ctr, Theoret Chem Sect, Chem Grp, Bombay 400085, Maharashtra, India	Nano Lett 8 (1): 13-19 Jan 2008
Binding of Dna Nucleobases and Nucleosides With Graphene	Varghese N, Mogera U, Govindaraj A, Das A, Maiti PK, Sood AK, Rao CNR	7	7	100.00	Jawaharlal Nehru Ctr Adv Sci Res, Chem & Phys Mat Unit, Dst Nanosci Unit, Jakkur Po, Bangalore 560064, Karnataka, India Indian Inst Sci, Dept Phys, Bangalore 560012, Karnataka, India.	Chemphyschem 10 (1): 206-210 Jan 12 2009
Adsorption of Phenolic Compounds on Low-Cost Adsorbents: A Review	Ahmaruzzaman M	1	1	100.00	Natl Inst Technol Silchar, Dept Chem, Silchar 788010, Assam, India.	Advan Colloid Interface Sci 143 (1-2): 48-67 Nov 4 2008

Title	Authors				Affiliation	Source
The Cms Experiment At the Cern Lhc	Bawa HS, Beri SB, Bhandari V, Bhatnagar V, Kaur M, Kohli JM, Kumar A, Singh B, Singh JB, Arora S, Bhattacharya S, S Chatterji S, Chauhan S, Choudhary BC, Gupta P, Jha M, Ranjan K, Shivpuri RK, Srivastava AK, Choudhury RK, Dutta D, Ghodgaonkar M, Kailas S, Kataria SK, Mohanty AK, Pant LM, Shukla P, Topkar A, Aziz T, Sunanda Banerjee, Bose S, Chendvankar S, Deshpande PV, Guchait M, Gurtu A, Maity M, Majumder G, Mazumdar K, Nayak A, Patil MR, Sharma S, Sudhakar K, Acharya BS, Sudeshna Banerjee, Bheesette S, Dugad S, Kalmani SD, Lakkireddi VR, Mondal NK, Panyam N, Verma P	51	3099	1.65	Panjab Univ, Chandigarh 160014, India. Univ Delhi, Delhi 110007, India. Bhabha Atom Res Ctr, Bombay 400085, Maharashtra, India. Tata Inst Fundamental Res Ehep, Bombay, Maharashtra, India.Visva Bharati Univ, Santini Ketan, W Bengal, India.	J Instrum 3: Art. No.-S08004 Aug 2008
Metal-Organic Framework Structures - How Closely Are They Related to Classical Inorganic Structures?	Natarajan S, Mahata P	2	2	100.00	Indian Inst Sci, Solid State & Struct Chem Unit, Framework Solids Lab, Bangalore 560012, Karnataka, India.	Chem Soc Rev 38 (8): 2304-2318 2009

Title	Authors				Affiliation	Reference
Alkyne Activation With Bronsted Acids, Iodine, or Gold Complexes, and Its Fate Leading to Synthetic Application	Patil NT	1	4	25.00	Indian Inst Chem Technol, Organ Chem Div 2, Hyderabad 500007, Andhra Pradesh, India.	Chem Commun (Camb). 2009 Sep 14; (34): 5075-87
Cuo Nanoparticles Catalyzed C–N, C–O, and C–S Cross-Coupling Reactions: Scope and Mechanism	Jammi S, Sakthivel S, Rout L, Mukherjee T, Mandal S, Mitra R, Saha P, Punniyamurthy T	8	8	100.00	Indian Inst Technol Guwahati, Dept Chem, Gauhati 781039, India	J Org Chem 74 (5): 1971-1976 Mar 6 2009
Adsorption of Basic Fuchsin Using Waste Materials-Bottom Ash and Deoiled Soya-As Adsorbents	Gupta VK, Mittal A, Gajbe V, Mittal J	4	4	100.00	Ind Inst Technol Roorkee, Dept Chem, Roorkee 247667, Uttar Pradesh, India. Maulana Azad Natl Inst Technol, Dept Chem, Bhopal 462007, Madhya Pradesh, India	J Colloid Interface Sci 319 (1): 30-39 Mar 1 2008
Ferromagnetism As A Universal Feature of Inorganic Nanoparticles	Sundaresan A, Rao CNR	2	2	100.00	Jawaharlal Nehru Ctr Adv Sci Res, Dept Sci & Technol Unit Nanosci, Bangalore 560064, Karnataka, India.	Nano Today 4 (1): 96-106 Feb 2009
Changes in the Electronic Structure and Properties of Graphene Induced by Molecular Charge-Transfer	Das B, Voggu R, Rout CS, Rao CNR	4	4	100.00	Indian Inst Sci, Jawaharlal Nehru Ctr Adv Sci Res, Chem & Phys Mat Unit, Bangalore 560064, Karnataka, India., Indian Inst Sci, Solid State & Struct Chem Unit, Bangalore 560012, Karnataka, India.	Chem Commun (41): 5155-5157 2008
Bonding and Structure Trends of Thiosemicarbazone Derivatives of Metals-An Overview	Lobana TS, Sharma R, Bawa G, Khanna S	4	4	100.00	Guru Nanak Dev Univ, Dept Chem, Amritsar 143005, Punjab, India	Coord Chem Rev 253 (7-8): 977-1055 Apr 2009

Title	Authors			%	Affiliation	Citation
On Two Novel Parameters For Validation of Predictive Qsar Models	Roy PP, Paul S, Mitra I, Roy K	4	4	100.00	Jadavpur Univ, Dept Pharmaceut Technol, Drug Theoret & Cheminformat Lab, Div Med & Pharmaceut Chem, Calcutta 700032, India	Molecules 14 (5): 1660-1701 May 2009
Nanoscale Particles For Polymer Degradation and Stabilization-Trends and Future Perspectives	Kumar AP, Depan D, Singh RP	3	4	75.00	Natl Chem Lab, Polymer Sci & Engn Div, Pune 411008, Maharashtra, India.	Prog Polym Sci 34 (6): 479-515 Jun 2009
Polymer-Supported Schiff Base Complexes in Oxidation Reactions	Gupta KC, Sutar AK	2	3	66.67	Indian Inst Technol, Dept Chem, Polymer Res Lab, Roorkee 247667, Uttar Pradesh, India	Coord Chem Rev 253 (13-14): 1926-1946 Jul 2009
Organocatalytic Reactions in Water	Raj M, Singh VK	2	2	100.00	Indian Inst Technol, Dept Chem, Kanpur 208016, Uttar Pradesh, India. Indian Inst Sci Educ & Res Bhopal, Bhopal 462023, India.	Chem Commun (44): 6687-6703 2009
Benzothiazinones Kill Mycobacterium Tuberculosis by Blocking Arabinan Synthesis	Nandi V, Bharath S, Gaonkar S, Shandil RK, Balasubramanian V, Balganesh T	7	37	18.92	Astrazeneca India, Bangalore, Karnataka, India.	Science 324 (5928): 801-804 May 8 2009
Resonance Energy Transfer Approach and A New Ratiometric Probe For Hg2+ in Aqueous Media and Living Organism	Suresh M, Mishra S, Mishra SK, Suresh E, Mandal AK, Shrivastav A, Das A	7	7	100.00	Cent Salt & Marine Chem Res Inst Csir, Bhavnagar 364002, Gujarat, India	Org Lett 11 (13): 2740-2743 Jul 2 2009
Nano Indium Oxide As A Recyclable Catalyst For C-S Cross-Coupling of Thiols With Aryl Halides Under Ligand Free Conditions	Reddy VP, Kumar AV, Swapna K, Rao KR	4	4	100.00	Indian Inst Chem Technol, Organ Chem Div 1, Hyderabad 500007, Andhra Pradesh, India.	Org Lett 11 (8): 1697-1700 Apr 16 2009

Title	Authors			%	Affiliation	Reference
Preparation of Graphene by the Rapid and Mild Thermal Reduction of Graphene Oxide Induced by Microwaves	Prakriti R Bangal	1	3	33.33	Indian Inst Chem Technol, Inorgan & Phys Chem Div, Hyderabad 500007, Andhra Pradesh, India.	Carbon 48 (4): 1146-1152 Apr 2010
Synthesis, Spectral Characterization, in Vitro Antibacterial, Antifungal and Cytotoxic Activities of Co(ii), Ni(ii) and Cu(ii) Complexes With 1, 2, 4-Triazole Schiff Bases	Bagihalli GB, Avaji PG, Patil SA, Badami PS	4	4	100.00	Karnatak Univ, Pg Dept Chem, Dharwad 580003, Karnataka, India. Shri Sharanabasaveswar Coll Sci, Dept Chem, Gulbarga 585102, India.	Eur J Med Chem 43 (12): 2639-2649 Dec 2008
Strong Broadband Optical Absorption in Silicon Nanowire Films	Davuluru A, Rapol U	2	12	16.67	Gen Elect John F Welch Technol Ctr, Bangalore, Karnataka, India.	J Nanophotonics 1: Art. No.-013552 2007
Graphene Analogues of Bn: Novel Synthesis and Properties	Nag A, Raidongia K, Hembram KPSS, Datta R, Waghmare UV, Rao CNR	6	6	100.00	Jawaharlal Nehru Ctr Adv Sci Res, Int Ctr Mat Sci, Theoret Sci Unit, Chem & Phys Mat Unit, Jakkur Po, Bangalore 560064, Karnataka, India.; Indian Inst Sci, Solid State & Struct Chem Unit, Bangalore 560012, Karnataka, India.	Acs Nano 4 (3): 1539-1544 Mar 2010
Biomedical Applications of Chitin and Chitosan Based Nanomaterials-A Short Review	Jayakumar, R, Menon, D, Manzoor, K, Nair, SV	4	5	80.00	Amrita Center For Nanosciences, Amrita Institute, Amrita Vishwa Vidyapeetham University, Kochi 682041, India	Carbohyd Polym 82 (2): 227-232 Sep 5 2010
Biological Synthesis of Metal Nanoparticles by Microbes	Narayanan KB, Sakthivel N	2	2	100.00	Pondicherry Univ, Dept Biotechnol, Kalapet 605014, Puducherry, India.	Advan Colloid Interface Sci 156 (1-2): 1-13 Apr 22 2010
A Bond by Any Other Name	Desiraju GR	1	1	100.00	Indian Inst Sci, Solid State & Struct Chem Unit, Bangalore 560012, Karnataka, India.	Angew Chem Int Ed 50 (1): 52-59 2011

Title	Authors			%	Affiliation	Source
Full Implementation and Benchmark Studies of Mukherjee's State-Specific Multireference Coupled-Cluster Ansatz	Das S, Mukherjee D,	2	3	66.67	Indian Assoc Cultivat Sci, Raman Ctr Atom Mol & Opt Sci, Calcutta 700032, W Bengal, India.	J Chem Phys 132 (7): Art. No.-074103 Feb 21 2010
Mos2 and Ws2 Analogues of Graphene	Matte HSSR, Gomathi A, Manna AK, Late DJ, Datta R, Pati SK, Rao CNR	7	7	100.00	Jawaharlal Nehru Ctr Adv Sci Res, Chem & Phys Mat Unit, Theoret Sci Unit, Jakkur Po, Bangalore 560064, Karnataka, India.	Angew Chem Int Ed 49 (24): 4059-4062 2010
Ultrathin Planar Graphene Supercapacitors	Srivastava A	1	11	9.09	Banaras Hindu Univ, Dept Phys, Varanasi 221005, Uttar Pradesh, India.	Nano Lett 11 (4): 1423-1427 Apr 2011
Block-Copolymer-Nanowires With Nanosized Domain Segregation and High Charge Mobilities As Stacked P/N Heterojunction Arrays For Repeatable Photocurrent Switching	Acharya S	1	9	11.11	Iacs, Calcutta 700032, India.	J Am Chem Soc 131 (50): 18030-+ Dec 23 2009
Adsorption Studies on the Removal of Hexavalent Chromium From Aqueous Solution Using A Low Cost Fertilizer Industry Waste Material	Gupta VK, Nayak A	2	3	66.67	Indian Inst Technol, Dept Chem, Roorkee 247667, Uttar Pradesh, India.; Kldavpg Coll, Dept Chem, Roorkee, Uttar Pradesh, India.	J Colloid Interface Sci 342 (1): 135-141 Feb 1 2010
Photochemical Green Synthesis of Calcium-Alginate-Stabilized Ag and Au Nanoparticles and Their Catalytic Application to 4-Nitrophenol Reduction	Saha S, Pal A, Kundu S, Basu S, Pal T	5	5	100.00	Indian Inst Technol, Dept Civil Engn, Kharagpur 721302, W Bengal, India.	Langmuir 26 (4): 2885-2893 Feb 16 2010

Title	Authors				Affiliation	Reference
Assessing the Strengths and Weaknesses of Various Types of Pre-Treatments of Carbon Nanotubes on the Properties of Polymer/Carbon Nanotubes Composites: A Critical Review	Khare RA,	1	3	33.33	Indian Inst Technol, Dept Met Engn & Mat Sci, Bombay 400076, Maharashtra, India.	Polymer 51 (5): 975-993 Mar 2 2010
2-Iodoxybenzoic Acid (Ibx): An Efficient Hypervalent Iodine Reagent	Ajay Harad, Hari Pati	2	4	50.00	Invent Pharma Pvt Ltd, Navi Mumbai 400701, Maharashtra, India; Sambalpur Univ, Dept Chem, Sambalpur 768019, Orissa, India.	Tetrahedron 66 (39): 7659-7706 Sep 25 2010
Solvation Dynamics in Ionic Liquids: What We Have Learned From the Dynamic Fluorescence Stokes Shift Studies	Samanta A	1	1	100.00	Univ Hyderabad, Sch Chem, Hyderabad 500046, Andhra Pradesh, India.	J Phys Chem Lett 1 (10): 1557-1562 May 20 2010
Synthesis, Structure, and Magnetic Properties of Cobalt(Ii) Coordination Polymers From A New Tripodal Carboxylate Ligand: Weak Ferromagnetism and Metamagnetism	Lama P, Aijaz A, Bharadwaj PK	3	4	75.00	Indian Inst Technol, Dept Chem, Kanpur 208016, Uttar Pradesh, India.	Cryst Growth Des 10 (1): 283-290 Jan 2010
The Many Facets of Adenine: Coordination, Crystal Patterns, and Catalysis	Verma S, Mishra AK, Kumar A	3	3	100.00	Indian Inst Technol, Dept Chem, Kanpur 208016, Uttar Pradesh, India.	Account Chem Res 43 (1): 79-91 Jan 2010
Solution Phase Epitaxial Self-Assembly and High Charge-Carrier Mobility Nanofibers of Semiconducting Molecular Gelators	Prasanthkumar S, Saeki A, Seki S, Ajayaghosh A	4	4	100.00	Photosciences and Photonics Group, Chemical Sciences and Technology Division, National Institute For Interdisciplinary Science and Technology (Niist), Csir, Trivandrum 695019, India.	J Am Chem Soc 132 (26): 8866-7 Jul 7 2010

Title	Authors			%	Address	Reference
Noncovalent Functionalization, Exfoliation, and Solubilization of Graphene in Water by Employing A Fluorescent Coronene Carboxylate	Ghosh A, Rao KV, George SJ, Rao CNR	4	4	100.00	Jncasr, New Chem Unit, Bangalore 560064, Karnataka, India.	Chem-Eur J 16 (9): 2700-2704 2010
Functional and Multifunctional Nanoparticles For Bioimaging and Biosensing	Jana NR	1	4	25.00	Indian Assoc Cultivat Sci, Ctr Adv Mat, Calcutta 700032, India.	Langmuir 26 (14): 11631-11641 Jul 20 2010
Nucleic Acid Based Molecular Devices	Krishnan Y	1	2	50.00	Tata Inst Fundamental Res, Natl Ctr Biol Sci, Gkvk, Bangalore 560065, Karnataka, India.	Angew Chem Int Ed 50 (14): 3124-3156 2011
Naphthalimide Appended Rhodamine Derivative: Through Bond Energy Transfer For Sensing of Hg2+ Ions	Kumar M, Kumar N, Bhalla V, Singh H, Sharma PR, Kaur T	6	6	100.00	Guru Nanak Dev Univ, Dept Chem, Ugc Ctr Adv Studies 1, Amritsar, Punjab, India. Indian Inst Integrat Med, Dept Canc Pharmacol, Jammu 180001	Org Lett 13 (6): 1422-1425 Mar 18 2011
Amino Functionalized Zeolitic Tetrazolate Framework (Ztf) With High Capacity For Storage of Carbon Dioxide	Panda T, Pachfule P, Banerjee R	3	5	60.00	Natl Chem Lab, Phys Mat Chem Div, Pune 411008, Maharashtra, India.	Chem Commun 47 (7): 2011-2013 2011
Optical Chemosensor For Ag+, Fe3+, and Cysteine: Information Processing At Molecular Level	Kumar M, Kumar R, Bhalla V	3	3	100.00	Guru Nanak Dev Univ, Dept Chem, Ugc Ctr Adv Studies 1, Amritsar, Punjab, India	Org Lett 13 (3): 366-369 Feb 4 2011
From Short Conjugated Oligomers to Conjugated Polymers. Lessons From Studies on Long Conjugated Oligomers	Zade SS	1	3	33.33	Indian Inst Sci Educ & Res, Dept Chem Sci, Nadia 741252, West Bengal, India.	Account Chem Res 44 (1): 14-24 Jan 2011

Title	Authors				Affiliation	Citation
Structural Chemistry of Peptides Containing Backbone Expanded Amino Acid Residues: Conformational Features of Beta, Gamma, and Hybrid Peptides	Vasudev PG, Chatterjee S, Shamala N, Balaram P	4	4	100.00	Indian Inst Sci, Dept Phys, Bangalore 560012, Karnataka, India.	Chem Rev 111 (2): 657-687 Feb 2011
Recent Advances in Sample Preparation Techniques For Effective Bioanalytical Methods	Jignesh Kotecha	1	4	25.00	Torrent Pharmaceut Ltd, Gandhinagar 382428, Gujarat, India.	Biomed Chromatogr 25 (1-2): 199-217 Sp. Iss. Si Jan-Feb 2011
Role of Organic Fluorine in Crystal Engineering	Chopra D, Row TNG	2	2	100.00	Harish Chandra Res Inst, Allahabad 211019, Uttar Pradesh, India. Tata Inst Fundamental Res, Dept Theoret Phys, Bombay 400005, Maharashtra, India.	Crystengcomm 13 (7): 2175-2186 2011
A Review on the Mechanical and Electrical Properties of Graphite and Modified Graphite Reinforced Polymer Composites	Sengupta R, Bhattacharya M, Bhowmick AK	3	4	75.00	Indian Inst Technol, Patna 800013, Bihar, India	Prog Polym Sci 36 (5): 638-670 Sp. Iss. Si May 2011
Adsorption Thermodynamics, Kinetics and Isosteric Heat of Adsorption of Malachite Green Onto Chemically Modified Rice Husk	Chowdhury S, Mishra R, Saha P, Kushwaha P	4	4	100.00	Natl Inst Technol Durgapur, Dept Biotechnol, Durgapur 713209, Wb, India.	Desalination 265 (1-3): 159-168 Jan 15 2011
Liquid-Crystal Nanoscience: An Emerging Avenue of Soft Self-Assembly	Bisoyi HK, Kumar S	2	2	100.00	Raman Res Inst, Bangalore 560080, Karnataka, India.	Chem Soc Rev 40 (1): 306-319 2011
Synthesis, Structure, Luminescent and Intramolecular Proton Transfer in Some Imidazole Derivatives	Jayabharathi J, Thanikachalam V, Srinivasan N, Saravanan K	4	4	100.00	Annamalai Univ, Dept Chem, Annamalainagar 608002, Tamil Nadu, India.	J Fluoresc 21 (2): 595-606 Mar 2011

Title	Authors				Affiliation	Citation
Coordination-Driven Self-Assembly of M3l2 Trigonal Cages From Preorganized Metalloligands Incorporating Octahedral Metal Centers and Fluorescent Detection of Nitroaromatics	Shanmugaraju S, Mukherjee PS	2	9	22.22	Indian Inst Sci, Dept Inorgan & Phys Chem, Bangalore 560012, Karnataka, India.	Inorg Chem 50 (4): 1506-1512 Feb 21 2011
Biological Water: A Critique	Pal SK	1	3	33.33	Sn Bose Natl Ctr Basic Sci, Dept Chem Biol & Macromol Sci, Unit Nano Sci & Technol, Calcutta 700098, India.	Chem Phys Lett 503 (1-3): 1-11 Feb 8 2011
Chemical Storage of Hydrogen in Few-Layer Graphene	Subrahmanyam KS, Kumar P, Maitra U, Govindaraj A, Hembram Kpss, Waghmare UV, Rao CNR	7	7	100.00	Jawaharlal Nehru Ctr Adv Sci Res, Int Ctr Mat Sci, Bangalore 560064, Karnataka, India.	Proc Nat Acad Sci USA 108 (7): 2674-2677 Feb 15 2011
Definition of the Hydrogen Bond (Iupac Recommendations 2011)	Arunan E, Desiraju GR	2	14	14.29	Indian Inst Sci, Dept Inorgan & Phys Chem, Bangalore 560012, Karnataka, India.	Pure Appl Chem 83 (8): 1637-1641 2011
Fluorescent Gold Clusters As Nanosensors For Copper Ions in Live Cells	Durgadas CV, Sharma CP, Sreenivasan K	3	3	100.00	Sree Chitra Tirunal Inst Med Sci & Technol, Biosurface Technol Div, Biomed Technol Wing, Trivandrum 695012, Kerala, India.	Analyst 136 (5): 933-940 2011
Rhodamine Hydrazone Derivatives As Hg2+ Selective Fluorescent and Colorimetric Chemosensors and Their Applications to Bioimaging and Microfluidic System	Swamy KMK	1	9	11.11	Vl Coll Pharm, Dept Pharmaceut Chem, Raichur, India.	Analyst 136 (7): 1339-1343 2011

Title	Authors				Affiliation	Citation
Immobilization of Bio-Macromolecules on Self-Assembled Monolayers: Methods and Sensor Applications	Samanta D, Sarkar A	2	2	100.00	Indian Assoc Cultivat Sci, Dept Organ Chem, Calcutta 700032, India	Chem Soc Rev 40 (5): 2567-2592 2011
Short-Peptide-Based Hydrogel: A Template For the in Situ Synthesis of Fluorescent Silver Nanoclusters by Using Sunlight	Adhikari B, Banerjee A	2	2	100.00	Indian Assoc Cultivat Sci, Dept Biol Chem, Calcutta 700032, India.	Chem-Eur J 16 (46): 13698-13705 Dec 2010
Nature of Cl Center Dot Center Dot Center Dot Cl Intermolecular Interactions Via Experimental and Theoretical Charge Density Analysis: Correlation of Polar Flattening Effects With Geometry	Hathwar VR, Row TNG	2	2	100.00	Indian Inst Sci, Solid State & Struct Chem Unit, Bangalore 560012, Karnataka, India.	J Phys Chem A 114 (51): 13434-13441 Dec 30 2010
Insight Into Adsorption Equilibrium, Kinetics and Thermodynamics of Malachite Green Onto Clayey Soil of Indian Origin	Saha P, Chowdhury S, Gupta S, Kumar I	4	4	100.00	Natl Inst Technol Durgapur, Dept Biotechnol, Durgapur 713209, Wb, India.	Chem Eng J 165 (3): 874-882 Dec 15 2010
Doping Cu in Semiconductor Nanocrystals: Some Old and Some New Physical Insights	Srivastava BB, Jana S, Pradhan N	3	3	100.00	Indian Assoc Cultivat Sci, Ctr Adv Mat, Jadavpur 700032, India.	J Am Chem Soc 133 (4): 1007-1015 Feb 2 2011
Sequential One-Pot Combination of Multi-Component and Multi-Catalysis Cascade Reactions: An Emerging Technology in Organic Synthesis	Ramachary DB, Jain S	2	2	100.00	Univ Hyderabad, Sch Chem, Hyderabad 500046, Andhra Pradesh, India.	Org Biomol Chem 9 (5): 1277-1300 2011
Ion Beam Induced Surface and Interface Engineering	Jain IP, Agarwal G	2	2	100.00	Univ Rajasthan, Mat Sci Lab, Ctr Nonconvent Energy Resources, Jaipur 302004, Rajasthan, India.	Surf Sci Rep 66 (3-4): 77-172 Mar 2011

Title	Authors			%	Affiliation	Journal
Synthesis and Characterization of Co0.8zn0.2fe2o4 Nanoparticles	Gonsalves LR, Verenkar VMS	2	2	66.67	Goa Univ, Dept Chem, Taleigao Plateau 403206, Goa, India.	J Therm Anal Calorim 104 (3): 869-873 Jun 2011
Defining the Hydrogen Bond: An Account (Iupac Technical Report)	Arunan E, Desiraju GR	2	14	14.29	Indian Inst Sci, Bangalore 560012, Karnataka, India.	Pure Appl Chem 83 (8): 1619-1636 2011
Luminescent Quantum Clusters of Gold in Transferrin Family Protein, Lactoferrin Exhibiting Fret	Xavier PL, Chaudhari K, Verma PK, Pal SK, Pradeep T	5	5	100.00	Indian Inst Technol, Madras 600036, Tamil Nadu, India. Satyendra Nath Bose Natl Ctr Basic Sci, Dept Chem Biol & Macromol Sci, Unit Nanosci & Technol, Calcutta 700098, India.	Nanoscale 2 (12): 2769-2776 Nov 2010
A One-Pot Catalysis: The Strategic Classification With Some Recent Examples	Patil NT, Shinde VS, Gajula B	3	3	100.00	Indian Inst Chem Technol, Csic, Hyderabad 500007, Andhra Pradesh, India.	Org Biomol Chem 10 (2): 211-224 2012
Synthesis of 1, 2, 3-Triazole-Fused Heterocycles Via Intramolecular Azide-Alkyne Cycloaddition Reactions	Majumdar KC, Ray K	2	2	100.00	Univ Kalyani, Dept Chem, Kalyani 741235, W Bengal, India	Synthesis-Stuttgart (23): 3767-3783 Dec 2011
A Spin-Adapted Size-Extensive State-Specific Multi-Reference Perturbation Schemes. I. Formal Developments	Mukherjee D	1	4	25.00	Indian Assoc Cultivat Sci, Raman Ctr Atom Mol & Opt Sci, Jadavpur Kolkata 700032, India	J Chem Phys 136 (2): Art. No.-024105 Jan 14 2012
Visible Light Photoredox Catalysis: Generation and Addition of N-Aryltetrahydroisoquinoline-Derived Alpha-Amino Radicals to Michael Acceptors	Jadhav D, Pandey G	2	4	50.00	Natl Chem Lab, Div Organ Chem, Dr Homi Bhabha Rd, Pune 411008, Maharashtra, India	Org Lett 14 (3): 672-675 Feb 3 2012

Title	Authors				Affiliation	Citation
Synthesis of the First Heterometalic Star-Shaped Oxido-Bridged Mncu3 Complex and Its Conversion Into Trinuclear Species Modulated by Pseudohalides (N-3(-), Ncs- and Nco-): Structural Analyses and Magnetic Properties	Biswas S, Naiya S, Ghosh A	3	4	75.00	Univ Calcutta, Univ Coll Sci, Dept Chem, Calcutta 700009, India.	Dalton Trans 41 (2): 462-473 2012
A Spin-Adapted Size-Extensive State-Specific Multi-Reference Perturbation Theory With Various Partitioning Schemes. Ii. Molecular Applications	Mukherjee D	1	4	25.00	Indian Assoc Cultivat Sci, Raman Ctr Atom Mol & Opt Sci, Jadavpur Kolkata 700032, India.	J Chem Phys 136 (2): Art. No.-024105 Jan 14 2012
Highly Selective Fluorescence Turn-On Chemodosimeter Based on Rhodamine For Nanomolar Detection of Copper Ions	Kumar M, Kumar N, Bhalla V, Sharma PR, Kaur T	5	5	100.00	Guru Nanak Dev Univ, Dept Chem, Ugc Sponsored Ctr Adv Studies 1, Amritsar 143005, Punjab, India. Indian Inst Integrat Med, Dept Canc Pharmacol, Jammu 180001, India.	Org Lett 14 (1): 406-409 Jan 6 2012
Strategies and Tactics in Olefin Metathesis	Kotha S, Dipak MK	2	2	100.00	Indian Inst Technol, Dept Chem, Bombay 400076, Maharashtra, India.	Tetrahedron 68 (2): 397-421 Jan 14 2012
From Anomalies in Neat Liquid to Structure, Dynamics and Function in the Biological World	Bagchi B	1	1	100.00	Indian Inst Sci, Solid State & Struct Chem Unit, Bangalore 12, Karnataka, India.	Chem Phys Lett 529: 1-9 Mar 9 2012
Self-Assembled Gelators For Organic Electronics	Babu SS, Prasanthkumar S, Ajayaghosh A	3	3	100.00	Niist, Photosci & Photon Grp, Chem Sci & Technol Div, Csir, Trivandrum, Kerala, India.	Angew Chem Int Ed 51 (8): 1766-1776 2012

Title	Authors				Affiliation	Reference
Structure-Activity Relationship Study of Copper(II) Complexes With 2-Oxo-1, 2-Dihydroquinoline-3-Carbaldehyde (4'-Methylbenzoyl) Hydrazone: Synthesis, Structures, Dna and Protein Interaction Studies, Antioxidative and Cytotoxic Activity	Raja DS, Natarajan K	2	3	66.67	Bharathiar Univ, Dept Chem, Coimbatore 641046, Tamil Nadu, India.	J Biol Inorg Chem 17 (2): 223-237 Feb 2012
Analysis of in Vitro Bioactivity Data Extracted From Drug Discovery Literature and Patents: Ranking 1654 Human Protein Targets by Assayed Compounds and Molecular Scaffolds	Boppana K, Jagarlapudi Sarma ARP	2	4	50.00	Gvk Biosci Pvt Ltd, Hyderabad 500037, Andhra Pradesh, India.	J Cheminformatics 3: Art. No.-14 May 13 2011
Solvent-Free Sonochemical One-Pot Three-Component Synthesis of 2H-Indazolo[2, 1-B]Phthalazine-1, 6, 11-Triones and 1H-Pyrazolo[1, 2-B] Phthalazine-5, 10-Diones	Shukla G, Verma RK, Verma GK, Singh MS	4	4	100.00	Banaras Hindu Univ, Fac Sci, Dept Chem, Varanasi 221005, Uttar Pradesh, India.	Tetrahedron Lett 52 (52): 7195-7198 Dec 28 2011
Supramolecular Coordination: Self-Assembly of Finite Two- and Three-Dimensional Ensembles	Partha Sarathi Mukherjee	1	3	33.33	Department of Inorganic and Physical Chemistry, Indian Institute of Science, Bangalore 560012, India	Chem Rev 111 (11): 6810-6918 Nov 2011

(d) Clinical Medicine

Title	Indian Authors	No. of Indian Authors	Total Author	Ratio I/F*	Institue Name	Source
Appropriate Body-Mass Index For Asian Populations and Its Implications For Policy and Intervention Strategies	Anura Kurpad, Srinath K Reddy, Yajnik CS	3	28	10.71	King Edward Mem Hosp, Diabet Unit, Pune, Maharashtra, India. All India Inst Med Sci, New Delhi, India. St Johns Med Coll, Inst Populat Hlth & Clin Res, Bangalore, Karnataka, India.	Lancet 363 (9403): 157-163 Jan 10 2004
Dabigatran Versus Warfarin in Patients With Atrial Fibrillation.	Xavier D	1	20	5.00	St Johns Natl Acad Hlth Sci, Bangalore, Karnataka, India.	N Engl J Med 361 (12): 1139-1151 Sep 17 2009
Gefitinib Plus Best Supportive Care in Previously Treated Patients With Refractory Advanced Non-Small-Cell Lung Cancer: Results From A Randomised, Placebo-Controlled, Multicentre Study (Iressa Survival Evaluation in Lung Cancer)	Purvish Parikh	1	11	9.09	Tata Mem Hosp, Bombay 400012, Maharashtra, India.	Lancet 366 (9496): 1527-1537 Oct-Nov 2005
Phase Iii Study Comparing Cisplatin Plus Gemcitabine With Cisplatin Plus Pemetrexed in Chemotherapy-Naive Patients With Advanced-Stage Non-Small-Cell Lung Cancer	Parikh P, Patil S, Raghunadharao Digumarti	3	20	15.00	Tata Mem Hosp, Mumbai 400012, Maharashtra, India. Nizams Inst Med Sci, Hyderabad, Andhra Pradesh, India. Bangalore Inst Oncol, Bangalore, Karnataka, India.	J Clin Oncol 26 (21): 3543-3551 Jul 20 2008

Title	Authors				Affiliation	Journal
Oncomine: A Cancer Microarray Database and Integrated Data-Mining Platform	Shanker K, Deshpande N	2	9	22.22	Inst Bioinformat, Bangalore, Karnataka, India	Neoplasia 6 (1): 1-6 Jan-Feb 2004
A Surgical Safety Checklist to Reduce Morbidity and Mortality in A Global Population.	Joseph S, Kumar A, Singh Chauhan H	3	15	20.00	St Stephens Hosp, New Delhi, India.	N Engl J Med 360 (5): 491-499 Jan 29 2009
Developmental Plasticity and Human Health	Deb D	1	15	6.67	Ctr Interdisciplinary Studies, Calcutta 700123, India.	Nature 430 (6998): 419-421 Jul 22 2004
Drotrecogin Alfa (Activated) For Adults With Severe Sepsis and A Low Risk of Death	Garg R	1	16	6.25	Metro Hosp & Heart Inst, Noida, Uttar Pradesh, India.	N Engl J Med 353 (13): 1332-1341 Sep 29 2005
Large-Scale Meta-Analysis of Cancer Microarray Data Identifies Common Transcriptional Profiles of Neoplastic Transformation and Progression	Shanker K, Deshpande N, Ghosh D, Pandey A,	4	9	44.44	Inst Bioinformat, Bangalore 560066, Karnataka, India.	Proc Nat Acad Sci USA 101 (25): 9309-9314 Jun 22 2004
Worldwide Distribution of Human Papillomavirus Types in Cytologically Normal Women in the International Agency For Research on Cancer Hpv Prevalence Surveys: A Pooled Analysis	Anh PTH	1	20	5.00	Christian Fellowship Community Hlth Area, Ambilikai, Tamil Nadu, India.	Lancet 366 (9490): 991-998 Sep 17 2005
Human Papillomavirus and Oral Cancer: The International Agency For Research on Cancer Multicenter Study	Balaram P, Rajkumar T, Sridhar H	3	23	13.04	Reg Canc Ctr, Trivandrum 695011, Kerala, India. Women India Assoc, Inst Canc, Madras, Tamil Nadu, India. Kidwai Mem Inst Oncol, Bangalore, Karnataka, India.	J Nat Cancer Inst 95 (23): 1772-1783 Dec 3 2003

Title	Authors			Affiliation	Citation
Molecular Predictors of Outcome With Gefitinib in A Phase Iii Placebo-Controlled Study in Advanced Non-Small-Cell Lung Cancer	Parikh P, Pereira JR	2	10.53	Tata Mem Hosp, Bombay 400012, Maharashtra, India.	J Clin Oncol 24 (31): 5034-5042 Nov 1 2006
Emergence of A New Antibiotic Resistance Mechanism in India, Pakistan, and the UK: A Molecular, Biological, and Epidemiological Study	Uma Chaudhary, Karthikeyan K Kumarasamy, Anil V Kumar, Madhu Sharma, Mandayam A Thirunarayan	5	16.13	Univ Madras, Dept Microbiol, Dr Alm Pg Ibms, Madras, Tamil Nadu, India. Apollo Gleneagles Hosp, Dept Microbiol, Calcutta, India. Apollo Hosp, Dept Microbiol, Madras, Tamil Nadu, India. Banaras Hindu Univ, Dept Microbiol, Inst Med Sci, Varanasi 221005, Uttar Pradesh, India.	Lancet Infect Dis 10 (9): 597-602 Sep 2010
Peritoneal Dialysis-Related Infections Recommendations: 2005 Update	Gupta A	1	6.67	Sanjay Gandhi Postgrad Inst Med Sci, Lucknow 226014, Uttar Pradesh, India.	Periton Dialysis Int 25 (2): 107-131 Mar-Apr 2005
The Indian Diabetes Prevention Programme Shows That Lifestyle Modification and Metformin Prevent Type 2 Diabetes in Asian Indian Subjects With Impaired Glucose Tolerance (Idpp-1)	Ramachandran A, Snehalatha C, Mary S, Mukesh B, Bhaskar AD, Vijay V, IDPP	7	100.00	Who, Collaborating Ctr Res Educ & Training Diabet, Mv Hosp Diabet, Diabet Res Ctr, Madras, Tamil Nadu, India.	Diabetologia 49 (2): 289-297 Feb 2006
Relation of Serial Changes in Childhood Body-Mass Index to Impaired Glucose Tolerance in Young Adulthood	Bhargava SK, Sachdev HS, Lakshmy R, Barker DJP, Biswas SKD, Ramji S, Reddy KS	7	70.00	E-6-12 Vasant Vihar, New Delhi 110057, India. Sunder Lal Jain Hosp, Dept Pediat, Delhi, India. Dept Pediat, New Delhi, India. All India Inst Med Sci, Dept Cardiol, New Delhi, India.Indian Council Med Res, New Delhi, India.	N Engl J Med 350 (9): 865-875 Feb 26 2004

Title	Author				Affiliation	Citation
Alcohol, Tobacco and Breast Cancer - Collaborative Reanalysis of Individual Data From 53 Epidemiological Studies, Including 58515 Women With Breast Cancer and 95067 Women Without the Disease	Gajalakshmi V	1	220	0.45	Chennai Cancer Inst, Madras, India.	Brit J Cancer 87 (11): 1234-1245 Nov 18 2002
Phase Iii Study of Letrozole Versus Tamoxifen As First-Line Therapy of Advanced Breast Cancer in Postmenopausal Women: Analysis of Survival and Update of Efficacy From the International Letrozole Breast Cancer Group	Bhatnagar A	1	20	5.00	Kidwai Mem Inst Oncol, Bangalore, Karnataka, India.	J Clin Oncol 21 (11): 2101-2109 Jun 1 2003
Effect of Glucose-Insulin-Potassium Infusion on Mortality in Patients With Acute St-Segment Elevation Myocardial Infarction - the Create-Ecla Randomized Controlled Trial	Prem Pais	1	14	7.14	St Johns Med Coll, Natl Acad Hlth Sci, Bangalore, Karnataka, India.	Jama-J Am Med Assn 293 (4): 437-446 Jan 26 2005
Update on the Pathophysiology and Classification of Von Willebrand Disease: A Report of the Subcommittee on Von Willebrand Factor	Srivastava A,	1	22	4.55	Christian Med Coll & Hosp, Dept Hematol, Vellore, Tamil Nadu, India.	J Thromb Haemost 4 (10): 2103-2114 Oct 2006

Title	Author				Affiliation	Reference
Oral Miltefosine For Indian Visceral Leishmaniasis	Sundar S, Jha TK, Thakur CP	3	8	37.50	Kala Azar Res Ctr, Brahmpura, Muzaffarpur, India. Banaras Hindu Univ, Inst Med Sci, Varanasi 221005, Uttar Pradesh, India. Balaji Utthan Sanastan, Patna, Bihar, India.	N Engl J Med 347 (22): 1739-1746 Nov 28 2002
Rapid Molecular Detection of Tuberculosis and Rifampin Resistance	Shenai S, Rodrigues C	2	19	10.53	Hinduja Natl Hosp, Bombay, Maharashtra, India. Med Res Ctr Hinduja, Bombay, Maharashtra, India.	N Engl J Med 363 (11): 1005-1015 Sep 9 2010
Maternal and Child Undernutrition 2 - Maternal and Child Undernutrition: Consequences For Adult Health and Human Capital	Sachdev HS	1	7	14.29	Sitaram Bhartia Inst Sci & Res, New Delhi, India.	Lancet 371 (9609): 340-357 Jan-Feb 2008
Low Dose Mifepristone and Two Regimens of Levonorgestrel For Emergency Contraception: A Who Multicentre Randomised Trial	Mittal Suneeta	1	17	5.88	All India Inst Med Sci, New Delhi, India.	Lancet 360 (9348): 1803-1810 Dec 7 2002
Maternal and Child Undernutrition 3 - What Works? Interventions For Maternal and Child Undernutrition and Survival	Sachdev HPS	1	11	9.09	Sitaram Bhartia Inst Sci & Res, New Delhi, India.	Lancet 371 (9610): 417-440 Feb 2 2008
Efficacy and Safety Comparison of Liraglutide, Glimepiride, and Placebo, All in Combination With Metformin, in Type 2 Diabetes the Lead (Liraglutide Effect and Action in Diabetes)-2 Study	Mitha IH	1	9	11.11	Seth Gs Med Coll, Bombay, Maharashtra, India. King Edward Mem Hosp, Bombay, Maharashtra, India.	Diabetes Care 32 (1): 84-90 Jan 2009

Title	Author				Institution	Citation
Micafungin Versus Liposomal Amphotericin B For Candidaemia and Invasive Candidosis: A Phase Iii Randomised Double-Blind Trial	Raghunadharao D, Sekhon JS, Ramasubramanian V	3	21	14.29	Nizams Inst Med Sci, Hyderabad, Andhra Pradesh, India. Dayanand Med Coll & Hosp, Ludhiana, Punjab, India. Apollo Hosp, Madras, Tamil Nadu, India.	Lancet 369 (9572): 1519-1527 May 5 2007
Variable Host-Pathogen Compatibility in Mycobacterium Tuberculosis	Sujatha Narayanan	1	13	7.69	Tuberculosis Res Ctr, Dept Immunol, Madras 600031, Tamil Nadu, India.	Proc Nat Acad Sci USA 103 (8): 2869-2873 Feb 21 2006
Micafungin Versus Caspofungin For Treatment of Candidemia and Other Forms of Invasive Candidiasis	Digumarti R, Talwar D	2	17	11.76	Nizam Inst Med Sci, Hyderabad, Andhra Pradesh, India.	Clin Infect Dis 45 (7): 883-893 Oct 1 2007
Maintenance Pemetrexed Plus Best Supportive Care Versus Placebo Plus Best Supportive Care For Non-Small-Cell Lung Cancer: A Randomised, Double-Blind, Phase 3 Study	Madhavan J	1	20	5.00	Reg Canc Ctr, Trivandrum 695011, Kerala, India.	Lancet 374 (9699): 1432-1440 Oct 24 2009
Clinical Efficacy of Sildenafil in Primary Pulmonary Hypertension - A Randomized, Placebo-Controlled, Double-Blind, Crossover Study	Sastry BKS, Narasimhan C, Reddy NK, Raju BS	4	4	100.00	Care Hosp, Dept Cardiol, Exhibit Rd, Nampally, Hyderabad 500001, Andhra Pradesh, India.	J Amer Coll Cardiol 43 (7): 1149-1153 Apr 7 2004

Telmisartan to Prevent Recurrent Stroke and Cardiovascular Events	Xavier D, Agarwal A, Agrawal RR, Anandan AM, Anandhi V, Babu GK, Bandishti S, Bhargava A, Bhargava N, Bharani A, Bhatt A, Chidambaram N, Dewan Y, Dinaker M, Joshi R, Joshi S, Kalanidhi A, Kalantri SP, Kothari S, Kumar A, Kumar P, Jain V, Mehndiratta MM, Mijar S, Mishra V, Murali S, Muralidharan RS, Murthy JMK, Nair R, Narayanan JT, Panwar RB, Patel P, Poncha F, Prasad VVR, Rath A, Reddy BCS, Rohatgi A, Roy AK, Sadanandham S, Salam A, Sarma GRK, Singh H, Singh Y, Shanmugasundaram S, Sharma S, Sivakumar S, Sundararajan R, Sundararajan T, Tukaram U, Umarani R, Varma S, Velmurugendran CU, Venkitachalam A, Verghese R, Vinayan KP, Vyas A, Wadia RS	57	1138	5.01	St Johns Med Coll, Bangalore, Karnataka, India.	N Engl J Med 359 (12): 1225-1237 Sep 18 2008
Pharmaceutical Approaches to Colon Targeted Drug Delivery Systems	Chourasia MK, Jain SK	2	2	100.00	Dr Hari Singh Gour Univ, Pharmaceut Res Projects Lab, Dept Pharmaceut Sci, Sagar, Mp, India.	J Pharm Pharm Sci 6 (1): 33-66 Jan-Apr 2003

Title	Authors				Institution	Reference
Hpv Screening For Cervical Cancer in Rural India	Bhagwan M Nene, Surendra S Shastri: Kasturi Jayant: Atul M Budukh, Sanjay Hingmire, Sylla G Malvi, Ranjit Thorat, Ashok Kothari, Roshan Chinoy, Rohini Kelkar, Shubhada Kane, Sangeetha Desai, Vijay R Keskar, Raghevendra Rajeshwarkar, Nandkumar Panse, Ketayun A Dinshaw	16	18	88.89	Nargis Dutt Mem Canc Hosp, Tata Mem Ctr, Rural Canc Project, Barshi, India. Tata Mem Hosp, Bombay, Maharashtra, India.	N Engl J Med 360 (14): 1385-1394 Apr 2 2009
Dasatinib Versus Imatinib in Newly Diagnosed Chronic-Phase Chronic Myeloid Leukemia	Agarwal U, Gangadharan VP, Mathew V, Narayan G, Prabhash K, Saikia T, Shah S	7	20	35.00	Hematooncol Clin, Ahmadabad, Gujarat, India.	N Engl J Med 362 (24): 2260-2270 Jun 17 2010
Aspirin and Extended-Release Dipyridamole Versus Clopidogrel For Recurrent Stroke	Pais P	1	34	2.94	St Johns Med Coll, Bangalore, Karnataka, India.	N Engl J Med 359 (12): 1238-1251 Sep 18 2008
Scarless Single Port Transumbilical Nephrectomy and Pyeloplasty: First Clinical Report	Rao PP, Desai MR, Mishra S	3	8	37.50	Muljibhai Patel Urol Hosp, Nadiad, India.	Bju Int 101 (1): 83-88 Jan 2008
Diabetes in Asia Epidemiology, Risk Factors, and Pathophysiology	Chittaranjan S Yajnik	1	7	14.29	Kem Hosp Res Ctr, Diabet Unit, Pune, Maharashtra, India.	Jama-J Am Med Assn 301 (20): 2129-2140 May 27 2009
Prevention of Hiv-1 Infection With Early Antiretroviral Therapy	Hakim James G, Kumwenda Johnstone	2	34	5.88	Yr Gaitonde Ctr Aids Res & Educ, Chennai, Tamil Nadu, India. Natl Aids Res Inst, Pune, Maharashtra, India.	N Engl J Med 365 (6): 493-505 Aug 11 2011

Title	Author				Affiliation	Journal
Apo B Versus Cholesterol in Estimating Cardiovascular Risk and in Guiding Therapy: Report of the Thirty-Person/Ten-Country Panel	Reddy KS	1	30	3.33	All India Inst Med Sci, New Delhi, India.	J Intern Med 259 (3): 247-258 Mar 2006
Safety of Coronary Sirolimus-Eluting Stents in Daily Clinical Practice - One-Year Follow-Up of the E-Cypher Registry	Seth A	1	12	8.33	Max Heart & Vasc Inst, New Delhi, India.	Circulation 113 (11): 1434-1441 Mar 21 2006
Mortality Associated With Aprotinin During 5 Years Following Coronary Artery Bypass Graft Surgery	Rajiv Juneja	1	13	7.69	Escorts Heart Inst, Dept Anesthesia, New Delhi, India.	Jama-J Am Med Assn 297 (5): 471-479 Feb 7 2007
International Network of Cancer Genome Projects	Bhan MK, Rao TS, Sarin R, Majumder P, Majumder PP	5	260	1.92	Natl Inst Biomed Genom, Kalyani 741251, W Bengal, India.Tata Mem Hosp, Adv Ctr Treatment, Res & Educ Canc, Navi Mumbai 41021, Maharashtra, India. Govt India, Minist Sci & Technol, Dept Biotechnol, New Delhi 110003, India.	Nature 464 (7291): 993-998 Apr 15 2010
Liraglutide Vs Insulin Glargine and Placebo in Combination With Metformin and Sulfonylurea Therapy in Type 2 Diabetes Mellitus (Lead-5 Met+Su): A Randomised Controlled Trial	Sethi BK	1	9	11.11	Care Hosp, Hyderabad, Andhra Pradesh, India.	Diabetologia 52 (10): 2046-2055 Oct 2009
T-Cell Accumulation and Regulated on Activation, Normal T Cell Expressed and Secreted Upregulation in Adipose Tissue in Obesity	Ghosh S	1	11	9.09	Natl Inst Nutr, Hyderabad 500007, Andhra Pradesh, India.	Circulation 115 (8): 1029-1038 Feb 27 2007

Title	Author(s)				Affiliation	Citation
Apixaban in Patients With Atrial Fibrillation	Prem Pais	1	30	3.33	St Johns Med Coll, Bangalore, Karnataka, India. Res Inst, Bangalore, Karnataka, India.	N Engl J Med 364 (9): 806-817 Mar 3 2011
Recent Advances on Surface Engineering of Magnetic Iron Oxide Nanoparticles and Their Biomedical Applications	Gupta AK, Naregalkar RR, Vaidya VD, Gupta M	4	4	100.00	Tpl, Torrent Res Ctr, Formulat & Dev Dept, Lab Nanoparticle Res, Village Bhat 382428, Gujarat, India.	Nanomedicine 2 (1): 23-39 Feb 2007
Eular/Pres Endorsed Consensus Criteria For the Classification of Childhood Vasculitides	Bagga A	1	12	8.33	All India Inst Med Sci, Dept Paediat, New Delhi, India.	Ann Rheum Dis 65 (7): 936-941 Jul 2006
Lack of Effectiveness of Cellulose Sulfate Gel For the Prevention of Vaginal Hiv Transmission	Solomon S, Krishnan AK, Pradeep BS,	3	14	21.43	Yr Gaitonde Ctr Aids Res & Educ Yrg Care, Madras, Tamil Nadu, India.St Johns Med Coll, Inst Populat Hlth & Clin Res, Bangalore, Karnataka, India.	N Engl J Med 359 (5): 463-472 Jul 31 2008
Apixaban Versus Warfarin in Patients With Atrial Fibrillation	Prem Pais	1	1151	0.09	St Johns Med Coll, Bangalore, Karnataka, India.	N Engl J Med 365 (11): 981-992 Sep 15 2011
Epidemiology of Antituberculosis Drug Resistance 2002-07: An Updated Analysis of the Global Project on Anti-Tuberculosis Drug Resistance Surveillance	Paramasivan CN	1	16	6.25	Tb Research Centre, Indian Council of Medical Research, Chennai, India	Lancet 373 (9678): 1861-1873 May-Jun 2009
Risk Factors For Ischaemic and Intracerebral Haemorrhagic Stroke in 22 Countries (The Interstroke Study): A Case-Control Study	Denis Xavier, Prem Pais	2	24 (total collaborators 283- investigators including 24)	#VALUE!	St Johns Med Coll & Res Inst, Bangalore, Karnataka, India.	Lancet 376 (9735): 112-123 Jul 10 2010

Title	Author(s)				Affiliation	Citation
Efficacy and Safety of Lapatinib As First-Line Therapy For Erbb2-Amplified Locally Advanced or Metastatic Breast Cancer	Doval DC, Nag S	2	14	14.29	Rajiv Gandhi Canc Inst & Res Ctr, New Delhi, India. Jehangir Hosp & Med Ctr, Pune, Maharashtra, India.	J Clin Oncol 26 (18): 2999-3005 Jun 20 2008
Global Mental Health 3 - Treatment and Prevention of Mental Disorders in Low-Income and Middle-Income Countries	Vikram Patel	1	10	10.00	Sangath Ctr, Alto Porvovim 403521, Goa, India.	Lancet 370 (9591): 991-1005 Sep 15 2007
Concise Review: Isolation and Characterization of Cells From Human Term Placenta: Outcome of the First International Workshop on Placenta Derived Stem Cells	Sankar V	1	26	3.85	Univ Madras, Dept Anat, Madras, Tamil Nadu, India.	Stem Cells 26 (2): 300-311 Feb 2008
Effect of Visual Screening on Cervical Cancer Incidence and Mortality in Tamil Nadu, India: A Cluster-Randomised Trial	Pulikkottil Okkuru Esmy, Rajamanickam Rajkumar, Rajaraman Swaminathan, Sivanandam Shanthakumari, Jacob Cherian	5	8	62.50	Christian Fellowship Community Hlth Ctr, Ambilikai, Tamil Nadu, India. Psg Inst Med Sci & Res, Dept Prevent Med, Coimbatore, Tamil Nadu, India. Canc Inst Wia, Div Epidemiol, Madras, Tamil Nadu, India. Psg Inst Med Sci & Res, Dept Pathol, Coimbatore, Tamil Nadu, India.	Lancet 370 (9585): 398-406 Aug 4 2007
A Phase Iii Study of Belatacept-Based Immunosuppression Regimens Versus Cyclosporine in Renal Transplant Recipients (Benefit Study)	Darji P	1	13	7.69	Sterling Hosp, Ahmadabad, Gujarat, India.	Am J Transplant 10 (3): 535-546 Mar 2010

Title	Authors			%	Affiliation	Source
Risk Factors For Early Myocardial Infarction in South Asians Compared With Individuals in Other Countries	Prashant Joshi, Prem Pais, Srinath Reddy, Prabhakaran Dorairaj	4	11	36.36	Dept Med, Nagpur, Maharashtra, India. St Johns Med Coll, Dept Med, Bangalore, Karnataka, India. All India Inst Med Sci, New Delhi, India	Jama-J Am Med Assn 297 (3): 286-294 Jan 17 2007
Single-Port-Access Nephrectomy and Other Laparoscopic Urologic Procedures Using A Novel Laparoscopic Port (R-Port)	Prashanth Rao, Pradeep Rao	2	3	66.67	Mamata Hosp, Bombay, Maharashtra, India.	Urology 72 (2): 260-263 Aug 2008
Grand Challenges in Chronic Non-Communicable Diseases	Ganguly N	1	19	5.26	Indian Council Med Res, New Delhi 110029, India.	Nature 450 (7169): 494-496 Nov 22 2007
Birth Weight and Risk of Type 2 Diabetes A Systematic Review	Bhargava SK, Sachdev HS	2	32	6.25	SI Jain Hosp, Dept Paediat, Delhi, India. Sitaram Bhartia Inst Sci & Res, Dept Paediat & Clin Epidemiol, New Delhi, India.	Jama-J Am Med Assn 300 (24): 2886-2897 Dec 24 2008
Efficacy and Safety of Dabigatran Compared With Warfarin At Different Levels of International Normalised Ratio Control For Stroke Prevention in Atrial Fibrillation: An Analysis of the Re-Ly Trial.	Prem Pais	1	14	7.14	St Johns Med Coll & Res Inst, Bangalore, Karnataka, India.	Lancet 376 (9745): 975-983 Sep 18 2010
A Treatment Planning Study Comparing Volumetric Arc Modulation With Rapidarc and Fixed Field Imrt For Cervix Uteri Radiotherapy	Dinshaw KA, Shrivastava SK, Mahantshetty U, Engineer R, Deshpande DD, Jamema SV	6	11	54.55	Tata Mem Hosp, Dept Radiat Oncol & Med Phys, Bombay 400012, Maharashtra, India.	Radiother Oncol 89 (2): 180-191 Nov 2008
Global Vitamin D Status and Determinants of Hypovitaminosis D	Mithal A	1	9	11.11	Indraprastha Apollo Hosp, New Delhi 110044, India.	Osteoporosis Int 20 (11): 1807-1820 Nov 2009

Title	Author				Affiliation	Citation
Trastuzumab Plus Anastrozole Versus Anastrozole Alone For the Treatment of Postmenopausal Women With Human Epidermal Growth Factor Receptor 2-Positive, Hormone Receptor-Positive Metastatic Breast Cancer: Results From the Randomized Phase Iii Tandem	Poonamalle P Bapsy, Ashok Vaid	2	12	16.67	Kidwai Mem Inst Oncol, Bangalore, Karnataka, India. Rajiv Gandhi Canc Inst, New Delhi, India.	J Clin Oncol 27 (33): 5529-5537 Nov 20 2009
Computationally Guided Photothermal Tumor Therapy Using Long-Circulating Gold Nanorod Antennas	Nanda Kishor Bandaru, Sarit K Das	2	7	28.57	Indian Inst Technol, Dept Mech Engn, Madras 600036, Tamil Nadu, India.	Cancer Res 69 (9): 3892-3900 May 1 2009
Global Mental Health 6 - Scale Up Services For Mental Disorders: A Call For Action	Sudipto Chatterjee, Patel V, Jacob KS, Thara R	4	40 (7+Group)	#VALUE!	Sangath Ctr, Alto Porvorim 403521, Goa, India. Christian Med Coll & Hosp, Vellore, Tamil Nadu, India. Schizophrenia Res Fdn, Madras, Tamil Nadu, India.	Lancet 370 (9594): 1241-1252 Oct 6 2007
Intravenous Platelet Blockade With Cangrelor During Pci	Jaspal Arneja	1	21 (307 champion platform Investigators)	#VALUE!	Arneja Heart Inst, Nagpur, Maharashtra, India.	N Engl J Med 361 (24): 2330-2341 Dec 10 2009
Platelet Inhibition With Cangrelor in Patients Undergoing Pci	Keyur H Parikh	1	20	5.00	Heart Care Clin, Ahmadabad, Gujarat, India.	N Engl J Med 361 (24): 2318-2329 Dec 10 2009
Chronic Obstructive Pulmonary Disease in Non-Smokers	Sundeep S Salvi	1	2	50.00	Chest Res Fdn, Pune 411014, Maharashtra, India	Lancet 374 (9691): 733-743 Aug-Sep 2009

Title	Authors				Affiliation	Citation
Gemcitabine Plus Paclitaxel Versus Paclitaxel Monotherapy in Patients With Metastatic Breast Cancer and Prior Anthracycline Treatment	Nag Shona M, Jagdev S Sekhon	2	12	16.67	Jehangir Hosp, Pune, Maharashtra, India. Dayanand Med Coll & Hosp, Ludhiana, Punjab, India.	J Clin Oncol 26 (24): 3950-3957 Aug 20 2008
Extended-Dose Nevirapine to 6 Weeks of Age For Infants to Prevent Hiv Transmission Via Breastfeeding in Ethiopia, India, and Uganda: An Analysis of Three Randomised Controlled Trials	Bhore AV, Bhosale R, Varadhrajan V, Gupte N, Sastry J, Suryavanshi N, Tripathy S, Chaudhary MA, Gupta A, Nayak U, Ram M, Shankar A	12	235	5.11	Bj Med Coll, Pune, Maharashtra, India. Pune Mit, Pune, Maharashtra, India. Natl Aids Res Inst, Pune, Maharashtra, India.	Lancet 372 (9635): 300-313 Jul-Aug 2008
Priority Actions For the Non-Communicable Disease Crisis	Shah Ebrahim, K Srinath Reddy	2	44 (for the Lancet NCD Action Group and the NCD Alliance)	#VALUE!	S Asia Network Chron Dis, New Delhi, India. Publ Hlth Fdn India, New Delhi, India.	Lancet 377 (9775): 1438-1447 Apr 23 2011
Vitamin B-12 and Folate Concentrations During Pregnancy and Insulin Resistance in the Offspring: The Pune Maternal Nutrition Study	Yajnik CS, Deshpande SS, Rao S, Fisher DJ, Bhat DS, Naik SS, Coyaji KJ, Joglekar CV, Joshi N, Lubree HG, Deshpande VU	11	15	73.33	Diabet Unit, 6Th Floor, Banoo Coyaji Bldg, Pune 411011, Maharashtra, India. Diabet Unit, Pune 411011, Maharashtra, India.Kem Hosp & Res Ctr, Diabet Unit, Pune, Maharashtra, India Agharkar Res Inst, Dept Biometry, Pune, Maharashtra, India	Diabetologia 51 (1): 29-38 Jan 2008
Health Professionals For A New Century: Transforming Education to Strengthen Health Systems in An Interdependent World	Srinath Reddy	1	19	5.26	Publ Hlth Fdn India, New Delhi, India.	Lancet 376 (9756): 1923-1958 Dec 4 2010

Title	Author				Affiliation	Citation
Global Burden of Acute Lower Respiratory Infections Due to Respiratory Syncytial Virus in Young Children: A Systematic Review and Meta-Analysis	Harish Nair	1	21	4.76	Publ Hlth Fdn India, New Delhi, India.	Lancet 375 (9725): 1545-1555 May 1 2010
Dissecting Lipid Raft Facilitated Cell Signaling Pathways in Cancer	Samir Kumar Patra	1	1	100.00	Canc Epigenet Res, Kalyani B-7-183, Nadia 741235, W Bengal, India., Canc Epigenet Res, Nadia 741235, W Bengal, India.	Bba-Rev Cancer 1785 (2): 182-206 Apr 2008
Chronic Diseases 5 - Prevention of Chronic Diseases: A Call to Action	Reddy S	1	5	20.00	Publ Hlth Fdn India, New Delhi, India.	Lancet 370 (9605): 2152-2157 Dec-Jan 2007
Key Gaps in the Knowledge of Plasmodium Vivax, A Neglected Human Malaria Parasite	Dhanpat K Kochar	1	7	14.29	Kothari Med & Res Inst, Bikaner, Rajasthan, India. Ag Hosp Bikaner, Bikaner, Rajasthan, India.	Lancet Infect Dis 9 (9): 555-566 Sep 2009
Snail and Slug Mediate Radioresistance and Chemoresistance by Antagonizing P53-Mediated Apoptosis and Acquiring A Stem-Like Phenotype in Ovarian Cancer Cells	Kurrey NK, Jalgaonkar SP, Joglekar AV, Ghanate AD, Chaskar PD, Doiphode RY, Bapat SA	7	7	100.00	Natl Ctr Cell Sci, Pune 411007, Maharashtra, India. Univ Pune, Inst Bioinformat & Biotechnol, Pune, Maharashtra, India.	Stem Cells 27 (9): 2059-2068 2009
Human Papillomavirus Genotype Attribution in Invasive Cervical Cancer: A Retrospective Cross-Sectional Worldwide Study	Asha Jain	1	68	1.47	Canc Prevent & Relief Soc, Raipur, Madhya Pradesh, India.	Lancet Oncol 11 (11): 1048-1056 Nov 2010

Title	Author		Count	%	Affiliation	Citation
New Insights Into the Management of Acne: An Update From the Global Alliance to Improve Outcomes in Acne Group	Raj Kubba	1	20	5.00	Delhi Dermatol Grp, New Delhi, India.	J Amer Acad Dermatol 60 (5): S1-S50 Suppl. S May 2009
Radial Versus Femoral Access For Coronary Angiography and Intervention in Patients With Acute Coronary Syndromes (Rival): A Randomised, Parallel Group, Multicentre Trial	Denis Xavier	1	18	5.56	St Johns Med Coll & Res Inst, Bangalore, Karnataka, India	Lancet 377 (9775): 1409-1420 Apr 23 2011
Efficacy and Safety of Belimumab in Patients With Active Systemic Lupus Erythematosus: A Randomised, Placebo-Controlled, Phase 3 Trial	Mathew Thomas	1	17	5.88	Kerala Inst Med Sci, Trivandrum, Kerala, India.	Lancet 377 (9767): 721-731 Feb-Mar 2011
Hiv Prevention, Treatment, and Care Services For People Who Inject Drugs: A Systematic Review of Global, Regional, and National Coverage	Atul Ambekar	1	9	11.11	National Drug Dependence Treatment Centre and Department of Psychiatry, All India Institute of Medical Sciences, New Delhi, India	Lancet 375 (9719): 1014-1028 Mar 20 2010
Peritoneal Dialysis-Related Infections Recommendations: 2010 Update	Amit Gupta	1	12	8.33	Sanjay Gandhi Postgrad Inst Med Sci, Lucknow, Uttar Pradesh, India.	Periton Dialysis Int 30 (4): 393-423 Jul 2010
Tuberculosis Control and Elimination 2010-50: Cure, Care, and Social Development	Chauhan LS	1	7	14.29	Minist Hlth & Family Welf, New Delhi, India.	Lancet 375 (9728): 1814-1829 May 22 2010

Title	Authors				Affiliation	Citation
Volumetric Modulated Arc Radiotherapy For Carcinomas of the Oro-Pharynx, Hypo-Pharynx and Larynx: A Treatment Planning Comparison With Fixed Field Imrt	Ghosh-Laskar S, Agarwal JP, Upreti RR, Budrukkar A, Murthy V, Deshpande DD, Shrivastava SK, Ketayun Ardeshir Dinshaw	8	12	66.67	Tata Mem Hosp, Mumbai 400012, Maharashtra, India.	Radiother Oncol 92 (1): 111-117 Jul 2009
Neratinib, An Irreversible Erbb Receptor Tyrosine Kinase Inhibitor, in Patients With Advanced Erbb2-Positive Breast Cancer	Ranade A, Badwe R	2	16	12.50	Deenanath Mangeshkar Hosp, Pune, Maharashtra, India. Tata Mem Hosp, Mumbai 400012, Maharashtra, India.	J Clin Oncol 28 (8): 1301-1307 Mar 10 2010
Longitudinal Study of the Assessment by Mri and Diffusion-Weighted Imaging of Tumor Response in Patients With Locally Advanced Breast Cancer Undergoing Neoadjuvant Chemotherapy	Sharma U, Danishad KKA, Seenu V, Jagannathan NR	4	4	100.00	All India Inst Med Sci, New Delhi 110029, India.	Nmr Biomed 22 (1): 104-113 Jan 2009
Transforming Growth Factor Beta Is Dispensable For the Molecular Orchestration of Th17 Cell Differentiation	Das A, Das G	2	9	22.22	Inst Mol Med, New Delhi 110020, India.Int Ctr Genet Engn & Biotechnol, Immunol Grp, New Delhi 110067, India.	J Exp Med 206 (11): 2407-2416 Oct 26 2009

Title	Authors				Affiliation	Citation
Acute-On-Chronic Liver Failure: Consensus Recommendations of the Asian Pacific Association For the Study of the Liver (Apasl)	Sarin SK, Kumar A, Chawla YK, Garg H, Madan K, Rastogi A, Sakhuja P, Shah S, Sharma BC, Sharma P, BR Thapa	11	27	40.74	Affiliated Univ Delhi, Gb Pant Hosp, Dept Gastroenterol, Jawahar Lal Nehru Rd, New Delhi 110002, India, Ilbs, Dept Hepatol, New Delhi 110070, India., Post Grad Inst Med Educ & Res, Dept Hepatol, Chandigarh, India; Ilbs, Dept Pathol, New Delhi 110070, India; Affiliated Univ Delhi, Gb Pant Hosp, Dept Pathol, New Delhi 110002, India.; Jaslok Hosp & Res Ctr, Dept Gastroenterol, Mumbai 400026, Maharashtra, India.; Post Grad Inst Med Educ & Res, Div Pediat Gastroenterol, Chandigarh, India.	Hepatol Int 3 (1): 269-282 Mar 2009
Gene Expression Profiling Reveals A New Classification of Adrenocortical Tumors and Identifies Molecular Predictors of Malignancy and Survival	Bertherat J	1	10	10.00	Hop Cochin, Ap Hp, Dept Endocrinol, Ctr Rare Adrenal Dis, Oncogenet Un, Dept Pathol, Unit Digest & Endocrine Surg, Cochin, Kerala, India	J Clin Oncol 27 (7): 1108-1115 Mar 1 2009
Feasibility, Diagnostic Accuracy, and Effectiveness of Decentralised Use of the Xpert Mtb/Rif Test For Diagnosis of Tuberculosis and Multidrug Resistance: A Multicentre Implementation Study	Joy S Michael, Kalaiselvan Sagadevan	2	22	9.09	Christian Med Coll & Hosp, Vellore, Tamil Nadu, India.	Lancet 377 (9776): 1495-1505 Apr-May 2011

Global Variation in the Prevalence and Severity of Asthma Symptoms: Phase Three of the International Study of Asthma and Allergies in Childhood (Isaac)	Shah J	1	6	16.67	Jaslok Hosp & Res Ctr, Bombay, Maharashtra, India.	Thorax 64 (6): 476-483 Jun 2009
Lasofoxifene in Postmenopausal Women With Osteoporosis	Usha Sriram	1	17	5.88	Associates Clin Endocrinol Educ & Res, Madras, Tamil Nadu, India.	N Engl J Med 362 (8): 686-696 Feb 25 2010
International Nosocomial Infection Control Consortium (Inicc) Report, Data Summary For 2003-2008, Issued June 2009	Subhash Kumar Todi	1	26	3.85	Amri Hosp, Calcutta, India.	Amer J Infect Control 38 (2): 95-U31 Mar 2010
Cardio-Renal Syndromes: Report From the Consensus Conference of the Acute Dialysis Quality Initiative	Sachin Soni	1	28	3.57	Mediciti Hosp, Div Nephrol, Hyderabad, Andhra Pradesh, India.	Eur Heart J 31 (6): 703-711C Mar 2010
The Metabolic Syndrome: Useful Concept or Clinical Tool? Report of A Who Expert Consultation	Ramachandran A	1	14	7.14	India Diabet Res Fdn, Chennai, Tamil Nadu, India.	Diabetologia 53 (4): 600-605 Apr 2010
Iof Position Statement: Vitamin D Recommendations For Older Adults	Mithal A	1	10	10.00	Indraprastha Apollo Hosp, New Delhi 110044, India.	Osteoporosis Int 21 (7): 1151-1154 Jul 2010
Diabetes in Asia	Ramachandran A, Snehalatha C	2	3	66.67	India Diabet Res Fdn, Madras 600008, Tamil Nadu, India.; Dr A Ramachandrans Diabet Hosp, Madras 600008, Tamil Nadu, India.	Lancet 375 (9712): 408-418 Jan-Feb 2010

Title	Authors				Affiliation	Reference
Coronary-Artery Bypass Surgery in Patients With Left Ventricular Dysfunction	Jain A, Prabhakaran D	2	22	9.09	Sal Hospital and Medical Institute, Ahmedabad, India; Center For Chronic Disease Control, New Delhi, India	N Engl J Med 364 (17): 1607-1616 Apr 28 2011
Dabigatran Compared With Warfarin in Patients With Atrial Fibrillation and Previous Transient Ischaemic Attack or Stroke: A Subgroup Analysis of the Re-Ly Trial	Denis Xavier	1	9	11.11	St Johns Med Coll, Dept Pharmacol, Bangalore, Karnataka, India.	Lancet Neurol 9 (12): 1157-1163 Dec 2010
Method of Delivery and Pregnancy Outcomes in Asia: The Who Global Survey on Maternal and Perinatal Health 2007-08	Roy M	1	23	4.35	Indian Council Med Res, New Delhi, India.	Lancet 375 (9713): 490-499 Feb 6 2010
Gemcitabine Sensitivity Can Be Induced in Pancreatic Cancer Cells Through Modulation of Mir-200 and Mir-21 Expression by Curcumin or Its Analogue Cdf	Subhash Padhye	1	9	11.11	Dr Dy Patil Univ, Pune, Maharashtra, India.	Cancer Res 70 (9): 3606-3617 May 1 2010
Genetic Variants in Novel Pathways Influence Blood Pressure and Cardiovascular Disease Risk	Vinay DG, Charles S Janipalli, Radha Mani K, Kranthi Kumar MV, Giriraj R Chandak, Smita R Kulkarni, Chittaranjan S Yajnik, Dorairaj Prabhakaran, Vikal Tripathy	9	337	2.67	Centre For Cellular and Molecular Biology (Ccmb), Council of Scientific and Industrial Research (Csir), Uppal Road, Hyderabad 500 007, India; Diabetes Unit, Kem Hospital and Research Centre, Rasta Peth, Pune-411011, Maharashtra, India; South Asia Network For Chronic Disease, Public Health Foundation of India, C-1/52, Sda, New Delhi 100016, India.	Nature 478 (7367): 103-109 Oct 6 2011

Title	Authors				Institution	Reference
Single-Dose Liposomal Amphotericin B For Visceral Leishmaniasis in India	Sundar S, Chakravarty J, Agarwal D, Rai M	4	5	80.00	Banaras Hindu Univ, Inst Med Sci, Dept Med, Kala Azar Med Res Ctr, Varanasi 221005, Uttar Pradesh, India.	N Engl J Med 362 (6): 504-512 Feb 11 2010
Pharmacological Inhibition of Gut-Derived Serotonin Synthesis Is A Potential Bone Anabolic Treatment For Osteoporosis	Padmanaban S Suresh, Rudraiah Medhamurthy, Anil K Balapure	3	14	21.43	Department of Molecular Reproduction, Development and Genetics, Indian Institute of Science, Bangalore, India; Tissue and Cell Culture Unit, Central Drug Research Institute, Lucknow, India.	Nature Med 16 (3): 308-U103 Mar 2010
Newborn-Care Training and Perinatal Mortality in Developing Countries	Goudar SS, Parida S, Tshefu A	3	18	16.67	Jawaharlal Nehru Med Coll, Belgaum, India. Sriramchandra Bhanja Med Coll, Cuttack, Orissa, India.	N Engl J Med 362 (7): 614-623 Feb 18 2010
Ticagrelor Versus Clopidogrel in Acute Coronary Syndromes in Relation to Renal Function Results From the Platelet Inhibition and Patient Outcomes (Plato) Trial	Keyur Parikh	1	16	6.25	the Heart Care Clinic, Ahmedabad, India	Circulation 122 (11): 1056-1067 Sep 14 2010
Adult and Child Malaria Mortality in India: A Nationally Representative Mortality Survey	Neeraj Dhingra, Vinod P Sharma, Raju M Jotkar	3	10	30.00	Natl Aids Control Org, New Delhi, India.; Indian Inst Technol, Ctr Rural Dev & Technol, New Delhi, India.; St Johns Res Inst, Bangalore, Karnataka, India.	Lancet 376 (9754): 1768-1774 Nov 20 2010

Title	Author				Affiliation	Citation
Effect of A Participatory Intervention With Women's Groups on Birth Outcomes and Maternal Depression in Jharkhand and Orissa, India: A Cluster-Randomised Controlled Trial	Prasanta Tripathy, Nirmala Nair, Rajendra Mahapatra, Shibanand Rath, Suchitra Rath, Rajkumar Gope, Dipnath Mahto, Rajesh Sinha, Vikram Patel	9	15	60.00	Ekjut, Ward 17, Plot 556B, Po Chakradharpur 833102, Jharkhand, India.	Lancet 375 (9721): 1182-1192 Apr 3 2010
India's Janani Suraksha Yojana, A Conditional Cash Transfer Programme to Increase Births in Health Facilities: An Impact Evaluation	Dandona L	1	6	16.67	Publ Hlth Fdn India, New Delhi, India.	Lancet 375 (9730): 2009-2023 Jun 5 2010
Genetic Characterization and Linkage Disequilibrium Estimation of A Global Maize Collection Using Snp Markers	Jianbing Yan	1	6	16.67	Int Crops Res Inst Semi Arid Trop, Dept Bioinformat, Hyderabad, Andhra Pradesh, India.	Plos One 4 (12): Art. No.-E8451 Dec 24 2009
A Research Agenda to Underpin Malaria Eradication	Chetan Chitnis	1	17	5.88	International Center For Genetic Engineering and Biotechnology, Delhi, India	Plos Med 8 (1): Art. No.-E1000406 Jan 2011
Colloidal Nanocarriers: A Review on Formulation Technology, Types and Applications Toward Targeted Drug Delivery	Mishra B, Patel BB, Tiwari S	3	3	100.00	Banaras Hindu Univ, Dept Pharmaceut, Inst Technol, Varanasi 221005, Uttar Pradesh, India	Nanomed-Nanotechnol Biol Med 6 (1): 9-24 Feb 2010
Voros Product, Noncommutative Schwarzschild Black Hole and Corrected Area Law	Mohan V	1	9	11.11	World Hlth Org Collaborating Ctr Noncommunicable, Madras Diabet Res Fdn, Madras, Tamil Nadu, India; World Hlth Org Collaborating Ctr Noncommunicable, Dr Mohans Diabet Special Ctr, Madras, Tamil Nadu, India.	Diabetes Care 33 (3): 580-582 Mar 2010

Title	Authors				Affiliation	Citation
Effect of Age on Outcome of Reduced-Intensity Hematopoietic Cell Transplantation For Older Patients With Acute Myeloid Leukemia in First Complete Remission or With Myelodysplastic Syndrome	Gupta PC, Pednekar MS, Ramadas K	3	44	6.82	Healis Sekhsaria Inst Publ Hlth, Bombay, Maharashtra, India. Reg Canc Ctr, Div Radiat Oncol, Trivandrum 695011, Kerala, India.	N Engl J Med 364 (8): 719-729 Feb 24 2011
Association Between Body-Mass Index and Risk of Death in More Than 1 Million Asians	Prem Pais	1	34	2.94	St. John's Medical College, Bangalore, India	N Engl J Med 365 (8): 699-708 Aug 25 2011
Apixaban With Antiplatelet Therapy After Acute Coronary Syndrome	Mahesh Desai	1	7	14.29	Department of Urology, Muljibhai Patel Urological Hospital, Nadiad, India	J Endourol 25 (1): 11-17 Jan 2011
The Clinical Research Office of the Endourological Society Percutaneous Nephrolithotomy Global Study: Indications, Complications, and Outcomes in 5803 Patients	Mahesh Desai, MD,	1	8	12.50	Muljibhai Patel Urol Hosp, Dept Urol, Nadiad, India.	J Endourol 25 (1): 11-17 Jan 2011
Effect of Nesiritide in Patients With Acute Decompensated Heart Failure	Mittal S	1	57	1.75	Escorts Heart Inst & Res Ctr, Dept Cardiol, New Delhi, India.	N Engl J Med 365 (1): 32-43 Jul 7 2011
Nf-Kappa B Addiction and Its Role in Cancer: 'One Size Does Not Fit All'	MM Chaturvedi	1	5	20.00	Univ Delhi, Dept Zool, Lab Chromatin Biol, Delhi 110007, India.	Oncogene 30 (14): 1615-1630 Apr 2011
Effectiveness of An Intervention Led by Lay Health Counsellors For Depressive and Anxiety Disorders in Primary Care in Goa, India (Manas): A Cluster Randomised Controlled Trial	Patel V, Chowdhary N, Naik S, Pednekar S, Chatterjee S, Bhat B	6	13	46.15	Sangath Ctr, Alto Porvorim 403521, Goa, India.	Lancet 376 (9758): 2086-2095 Dec 18 2010

Title	Authors				Affiliation	Journal
Targeted Therapy For High-Grade Glioma With the Tgf-Beta 2 Inhibitor Trabedersen: Results of A Randomized and Controlled Phase IIb Study	Venkataramana NK, Mahapatra AK, Suri A, Balasubramaniam A, Nair S	5	17	29.41	Manipal Hosp, Manipal Inst Neurol Disorders, Bangalore, Karnataka, India. Sanjay Gandhi Postgrad Inst Med Sci, Lucknow, Uttar Pradesh, India. All India Inst Med Sci, Dept Neurosurg, New Delhi, India. Natl Inst Mental Hlth & Neurosci, Bangalore 560029, Karnataka, India. Sree Chitra Tirunal Inst Med Sci & Technol, Dept Neurosurg, Thiruvananthapuram, Kerala, India.	Neuro-Oncology 13 (1): 132-142 Jan 2011
A 2-H Diagnostic Protocol to Assess Patients With Chest Pain Symptoms in the Asia-Pacific Region (Aspect): A Prospective Observational Validation Study	Ravi R Kasliwal, Manish Bansal	2	27	7.41	Medanta Medicity, Gurgaon, India.	Lancet 377 (9771): 1077-1084 Mar-Apr 2011
Teplizumab For Treatment of Type 1 Diabetes (Protege Study): 1-Year Results From A Randomised, Placebo-Controlled Trial	Sunil M Jain,	1	85	1.18	Totall Diabet Hormone Res Inst, Indore, Madhya Pradesh, India.	Lancet 378 (9790): 487-497 Aug 6 2011
Ethics and Best Practice Guidelines For Training Experiences in Global Health	Anant Bhan	1	15	6.67	Bioeth & Global Hlth, Pune, Maharashtra, India.	Amer J Trop Med Hyg 83 (6): 1178-1182 Dec 2010
Stillbirths 2 Stillbirths: Where? When? Why? How to Make the Data Count?	Kumar R	1	9	11.11	Postgrad Inst Med Educ & Res, Chandigarh 160012, India.	Lancet 377 (9775): 1448-1463 Apr 23 2011

Title	Author				Affiliation	Citation
Timing of Antiretroviral Therapy For Hiv-1 Infection and Tuberculosis	Kumarasamy N	1	31	3.23	Yr Gaitonde Ctr Aids Res & Educ, Voluntary Hlth Serv, Madras, Tamil Nadu, India	N Engl J Med 365 (16): 1482-1491 Oct 20 2011
Dronedarone in High-Risk Permanent Atrial Fibrillation	Narasimhan C, Xavier D	2	55	3.64	St Johns Med Coll, Bangalore, Karnataka, India. Care Hosp, Hyderabad, Andhra Pradesh, India.	N Engl J Med 365 (24): 2268-2276 Dec 15 2011
India: Towards Universal Health Coverage 7 Towards Achievement of Universal Health Care in India by 2020: A Call to Action	Srinath Reddy K, Vikram Patel, Vinod K Paul, Shiva Kumar AK, Lalit Dandona	5	6	83.33	Publ Hlth Fdn India, New Delhi, India. All India Inst Med Sci, New Delhi, India. Unicef India, New Delhi, India.	Lancet 377 (9767): 760-768 Feb-Mar 2011
Un High-Level Meeting on Non-Communicable Diseases: Addressing Four Questions	Reddy KS	1	14	7.14	Publ Hlth Fdn India, New Delhi, India	Lancet 378 (9789): 449-455 Jul-Aug 2011
The London Position Statement of the World Congress of Gastroenterology on Biological Therapy For Ibd With the European Crohn's and Colitis Organization: When to Start, When to Stop, Which Drug to Choose, and How to Predict Response	Sood A	1	27	3.70	Dayanand Med Coll & Hosp, Dept Gastroenterol, Ludhiana, Punjab, India.	Amer J Gastroenterol 106 (2): 199-212 Feb 2011
Resveratrol and Cellular Mechanisms of Cancer Prevention	Shukla Y, Singh R	2	2	100.00	Indian Inst Toxicol Res, Csir, Prote Lab, Lucknow 226001, Uttar Pradesh, India.	Ann N Y Acad Sci 1215: 1-8 2011

Title	Authors				Affiliation	Source
Selective Toxicity of Zno Nanoparticles Toward Gram-Positive Bacteria and Cancer Cells by Apoptosis Through Lipid Peroxidation	Mariappan Premanathan, Krishnamoorthy Karthikeyan, Govindasamy Manivannan	3	4	75.00	Mepco Schlenk Engn Coll, Dept Biotechnol, Sivakasi 608502, Tamil Nadu, India. Mepco Schlenk Engn Coll, Dept Nanosci & Technol, Sivakasi 608502, Tamil Nadu, India. Nmss Vellaichamy Nadar Coll, Dept Microbiol, Madurai, Tamil Nadu, India.	Nanomed-Nanotechnol Biol Med 7 (2): 184-192 Apr 2011
What Is New For An Old Molecule? Systematic Review and Recommendations on the Use of Resveratrol	Namasivayam Nalini, Yogeshwer Shukla	2	21	9.52	Annamalai Univ, Dept Biochem & Biotechnol, Annamalainagar 608002, Tamil Nadu, India.Indian Inst Toxicol Res, Prote Lab, Lucknow, Uttar Pradesh, India.	Plos One 6 (6): Art. No.-E19881 Jun 16 2011
Evaluation of the Xpert Mtb/ Rif Assay For the Diagnosis of Pulmonary Tuberculosis in A High Hiv Prevalence Setting	Mishra Hridesh, Sharma Surendra	2	13	15.38	All India Inst Med Sci, Dept Med, New Delhi 110029, India.	Amer J Respir Crit Care Med 184 (1): 132-140 Jul 1 2011
Efficacy and Safety of Entecavir Versus Adefovir in Chronic Hepatitis B Patients With Hepatic Decompensation: A Randomized, Open-Label Study	Shiv Kumar Sarin	1	15	6.67	Gb Pant Hosp, Dept Gastroenterol, New Delhi, India.	Hepatology 54.(1): 91-100 Jul 2011
Antimalarial Drug Resistance of Plasmodium Falciparum in India: Changes Over Time and Space	Dhillon GPS, Dash AP, Arora U, Meshnick SR, Valecha N	5	6	83.33	Indian Council Med Res, Natl Inst Malaria Res, New Delhi 110077, India. Minist Hlth & Family Welf, Natl Vector Borne Dis Control Programme, New Delhi, India.	Lancet Infect Dis 11 (1): 57-64 Jan 2011

Title	Authors				Affiliation	Reference
Who Guidelines For the Programmatic Management of Drug-Resistant Tuberculosis: 2011 Update	Sarin R	1	46	2.17	Lrs Inst Tb & Allied Dis, New Delhi, India.	Eur Resp J 38 (3): 516-528 Sep 2011
India: Towards Universal Health Coverage 2 Reproductive Health, and Child Health and Nutrition in India: Meeting the Challenge	Paul VK, Sachdev HS, Mavalankar D, Ramachandran P, Sankar MJ, Bhandari N, Sreenivas V, Sundararaman T, Govil D	9	11	81.82	All India Inst Med Sci, Dept Paediat, New Delhi 110029, India. Sitaram Bhartia Inst Sci & Res, New Delhi, India. Indian Inst Management, Ahmadabad 380015, Gujarat, India. Nutr Fdn India, New Delhi, India. Soc Appl Studies, New Delhi, India. Natl Hlth Syst Resources Ctr, New Delhi, India. Inst Hlth Management & Res, Jaipur, Rajasthan, India.	Lancet 377 (9762): 332-349 Jan 22 2011
Comparison of Short-Course Multidrug Treatment With Standard Therapy For Visceral Leishmaniasis in India: An Open-Label, Non-Inferiority, Randomised Controlled Trial	Sundar S, Sinha PK, Rai M, Verma DK, Nawin K, Alam S, Chakravarty J, Verma N, Pandey K, Kumari P, Lal CS, Arora R	12	20	60.00	6 Sk Gupta Nagar, Varanasi 221005, Uttar Pradesh, India. Banaras Hindu Univ, Inst Med Sci, Dept Med, Kala Azar Med Res Ctr, Varanasi 221005, Uttar Pradesh, India. Indian Council Med Res, Rajendra Mem Res Inst Med Sci, Patna, Bihar, India.	Lancet 377 (9764): 477-486 Feb 5 2011
Research Reporting Standards For Radioembolization of Hepatic Malignancies	Kulkarni S, Kulkarni A,	2	47	4.26	Tata Mem Hosp, Dept Radiol, Bombay 400012, Maharashtra, India.	J Vasc Interven Radiol 22 (3): 265-278 Mar 2011

Title	Author				Institution	Publication
Insulin Degludec, An Ultra-Long-Acting Basal Insulin, Once A Day or Three Times A Week Versus Insulin Glargine Once A Day in Patients With Type 2 Diabetes: A 16-Week, Randomised, Open-Label, Phase 2 Trial	Rao PV, Thomas N	2	11	18.18	Med Sci Univ, Nizams Inst, Hyderabad, Andhra Pradesh, India. Christian Med Coll & Hosp, Vellore, Tamil Nadu, India.	Lancet 377 (9769): 924-931 Mar 12 2011
Characteristics and Short-Term Prognosis of Perioperative Myocardial Infarction in Patients Undergoing Noncardiac Surgery A Cohort Study	Xavier D, Sigamani A	2	15	13.33	St Johns Res Inst, Bangalore, Karnataka, India	Ann Intern Med 154 (8): 523-W177 Apr 19 2011
Phase Iii, Open-Label, Randomized Study Comparing Concurrent Gemcitabine Plus Cisplatin and Radiation Followed by Adjuvant Gemcitabine and Cisplatin Versus Concurrent Cisplatin and Radiation in Patients With Stage Iib to Iva Carcinoma of the Cervix	Patel Firuza	1	11	9.09	Postgrad Inst Med Educ & Res, Chandigarh 160012, India.	J Clin Oncol 29 (13): 1678-1685 May 1 2011
Efficacy and Safety of Treatment With Rituximab For Difficult Steroid-Resistant and -Dependent Nephrotic Syndrome: Multicentric Report	Ashima Gulati, Aditi Sinha, Pankaj Hari, Amit K Dinda, Sonika Sharma, Rajendra N Srivastava, Arvind Bagga	7	9	77.78	All India Inst Med Sci, New Delhi 110029, India.	Clin J Am Soc Nephrol 5 (12): 2207-2212 Dec 2010
Regional Management Units For Marine Turtles: A Novel Framework For Prioritizing Conservation and Research Across Multiple Scales	Chowdhary BC	1	33	3.03	Wildlife Inst India, Dept Endangered Species Management, Dehra Dun, Uttarakhand, India.	Plos One 5 (12): Art. No.-E15465 Dec 17 2010

Title	Author				Affiliation	Journal
Maintenance Therapy With Pemetrexed Plus Best Supportive Care Versus Placebo Plus Best Supportive Care After Induction Therapy With Pemetrexed Plus Cisplatin For Advanced Non-Squamous Non-Small-Cell Lung Cancer (Paramount): A Double-Blind, Phase 3,...	Jayaprakash Madhavan	1	17	5.88	Jawaharlal Nehru Canc Hosp & Res Ctr, Bhopal, India.	Lancet Oncol 13 (3): 247-255 Mar 2012
Randomized, Double-Blind, Placebo-Controlled Phase Ii Study of Amg 386 Combined With Weekly Paclitaxel in Patients With Recurrent Ovarian Cancer	Nagarkar Raj V	1	20	5.00	Curie Manavata Canc Ctr, Nasik, Maharashtra, India.	J Clin Oncol 30 (4): 362-371 Feb 1 2012
Curcumin Analogue Cdf Inhibits Pancreatic Tumor Growth by Switching on Suppressor Micrornas and Attenuating Ezh2 Expression	Padhye S	1	11	9.09	Abeda Inamdar Senior Coll, Interdisciplinary Sci & Technol Res Acad, Pune, Maharashtra, India.	Cancer Res 72 (1): 335-345 Jan 1 2012
Ulipristal Acetate Versus Placebo For Fibroid Treatment Before Surgery	Manju P Jilla	1	13	7.69	Dr Jilla Hosp, Aurangabad, Maharashtra, India.	N Engl J Med 366 (5): 409-420 Feb 2 2012
Three-Year Outcomes From Benefit, A Randomized, Active-Controlled, Parallel-Group Study in Adult Kidney Transplant Recipients	Kothari J	1	18	5.56	Pd Hinduja Hosp, Bombay, Med Res Ctr, Bombay,	Am J Transplant 12 (1): 210-217 Jan 2012
Efficacy of Antidepressants As Analgesics: A Review	Dharmshaktu P, Tayal V, Kalra BS	3	3	100.00	Maulana Azad Med Coll, Dept Pharmacol, Bahadur Shah Zafar Marg, Delhi 110002, India.	J Clin Pharmacol 52 (1): 6-17 Jan 2012

Title	Authors				Affiliation	Citation
Nanoparticles: A Boon to Drug Delivery, Therapeutics, Diagnostics and Imaging	Parveen S, Misra R, Sahoo SK	3	3	100.00	Inst Life Sci, Lab Nanomed, Bhubaneswar, Orissa, India	Nanomed-Nanotechnol Biol Med 8 (2): 147-166 Feb 2012
Linear Pei Nanoparticles: Efficient Pdna/Sirna Carriers in Vitro and in Vivo	Goyal R, Tripathi SK, Tyagi S, Sharma A, Ram KR, Chowdhuri DK, Shukla Y, Kumar P, Gupta KC	9	9	100.00	Indian Inst Toxicol Res, Csir, Mahatma Gandhi Marg, Lucknow 226001, Uttar Pradesh, India. Univ Delhi, Csir, Inst Genom & Integrat Biol, Delhi 110007, India.	Nanomed-Nanotechnol Biol Med 8 (2): 167-175 Feb 2012
Dasatinib or Imatinib in Newly Diagnosed Chronic-Phase Chronic Myeloid Leukemia: 2-Year Follow-Up From A Randomized Phase 3 Trial (Dasision)	Agarwal MB,	1	20	5.00	Bombay Hosp & Med Res Ctr, Inst Med Sci, Bombay, Maharashtra, India.	Blood 119 (5): 1123-1129 Feb 2 2012
Magnetic Resonance Evaluation of Tubercular Lesion in Spine	Jain AK, Sreenivasan R, Saini NS, Kumar S, Jain S, Dhammi IK	6	6	100.00	A-10 Part B, Ashok Nagar, Ghaziabad Guru Teg Bahadur Hosp, New Delhi, Univ Delhi, Inst Nucl Med & Allied Sci, Dept Radiol, New Delhi	Int Orthop 36 (2): 261-269 Sp. Iss. Si Feb 2012
Rapid Diagnosis of Tuberculosis With the Xpert Mtb/Rif Assay in High Burden Countries: A Cost-Effectiveness Analysis	Joy S Michael, KR John	2	15	13.33	Christian Med Coll & Hosp, Vellore, Tamil Nadu, India. Natl Tb Program, Vellore, Tamil Nadu, India.	Plos Med 8 (11): Art. No.-E1001120 Nov 2011

(e) Computer Sciences

Title	Indian Authors	No. of Indian Authors	Total Author	Ratio I/F*	Institue Name	Source
Propred: Prediction of Hla-Dr Binding Sites	Singh H, Raghava Gps	2	2	100.00	Inst Microbial Technol, Bioinformat Ctr, Chandigarh 160036, India.	Bioinformatics 17 (12): 1236-1237 Dec 2001
Convergence and Diversity in Evolutionary Multiobjective Optimization	Kalyanmoy Deb	1	4	25.00	Indian Inst Technol, Dept Mech Engn, Kanpur 208016, Uttar Pradesh, India.	Evol Comput 10 (3): 263-282 Fal 2002
Analysis and Prediction of Dna-Binding Proteins and Their Binding Residues Based on Composition, Sequence and Structural Information	Ahmad S	1	3	33.33	Jamia Millia Islamia, Dept Biosci, New Delhi 110025, India. Aist, Computat Biol Res Ctr, Cbrc, Koto Ku, Tokyo 1350064, Japan.	Bioinformatics 20 (4): 477-486 Mar 1 2004
Full-Diversity, High-Rate Space-Time Block Codes From Division Algebras	Rajan BS, Shashidhar V	2	3	66.67	Anna Univ, Aukbc Res Ctr, Madras 600044, Tamil Nadu, India. Indian Inst Sci, Dept Elect Commun Engn, Bangalore 560012, Karnataka, India.	Ieee Trans Inform Theory 49 (10): 2596-2616 Oct 2003
Analysis of Transmit-Receive Diversity in Rayleigh Fading	Dighe PA, Mallik RK	2	3	66.67	Hellosoft Inc, Banjara Hills, Hyderabad 500034, Andhra Pradesh, India. Indian Inst Technol, Dept Elect Engn, New Delhi 110016, India.	Ieee Trans Commun 51 (4): 694-703 Apr 2003
Complex Minimax Programming Under Generalized Type-I Functions	Mishra SK	1	1	100.00	Govind Ballabh Pant Univ Agr & Technol, Coll Basic Sci & Humanities, Dept Math Stat & Comp Sci, Pantnagar 263145, Uttar Pradesh, India.	Comput Math Appl 50 (1-2): 1-11 Jul 2005

Title	Authors				Affiliation	Citation
Bounds For Multihop Relayed Communications in Nakagami-M Fading	Mallik RK	1	3	33.33	Indian Inst Technol, Dept Elect Engn, New Delhi 110016, India.	Ieee Trans Commun 54 (1): 18-22 Jan 2006
Multi-Objective Optimization of An Industrial Fluidized-Bed Catalytic Cracking Unit (Fccu) Using Genetic Algorithm (Ga) With the Jumping Genes Operator	Kasat RB, Gupta SK	2	2	100.00	Indian Inst Technol, Dept Chem Engn, Kanpur 208016, Uttar Pradesh, India.	Comput Chem Eng 27 (12): 1785-1800 Dec 15 2003
Soft Set Theory	Maji PK, Biswas R, Roy AR	3	3	100.00	Indian Inst Technol, Dept Math, Kharagpur 721302, W Bengal, India.	Comput Math Appl 45 (4-5): 555-562 Feb-Mar 2003
Analyzing Alternatives in Reverse Logistics For End-Of-Life Computers: Anp and Balanced Scorecard Approach	Ravi V, Shankar R, Tiwari MK	3	3	100.00	Indian Inst Technol, Dept Management Studies, New Delhi 110016, India. Natl Inst Foundry & Forge Technol, Dept Mfg Engn, Ranchi 834003, Bihar, India	Comput Ind Eng 48 (2): 327-356 Mar 2005
Mhcbn: A Comprehensive Database of Mhc Binding and Non-Binding Peptides	Bhasin M, Singh H, Raghava GPS	3	3	100.00	Inst Microbial Technol, Chandigarh, India	Bioinformatics 19 (5): 665-666 Mar 22 2003
A Fuzzy Goal Programming Approach For Vendor Selection Problem in A Supply Chain	Kumar M, Vrat P, Shankar R	3	3	100.00	Indian Inst Technol, Dept Management Sci, Haus Khas, New Delhi 110016, India; Indian Inst Technol, Roorkee 247667, Uttar Pradesh, India; Indian Inst Technol, Dept Mech Engn, New Delhi 110016, India.	Comput Ind Eng 46 (1): 69-85 Mar 2004
Modified Differential Evolution (Mde) For Optimization of Non-Linear Chemical Processes	Babu BV, Angira R	2	2	100.00	Birla Inst Technol & Sci, Dept Chem Engn, Pilani 333031, Rajasthan, India	Comput Chem Eng 30 (6-7): 989-1002 May 15 2006

Title	Authors				Affiliation	Journal
Explicit Space-Time Codes Achieving the Diversity-Multiplexing Gain Tradeoff	Raj K Kumar, Sameer A Pawar	2	5	40.00	Indian Inst Sci, Dept Elect Commun Engn, Bangalore 560012, Karnataka, India.	Ieee Trans Inform Theory 52 (9): 3869-3884 Sep 2006
Propred1: Prediction of Promiscuous Mhc Class-I Binding Sites	Singh H, Raghava GPS	2	2	100.00	Inst Microbial Technol, Bioinformat Ctr, Sector 39A, Chandigarh 160036, India.	Bioinformatics 19 (8): 1009-1014 May 22 2003
Pslpred: Prediction of Subcellular Localization of Bacterial Proteins	Bhasin M, Garg A, Raghava GPS	3	3	100.00	Inst Microbial Technol, Sector 39A, Chandigarh, India.	Bioinformatics 21 (10): 2522-2524 May 15 2005
Svm Based Method For Predicting Hla-Drb1(*)0401 Binding Peptides In An Antigen Sequence	Bhasin M, Raghava GPS	2	2	100.00	Inst Microbial Technol, Sector 39A, Chandigarh, India.	Bioinformatics 20 (3): 421-423 Feb 12 2004
An Evaluation of Human Protein-Protein Interaction Data in the Public Domain	Mathivanan S, Periaswamy B, Gandhi TKB, Kandasamy K, Suresh S, Mohmood R, Ramachandra YL	7	8	87.50	Inst Bioinformat, Bangalore, Karnataka, India.; Kuvempu Univ, Dept Biotechnol, Shankaraghatta, Karnataka, India.	Bmc Bioinformatics 7: Art. No.-S19 Suppl. 5 2006
Pssm-Based Prediction of Dna Binding Sites in Proteins	Ahmad S	1	2	50.00	Jamia Millia Islamia Univ, Dept Biosci, New Delhi 110025, India.	Bmc Bioinformatics 6: Art. No.-33 Feb 19 2005
Managing Cross-Cultural Issues in Global Software Outsourcing	Krishna S	1	3	33.33	Indian Inst Management, Software Enterprise Management Program, Bangalore, Karnataka, India.	Commun Acm 47 (4): 62-+ Apr 2004
A Genetic Algorithms Based Multi-Objective Neural Net Applied to Noisy Blast Furnace Data	Chakraborti N	1	3	33.33	Department of Metallurgical & Materials Engineering, Indian Institute of Technology, Kharagpur 721302, West Bengal, India	Appl Soft Comput 7 (1): 387-397 Jan 2007

Title	Authors			%	Affiliation	Reference
Multi-Objective Particle Swarm Optimization With Time Variant Inertia and Acceleration Coefficients	Tripathi PK, Bandyopadhyay S, Pal SK	3	3	100.00	Indian Stat Inst, Machine Intelligence Unit, 203 Bt Rd, Calcutta 700108, India.	Inform Sciences 177 (22): 5033-5049 Nov 15 2007
Hybrid Neural Network Models For Hydrologic Time Series Forecasting	Jain A, Kumar AM	2	2	100.00	Indian Inst Technol, Dept Civil Engn, Kanpur 208016, Uttar Pradesh, India.	Appl Soft Comput 7 (2): 585-592 Mar 2007
Support Vector Machine-Based Method For Predicting Subcellular Localization of Mycobacterial Proteins Using Evolutionary Information and Motifs	Rashid M, Saha S, Raghava GPS	3	3	100.00	Bioinformat Ctr, Inst Microbial Technol, Chandigarh, India.	Bmc Bioinformatics 8: Art. No.-337 Sep 13 2007
Flux Balance Analysis of Biological Systems: Applications and Challenges	Raman K, Chandra N	2	2	100.00	Indian Inst Sci, Bioinformat Ctr, Bangalore 560012, Karnataka, India.	Brief Bioinform 10 (4): 435-449 Jul 2009
A Novel Feature Representation Method Based on Chou's Pseudo Amino Acid Composition For Protein Structural Class Prediction	Sahu SS, Panda G	2	2	100.00	Natl Inst Technol, Sch Elect Sci, Rourkela, Orissa, India.	Comput Biol Chem 34 (5-6): 320-327 Dec 2010
An Artificial Bee Colony Algorithm For the Leaf-Constrained Minimum Spanning Tree Problem	Singh A	1	1	100.00	Univ Hyderabad, Dept Comp & Informat Sci, Hyderabad 500046, Andhra Pradesh, India	Appl Soft Comput 9 (2): 625-631 Mar 2009
Strain Smoothing in Fem and Xfem	Natarajan S	1	8	12.50	Ge Aviat India Technol Ctr, Bangalore, Karnataka, India.	Comput Struct 88 (23-24): 1419-1443 Sp. Iss. Si Dec 2010
Methods Used For the Development of Neural Networks For the Prediction of Water Resource Variables in River Systems: Current Status and Future Directions	Jain A, Sudheer KP	2	4	50.00	Indian Inst Technol, Dept Civil Engn, Kanpur 208016, Uttar Pradesh, India. Indian Inst Technol, Dept Civil Engn, Madras 600036, Tamil Nadu, India.	Environ Modell Softw 25 (8): 891-909 Aug 2010

Title	Authors				Affiliation	Citation
Generalised Fuzzy Soft Sets	Majumdar P, Samanta SK	2	2	100.00	Muc Womens Coll, Dept Math, Burdwan 713104, W Bengal, India. Visva Bharati, Dept Math, Santini Ketan 731235, W Bengal, India	Comput Math Appl 59 (4): 1425-1432 Feb 2010
Bioinformatics Analysis of Functional Relations Between Cnps Regions	Dave K	1	2	50.00	Sardar Patel Univ, Gh Patel Pg Dept Comp Sci & Technol, Vallabh Vidyanagar 388120, Gujarat, India.	Curr Bioinform 6 (1): 122-128 Mar 2011
A Survey on Security Issues in Service Delivery Models of Cloud Computing	Subashini S, Kavitha V	2	2	100.00	Anna Univ Tirunelveli, Tirunelveli 627007, Tn, India	J Netw Comput Appl 34 (1): 1-11 Jan 2011
Complex Network and Gene Ontology in Pharmacology Approaches: Mapping Natural Compounds on Potential Drug Target Colon Cancer Network	Bhattacharjee B, Jayadeepa RM, Talambedu U, Banerjee S, Joshi J, Mole JP, Samuel J, Middha SK	8	8	100.00	Inst Computat Biol, -2 Karnataka Maharani Lakshmi Ammani Coll Women, Karanataka Indian Stat Inst, Calcutta Inst Genom & Integrat Biol, Delhi Karunya Univ, Coimbatore, Tamil Nadu	Curr Bioinform 6 (1): 44-52 Mar 2011
Coupled Common Fixed Point Theorems For A Pair of Commuting Mappings in Partially Ordered Complete Metric Spaces	Hemant Kumar Nashine	1	2	50.00	Disha Inst Management & Technol, Chhatisgarh-2	Comput Math Appl 62 (4): 1984-1993 Aug 2011
On Averaging Operators For Atanassov's Intuitionistic Fuzzy Sets	Goswami DP, Mukherjee UK, Pal NR	3	5	60.00	Jadavpur Univ, Sch Biosci & Engn, Calcutta, India. Sarat Centenary Coll, Hooghly, India. Indian Stat Inst, Calcutta 700035, W Bengal, India.	Inform Sciences 181 (6): 1116-1124 Mar 15 2011

Title	Indian Authors	No. of Indian Authors	Total Author	Ratio I/F*	Institute Name	Source
Peristaltic Transport of Fractional Maxwell Fluids in Uniform Tubes: Applications in Endoscopy	Dharmendra Tripathi	1	1	100.00	Birla Inst Technol & Sci Pilani, Dept Math, Hyderabad 500078, Andhra Pradesh, India.	Comput Math Appl 62 (3): 1116-1126 Sp. Iss. Si Aug 2011
Fractional Bloch Equation With Delay	Bhalekar S, Daftardar-Gejji V	2	4	50.00	Univ Pune, Dept Math, Pune 411007, Maharashtra, India.	Comput Math Appl 61 (5): 1355-1365 Mar 2011
Mayavi: 3D Visualization of Scientific Data	Ramachandran P,	1	2	50.00	Iit Bombay, Dept Aerosp Engn, Bombay, Maharashtra, India.	Comput Sci Eng 13 (2): 40-50 Mar-Apr 2011
Hsp90/Cdc37 Chaperone/Co-Chaperone Complex, A Novel Junction Anticancer Target Elucidated by the Mode of Action of Herbal Drug Withaferin A	Grover A, Shandilya A, Agrawal V, Pratik P, Bhasme D, Bisaria VS, Sundar D	7	7	100.00	Indian Inst Technol Iit Delhi, Dept Biochem Engn & Biotechnol, New Delhi 110016, India.	Bmc Bioinformatics 12: Art. No.-S30 Suppl. 1 Feb 15 2011
Discribinate: A Rapid Method For Accurate Taxonomic Classification of Metagenomic Sequences	Ghosh TS, Haque M, Mande SS	3	3	100.00	Tata Consultancy Serv, Innovat Labs, Biosci Div, 1 Software Units Layout, Hyderabad 500081, Andhra Pradesh, India	Bmc Bioinformatics 11: Art. No.-S14 Suppl. 7 Oct 15 2010

(f) Economics and Business

Title	Indian Authors	No. of Indian Authors	Total Author	Ratio I/F*	Institute Name	Source
Environmental Kuznets Curve Hypothesis: A Survey	Dinda S	1	1	100.00	Indian Stat Inst, Econ Res Unit, Calcutta 108, India. Sr Fatepuria Coll, Murshidabad, India.	Ecol Econ 49 (4): 431-455 Aug 1 2004
Green Supply-Chain Management: A State-Of-The-Art Literature Review	Srivastava SK	1	1	100.00	Management Dev Inst, Post Box 60, Sukhrali 122001, Gurgaon, India	Int J Manag Rev 9 (1): 53-80 Mar 2007

Title	Indian Authors	No. of Indian Authors	Total Author	Ratio I/F*	Institue Name	Source
Panel Data Models With Spatially Correlated Error Components	Kapoor M	1	3	33.33	Isb, Hyderabad 500019, Andhra Pradesh, India.	J Econometrics 140 (1): 97-130 Sep 2007
Jatropha Plantations For Biodiesel in Tamil Nadu, India Viability, Livelihood Trade-Offs, and Latent Conflict	Lele S	1	2	50.00	Atree, Ctr Environm & Dev, Bangalore, Karnataka, India.	Ecol Econ 70 (2): 189-195 Dec 15 2010

(g) Engineering

Title	Indian Authors	No. of Indian Authors	Total Author	Ratio I/F*	Institue Name	Source
Geant4-A Simulation Toolkit	Banerjee S	1	127	0.79	Tata Inst Fundamental Res, Bombay	Nucl Instrum Meth Phys Res A 506 (3): 250-303 Jul 1 2003
A Fast and Elitist Multiobjective Genetic Algorithm: Nsga-Ii	Deb K, Pratap A, Agarwal S, Meyarivan T	4	4	100.00	Indian Inst Technol, Kanpur	Ieee Trans Evol Computat 6 (2): 182-197 Apr 2002
Pyrolysis of Wood/Biomass For Bio-Oil: A Critical Review	Mohan D	1	3	33.33	Ind Toxicol Res Ctr, Environm Chem Div, Lucknow 226001, Uttar Pradesh, India.	Energ Fuel 20 (3): 848-889 May 2006
The Belle Dectetor	Hayashil H, Miyatayashi K, Noguchi S, Wang CH	4	343	1.17	Panjab Univ, Chandigarh 160014, India. Utkal Univ, Bhubaneswar 751004, Orissa. India	Nucl Instrum Meth Phys Res A 479 (1): 117-232 Feb 21 2002
Arsenic Removal From Water/Wastewater Using Adsorbents - A Critical Review	Mohan D	1	2	50.00	Ind Toxicol Res Ctr, Environm Chem Div, Lucknow 226001, Uttar Pradesh, India.	J Hazard Mater 142 (1-2): 1-53 Apr 2 2007

Title	Author				Affiliation	Source
Biofuels (Alcohols and Biodiesel) Applications As Fuels For Internal Combustion Engines	Agarwal AK	1	1	100.00	Indian Inst Technol, Dept Mech Engn, Kanpur 208016, Uttar Pradesh, India.	Prog Energ Combust Sci 33 (3): 233-271 Jun 2007
Temperature Dependence of Thermal Conductivity Enhancement For Nanofluids	Das SK	1	4	25.00	Indian Inst Technol, Dept Mech Engn, Heat Transfer & Thermal Power Lab, Madras 600036, Tamil Nadu, India.	J Heat Transfer 125 (4): 567-574 Aug 2003
Pool Boiling Characteristics of Nano-Fluids	Das SK	1	3	33.33	Indian Inst Technol, Dept Mech Engn, Heat Transfer & Thermal Power Lab, Madras 600036, Tamil Nadu, India.	Int J Heat Mass Transfer 46 (5): 851-862 Feb 2003
Activated Carbons and Low Cost Adsorbents For Remediation of Tri- and Hexavalent Chromium From Water	Mohan D	1	2	50.00	Ind Toxicol Res Ctr, Environm Chem Div, Lucknow 226001, Uttar Pradesh, India.	J Hazard Mater 137 (2): 762-811 Sep 21 2006
The Role of Sawdust in the Removal of Unwanted Materials From Water	Dubey P	1	5	20.00	Vikram Univ, Ujjain, Madhya Pradesh, India.	J Hazard Mater 95 (1-2): 137-152 Nov 11 2002
Vortex-Induced Vibrations	Govardhan R	1	2	50.00	Indian Inst Sci, Mech Engn Dept, Bangalore, Karnataka, India.	Annu Rev Fluid Mech 36: 413-455 2004

Title	Authors				Affiliation	Publication
The Upgraded Do Detector	Reddy LV, Rani KJ, Narasimhan VS, Krishnaswamyp MR, Kalmani SD, Gupta A, Ranjan K, Beri SB, Bhatnagar V, Kaur R, Kohli JM, Chowdhary B, Kumar A, Naimuddin M, Mal PK, Nagaraj P, Rao MVS, Mondal NK, Dugal SR, Chandra A, Chakrabarti S, Banerjee S, Banerjee P, Acharya BS, Shivpuri RK, Satyanarayana B, Shankar HC, Vishwanath PR	28	356	7.87	Panjab Univ, Chandigarh 160014, India. Univ Delhi, Delhi 110007, India. Tata Inst Fundamental Res, Ehep, Bombay 400005, Maharashtra, India.	Nucl Instrum Meth Phys Res A 565 (2): 463-537 Sep 15 2006
Energy-Aware Wireless Microsensor Networks	Raghunathan V	1	4	25.00	Indian Inst Technol, Madras 600036, Tamil Nadu, India.	Ieee Signal Process Mag 19 (2): 40-50 Mar 2002
Phenix Detector Overview	Chand P, Choudhury RK, Dinesh BV, Dutta D, Gupta SK, Kapoor SS, Mohanty AK, Ojha ID, Singh CP, Singh V, Tuli SK	11	126	8.73	Banaras Hindu Univ, Dept Phys, Varanasi 221005, Uttar Pradesh, India. Bhabha Atom Res Ctr, Bombay 400085, Maharashtra, India.	Nucl Instrum Meth Phys Res A 499 (2-3): 469-479 Mar 1 2003
Analytic Hierarchy Process: An Overview of Applications	Vaidya OS, Kumar S	2	2	100.00	Natl Inst Ind Engn, Bombay 400087, Maharashtra, India. Army Inst Technol, Dept Mech Engn, Pune 411015, Maharashtra, India.	Eur J Oper Res 169 (1): 1-29 Feb 16 2006
Natural Convection of Nano-Fluids	Das SK	1	2	50.00	Indian Inst Technol, Dept Mech Engn, Heat Transfer & Thermal Power Lab, Madras 600036, Tamil Nadu, India.	Heat Mass Transfer 39 (8-9): 775-784 Sep 2003

Title	Authors				Affiliation	Source
Solar/Uv-Induced Photocatalytic Degradation of Three Commercial Textile Dyes	Sakthivel S, Arabindoo B, Murugesan V	3	5	60.00	Anna Univ, Dept Chem, Chennai 25, India.	J Hazard Mater 89 (2-3): 303-317 Jan 28 2002
Detector Description and Performance For the First Coincidence Observations Between Ligo and Geo	Dhurandar S, Nayak R, Sengupta AS	3	372	0.81	Interuniv Ctr Astron & Astrophys, Pune 411007, Maharashtra, India.	Nucl Instrum Meth Phys Res A 517 (1-3): 154-179 Jan 21 2004
Utilization of Industrial Waste Products As Adsorbents For the Removal of Dyes	Jain AK, Gupta VK, Bhatnagar A, Suhas	4	4	100.00	Indian Inst Technol, Dept Chem, Roorkee 247667, Uttar Pradesh, India.	J Hazard Mater 101 (1): 31-42 Jul 4 2003
Properties and Use of Jatropha Curcas Oil and Diesel Fuel Blends in Compression Ignition Engine	Pramanik K	1	1	100.00	Reg Engn Coll, Dept Chem Engn, Warangal 506004, Andhra Pradesh, India.	Renewable Energy 28 (2): 239-248 Feb 2003
A New Proof For the Existence of Mutually Unbiased Bases	Bandyopadhyay S	1	4	25.00	Bose Inst, Dept Phys, Calcutta 700009, W Bengal, India.	Algorithmica 34 (4): 512-528 Dec 2002
Biodegradation Aspects of Polycyclic Aromatic Hydrocarbons (Pahs): A Review	Haritash AK, Kaushik CP	2	2	100.00	Univ Delhi, Delhi Coll Engn, Dept Civil & Environm Engn, Delhi 110007, India. Guru Jambheshwar Univ Sci & Technol, Dept Environm Sci & Engn, Hisar, Haryana, India.	J Hazard Mater 169 (1-3): 1-15 Sep 30 2009
Use of Vegetable Oils As Ic Engine Fuels - A Review	Ramadhas AS, Jayaraj S, Muraleedharan C	3	3	100.00	Natl Inst Technol Calicut, Dept Engn Mech, Calicut Rec Po, Calicut 673601, Kerala, India.	Renewable Energy 29 (5): 727-742 Apr 2004
Thin-Film Solar Cells: An Overview	Chopra KL, Dutta V	2	3	66.67	Indian Inst Technol, Ctr Energy Studies, Photovolta Lab, New Delhi 110016, India.	Prog Photovoltaics 12 (2-3): 69-92 Mar-May 2004

Title	Authors				Affiliation	Citation
Studies on Photodegradation of Two Commercial Dyes in Aqueous Phase Using Different Photocatalysts	Kansal SK, Singh M, Sud D	3	3	100.00	Panjab Univ, Dept Chem Engn & Technol, Chandigarh 160014, India. Sant Longowal Inst Engn & Technol, Dept Chem Technol, Sangrur 148106, Punjab, India.	J Hazard Mater 141 (3): 581-590 Mar 22 2007
Evolutionary Programming Techniques For Economic Load Dispatch	Sinha N, Chakrabarti R, Chattopadhyay RK	3	3	100.00	Jadavpur Univ, Dept Elect Engn, Calcutta 700032, W Bengal, India.	Ieee Trans Evol Computat 7 (1): 83-94 Feb 2003
Photocatalytic Degradation of Model Textile Dyes in Wastewater Using Zno As Semiconductor Catalyst	Chakrabarti S, Dutta BK	2	2	100.00	Univ Calcutta, Dept Chem Engn, 92 Acharya Pc Rd, Calcutta 700009, W Bengal, India	J Hazard Mater 112 (3): 269-278 Aug 30 2004
Performance Evaluation of Some Clustering Algorithms and Validity Indices	Bandyopadhyay S	1	2	50.00	Indian Stat Inst, Machine Intelligence Unit, Calcutta 700108, W Bengal, India.	Ieee Trans Patt Anal Mach Int 24 (12): 1650-1654 Dec 2002
Chromium(Vi) Adsorption From Aqueous Solution by Hevea Brasilinesis Sawdust Activated Carbon	Karthikeyan T, Rajgopal S, Miranda LR	3	3	100.00	Anna Univ, Dept Chem Engn, Alagappa Col Technol, Sardar Patel Rd, Madras 600025, Tamil Nadu, India	J Hazard Mater 124 (1-3): 192-199 Sep 30 2005
Adsorptive Removal of Direct Azo Dye From Aqueous Phase Onto Coal Based Sorbents: A Kinetic and Mechanistic Study	Mohan SV, Rao NC, Karthikeyan J	3	3	100.00	Indian Inst Chem Technol, Biochem & Environm Engn Grp, Hyderabad 500007, Andhra Pradesh, India. Sri Venkateswara Univ, Dept Civil Engn, Tirupati 517502, Andhra Pradesh, India	J Hazard Mater 90 (2): 189-204 Mar 1 2002
Removal of Cr6+ and Ni2+ From Aqueous Solution Using Bagasse and Fly Ash	Rao M, Parwate AV, Bhole AG	3	3	100.00	Coll Engn, Dept Civil Engn, Badnera 444701, Maharashtra, India. Nagpur Univ, Laxminarayan Inst Technol, Nagpur 440010, Maharashtra, India	Waste Management 22 (7): 821-830 2002

Title	Authors				Address	Source
Removal of Heavy Metal Ions From Aqueous Solutions Using Carbon Aerogel As An Adsorbent	Meena AK, Mishra GK, Rai PK, Rajagopal C, Nagar PN	5	5	100.00	Ctr Fire Explos & Environm Safety, Brig Sk Mazumdar Rd, Delhi 110054, India. Univ Rajasthan, Dept Chem, Jaipur 302004, Rajasthan, India.	J Hazard Mater 122 (1-2): 161-170 Jun 30 2005
Unsupervised Feature Selection Using Feature Similarity	Mitra P, Murthy CA, Pal SK	3	3	100.00	Indian Stat Inst, Inst Machine Intelligence, Calcutta 700035, W Bengal, India	Ieee Trans Patt Anal Mach Int 24 (3): 301-312 Mar 2002
Biohydrogen As A Renewable Energy Resource - Prospects and Potentials	Kotay SM, Das D	2	2	100.00	Indian Inst Technol, Dept Biotechnol, Kharagpur 721302, W Bengal, India	Int J Hydrogen Energ 33 (1): 258-263 Jan 2008
Biodiesel Preparation by Lipase-Catalyzed Transesterification of Jatropha Oil	Shah S, Sharma S, Gupta MN	3	3	100.00	Indian Inst Technol, Dept Chem, Hauz Khas, New Delhi 110016, India	Energ Fuel 18 (1): 154-159 Jan-Feb 2004
A Neuro-Fuzzy Computing Technique For Modeling Hydrological Time Series	Nayak PC, Sudheer KP, Rangan DM, Ramasastri KS	4	4	100.00	Natl Inst Hydrol, Deltaic Reg Ctr, Kakinada 533003, India. Indian Inst Technol, Dept Civil Engn, Madras 600036, Tamil Nadu, India. Natl Inst Hydrol, Roorkee 247667, Uttar Pradesh, India.	J Hydrol 291 (1-2): 52-66 May 31 2004
Biosorption of Lead From Aqueous Solutions by Green Algae Spirogyra Species: Kinetics and Equilibrium Studies	Gupta VK, Rastogi A	2	2	100.00	Indian Inst Technol, Dept Chem, Roorkee 247667, Uttar Pradesh, India.	J Hazard Mater 152 (1): 407-414 Mar 21 2008
Performance of Polymer Electrolyte Membrane Fuel Cells With Carbon Nanotubes As Oxygen Reduction Catalyst Support Material	Shaijumon MM, Ramaprabhu S	2	4	50.00	Indian Inst Technol, Dept Phys, Alternat Energy Technol Lab, Madras 600036, Tamil Nadu, India	J Power Sources 140 (2): 250-257 Feb 2 2005

Title	Authors				Affiliation	Reference
Removal of Lead(II) by Adsorption Using Treated Granular Activated Carbon: Batch and Column Studies	Goel J, Kadirvelu K, Rajagopal C, Garg VK	4	4	100.00	Def R&D Org, Ctr Fire Explosives & Environm Safety, Brig Sk Majumdar Marg, Delhi 110054, India. Guru Jambheshwar Univ, Dept Environm Sci & Engn, Hisar 125001, Haryana, India.	J Hazard Mater 125 (1-3): 211-220 Oct 17 2005
Safety Mechanisms in Lithium-Ion Batteries	Balakrishnan PG, Ramesh R, Kumar TP	3	3	100.00	Cent Electrochem Res Inst, Electrochem Power Syst Div, Karaikkudi 630006, Tamil Nadu, India	J Power Sources 155 (2): 401-414 Apr 21 2006
Laboratory Based Approaches For Arsenic Remediation From Contaminated Water: Recent Developments	Mondal P, Majumder CB, Mohanty B	3	3	100.00	Indian Inst Technol, Dept Chem Engn, Roorkee 247667, Uttar Pradesh, India	J Hazard Mater 137 (1): 464-479 Sep 1 2006
A New Beam Finite Element For the Analysis of Functionally Graded Materials	Chakraborty A, Gopalakrishnan S	2	3	66.67	Indian Inst Sci, Dept Aeronaut Engn, Bangalore 560012, Karnataka, India	Int J Mech Sci 45 (3): 519-539 Mar 2003
Ant-Colony Algorithms For Permutation Flowshop Scheduling to Minimize Makespan/Total Flowtime of Jobs	Rajendran C	1	2	50.00	Indian Inst Technol, Dept Humanities & Social Sci, Ind Engn & Mangement Div, Madras 600036, Tamil Nadu, India	Eur J Oper Res 155 (2): 426-438 Jun 1 2004
Characterization of Mesoporous Rice Husk Ash (Rha) and Adsorption Kinetics of Metal Ions From Aqueous Solution Onto Rha	Srivastava VC, Mall ID, Mishra IM	3	3	100.00	Indian Inst Technol, Dept Chem Engn, Roorkee 247667, Uttar Pradesh, India	J Hazard Mater 134 (1-3): 257-267 Jun 30 2006
Performance and Emission Evaluation of A Diesel Engine Fueled With Methyl Esters of Rubber Seed Oil	Ramadhas AS, Muraleedharan C, Jayaraj S	3	3	100.00	Natl Inst Technol, Dept Mech Engn, Calicut 673601, Kerala, India	Renewable Energy 30 (12): 1789-1800 Oct 2005

Title	Authors				Location	Journal
A Review of Advanced High Performance, Insensitive and Thermally Stable Energetic Materials Emerging For Military and Space Applications	Sikder AK, Sikder N	2	2	100.00	High Energy Mat Res Lab, Pune 411021, Maharashtra, India	J Hazard Mater 112 (1-2): 1-15 Aug 9 2004
Bankruptcy Prediction in Banks and Firms Via Statistical and Intelligent Techniques - A Review	Kumar PR, Ravi V	2	2	100.00	Inst Dev & Res Banking Technol, Castle Hills Rd 1, Masab Tank, Hyderabad 500057, Andhra Pradesh, India	Eur J Oper Res 180 (1): 1-28 Jul 1 2007
A Comprehensive Model of Pmos Nbti Degradation	Mahapatra S	1	2	50.00	Iit, Dept Elect Engn, Bombay, Maharashtra, India.	Microelectron Rel 45 (1): 71-81 Jan 2005
Removal of Mercury(Ii) From Aqueous Solutions and Chlor-Alkali Industry Effluent by Steam Activated and Sulphurised Activated Carbons Prepared From Bagasse Pith: Kinetics and Equilibrium Studies	Krishnan KA, Anirudhan TS	2	2	100.00	Univ Kerala, Dept Chem, Trivandrum 695581, Kerala, India.	J Hazard Mater 92 (2): 161-183 May 27 2002
A Review of Dna Functionalized/Grafted Carbon Nanotubes and Their Characterization	Daniel S, Rao TP, Rao KS,	3	7	42.86	Csir, Reg Res Lab, Inorgan Mat Grp, Trivandrum 695019, Kerala, India.Sri Venkateswara Univ, Dept Environm Sci, Tirupati 517502, Andhra Pradesh, India.	Sensor Actuator B-Chem 122 (2): 672-682 Mar 26 2007
Biosorption Mechanism of Nine Different Heavy Metals Onto Biomatrix From Rice Husk	Krishnani KK	1	4	25.00	Cent Inst Brackishwater Aquaculture, Madras 600028, Tamil Nadu, India.	J Hazard Mater 153 (3): 1222-1234 May 30 2008
Estimating Evapotranspiration Using Artificial Neural Network	Kumar M, Raghuwanshi NS, Singh R,	3	5	60.00	Indian Inst Technol, Dept Agr & Food Engn, Kharagpur 721302, W Bengal, India.	J Irrig Drain Eng-Asce 128 (4): 224-233 Jul-Aug 2002

Title	Authors			%	Affiliation	Citation
Adsorption of Acid Dye Onto Organobentonite	Baskaralingam P, Pulikesi M, Elango D, Ramamurthi V, Sivanesan S	5	5	100.00	Anna Univ, Ac Coll Technol, Dept Chem Engn, Madras 6000025, Tamil Nadu, India. Hindustan Coll Engn, Dept Civil Engn, Madras 21, Tamil Nadu, India.	J Hazard Mater 128 (2-3): 138-144 Feb 6 2006
Comparison of Biohydrogen Production Processes	Manish S, Banerjee R	2	2	100.00	Indian Inst Technol, Bombay 400076, Maharashtra, India.	Int J Hydrogen Energ 33 (1): 279-286 Jan 2008
Hybrid Pso-Sqp For Economic Dispatch With Valve-Point Effect	Victoire TAA, Jeyakumar AE	2	2	100.00	Karunya Inst Technol, Dept Elect & Elect Engn, Coimbatore 641114, Tamil Nadu, India., Anna Univ, Dept Elect & Elect Engn, Coimbatore 641013, Tamil Nadu, India.	Elec Power Syst Res 71 (1): 51-59 Sep 2004
Heat Transfer in Stag Nation-Point Flow Towards A Stretching Sheet	Mahapatra TR, Gupta AS	2	2	100.00	Indian Inst Technol, Dept Math, Kharagpur 721302, W Bengal, India	Heat Mass Transfer 38 (6): 517-521 Jun 2002
A Physical Snowtack Model For the Swiss Avalanche Warning Part Ii: Snow Microstructure	Satyawali P	1	5	20.00	Snow & Avalance Study Estab, Manali, India.	Cold Reg Sci Technol 35 (3): 147-167 Nov 2002
An Alkaline Direct Borohydride Fuel Cell With Hydrogen Peroxide As Oxidant	Choudhury NA, Raman RK, Sampath S, Shukla AK	4	4	100.00	Cent Electrochem Res Inst, Karaikkudi 630006, Tamil Nadu, India, Indian Inst Sci, Solid State & Struct Chem Unit, Bangalore 560012, Karnataka, India	J Power Sources 143 (1-2): 1-8 Apr 27 2005
Ft Raman and Ft Ir Spectra, Vibrational Assignments and Density Functional Studies of 5-Bromo-2-Nitropyridine	Sundaraganesan N, Saleem H,	2	5	40.00	Annamalai Univ, Dept Engn Phys, Annamalainagar 608002, Tamil Nadu, India.	Spectrochim Acta Pt A-Mol Bio 61 (13-14): 2995-3001 Oct 2005

Title	Author				Affiliation	Citation
Advances in Biological Hydrogen Production Processes	Das D,	1	4	25.00	Indian Inst Technol, Dept Biotechnol, Kharagpur 721302, W Bengal, India.	Int J Hydrogen Energ 33 (21): 6046-6057 Nov 2008
Identification of Humans Using Gait	Rajagopalan AN	1	7	14.29	Indian Inst Technol, Dept Elect Engn, Madras 600036, Tamil Nadu, India.	Ieee Trans Image Processing 13 (9): 1163-1173 Sep 2004
Removal of Hexavalent Chromium From Aqueous Solution by Agricultural Waste Biomass	Garg UK, Kaur MP, Garg VK, Sud D	4	4	100.00	St Longowal Inst Engn & Technol, Dept Chem, Longowal 148106, Punjab, India; Guru Jambheshwar Univ Sci & Technol, Dept Environm Sci & Engn, Hisar, Haryana, India.	J Hazard Mater 140 (1-2): 60-68 Feb 9 2007
The Compass Experiment At Cern	L Sinha, SS Dasgupta	2	318	0.63	Univ Burdwan, Burdwan 713104, W Bengal, India.; Matrivani Inst Expt Res & Educ, Calcutta 700030, W Bengal, India.	Nucl Instrum Meth Phys Res A 577 (3): 455-518 Jul 11 2007
Hydrothermal Synthesis of Highly Crystalline Zno Nanoparticles: A Competitive Sensor For Lpg and Etoh	Baruwati B, Kumar DK, Manorama SV	3	3	100.00	Indian Inst Chem Technol, Nanomat Lab, Inorgan & Phys Chem Div, Hyderabad 500007, Andhra Pradesh, India.	Sensors and Actuators B: Chemical Volume 119, Issue 2, 7 December 2006, Pages 676-682
Dependence of Optimum Nafion Content in Catalyst Layer on Platinum Loading	Sasikumar G	1	3	33.33	Spic Sci Fdn, Energy Res Ctr, Madras 600032, Tamil Nadu, India.	J Power Sources 132 (1-2): 11-17 May 20 2004
Pool Boiling of Nano-Fluids on Horizontal Narrow Tubes	Das SK	1	3	33.33	Heat Transfer & Thermal Power Lab, Dept Mech Engn, Madras 600036, Tamil Nadu, India.	J Power Sources 132 (1-2): 11-17 May 20 2004

Title	Authors			Institution	%	Journal
Removal of Copper and Cadmium From the Aqueous Solutions by Activated Carbon Derived From Ceiba Pentandra Hulls	M Madhava Raoa, G Purna Chandra Raoa, K Seshaiah	3	4	Sri Venkateswara Univ, Dept Chem, Tirupati 517502, Andhra Pradesh, India.	75.00	J Hazard Mater 129 (1-3): 123-129 Feb 28 2006
Estimating Actual Evapotranspiration From Limited Climatic Data Using Neural Computing Technique	Sudheer KP, Gosain AK, Ramasastri KS	3	3	Delta Reg Ctr, Natl Inst Hydrol, Siddartha Nagar, Kakinada, India.Delta Reg Ctr, Natl Inst Hydrol, Kakinada, India.Indian Inst Technol, Dept Civil Engn, Delhi, India.Natl Hydrol Res Inst, Roorkee, Uttar Pradesh, India.	100.00	J Irrig Drain Eng-Asce 129 (3): 214-218 May-Jun 2003
Sensors - An Effective Approach For the Detection of Explosives	Singh S	1	1	Cent Mech Engn Res Inst, Durgapur 713209, W Bengal, India.	100.00	J Hazard Mater 144 (1-2): 15-28 Jun 1 2007
Trivalent Chromium Removal From Wastewater Using Low Cost Activated Carbon Derived From Agricultural Waste Material and Activated Carbon Fabric Cloth	Mohan D, Singh KP, Singh VK	3	3	Ind Toxicol Res Ctr, Environm Chem Div, Lucknow 226001, Uttar Pradesh, India.	100.00	J Hazard Mater 135 (1-3): 280-295 Jul 31 2006
Hydrogen Storage in Mg: A Most Promising Material	Jain IP, Lal C, Jain A	3	3	Univ Rajasthan, Ctr Nonconvent Energy Resources, Jaipur 302004, Rajasthan, India.	100.00	Int J Hydrogen Energ 35 (10): 5133-5144 Sp. Iss. Si May 2010
Heat Transfer Augmentation in A Two-Sided Lid-Driven Differentially Heated Square Cavity Utilizing Nanofluids	Tiwari RK, Das MK	2	2	Indian Inst Technol, Dept Mech Engn, Gauhati 781039, Assam, India.	100.00	Int J Heat Mass Transfer 50 (9-10): 2002-2018 May 2007

Title	Authors				Affiliation	Journal Reference
Advances in Science and Technology of Modern Energetic Materials: An Overview	Badgujar DM, Talawar MB, Asthana SN, Mahulikar PP	4	4	100.00	N Maharashtra Univ, Sch Chem Sci, Jalgaon 425001, Maharashtra, India., High Energy Mat Res Lab, Pune 41102, Maharashtra, India.	J Hazard Mater 151 (2-3): 289-305 Mar 1 2008
Image Encryption Using Chaotic Logistic Map	Pareek NK, Patidar V, Sud KK	3	3	100.00	Mls Univ, Ctr Comp, Udaipur 313002, Rajasthan, India.	Image Vision Comput 24 (9): 926-934 Sep 1 2006
Adsorption Kinetics of Removal of A Toxic Dye, Malachite Green, From Wastewater by Using Hen Feathers	Mittal A	1	1	100.00	Maulana Azad Natl Inst Technol, Dept Appl Chem, Bhopal 462007, India.	J Hazard Mater 133 (1-3): 196-202 May 20 2006
Potential Benefits and Risks of Land Application of Sewage Sludge	Singh RP, Agrawal M	2	2	100.00	Banaras Hindu Univ, Dept Bot, Varanasi 221005, Uttar Pradesh, India.	Waste Management 28 (2): 347-358 2008
Sorption and Desorption Studies of Chromium(Vi) From Nonviable Cyanobacterium Nostoc Muscorum Biomass	Gupta VK, Rastogi A	2	2	100.00	Indian Inst Technol, Dept Chem, Roorkee 247667, Uttar Pradesh, India.	J Hazard Mater 154 (1-3): 347-354 Jun 15 2008
Dynamic Analysis of Flexible Manipulators, A Literature Review	Dwivedy SK	1	2	50.00	Indian Inst Technol, Dept Engn Mech, Gauhati 781039, India.	Mech Mach Theor 41 (7): 749-777 Jul 2006
Modified Zinc Oxide Thick Film Resistors As Nh3 Gas Sensor	Wagh MS, Jain GH, Patil DR, Patil SA, Patil LA	5	5	100.00	Pratap Coll, Pg Dept Phys, Mat Res Lab, Amalner 425401, Maharashtra, India	Sensor Actuator B-Chem 115 (1): 128-133 May 23 2006
Alternate Synthetic Strategy For the Preparation of Cds Nanoparticles and Its Exploitation For Water Splitting	Sathish M, Viswanathan B, Viswanath RP	3	3	100.00	Indian Inst Technol, Dept Chem, Madras 600036, Tamil Nadu, India	Int J Hydrogen Energ 31 (7): 891-898 Jun 2006
Unit Commitment - A Bibliographical Survey	Padhy NP	1	1	100.00	Indian Inst Technol, Dept Elect Engn, Roorkee 247667, Uttar Pradesh, India.	

Title	Authors				Affiliation	Reference
Design and Development of the Ahwr - the Indian Thorium Fuelled Innovative Nuclear Reactor	Sinha RK, Kakodkar A	2	2	100.00	Bhabha Atom Res Ctr, Bombay 400085, Maharashtra, India	Nucl Eng Des 236 (7-8): 683-700 Apr 2006
Biosorption of Nickel(Ii) Ions Onto Sargassum Wightii: Application of Two-Parameter and Three-Parameter Isotherm Models	Vijayaraghavan K, Padmesh TVN, Palanivelu K, Velan M	4	4	100.00	Anna Univ, Madras 600025, Tamil Nadu, India.	J Hazard Mater 133 (1-3): 304-308 May 20 2006
Ripl - Reference Input Parameter Library For Calculation of Nuclear Reactions and Nuclear Data Evaluations	S Kailas	1	23	4.35	Bhabha Atom Res Ctr, Bombay 400085, Maharashtra, India.	Nucl Data Sheets 110 (12): 3107-3213 Sp. Iss. Si Dec 2009
Peristaltic Transport in An Asymmetric Channel With Heat Transfer - A Note	Srinivas S, Kothandapani M	2	2	100.00	Vit Univ, Sch Sci & Humanities, Vellore 632014, Tamil Nadu, India. Ama Coll Engn, Dept Math, Vadamavandal 604410, Tamil Nadu, India.	Int Commun Heat Mass Trans 35 (4): 514-522 Apr 2008
Equilibrium and Kinetic Modelling of Cadmium(Ii) Biosorption by Nonliving Algal Biomass Oedogonium Sp From Aqueous Phase	Gupta VK, Rastogi A	2	2	100.00	Indian Inst Technol Roorkee, Dept Chem, Roorkee 247667, Uttar Pradesh, India.	J Hazard Mater 153 (1-2): 759-766 May 2008
A New Particle Swarm Optimization Solution to Nonconvex Economic Dispatch Problems	Selvakumar AI, Thanushkodi K	2	2	100.00	Karunya Deemed Univ, Dept Elect Sci, Coimbatore 641114, Tamil Nadu, India. Govt Coll Technol, Dept Elect Engn, Coimbatore 641114, Tamil Nadu, India.	Ieee Trans Power Syst 22 (1): 42-51 Feb 2007
Heart Rate Variability: A Review	Joseph KP	1	5	20.00	Natl Inst Technol Calicut, Calicut 673601, Kerala, India.	Med Biol Eng Comput 44 (12): 1031-1051 Dec 2006

Title	Authors				Affiliation	Citation
Use of Waste Materials - Bottom Ash and De-Oiled Soya, As Potential Adsorbents For the Removal of Amaranth From Aqueous Solutions	Mittal A, Kurup L, Gupta VK	3	3	100.00	Indian Inst Technol, Dept Chem, Roorkee 247667, Uttar Pradesh, India. Maulana Azad Natl Inst Technol, Dept Appl Chem, Bhopal 462007, India.	J Hazard Mater 117 (2-3): 171-178 Jan 31 2005
Aggregation Behavior of Quaternary Salt Based Cationic Surfactants	Mata J, Varade D, Bahadur P	3	3	100.00	S Gujarat Univ, Dept Chem, Surat 395007, India. Sarvajanik Coll Engn & Technol, Dept Chem Engn, Surat 395001, India.	Thermochim Acta 428 (1-2): 147-155 Apr 2005
High-Performance Dual-Gate Carbon Nanotube Fets With 40-Nm Gate Length	Yu-Ming Lin, Joerg Appenzeller, Zhihong Chen, Zhi-Gang Chen, Hui-Ming Cheng, and Phaedon Avouris	1	6	16.67	Inst Met Sci & Technol, Shenyang Natl Lab Mat Sci, New Delhi 110016, India.	Ieee Electron Dev Lett 26 (11): 823-825 Nov 2005
Environmentally Compatible Next Generation Green Energetic Materials (Gems	Talawar MB, Sivabalan R, Mukundan T, Muthurajan H, Sikder AK, Gandhe BR, Rao AS	7	7	100.00	High Energy Mat Res Lab, Pune 411021, Maharashtra, India.	J Hazard Mater 161 (2-3): 589-607 Jan 30 2009
A Possibilistic Fuzzy C-Means Clustering Algorithm	Pal NR, Pal K	2	4	50.00	Indian Stat Inst, Elect & Commun Sci Unit, Calcutta 700108, W Bengal, India. Salt Lake Elect Complex, Inst Engn & Management, Calcutta 700091, W Bengal, India.	Ieee Trans Fuzzy Syst 13 (4): 517-530 Aug 2005
Catalytic Hydrolysis of Sodium Borohydride by A Novel Nickel-Cobalt-Boride Catalyst	Ingersoll JC, Mani N	2	4	50.00	Aartral Energy Res Org, 42-9 Avvai Nagar, Kannagi St, Madras 600094, Tamil Nadu, India.	J Power Sources 173 (1): 450-457 Nov 8 2007

Title	Authors			%	Affiliation	Journal
Exploring the Impact of Size of Training Sets For the Development of Predictive Qsar Models	Roy PP, Leonard JT, Roy K	3	3	100.00	Jadavpur Univ, Dept Pharmaceut Technol, Div Med Chem & Pharmaceut, Drug Theoret & Cheminformat Lab, Calcutta 700032, India.; Km Coll Pharm, Dept Pharmaceut Chem, Madurai 625107, Tamil Nadu, India.	Chemometrics and Intelligent Laboratory Systems
Steam Reforming of Ethanol For Production of Hydrogen Over Ni/Ceo2-Zro2 Catalyst: Effect of Support and Metal Loading	Biswas P, Kunzru D	2	2	100.00	Indian Inst Technol, Dept Chem Engn, Kanpur 208016, Uttar Pradesh, India.	Int J Hydrogen Energ 32 (8): 969-980 Jun 2007
Studies on the Adsorption Kinetics and Isotherms For the Removal and Recovery of Methyl Orange From Wastewaters Using Waste Materials	Mittal A, Malviya A, Kaur D, Mittal J, Kurup L	5	5	100.00	Rofel Shri Gm Bilakhia Coll Appl Sci, Vapi Namcha Rd, Pob 61, Vapi 396191, India., Maulana Azad Natl Inst Technol, Dept Appl Chem, Bhopal 462007, Mp, India.	J Hazard Mater 148 (1-2): 229-240 Sep 5 2007
Peristaltic Transport of A Jeffrey Fluid Under the Effect of Magnetic Field in An Asymmetric Channel	Kothandapani M, Srinivas S	2	3	66.67	Vit Univ, Div Appl Math, Vellore 632014, Tamil Nadu, India.	Int J Non-Linear Mech 43 (9): 915-924 Nov 2008
Rare Isotopes Investigation At Gsi (Rising) Using Gamma-Ray Spectroscopy At Relativistic Energies	Muralithar S	1	72	1.39	Nucl Sci Ctr, New Delhi 110067, India.	Nucl Instrum Meth Phys Res A 537 (3): 637-657 Feb 1 2005
Lpg Sensing Properties of Zno Films Prepared by Spray Pyrolysis Method: Effect of Molarity of Precursor Solution	Shinde VR, Gujar TP, Lokhande CD	3	3	100.00	Shivaji Univ, Dept Phys, Thin Film Phys Lab, Kolhapur 416004, Maharashtra, India.	Sensor Actuator B-Chem 120 (2): 551-559 Jan 10 2007

Title	Authors				Affiliation	Citation
Removal of Reactofix Golden Yellow 3 Rfn From Aqueous Solution Using Wheat Husk - An Agricultural Waste	Gupta VK, Jain R, Varshney S	3	3	100.00	Indian Inst Technol, Dept Chem, Roorkee 247667, Uttar Pradesh, India., Jiwaji Univ, Dept Environm Chem, Gwalior 474011, India.	J Hazard Mater 142 (1-2): 443-448 Apr 2 2007
Characterization and Effect of Using Rubber Seed Oil As Fuel in the Compression Ignition Engines	Ramadhas AS, Jayaraj S, Muraleedharan C	3	3	100.00	Natl Inst Technol Calicut, Dept Mech Engn, Calicut 673601, Kerala, India.	Renewable Energy 30 (5): 795-803 Apr 2005
Studies on Distribution and Fractionation of Heavy Metals in Gomti River Sediments - A Tributary of the Ganges, India	Singh KP, Mohan D, Singh VK, Malik A	4	4	100.00	Ind Toxicol Res Ctr, Environm Chem Div, Pob 80, Mahatma Gandhi Marg, Lucknow 226001, Uttar Pradesh, India.	J Hydrol 312 (1-4): 14-27 Oct 10 2005
Removal of Basic Yellow Dye From Aqueous Solution by Sorption on Green Alga Caulerpa Scalpelliformis	Aravindhan R, Rao JR, Nair BU	3	3	100.00	Cent Leather Res Inst, Chem Lab, Madras 600020, Tamil Nadu, India	J Hazard Mater 142 (1-2): 68-76 Apr 2 2007
The Influence of Heat and Mass Transfer on Mhd Peristaltic Flow Through A Porous Space With Compliant Walls	Srinivas S, Kothandapani M	2	2	100.00	Vit Univ, Div Appl Math, Vellore 632014, Tamil Nadu, India.	Appl Math Comput 213 (1): 197-208 Jul 1 2009
Differential Evolution Using A Neighborhood-Based Mutation Operator	S Das and A Konar	2	4	50.00	Jadavpur Univ, Dept Elect & Telecommun Engn, Calcutta 700032, India.	Ieee Trans Evol Computat 13 (3): 526-553 Jun 2009
Cholesterol Biosensor Based on Cholesterol Esterase, Cholesterol Oxidase and Peroxidase Immobilized Onto Conducting Polyaniline Films	Singh S, Solanki PR, Pandey MK, Malhotra BD	4	4	100.00	Natl Phys Lab, Biomol Elect & Conducting Polymer Res Grp, Dr Ks Krishnan Marg, New Delhi 110012, India.	Sensor Actuator B-Chem 115 (1): 534-541 May 23 2006

Title	Authors			%	Affiliation	Citation
Pesticide Use and Application: An Indian Scenario	Abhilash PC, Singh N	2	2	100.00	Natl Bot Res Inst, Council Sci & Ind Res, Ecoauditing Grp, Rana Pratap Marg, Lucknow 226001, Uttar Pradesh, India.	J Hazard Mater 165 (1-3): 1-12 Jun 15 2009
A Simulated Annealing-Based Multiobjective Optimization Algorithm: Amosa	Bandyopadhyay S, Saha S, Maulik U, Deb K	4	4	100.00	Indian Stat Inst, Machine Intelligence Unit, Calcutta 700108, India. Jadavpur Univ, Dept Comp Sci & Engn, Calcutta 700032, India. Indian Inst Technol, Dept Mech Engn, Kangal, Kanpur 208016, Uttar Pradesh, India.	Ieee Trans Evol Computat 12 (3): 269-283 Jun 2008
Peristaltic Flow and Heat Transfer in A Vertical Porous Annulus, With Long Wave Approximation	Radhakrishnamacharya G, Radhakrishnamurty V	2	3	66.67	Natl Inst Technol, Dept Math & Humanities, Warangal 506004, Andhra Pradesh, India. Sir Cr Reddy Coll Engn, Dept Math, Eluru 534007, India.	Int J Non-Linear Mech 42 (5): 754-759 Jun 2007
Removal of Congo Red Using Activated Carbon and Its Regeneration	Purkait MK, Maiti A, Dasgupta S, De S	4	4	100.00	Indian Inst Technol, Dept Chem Engn, Kharagpur 721302, W Bengal, India.	J Hazard Mater 145 (1-2): 287-295 Jun 25 2007
Biosorption of Chromium and Nickel by Heavy Metal Resistant Fungal and Bacterial Isolates	Congeevaram S, Dhanarani S, Dexilin M, Thamaraiselvi K	4	5	80.00	Bharathidasan Univ, Dept Ecobiotechnol, Tiruchchirappalli 620024, Tamil Nadu, India.	J Hazard Mater 146 (1-2): 270-277 Jul 19 2007
Using Pseudo Amino Acid Composition to Predict Protein Subnuclear Localization: Approached With Pssm	Mundra P, Kumar M, Kumar KK, Jayaraman VK, Kulkarni BD	5	5	100.00	Natl Chem Lab, Chem Engn & Proc Dev Div, Dr Homi Bhabha Rd, Pune 411008, Maharashtra, India.	Pattern Recognition Lett 28 (13): 1610-1615 Oct 1 2007

Title	Authors				Affiliation	Source
Electrochemical Oxidation of Textile Wastewater and Its Reuse	Mohan N, Balasubramanian N, Basha CA	3	3	100.00	Anna Univ, Ac Coll Technol, Dept Chem Engn, Madras 600025, Tamil Nadu, India.; Def Res & Dev Org, Ctr Environm & Explos Safety, Delhi 54, India.; Cent Electrochem Res Inst, Pollut Control Div, Karaikkudi 630006, Tamil Nadu, India.	J Hazard Mater 147 (1-2): 644-651 Aug 17 2007
Metal Decorated Graphene Nanosheets As Immobilization Matrix For Amperometric Glucose Biosensor	Baby TT, Aravind SSJ, Arockiadoss T, Rakhi RB, Ramaprabhu S	5	5	100.00	Indian Inst Technol Madras, Dept Phys, Nano Funct Mat Technol Ctr, Alternat Energy & Nanotechnol Lab, Madras 600036, Tamil Nadu, India.	Sensor Actuator B-Chem 145 (1): 71-77 Mar 4 2010
A Review on the Utilization of Fly Ash	Ahmaruzzaman M	1	1	100.00	Natl Inst Technol Silchar, Dept Chem, Silchar 788010, Assam, India.	Prog Energ Combust Sci 36 (3): 327-363 Jun 2010
Biosorption of Hexavalent Chromium by Raw and Acid-Treated Green Alga Oedogonium Hatei From Aqueous Solutions	Gupta VK, Rastogi A	2	2	100.00	Indian Inst Technol Roorkee, Dept Chem, Roorkee 247667, Uttar Pradesh, India.	J Hazard Mater 163 (1): 396-402 Apr 15 2009
Kinetics of Two-Stage Fermentation Process For the Production of Hydrogen	Nath K, Muthukumar M, Kumar A, Das D	4	4	100.00	Indian Inst Technol, Dept Biotechnol, Fermentat Technol Lab, Kharagpur 721302, W Bengal, India.; Gh Patel Coll Engn & Technol, Dept Chem Engn, Vallabh Vidyanagar 388120, Gujarat, India.	Int J Hydrogen Energ 33 (4): 1195-1203 Feb 2008
Hexavalent Chromium Removal From Wastewater Using Aniline Formaldehyde Condensate Coated Silica Gel	Kumar PA, Ray M, Chakraborty S	3	3	100.00	Indian Inst Technol, Dept Chem, Gauhati 781039, Assam, India.	J Hazard Mater 143 (1-2): 24-32 May 8 2007

Evaluation of Removal Efficiency of Fluoride From Aqueous Solution Using Quick Lime	Islam M, Patel RK	2	2	100.00	Natl Inst Technol, Dept Chem, Rourkela 769008, India.	J Hazard Mater 143 (1-2): 303-310 May 8 2007
Modeling the Metrics of Lean, Agile and Leagile Supply Chain: An Anp-Based Approach	Agarwal A, Shankar R, Tiwari MK	3	3	100.00	Indian Inst Technol, Dept Management Studies, New Delhi 110016, India.; Natl Inst Forged & Foundry Technol, Dept Mfg Engn, Ranchi 834003, Bihar, India.	Eur J Oper Res 173 (1): 211-225 Aug 16 2006
Adsorption of Arsenate on Synthetic Goethite From Aqueous Solutions	Lakshmipathiraj P, Narasimhan BR, Prabhakar S, Raju GB	4	4	100.00	Natl Met Lab Madras Ctr, Csir Madras Complex, Madras 600113, Tamil Nadu, India.	J Hazard Mater 136 (2): 281-287 Aug 21 2006
Cellulase Production Using Biomass Feed Stock and Its Application in Lignocellulose Saccharification For Bio-Ethanol Production	Sukumaran RK, Singhania RR, Mathew GM, Pandey A	4	4	100.00	Csir, Div Biotechnol, Natl Inst Interdisciplinary Sci & Technol, Trivandrum 695019, Kerala, India.	Renewable Energy 34 (2): 421-424 Feb 2009
Catalysis For Nox Abatement	Roy S, Hegde MS, Madras G	3	3	100.00	Indian Inst Sci, Solid State & Struct Chem Unit, Bangalore 560012, Karnataka, India.	Appl Energ 86 (11): 2283-2297 Nov 2009
Performance and Emissions Characteristics of Jatropha Oil (Preheated and Blends) in A Direct Injection Compression Ignition Engine	Agarwal D, Agarwal AK	2	2	100.00	Indian Inst Technol, Environm Engn & Management Program, Kanpur 208016, Uttar Pradesh, India.	Appl Therm Eng 27 (13): 2314-2323 Sep 2007
Adsorption of Malachite Green on Groundnut Shell Waste Based Powdered Activated Carbon	Malik R, Ramteke DS, Wate SR	3	3	100.00	Natl Environm Engn Res Inst, Environm Imapct & Risk Assesment Div, Nagpur 440020, Maharashtra, India.	Waste Management 27 (9): 1129-1138 2007

Title	Authors				Affiliation	Reference
Heterocontact Type Cuo-Modified Sno2 Sensor For the Detection of A Ppm Level H2s Gas At Room Temperature	Patil LA, Patil DR	2	2	100.00	Pratap Coll, Dept Phys, Mat Res Lab, Amalner 425401, Maharashtra, India.	Sensor Actuator B-Chem 120 (1): 316-323 Dec 14 2006
Adsorption Characteristics of Brilliant Green Dye on Kaolin	Nandi BK, Goswami A, Purkait MK	3	3	100.00	Indian Inst Technol Guwahati, Dept Chem Engn, Gauhati 781039, Assam, India.	J Hazard Mater 161 (1): 387-395 Jan 15 2009
Electrical and Humidity Sensing Properties of Polyaniline/Wo3 Composites	Parvatikar N, Jain S, Khasim S, Revansiddappa M, Bhoraskar SV, Prasad MCNA	6	6	100.00	Gulbarga Univ, Dept Chem, Gulbarga 585106, Karnataka, India. Univ Pune, Dept Phys, Ctr Adv Res Mat Sci & Solid State Phys, Pune, Maharashtra, India.	Sensor Actuator B-Chem 114 (2): 599-603 Apr 26 2006
Liquid Phase Adsorption of Crystal Violet Onto Activated Carbons Derived From Male Flowers of Coconut Tree	Senthilkumaar S, Kalaamani P, Subburaam CV	3	3	100.00	Psg Coll Technol, Fac Engn, Dept Chem, Coimbatore 641004, Tamil Nadu, India. Ngm Coll, Dept Chem, Pollachi 642001, India. Bharathiar Univ, Dept Environm Sci, Coimbatore 641046, Tamil Nadu, India.	J Hazard Mater 136 (3): 800-808 Aug 25 2006
Guava (Psidium Guaiava) Leaf Powder: Novel Adsorbent For Removal of Methylene Blue From Aqueous Solutions	Ponnusami V, Vikram S, Srivastava SN	3	3	100.00	Sastra Univ, Dept Chem Engn, Sch Chem & Biotechnol, Thirumalaisamudram 613402, Thanjavur, India.	J Hazard Mater 152 (1): 276-286 Mar 21 2008
H2s Sensors Based on Tungsten Oxide Nanostructures	Rout CS, Hegde M, Rao CNR	3	3	100.00	Jawaharlal Nehru Ctr Adv Sci Res, Chem & Phys Mat Unit, Dst Unit Nanosci, Bangalore 560064, Karnataka, India.	Sensor Actuator B-Chem 128 (2): 488-493 Jan 15 2008

Title	Author				Address	Reference
Biosynthesis of Au, Ag and Au-Ag Nanoparticles Using Edible Mushroom Extract	Philip D	1	1	100.00	Mar Ivanios Coll, Dept Phys, Thiruvananthapuram 695015, Kerala, India.	Spectrochim Acta Pt A-Mol Bio 73 (2): 374-381 Jul 15 2009
Removal of Cadmium (Ii) From Aqueous Solutions by Adsorption on Agricultural Waste Biomass	Garg U, Kaur MP, Jawa GK, Sud D, Garg VK	5	5	100.00	Guru Jambheshwar Univ Sci & Technol, Dept Environm Sci & Engn, Hisar 125001, Haryana, India., Sant Longowal Inst Engn & Technol, Dept Chem, Longowal 148106, Punjab, India.	J Hazard Mater 154 (1-3): 1149-1157 Jun 15 2008
Chlorination Byproducts, Their Toxicodynamics and Removal From Drinking Water	Gopal K, Dubey SP	2	4	50.00	Ind Toxicol Res Ctr, Aquat Toxicol Div, Pob 80, Mg Marg, Lucknow 226001, Uttar Pradesh, India.	J Hazard Mater 140 (1-2): 1-6 Feb 9 2007
Equilibrium, Kinetic and Thermodynamic Studies of Adsorption of Fluoride Onto Plaster of Paris	Gopal V, Elango KP	2	2	100.00	Deemed Univ, Gandhigram Rural Inst, Dept Chem, Gandhigram 624302, India. Tamil Nadu Water Supply & Drainage Board, Dist Water Testing Lab, Theni 625531, India.	J Hazard Mater 141 (1): 98-105 Mar 6 2007
Gas Diffusion Layer For Proton Exchange Membrane Fuel Cells-A Review	Cindrella L	1	7	14.29	Natl Inst Technol, Dept Chem, Tiruchchirappalli 620015, Tamil Nadu, India.	J Power Sources 194 (1): 146-160 Sp. Iss. Si Oct 20 2009
Studies on Tio2/Zno Photocatalysed Degradation of Lignin	Kansal SK, Singh M, Sud D	3	3	100.00	Panjab Univ, Dept Chem Engn & Technol, Chandigarh 160014, India. St Longowal Inst Engn & Technol, Dept Chem Technol, Sangrur 148106, Punjab, India.	J Hazard Mater 153 (1-2): 412-417 May 2008

Title	Authors				Institution	Source
Fiber-Optic Sensors Based on Surface Plasmon Resonance: A Comprehensive Review	Sharma AK, Jha R, Gupta BD	3	3	100.00	Indian Inst Technol, New Delhi 110016, India.	Ieee Sens J 7 (7-8): 1118-1129 Jul-Aug 2007
Neural Networks and Statistical Techniques: A Review of Applications	Paliwal M, Kumar UA	2	2	100.00	Indian Inst Technol, Shailesh J Mehta Sch Management, Bombay 400076, Maharashtra, India.	Expert Syst Appl 36 (1): 2-17 Jan 2009
Sensors For 5-Hydroxytryptamine and 5-Hydroxyindole Acetic Acid Based on Nanomaterial Modified Electrodes	Goyal RN, Gupta VK, Singh SP, Sharma RA	4	5	80.00	Indian Inst Technol, Dept Chem, Roorkee 247667, Uttar Pradesh, India.	Sensor Actuator B-Chem 134 (2): 816-821 Sep 25 2008
Effect of Particle Size on the Convective Heat Transfer in Nanofluid in the Developing Region	Anoop KB, Sundararajan T, Das SK	3	3	100.00	Indian Inst Technol, Dept Mech Engn, Heat Transfer & Thermal Power Lab, Madras 600036, Tamil Nadu, India.	Int J Heat Mass Transfer 52 (9-10): 2189-2195 Apr 2009
Photo Degradation of Methyl Orange An Azo Dye by Advanced Fenton Process Using Zero Valent Metallic Iron: Influence of Various Reaction Parameters and Its Degradation Mechanism	Devi LG, Kumar SG, Reddy KM, Munikrishnappa C	4	4	100.00	Bangalore Univ, Dept Post Grad Studies Chem, Cent Coll City Campus, Dr Br Ambedkar Veedi, Bangalore 560001, Karnataka, India.	J Hazard Mater 164 (2-3): 459-467 May 30 2009
On the Application and Extension of System Signatures in Engineering Reliability	Bhattacharya D	1	4	25.00	Visva Bharati Univ, Dept Stat, Santini Ketan, W Bengal, India.	Nav Res Log 55 (4): 313-327 Jun 2008
Dielectric Properties of Epoxy Nanocomposites	Singha S, Thomas MJ	2	2	100.00	Indian Inst Sci, Dept Elect Engn, High Voltage Lab, Bangalore 560012, Karnataka, India.	Ieee Trans Dielect Electr in 15 (1): 12-23 Feb 2008

Title	Authors				Affiliation	Citation
Adsorptive Removal of Heavy Metals From Aqueous Solution by Treated Sawdust (Acacia Arabica)	Meena AK, Kadirvelu K, Mishra GK, Rajagopal C, Nagar PN	5	5	100.00	Ctr Fire Explos & Environm Safety, Def R&D Org, Brig Sk Mazumdar Rd, Delhi 110054, India., Univ Rajasthan, Dept Chem, Jaipur 302004, Rajasthan, India.	J Hazard Mater 150 (3): 604-611 Feb 11 2008
Liquid-Phase Adsorption of Phenols Using Activated Carbons Derived From Agricultural Waste Material	Singh KP, Malik A, Sinha S, Ojha P	4	4	100.00	Ind Toxicol Res Ctr, Environm Chem Sect, Lucknow 226001, Uttar Pradesh, India. Natl Bot Res Inst, Lucknow 226001, Uttar Pradesh, India.	J Hazard Mater 150 (3): 626-641 Feb 11 2008
Biosorption of Chromium Species by Aquatic Weeds: Kinetics and Mechanism Studies	Elangovan R, Philip L, Chandraraj K	3	3	100.00	Indian Inst Technol, Dept Biotechnol, Madras 600036, Tamil Nadu, India.	J Hazard Mater 152 (1): 100-112 Mar 21 2008
Enhanced Fluoride Sorption by Mechanochemically Activated Kaolinites	Meenakshi S, Sundaram CS, Sukumar R	3	3	100.00	Gandhigram Rural Univ, Dept Chem, Gandhigram 624302, Tamil Nadu, India. Karaikal Polytech Coll, Dept Sci & Humanities, Karaikal 609609, Puducherry, India. Chem Sci & Technol, Reg Res Lab, Trivandrum 695019, Kerala, India.	J Hazard Mater 153 (1-2): 164-172 May 2008
Experimental Investigations of Performance and Emissions of Karanja Oil and Its Blends in A Single Cylinder Agricultural Diesel Engine	Agarwal AK, Rajamanoharan K	2	2	100.00	Indian Inst Technol, Dept Mech Engn, Kanpur 208016, Uttar Pradesh, India.	Appl Energ 86 (1): 106-112 Jan 2009

Title	Authors				Affiliation	Source
Hydrogen the Fuel For 21St Century	Jain IP	1	1	100.00	Univ Rajasthan, Ctr Nonconvent Energy Resources, 14 Vigyan Bhavan, Jaipur 302004, Rajasthan, India.	Int J Hydrogen Energ 34 (17): 7368-7378 Sp. Iss. Si Sep 2009
Pt-Ru/Multi-Walled Carbon Nanotubes As Electrocatalysts For Direct Methanol Fuel Cell	Jha N, Reddy ALM, Shaijumon MM, Rajalakshmi N, Ramaprabhu S	5	5	100.00	Indian Inst Technol, Dept Phys, Alternat Energy Technol Lab, Madras 600036, Tamil Nadu, India. Ctr Fuel Cell Technol, Madras 601302, Tamil Nadu, India.	Int J Hydrogen Energ 33 (1): 427-433 Jan 2008
Chemical Hydrides: A Solution to High Capacity Hydrogen Storage and Supply	Biniwale RB, Rayalu S, Devotta S	3	4	75.00	Natl Environm Engn Res Inst, Nagpur 440020, Maharashtra, India	Int J Hydrogen Energ 33 (1): 360-365 Jan 2008
Adsorption of Cr(Vi) From Aqueous Solutions by Spent Activated Clay	Sharma YC	1	3	33.33	Banaras Hindu Univ, Inst Technol, Dept Appl Chem, Environm Engn & Res Labs, Varanasi 221005, Uttar Pradesh, India.	J Hazard Mater 155 (1-2): 65-75 Jun 30 2008
Study of Heat Transfer Augmentation in A Differentially Heated Square Cavity Using Copper-Water Nanofluid	Santra AK, Sen S, Chakraborty N	3	3	100.00	Jadavpur Univ, Dept Mech Engn, Calcutta 700032, India.	Int J Therm Sci 47 (9): 1113-1122 Sep 2008
Cyanide in Industrial Wastewaters and Its Removal: A Review on Biotreatment	Dash RR, Gaur A, Balomajumder C	3	3	100.00	Natl Inst Technol Hamirpur, Dept Civil Engn, Hamirpur 177005, Hp, India. Indian Inst Technol Roorkee, Dept Chem Engn, Roorkee 247667, Uttar Pradesh, India	J Hazard Mater 163 (1): 1-11 Apr 15 2009
Distillery Spent Wash: Treatment Technologies and Potential Applications	Mohana S, Acharya BK, Madamwar D	3	3	100.00	Sardar Patel Univ, Brd Sch Biosci, Vallabh Vidyanagar 388120, Gujarat, India.	J Hazard Mater 163 (1): 12-25 Apr 15 2009

Title	Authors				Affiliation	Source
An 80-Tile Sub-100-W Teraflops Processor in 65-Nm Cmos	Singh A, Jacob T, Jain S, Erraguntla V	4	15	26.67	Intel Corp, Microprocessor Technol Lab, Bangalore 560037, Karnataka, India.	Ieee J Solid-State Circuits 43 (1): 29-41 Jan 2008
Electrochemical Sensor For the Determination of Dopamine in Presence of High Concentration of Ascorbic Acid Using A Fullerene-C-60 Coated Gold Electrode	Goyal RN, Gupta VK, Bachheti N, Sharma RA	4	4	100.00	Indian Inst Technol, Dept Chem, Roorkee 247667, Uttar Pradesh, India	Electroanal 20 (7): 757-764 Apr 2008
Reliability Issues in Photovoltaic Power Processing Systems	Veerachary M	1	5	20.00	Indian Inst Technol, Dept Elect Engn, New Delhi 110016, India.	Ieee Trans Ind Electron 55 (7): 2569-2580 Jul 2008
A Survey on Cascaded Multilevel Inverters	Gopakumar K	1	4	25.00	Indian Inst Sci, Ctr Elect Design & Technol, Bangalore 560012, Karnataka, India.	Ieee Trans Ind Electron 57 (7): 2197-2206 Jul 2010
Ecofriendly Biodegradation and Detoxification of Reactive Red 2 Textile Dye by Newly Isolated Pseudomonas Sp Suk1	Kalyani DC, Telke AA, Dhanve RS, Jadhav JP	4	4	100.00	Shivaji Univ, Dept Biochem, Kolhapur 416004, Maharashtra, India	J Hazard Mater 163 (2-3): 735-742 Apr 30 2009
Removal of Acid Violet 17 From Aqueous Solutions by Adsorption Onto Activated Carbon Prepared From Sunflower Seed Hull	Thinakaran N, Baskaralingam P, Pulikesi M, Panneerselvam P, Sivanesan S	5	5	100.00	St Josephs Coll Engn, Dept Chem, Madras 600119, Tamil Nadu, India. Velammal Engn Coll, Dept Chem, Madras 600066, Tamil Nadu, India. Anna Univ, Ac Coll Technol, Dept Chem Engn, Madras 600025, Tamil Nadu, India.	J Hazard Mater 151 (2-3): 316-322 Mar 1 2008
Al-Doped Zno Thin Films As Methanol Sensors	Sahay PP, Nath RK	2	2	100.00	Motilal Nehru Natl Inst Technol, Dept Phys, Allahabad 211004, Uttar Pradesh, India. Natl Inst Technol, Dept Phys, Silchar 788100, India.	Sensor Actuator B-Chem 134 (2): 654-659 Sep 25 2008

Title	Authors			%	Affiliation	Source
Biodiesel Development From Rice Bran Oil: Transesterification Process Optimization and Fuel Characterization	Sinha S, Agarwal AK, Garg S	3	3	100.00	Indian Inst Technol, Dept Mech Engn, Kanpur 208016, Uttar Pradesh, India	Energ Conv Manage 49 (5): 1248-1257 May 2008
Adaptive Particle Swarm Optimization Approach For Static and Dynamic Economic Load Dispatch	Panigrahi BK, Pandi VR	2	3	66.67	Indian Inst Technol, Dept Elect Engn, New Delhi 110016, India	Energ Conv Manage 49 (6): 1407-1415 Jun 2008
Copper Sulphide (Cuxs) As An Ammonia Gas Sensor Working At Room Temperature	Sagade AA, Sharma R	2	2	100.00	Dr Babasaheb Ambedkar Marathwada Univ, Dept Phys, Thin Film & Nanotechnol Lab, Aurangabad 431004, Maharashtra, India.	Sensor Actuator B-Chem 133 (1): 135-143 Jul 28 2008
Biology and Genetic Improvement of Jatropha Curcas L.: A Review	Divakara BN, Upadhyaya HD, Wani SP, Gowda CLL	4	4	100.00	Int Crops Res Inst Semi Arid Trop, Hyderabad 502324, Andhra Pradesh, India	Appl Energ 87 (3): 732-742 Mar 2010
Application of Taguchi and Response Surface Methodologies For Surface Roughness in Machining Glass Fiber Reinforced Plastics by Pcd Tooling	Palanikumar K	1	1	100.00	Deemed Univ, Sathysbama Inst Sci & Technol, Dept Mech & Prod Engn, Madras 119, Tamil Nadu, India	Int J Adv Manuf Technol 36 (1-2): 19-27 Feb 2008
Hydrogen Adsorption Properties of Single-Walled Carbon Nanotube - Nanocrystalline Platinum Composites	Reddy ALM, Ramaprabhu S	2	2	100.00	Indian Inst Technol, Dept Phys, Alternat Energy Technol Lab, Madras 600036, Tamil Nadu, India	Int J Hydrogen Energ 33 (3): 1028-1034 Feb 2008

Title	Authors				Affiliation	Citation
Defluoridation Chemistry of Synthetic Hydroxyapatite At Nano Scale: Equilibrium and Kinetic Studies	Sundaram CS, Viswanathan N, Meenakshi S	3	3	100.00	Polytecch Coll, Dept Sci & Humanities, Karaikal 609609, Puducherry, India. Gandhigram Rural Univ, Dept Chem, Gandhigram 624302, Tamil Nadu, India	J Hazard Mater 155 (1-2): 206-215 Jun 30 2008
Performance and Emission Characteristics of A Di Compression Ignition Engine Operated on Honge, Jatropha and Sesame Oil Methyl Esters	Banapurmath NR, Tewari PG, Hosmath RS	3	3	100.00	Bvb Coll Engn & Technol, Dept Mech Engn, Hubli 580031, India. Kels Cet, Dept Mech Engn, Belgaum, India.	Renewable Energy 33 (9): 1982-1988 Sep 2008
Self-Organizing Hierarchical Particle Swarm Optimization For Nonconvex Economic Dispatch	Chaturvedi KT, Pandit M, Srivastava L	3	3	100.00	Mits, Dept Elect Engn, Gwalior, India.	Ieee Trans Power Syst 23 (3): 1079-1087 Aug 2008
Biodiesel Production Process Optimization and Characterization to Assess the Suitability of the Product For Varied Environmental Conditions	Eevera T, Rajendran K, Saradha S	3	3	100.00	Periyar Maniammai Univ, Dept Biotechnol, Thanjavur 613403, Tamil Nadu, India	Renewable Energy 34 (3): 762-765 Mar 2009
Biosorption of Cadmium (Ii) and Lead (Ii) From Aqueous Solutions Using Mushrooms: A Comparative Study	Vimala R, Das N	2	2	100.00	Vit Univ, Sch Biotechnol Chem & Biomed Engn, Vellore 632014, Tamil Nadu, India	J Hazard Mater 168 (1): 376-382 Aug 30 2009
Computer-Vision-Based Fabric Defect Detection: A Survey	Kumar A	1	1	100.00	Indian Inst Technol, Dept Elect Engn, New Delhi 110016, India	Ieee Trans Ind Electron 55 (1): 348-363 Jan 2008
Design of Dual-Band Bandpass Filters Using Stub-Loaded Open-Loop Resonators	Mondal P, Mandal MK	2	2	100.00	Indian Inst Technol, Dept Elect & Elect Commun Engn, Kharagpur 721302, W Bengal, India	Ieee Trans Microwave Theory 56 (1): 150-155 Jan 2008

Title	Authors			%	Affiliation	Citation
Room-Temperature H2s Gas Sensing At Ppb Level by Single Crystal In2o3 Whiskers	Kaur M, Jain N, Sharma K, Bhattacharya S, Roy M, Tyagi AK, Gupta SK, Yakhmi JV	8	8	100.00	Bhabha Atom Res Ctr, Div Chem, Bombay 400085, Maharashtra, India. Amity Inst Nanotechnol, Noida 201301, India.	456-461 Aug 12 2008
Comparison of Various Advanced Oxidation Processes For the Degradation of 4-Chloro-2 Nitrophenol	Saritha P, Aparna C, Himabindu V	3	4	75.00	Jawaharlal Nehru Technol Univ, Ctr Environm, Inst Sci & Technol, Hyderabad 500072, Andhra Pradesh, India	J Hazard Mater 149 (3): 609-614 Nov 19 2007
Removal of Fluoride From Aqueous Solution Using Protonated Chitosan Beads	Viswanathan N, Sundaram CS, Meenakshi S	3	3	100.00	Gandhigram Rural Univ, Dept Chem, Gandhigram 624302, Tamil Nadu, India. Karaikal Polytech Coll, Dept Sci & Humanities, Karaikal 609609, Puducherry, India	J Hazard Mater 161 (1): 423-430 Jan 15 2009
Arsenic Removal Using Hydrous Nanostructure Iron(Iii)-Titanium(Iv) Binary Mixed Oxide From Aqueous Solution	Gupta K, Ghosh UC	2	2	100.00	Presidency Coll, Dept Chem, Calcutta 700073, W Bengal, India.	J Hazard Mater 161 (1): 423-430 Jan 15 2009
Fourier Transform Infrared and Ft-Raman Spectra, Assignment, Ab Initio, Dft and Normal Co-Ordinate Analysis of 2-Chloro-4-Methylaniline and 2-Chloro-6-Methylaniline	Arjunan V, Mohan S	2	2	100.00	Kanchi Mamunivar Ctr Postgrad Studies, Dept Chem, Puducherry 605008, India. Prist Univ, Ctr Res & Dev, Vallam 613403, Thanjavur, India.	Spectrochim Acta Pt A-Mol Bio 72 (2): 436-444 Mar 2009
Municipal Solid Waste Management in Indian Cities - A Review	Sharholy M, Ahmad K, Mahmood G, Trivedi RC	4	4	100.00	Jamia Millia Islamia, Dept Civil Engn, New Delhi 110025, India.Cent Pollut Control Board, New Delhi 110092, India.	Waste Management 28 (2): 459-467 2008

Title	Authors				Affiliation	Citation
Electrochemical Properties of Doped Lithium Titanate Compounds and Their Performance in Lithium Rechargeable Batteries	Murali K	1	2	50.00	Cent Electrochem Res Inst, Electrochem Mat Sci Div, Karaikkudi 630006, Tamil Nadu, India.	J Power Sources 176 (1): 332-339 Jan 21 2008
Effects of Adsorbent Dose, Its Particle Size and Initial Arsenic Concentration on the Removal of Arsenic, Iron and Manganese From Simulated Ground Water by Fe^{3+} Impregnated Activated Carbon	Mondal P, Majumder CB, Mohanty B	3	3	100.00	Ind Technol Inst, Dept Chem Engn, Roorkee 247667, Uttar Pradesh, India	J Hazard Mater 150 (3): 695-702 Feb 11 2008
Wastewater Treatment Using Low Cost Activated Carbons Derived From Agricultural Byproducts - A Case Study	Mohan D, Singh KR, Singh VK	3	3	100.00	Ind Toxicol Res Ctr, Div Environm Chem, Lucknow 226001, Uttar Pradesh, India.	J Hazard Mater 152 (3): 1045-1053 Apr 15 2008
Studies on the Removal of Pb(ii) From Wastewater by Activated Carbon Developed From Tamarind Wood Activated With Sulphuric Acid	Singh CK, Sahu JN, Mahalik KK, Mohanty CR, Mohan BR, Meikap BC	6	6	100.00	Indian Inst Technol, Dept Chem Engn, Kharagpur 721302, W Bengal, India. Orissa State Pollut Control Board, Bubaneswar, Orissa, India. Natl Inst Technol, Dept Chem Engn, Surathkal, India	J Hazard Mater 153 (1-2): 221-228 May 2008
On the Effective Atomic Number and Electron Density: A Comprehensive Set of Formulas For All Types of Materials and Energies Above 1 Kev	Manohara SR, Hanagodimath SM, Thind KS	3	4	75.00	Gulbarga Univ, Dept Phys, Gulbarga 585106, Karnataka, India. Guru Nanak Dev Univ, Dept Phys, Amritsar 143005, Punjab, India.	Nucl Instrum Meth Phys Res B 266 (18): 3906-3912 Sep 2008

Title	Authors			%	Affiliation	Journal
Poly(Methylmethacrylate) Grafted Chitosan: An Efficient Adsorbent For Anionic Azo Dyes	Singh V, Sharma AK, Tripathi DN, Sanghi R	4	4	100.00	Univ Allahabad, Dept Chem, Allahabad 211002, Uttar Pradesh, India. Indian Inst Technol, So Labs 302, Facil Ecol & Analyt Testing, Kanpur 208016, Uttar Pradesh, India.	J Hazard Mater 161 (2-3): 955-966 Jan 30 2009
Electrocatalytic Activity of Binary and Ternary Composite Films of Pd, Mwcnt and Ni, Part Ii: Methanol Electrooxidation in 1 M Koh	Singh RN, Singh A, Anindita	3	3	100.00	Banaras Hindu Univ, Fac Sci, Dept Chem, Varanasi 221005, Uttar Pradesh, India.	Int J Hydrogen Energ 34 (4): 2052-2057 Feb 2009
Improving the Low Temperature Properties of Biodiesel Fuel	Bhale PV, Deshpande NV, Thombre SB	3	3	100.00	Visvesvaraya Natl Inst Technol, Nagpur 440011, Maharashtra, India.	Renewable Energy 34 (3): 794-800 Mar 2009
Removal of Malachite Green From Dye Wastewater Using Neem Sawdust by Adsorption	Khattri SD, Singh MK	2	2	100.00	Banaras Hindu Univ, Fac Sci, Dept Chem, Varanasi 221005, Uttar Pradesh, India. Ideal Inst Technol, Dept Chem, Ghaziabad, Up, India.	J Hazard Mater 167 (1-3): 1089-1094 Aug 15 2009
Advances in Biohydrogen Production Processes: An Approach Towards Commercialization	Das D	1	1	100.00	Indian Inst Technol, Dept Biotechnol, Kharagpur 721302, W Bengal, India	Int J Hydrogen Energ 34 (17): 7349-7357 Sp. Iss. Si Sep 2009
Water Purification Using Magnetic Assistance: A Review	Ambashta RD	1	2	50.00	Bhabha Atom Res Ctr, Backend Technol Dev Div, Bombay 400085, Maharashtra, India	J Hazard Mater 180 (1-3): 38-49 Aug 15 2010
Removal of Some Metal Ions by Activated Carbon Prepared From Phaseolus Aureus Hulls	Rao MM, Ramana DK, Seshaiah K	3	5	60.00	Sri Venkateswara Univ, Dept Chem, Analyt & Environm Chem Div, Tirupati 517502, Andhra Pradesh, India.	J Hazard Mater 166 (2-3): 1006-1013 Jul 30 2009

Title	Authors			Affiliation	Source
Peristaltic Transport of A Newtonian Fluid in A Vertical Asymmetric Channel With Heat Transfer and Porous Medium	Srinivas S, Gayathri R	2	2	100.00	Vit Univ, Div Appl Math, Vellore 632014, Tamil Nadu, India.	Appl Math Comput 215 (1): 185-196 Sep 1 2009
Influence of Metal Contaminants on Oxidation Stability of Jatropha Biodiesel	Sarin A, Arora R, Singh NP, Sharma M, Malhotra RK	5	5	100.00	Amritsar Coll Engn & Technol, Dept Appl Sci, Amritsar 143001, Punjab, India. Punjab Tech Univ, Jalandhar, India. Indian Oil Corp Ltd, R&D Ctr, Faridabad 121007, India.	Energy 34 (9): 1271-1275 Sep 2009
Biosorption of Arsenic From Aqueous Solution Using Agricultural Residue 'Rice Polish'	Ranjan D, Talat M, Hasan SH	3	3	100.00	Banaras Hindu Univ, Fac Sci, Dept Biochem, Varanasi 221005, Uttar Pradesh, India.	J Hazard Mater 166 (2-3): 1050-1059 Jul 30 2009
Study of Heat Transfer Due to Laminar Flow of Copper-Water Nanofluid Through Two Isothermally Heated Parallel Plates	Santra AK, Sen S, Chakraborty N	3	3	100.00	Jadavpur Univ, Calcutta 700032, India.	Int J Therm Sci 48 (2): 391-400 Feb 2009
Silicotungstic Acid Stabilized Pt-Ru Nanoparticles Supported on Carbon Nanofibers Electrodes For Methanol Oxidation	Maiyalagan T	1	1	100.00	Vit Univ, Sch Sci & Humanities, Dept Chem, Vellore 632014, Tamil Nadu, India.	Int J Hydrogen Energ 34 (7): 2874-2879 Apr 2009
The Murchison Widefield Array: Design Overview	A Deshpande, K Deepak, MR Gopalakrishna, PA Kamini, S Madhavi, D Prabu, A Roshi, NU Shankar, KS Srivani	9	47	19.15	Raman Res Inst, Bangalore 560080, Karnataka, India.	Proc Ieee 97 (8): 1497-1506 Aug 2009
Process Integration of Organic Rankine Cycle	Desai NB, Bandyopadhyay S	2	2	100.00	Indian Inst Technol, Dept Energy Sci & Engn, Bombay 400076, Maharashtra, India.	Energy 34 (10): 1674-1686 Oct 2009

Title	Authors				Affiliation	Source
B-Jet Identification in the D0 Experiment	Shivpuri RK, Rajan K, Adubey, Dutt S, Beri SB, Acharya BS, Bhatnagar V, Koholi JM, Choudhary B, Naimuddin M, Nayyar R, Banerjee S, Mondal NK	13	455	2.86	Panjab Univ, Chandigarh 160014, India. Univ Delhi, Delhi 110007, India. Tata Inst Fundamental Res, Mumbai 400005, Maharashtra, India.	Nucl Instrum Meth Phys Res A 620 (2-3): 490-517 Aug 11 2010
Preconcentration and Determination of Trace Metal Ions From Aqueous Samples by Newly Developed Gallic Acid Modified Amberlite Xad-16 Chelating Resin	Sharma RK, Pant P	2	2	100.00	Univ Delhi, Dept Chem, Green Chem Network Ctr, Delhi 110007, India.	J Hazard Mater 163 (1): 295-301 Apr 15 2009
Fixed-Bed Column Study For Hexavalent Chromium Removal and Recovery by Short-Chain Polyaniline Synthesized on Jute Fiber	Kumar PA, Chakraborty S	2	2	100.00	Indian Inst Technol, Dept Civil Engn, Gauhati 781039, Assam, India	J Hazard Mater 162 (2-3): 1086-1098 Mar 15 2009
Spectroscopic Analysis and Dft Calculations of A Food Additive Carmoisine	Snehalatha M, Ravikumar C, Joe IH, Sekar N, Jayakumar VS	5	5	100.00	Mar Ivanios Coll, Dept Phys, Ctr Mol & Biophys Res, Thiruvananthapuram 695015, Kerala, India. Univ Bombay, Inst Chem Technol, Bombay 400019, Maharashtra, India. Christian Coll, Dept Phys, Kattakada, Kerala, India.	Spectrochim Acta Pt A-Mol Bio 72 (3): 654-662 Apr 2009
Synthesis, Characterization and Removal of Cd(li) Using Cd(li)-Ion Imprinted Polymer	Singh DK, Mishra S	2	2	100.00	Harcourt Butler Technol Inst, Dept Chem, Analyt Res Lab, Kanpur 208002, Uttar Pradesh, India.	J Hazard Mater 164 (2-3): 1547-1551 May 30 2009

Title	Authors			%	Affiliation	Reference
Removal of Lead and Zinc Ions From Water by Low Cost Adsorbents	Mishra PC, Patel RK	2	2	100.00	Piet, Dept Chem, Rourkela 770034, Orissa, India. Natl Inst Technol, Dept Chem, Rourkela 769008, India	J Hazard Mater 168 (1): 319-325 Aug 30 2009
Response Surface Modeling and Optimization of Chromium(Vi) Removal From Aqueous Solution Using Tamarind Wood Activated Carbon in Batch Process	Sahu JN, Acharya J, Meikap BC	3	3	100.00	Indian Inst Technol, Dept Chem Engn, Po Kharagpur Technol, Kharagpur 721302, W Bengal, India. Indian Inst Technol, Dept Chem Engn, Kharagpur 721302, W Bengal, India. Rgtu, Sch Energy & Environm Management, Bhopal, Madhya Pradesh, India.	J Hazard Mater 172 (2-3): 818-825 Dec 30 2009
Analytical Expression For Electrical Efficiency of Pv/T Hybrid Air Collector	Dubey S, Sandhu GS, Tiwari GN	3	3	100.00	Indian Inst Technol, Ctr Energy Studies, New Delhi 110016, India	Appl Energ 86 (5): 697-705 May 2009
The Sorption of Lead(Ii) Ions on Rice Husk Ash	Naiya TK, Bhattacharya AK, Mandal S, Das SK	4	4	100.00	Univ Calcutta, 92 A P C Rd, Calcutta 700009, India. Natl Inst Tech Teachers Training & Res, Calcutta 700106, India.	J Hazard Mater 163 (2-3): 1254-1264 Apr 30 2009
Low Band Gap Vinylene Compounds With Triphenylamine and Benzothiadiazole Segments For Use in Photovoltaic Cells	Suresh P, Balraju P, Sharma GD	3	5	60.00	Jai Narain Vyas Univ, Mol Elect & Optoelect Device Lab, Dept Phys, Jodhpur 342005, Rajasthan, India. Jaipur Engn Coll, Jaipur, Rajasthan, India	Org Electron 10 (7): 1320-1333 Nov 2009
Studies on Sorption of Some Geomaterials For Fluoride Removal From Aqueous Solutions	Sujana MG, Pradhan HK, Anand S	3	3	100.00	Inst Minerals & Mat Technol, Bhubaneswar 751013, Orissa, India	J Hazard Mater 161 (1): 120-125 Jan 15 2009

Title	Authors				Affiliation	Reference
Removal of Cr(Vi) From Aqueous Solutions Using Pre-Consumer Processing Agricultural Waste: A Case Study of Rice Husk	Bansal M, Garg U, Singh D, Garg VK	4	4	100.00	Guru Jambheshwar Univ Sci & Technol, Dept Environm Sci & Engn, Hisar 125001, Haryana, India. Natl Inst Technol, Dept Civil Engn, Kurukshetra 136119, Haryana, India. St Longowal Inst Engn & Technol, Dept Chem, Longowal 148106, Punjab, India	J Hazard Mater 162 (1): 312-320 Feb 15 2009
Chromium(Vi) Removal From Aqueous System Using Helianthus Annuus (Sunflower) Stem Waste	Jain M, Garg VK, Kadirvelu K	3	3	100.00	Guru Jambheshwar Univ Sci & Technol, Dept Environm Sci & Engn, Hisar 125001, Haryana, India. Def Res & Dev Org, Cfees, Delhi 110054, India	J Hazard Mater 162 (1): 365-372 Feb 15 2009
Recent Progress in the Development of Nano-Structured Conducting Polymers/Nanocomposites For Sensor Applications	Rajesh, Ahuja T, Kumar D	3	3	100.00	Csir, Special Inst Area, Delhi 110067, India. Univ Delhi, Delhi Coll Engn, Dept Appl Chem, Delhi 110042, India.	Sensor Actuator B-Chem 136 (1): 275-286 Feb 2 2009
Jute Stick Powder As A Potential Biomass For the Removal of Congo Red and Rhodamine B From Their Aqueous Solution	Panda GC, Das SK, Guha AK	3	3	100.00	Indian Assoc Cultivat Sci, Dept Biol Chem, Calcutta 700032, India.	J Hazard Mater 164 (1): 374-379 May 15 2009
Nano Titanium Oxide Catalyst Support For Proton Exchange Membrane Fuel Cells	Rajalakshmi N, Lakshmi N, Dhathathreyan KS	3	3	100.00	Ctr Fuel Cell Technol Arci, Madras 600100, Tamil Nadu, India.	Int J Hydrogen Energ 33 (24): 7521-7526 Dec 2008

Title	Authors				Affiliation	Reference
Simultaneous Determination of Ascorbic Acid, Dopamine and Uric Acid Using Polystyrene Sulfonate Wrapped Multiwalled Carbon Nanotubes Bound to Graphite Electrode Through Layer-By-Layer Technique	Manjunatha R, Suresh GS, Melo JS, D'souza SF, Venkatesha TV	5	5	100.00	Ssmrv Degree Coll, Chem Res Ctr, Bangalore 560041, Karnataka, India. Bhabha Atom Res Ctr, Nucl Agr & Biotechnol Div, Bombay 400085, Maharashtra, India. Kuvempu Univ, Dept Chem, Jnanasahyadri 577451, Shimoga, India.	Sensor Actuator B-Chem 145 (2): 643-650 Mar 19 2010
Experimental Investigations and Theoretical Determination of Thermal Conductivity and Viscosity of Al2o3/Water Nanofluid	Chandrasekar M, Suresh S, Bose AC	3	3	100.00	Natl Inst Technol, Nanomat Lab, Tiruchchirappalli 620015, Tamil Nadu, India.	Exp Therm Fluid Sci 34 (2): 210-216 Feb 2010
Ftir and Ftraman Spectra, Assignments, Ab Initio Hf and Dft Analysis of 4-Nitrotoluene	Ramalingam S, Periandy S, Govindarajan M, Mohan S	4	4	100.00	Avc Coll Autonomous, Dept Phys, Mayiladuthurai 609305, India. Thagore Arts Coll, Dept Phys, Pondicherry, India. Avvaiyar Govt Coll Women, Dept Phys, Karaikal, India. Prist Univ, Ctr Res & Dev, Thanjavur, India.	Spectrochim Acta Pt A-Mol Bio 75 (4): 1308-1314 Apr 2010
Ftir and Ftraman Spectroscopic Investigation of 2-Bromo-4-Methylaniline Using Ab Initio Hf and Dft Calculations	Ramalingam S, Periandy S, Narayanan B, Mohan S	4	4	100.00	Avc Coll Autonomous, Dept Phys, Mappadugai 609305, Mayiladuthurai, India. Tagore Arts Coll, Dept Phys, Pondicherry, India. Prist Univ, Ctr Res & Dev, Thanjavur, India.	Spectrochim Acta Pt A-Mol Bio 76 (1): 84-92 Jun 2010
Effectiveness-Thermal Resistance Method For Heat Exchanger Design and Analysis	Shah RK	1	4	25.00	Indian Inst Technol, Dept Energy Sci & Engn, Bombay 400076, Maharashtra, India.	Int J Heat Mass Transfer 53 (13-14): 2877-2884 Jun 2010

Title	Authors				Affiliation	Source
Recent Trends on the Development of Photobiological Processes and Photobioreactors For the Improvement of Hydrogen Production	Dasgupta CN, Gilbert JJ, Das D	3	7	42.86	Indian Inst Technol, Dept Biotechnol, Kharagpur 721302, W Bengal, India.	Int J Hydrogen Energ 35 (19): 10218-10238 Sp. Iss. Si Oct 2010
Differential Evolution: A Survey of the State-Of-The-Art	Das S	1	2	50.00	Jadavpur Univ, Dept Elect & Telecommun Engn, Calcutta 700032, W Bengal, India.	Ieee Trans Evol Computat 15 (1): 27-54 Feb 2011
Lead(ii) Adsorption From Aqueous Solutions by Raw and Activated Charcoals of Melocanna Baccifera Roxburgh (Bamboo)-A Comparative Study	Lalhruaitluanga H, Jayaram K, Prasad MNV, Kumar KK	4	4	100.00	Univ Hyderabad, Dept Plant Sci, Hyderabad 500046, Andhra Pradesh, India.	J Hazard Mater 175 (1-3): 311-318 Mar 15 2010
Bio-Electrochemical Treatment of Distillery Wastewater in Microbial Fuel Cell Facilitating Decolorization and Desalination Along With Power Generation	Mohanakrishna G, Mohan SV, Sarma PN	3	3	100.00	Indian Inst Chem Technol, Bioengn & Environm Ctr Beec, Hyderabad 500607, Andhra Pradesh, India.	J Hazard Mater 177 (1-3): 487-494 May 15 2010
Molecularly Imprinted Electrochemical Sensors	Suryanarayanan V	1	3	33.33	Cent Electro Chem Res Inst, Electroorgan Div, Karaikkudi 630006, Tamil Nadu, India.	Electroanal 22 (16): 1795-1811 Aug 2010
Peristaltic Flow of Viscoelastic Fluid With Fractional Maxwell Model Through A Channel	Tripathi D, Pandey SK, Das S	3	3	100.00	Banaras Hindu Univ, Inst Technol, Dept Appl Math, Varanasi 221005, Uttar Pradesh, India.	Appl Math Comput 215 (10): 3645-3654 Jan 15 2010
Feature Selection For Classification of Hyperspectral Data by Svm	Pal M	1	2	50.00	Natl Inst Technol, Dept Civil Engn, Kurukshetra 136119, Haryana, India.	Ieee Trans Geosci Remot Sen 48 (5): 2297-2307 May 2010

Title	Authors				Affiliation	Reference
Murraya Koenigii Leaf-Assisted Rapid Green Synthesis of Silver and Gold Nanoparticles	Philip D, Unni C, Aromal SA, Vidhu VK	4	4	100.00	Mar Ivanios Coll, Dept Phys, Thiruvananthapuram 695015, Kerala, India. Tkm Coll Engn, Dept Elect & Commun, Kollam 691005, India.	Spectrochim Acta Pt A-Mol Bio 78 (2): 899-904 Feb 2011
A Comparative Investigation on Adsorption Performances of Mesoporous Activated Carbon Prepared From Waste Rubber Tire and Activated Carbon For A Hazardous Azo Dye-Acid Blue 113	Gupta VK, Gupta B, Rastogi A, Agarwal S, Nayak A	5	5	100.00	Indian Inst Technol, Dept Chem, Roorkee 247667, Uttar Pradesh, India. Kldav Pg Coll, Dept Chem, Roorkee, Uttar Pradesh, India. Jiwaji Univ, Sch Studies Chem, Gwalior 474011, Mp, India.	J Hazard Mater 186 (1): 891-901 Feb 15 2011
Fluorescence Resonance Energy Transfer From A Bio-Active Imidazole Derivative 2-(1-Phenyl-1H-Imidazo[4,5-F][1, 10]Phenanthrolin-2-Yl)Phenol to A Bioactive Indoloquinolizine System	Jayabharathi J, Thanikachalam V, Perumal MV, Srinivasan N	4	4	100.00	Annamalai Univ, Dept Chem, Annamalainagar 608002, Tamil Nadu, India.	Spectrochim Acta Pt A-Mol Bio 79 (1): 236-244 Jun 2011
A Kinetic Model For the Selective Catalytic Reduction of Nox With Nh3 Over An Fe-Zeolite Catalyst	Gopinath A, Olsson L	2	4	50.00	Gen Motors R&D India Sci Lab, Bangalore 560066, Karnataka, India.	Ind Eng Chem Res 49 (1): 39-52 Jan 6 2010
Biosorption of Pb2+ From Aqueous Solutions by Moringa Oleifera Bark: Equilibrium and Kinetic Studies	Reddy DHK, Seshaiah K, Reddy AVR	3	5	60.00	Sri Venkateswara Univ, Dept Chem, Analyt & Environm Chem Div, Tirupati 517502, Andhra Pradesh, India. Bhabha Atom Res Ctr, Div Analyt Chem, Bombay 400085, Maharashtra, India.	J Hazard Mater 174 (1-3): 831-838 Feb 15 2010

Title	Authors				Institution	Reference
Removal of Cu(Ii) From Aqueous Solutions Using Chemically Modified Chitosan	Kannamba B, Reddy KL, Apparao BV	3	3	100.00	Natl Inst Technol Warangal, Dept Chem, Warangal 506004, Andhra Pradesh, India.	J Hazard Mater 175 (1-3): 939-948 Mar 15 2010
Qsar Modeling of Toxicity of Diverse Organic Chemicals to Daphnia Magna Using 2D and 3D Descriptors	Kar S, Roy K	2	2	100.00	Jadavpur Univ, Dept Pharmaceut Technol, Drug Theoret & Cheminformat Lab, Calcutta 700032, India.	J Hazard Mater 177 (1-3): 344-351 May 15 2010
Removal of Malachite Green From Aqueous Solution by Activated Carbon Prepared From the Epicarp of Ricinus Communis by Adsorption	Santhi T, Manonmani S, Smitha T	3	3	100.00	Karpagam Univ, Dept Chem, Coimbatore 641021, Tamil Nadu, India. Psg Coll Arts & Sci, Dept Chem, Coimbatore 641014, Tamil Nadu, India	J Hazard Mater 179 (1-3): 178-186 Jul 15 2010
A Review on Electrochemical Double-Layer Capacitors	Sharma P, Bhatti TS	2	2	100.00	Indian Inst Technol, Ctr Energy Studies, New Delhi 110016, India.	Energ Conv Manage 51 (12): 2901-2912 Dec 2010
Slip Effects on Boundary Layer Stagnation-Point Flow and Heat Transfer Towards A Shrinking Sheet	Bhattacharyya K, Mukhopadhyay S, Layek GC	3	3	100.00	Univ Burdwan, Dept Math, Burdwan 713104, W Bengal, India.	Int J Heat Mass Transfer 54 (1-3): 308-313 Jan 15 2011
Reduced Graphene Oxide-Metal/Metal Oxide Composites: Facile Synthesis and Application in Water Purification	Sreeprasad TS, Maliyekkal SM, Lisha KP, Pradeep T	4	4	100.00	Indian Inst Technol, Dst Unit Nanosci, Dept Chem & Sophisticated Analyt Instrument Facil, Madras 600036, Tamil Nadu, India.	J Hazard Mater 186 (1): 921-931 Feb 15 2011
A Physiochemical Study of Excited State Intramolecular Proton Transfer Process Luminescent Chemosensor by Spectroscopic Investigation Supported by Ab Initio Calculations	Jayabharathi J, Thanikachalam V, Jayamoorthy K, Perumal MV	4	4	100.00	Annamalai Univ, Dept Chem, Annamalainagar 608002, Tamil Nadu, India.	Spectrochim Acta Pt A-Mol Bio 79 (1): 6-16 Jun 2011

Title	Author				Affiliation	Citation
Role of Organic Amendments on Enhanced Bioremediation of Heavy Metal(Loid) Contaminated Soils	Paneerselvam P	1	6	16.67	Indian Inst Hort Res, Bangalore 560089, Karnataka, India.	J Hazard Mater 185 (2-3): 549-574 Jan 30 2011
Developments in Biobutanol Production: New Insights	Kumar M, Gayen K	2	2	100.00	Indian Inst Technol Gandhinagar, Dept Chem Engn, Ahmadabad 382424, Gujarat, India.	Appl Energ 88 (6): 1999-2012 Jun 2011
Synthesis and Characterization of Alumina-Coated Carbon Nanotubes and Their Application For Lead Removal	Gupta VK, Agarwal S,	2	3	66.67	Jiwaji Univ, Dept Chem, Gwalior, Mp, India.Indian Inst Technol Roorkee, Dept Chem, Roorkee 247667, Uttar Pradesh, India.	J Hazard Mater 185 (1): 17-23 Jan 15 2011
Further Exploring R(M)(2) Metrics For Validation of Qspr Models	Ojha PK, Mitra I, Das RN, Roy K	4	4	100.00	Jadavpur Univ, Dept Pharmaceut Technol, Div Med & Pharmaceut Chem, Drug Theoret & Cheminformat Lab, Calcutta 700032, India.	Chemometr Intell Lab Syst 107 (1): 194-205 May 2011
Synthesis and Drug-Delivery Behavior of Chitosan-Functionalized Graphene Oxide Hybrid Nanosheets	Mishra S	1	9	11.11	N Maharastra Univ, Dept Chem Technol, Jalgaon 425001, India. Natl Chem Lab, Div Polymer Sci & Engn, Pune 411008, Maharashtra, India.	Macromol Mater Eng 296 (2): 131-140 Feb 14 2011
Peristaltic Transport of A Viscoelastic Fluid in A Channel	Tripathi D	1	1	100.00	Birla Inst Technol & Sci Pilani, Math Grp, Hyderabad 500078, Andhra Pradesh, India.	Acta Astronaut 68 (7-8): 1379-1385 Apr-May 2011

Title	Authors				Affiliation	Source
Applications of the Moora Method For Decision Making in Manufacturing Environment	Chakraborty S	1	1	100.00	Jadavpur Univ, Dept Prod Engn, Calcutta 700032, W Bengal, India.	Int J Adv Manuf Technol 54 (9-12): 1155-1166 Jun 2011
Parametric Optimization and Performance Analysis of A Waste Heat Recovery System Using Organic Rankine Cycle	Roy JP, Mishra MK, Misra A	3	3	100.00	Ntpc Ltd Kahalgaon, Pts, Bhagalpur 813214, India. Birla Inst Technol, Dept Mech Engn, Ranchi 835215, Bihar, India.	Energy 35 (12): 5049-5062 Dec 2010
Effects of Suction/Blowing on Steady Boundary Layer Stagnation-Point Flow and Heat Transfer Towards A Shrinking Sheet With Thermal Radiation	Bhattacharyya K, Layek GC	2	2	100.00	Univ Burdwan, Dept Math, Burdwan 713104, W Bengal, India.	Int J Heat Mass Transfer 54 (1-3): 302-307 Jan 15 2011
Existence and Global Stability Analysis of Equilibrium of Fuzzy Cellular Neural Networks With Time Delay in the Leakage Term Under Impulsive Perturbations	Rakkiyappan R, Balasubramaniam P	2	3	66.67	Gandhigram Rural Univ, Dept Math, Gandhigram 624302, Tamil Nadu, India.	J Franklin Inst-Eng Appl Math 348 (2): 135-155 Mar 2011
A Shunt Active Power Filter With Enhanced Performance Using Ann-Based Predictive and Adaptive Controllers	Bhattacharya A, Chakraborty C	2	2	100.00	Indian Inst Technol, Dept Elect Engn, Kharagpur 721302, W Bengal, India.	Ieee Trans Ind Electron 58 (2): 421-428 Feb 2011
Momentum and Heat Transfer From A Square Cylinder in Power-Law Fluids	Rao PK, Sahu AK, Chhabra RP	3	3	100.00	Indian Inst Technol, Dept Chem Engn, Kanpur 208016, Uttar Pradesh, India	Int J Heat Mass Transfer 54 (1-3): 390-403 Jan 15 2011
Cation-Anion-Water Interactions in Aqueous Mixtures of Imidazolium Based Ionic Liquids	Singh T, Kumar A	2	2	100.00	Csir, Salt & Marine Chem Div, Cent Salt & Marine Chem Res Inst, Bhavnagar 364002, Gujarat, India.	Vib Spectrosc 55 (1): 119-125 Jan 18 2011

Title	Authors				Affiliation	Journal
Preconcentration and Separation of Copper, Nickel and Zinc in Aqueous Samples by Flame Atomic Absorption Spectrometry After Column Solid-Phase Extraction Onto Mwcnts Impregnated With D2ehpa-Topo Mixture	Vellaichamy S, Palanivelu K	2	2	100.00	Anna Univ, Ctr Environm Studies, Madras 600025, Tamil Nadu, India.	J Hazard Mater 185 (2-3): 1131-1139 Jan 30 2011
Physiochemical Properties of Organic Nonlinear Optical Crystal From Combined Experimental and Theoretical Studies	Jayabharathi J, Thanikachalam V, Saravanan K, Srinivasan N, Perumal MV	5	5	100.00	Annamalai Univ, Dept Chem, Annamalainagar 608002, Tamil Nadu, India.	Spectrochim Acta Pt A-Mol Bio 78 (2): 794-802 Feb 2011
Single Sensor For Two Metal Ions: Colorimetric Recognition of Cu2+ and Fluorescent Recognition of Hg2+	Nandhakumar R	1	4	25.00	Karunya Univ, Dept Chem, Coimbatore 641114, Tamil Nadu, India.	Spectrochim Acta Pt A-Mol Bio 78 (3): 1168-1172 Mar 2011
Trends in Reference Crop Evapotranspiration Over Iran	Jhajharia D	1	5	20.00	Ne Reg Inst Sci & Technol, Dept Agr Engn, Itanagar 791109, Arunachal Prade, India.	J Hydrol 399 (3-4): 422-433 Mar 18 2011
Coupled Fixed Point Results in Generalized Metric Spaces	Choudhury BS, Maity P	2	2	100.00	Bengal Engn & Sci Univ, Dept Math, Howrah 711103, W Bengal, India.	Math Comput Modelling 54 (1-2): 73-79 Jul 2011
Best Proximity Points: Global Optimal Approximate Solutions	Basha SS	1	1	100.00	Anna Univ, Dept Math, Madras 600025, Tamil Nadu, India.	J Global Optim 49 (1): 15-21 Jan 2011
Experimental Study of Combustion and Emission Characteristics of Ethanol Fuelled Port Injected Homogeneous Charge Compression Ignition (Hcci) Combustion Engine	Maurya RK, Agarwal AK	2	2	100.00	Indian Inst Technol, Engine Res Lab, Dept Mech Engn, Kanpur 208016, Uttar Pradesh, India.	Appl Energ 88 (4): 1169-1180 Apr 2011

Title	Authors				Affiliation	Citation
Laminar Natural Convection From A Horizontal Cylinder in Power-Law Fluids	Prhashanna A, Chhabra RP	2	2	100.00	Indian Inst Technol, Dept Chem Engn, Kanpur 208016, Uttar Pradesh, India.	Ind Eng Chem Res 50 (4): 2424-2440 Feb 16 2011
Mangifera Indica Leaf-Assisted Biosynthesis of Well-Dispersed Silver Nanoparticle	Philip D	1	1	100.00	Mar Ivanios Coll, Dept Phys, Thiruvananthapuram 695015, Kerala, India.	Spectrochim Acta Pt A-Mol Bio 78 (1): 327-331 Jan 2011
Optimizing Adsorption of Crystal Violet Dye From Water by Magnetic Nanocomposite Using Response Surface Modeling Approach	Singh KP, Gupta S, Singh AK, Sinha S	4	4	100.00	Indian Inst Toxicol Res, Div Environm Chem, Csir, Lucknow 226001, Uttar Pradesh, India. Natl Bot Res Inst, Csir, Lucknow 226001, Uttar Pradesh, India	J Hazard Mater 186 (2-3): 1462-1473 Feb 28 2011
Mechanistic Investigation on Binding Interaction of Bioactive Imidazole With Protein Bovine Serum Albumin-A Biophysical Study	Jayabharathi J, Thanikachalam V, Perumal MV	3	3	100.00	Annamalai Univ, Dept Chem, Annamalainagar 608002, Tamil Nadu, India.	Spectrochim Acta Pt A-Mol Bio 79 (3): 502-507 Aug 2011
Spectrofluorometric Studies on the Binding Interaction of Bioactive Imidazole With Bovine Serum Albumin: A Dft Based Esipt Process	Jayabharathi J, Thanikachalam V, Saravanan K, Perumal MV	1	4	25.00	Annamalai Uni., Tamil Nadu	Spectrochim Acta Pt A-Mol Bio 79 (5): 1240-1246 Sep 2011
Flow Over and Forced Convection Heat Transfer in Newtonian Fluids From A Semi-Circular Cylinder	Chandra A, Chhabra RP	2	2	100.00	Indian Inst Technol, Dept Chem Engn, Kanpur 208016, Uttar Pradesh, India.	Int J Heat Mass Transfer 54 (1-3): 225-241 Jan 15 2011
Biomimetic Sequestration of Co2 and Reformation to Caco3 Using Bovine Carbonic Anhydrase Immobilized on Sba-15	Muthukaruppan Alagar, Mari Vinoba	2	6	33.33	Anna Univ, Dept Chem Engn, Madras 600025, Tamil Nadu, India.	Energ Fuel 25: 438-445 Jan 2011

Title	Authors				Institution	Journal
Studies on Fractional Order Differentiators and Integrators: A Survey	Krishna BT	1	1	100.00	Gitam Univ, Dept Ece, Visakhapatnam, Andhra Pradesh, India.	Signal Process 91 (3): 386-426 Sp. Iss. Si Mar 2011
Decolourization of Methyl Orange Using Fenton-Like Mesoporous Fe2o3-Sio2 Composite	Panda N, Sahoo H, Mohapatra S	3	3	100.00	Nit, Orrisa	J Hazard Mater 185 (1): 359-365 Jan 15 2011
Optimization of Orange G Dye Adsorption by Activated Carbon of Thespesia Populnea Pods Using Response Surface Methodology	Arulkumar M, Sathishkumar P, Palvannan T	3	3	100.00	Periyar Univ, Tamil Nadu	J Hazard Mater 186 (1): 827-834 Feb 15 2011
Adsorption of Hydrogen Molecules on the Alkali Metal Ion Decorated Boric Acid Clusters: A Density Functional Theory Investigation	Prakash M, Elango M, Subramanian V	3	3	100.00	Csir, Cent Leather Res Inst, Chem Lab, Madras 600020, Tamil Nadu, India.	Int J Hydrogen Energ 36 (6): 3922-3931 Mar 2011
Modeling and Evaluation of Chromium Remediation From Water Using Low Cost Bio-Char, A Green Adsorbent	Mohan D, Rajput S, Singh VK,	3	5	60.00	Jnu, Delhi	J Hazard Mater 188 (1-3): 319-333 Apr 15 2011
Phytosynthesis of Au, Ag and Au-Ag Bimetallic Nanoparticles Using Aqueous Extract and Dried Leaf of Anacardium Occidentale	Sheny DS, Mathew J, Philip D	3	3	100.00	Mar Ivanios Coll, Kerela-2	Spectrochim Acta Pt A-Mol Bio 79 (1): 254-262 Jun 2011
A Printed 2.4 Ghz/5.8 Ghz Dual-Band Monopole Antenna With A Protruding Stub in the Ground Plane For Wlan and Rfid Applications	Panda JR, Kshetrimayum RS	2	2	100.00	Indian Inst Technol, Dept Elect & Elect Engn, Gauhati 781039, India.	

Title	Authors				Affiliation	Source
Influence of Operational Parameters on Photodegradation of Acid Black 1 With Zno	Krishnakumar B, Selvam K, Velmurugan R, Swaminathan M	4	4	100.00	Annamalai Univ, Dept Chem, Annamalainagar 608002, Tamil Nadu, India.	Desalin Water Treat 24 (1-3): 132-139 Dec 2010
Studies on Metal Oxide Nanoparticles Catalyzed Sodium Aluminum Hydride	Pukazhselvan D, Hudson MSL, Sinha ASK, Srivastava ON	4	4	100.00	Banaras Hindu Univ, Dept Phys, Unit Nanosci & Technol, Varanasi 221005, Uttar Pradesh, India.	Energy 35 (12): 5037-5042 Dec 2010
Hydro-Magnetic Combined Convection in A Lid-Driven Cavity With Sinusoidal Boundary Conditions on Both Sidewalls	Malleswaran A	1	4	25.00	Kandaswami Kandars Coll, P Velur 638182, Tamil Nadu, India. Ks Rangasamy Coll Technol, Tiruchengode 637215, Tamil Nadu, India.	Int J Heat Mass Transfer 54 (1-3): 512-525 Jan 15 2011
Biocatalytic Production of Biodiesel From Cottonseed Oil: Standardization of Process Parameters and Comparison of Fuel Characteristics	Chattopadhyay S, Karemore A, Das S, Sen R	4	5	80.00	Indian Inst Technol, Dept Biotechnol, Kharagpur 721302, W Bengal, India.	Appl Energ 88 (4): 1251-1256 Apr 2011
Lpg Sensing Properties of Pd-Sensitized Vertically Aligned Zno Nanorods	Gurav KV, Deshmukh PR, Lokhande CD	3	3	100.00	Shivaji Univ, Dept Phys, Thin Film Phys Lab, Kolhapur 416004, Ms, India.	Sensor Actuator B-Chem 151 (2): 365-369 Jan 28 2011
Collective Hydrodynamics of Swimming Microorganisms: Living Fluids	Subramanian G	1	2	50.00	Incasr, Engn Mech Unit, Bangalore 560064, Karnataka, India.	Annu Rev Fluid Mech 43: 637-659 2011
Synthesis of Surface Imprinted Nanospheres For Selective Removal of Uranium From Simulants of Sambhar Salt Lake and Ground Water	Milja TE, Prathish KP, Rao TP	3	3	100.00	Csir, Cstd, Niist, Trivandrum 695019, Kerala, India.	J Hazard Mater 188 (1-3): 384-390 Apr 15 2011

Title	Authors				Affiliation	Reference
Products of Multiplication Composition and Differentiation Operators on Weighted Bergman Spaces	Sharma AK, Bhat A	2	3	66.67	Shri Mata Vaishno Devi Univ, Sch Math, Katra 182320, J&K, India.	Appl Math Comput 217 (20): 8115-8125 Jun 15 2011
Physicochemical Studies of Molecular Hyperpolarizability of Imidazole Derivatives	Jayabharathi J, Thanikachalam V, Srinivasan N, Perumal MV, Jayamoorthy K	5	5	100.00	Annamalai Univ, Dept Chem, Annamalainagar 608002, Tamil Nadu, India.	Spectrochim Acta Pt A-Mol Bio 79 (1): 137-147 Jun 2011
Performance Analysis of An Organic Rankine Cycle With Superheating Under Different Heat Source Temperature Conditions	Roy JP, Mishra MK, Misra A	3	3	100.00	Ntpc Ltd Kahalgaon, Kahalgaon Stpp, Qtr C-150, Bhagalpur 813214, India. Ntpc Ltd Kahalgaon, Kahalgaon Stpp, Bhagalpur 813214, India.	Appl Energ 88 (9): 2995-3004 Sep 2011
Graphene-Multi Walled Carbon Nanotube Hybrid Electrocatalyst Support Material For Direct Methanol Fuel Cell	Jha N, Jafri RI, Rajalakshmi N, Ramaprabhu S	4	4	100.00	Indian Inst Technol, Dept Phys, Nfmtc, Aenl, Madras 600036, Tamil Nadu, India. Ctr Fuel Cell Technol Arci, Madras 600113, Tamil Nadu, India.	Int J Hydrogen Energ 36 (12): 7284-7290 Jun 2011
An Evaluative Report and Challenges For Fermentative Biohydrogen Production	Sinha P, Pandey A	2	2	100.00	Univ Allahabad, Nanotechnol & Mol Biol Lab, Ctr Biotechnol, Allahabad 211002, Uttar Pradesh, India	Int J Hydrogen Energ 36 (13): 7460-7478 Jul 2011
Artificial Neural Networks in Hardware A Survey of Two Decades of Progress	Misra J,	1	2	50.00	Hts Res, 151-1 Doraisanipalya, Bannerghatta Rd, Bangalore 560076, Karnataka, India	Neurocomputing 74 (1-3): 239-255 Sp. Iss. Si Dec 2010
New Hybrid Multilevel Inverter Fed Induction Motor Drive - A Diagnostic Study	Ramani K, Krishnan A	2	2	100.00	Ks Rangasamy Coll Technol, Dept Elect & Elect Engn, Tiruchengode, Tamilnadu, India	Int Rev Electr Eng-Iree 5 (6): 2562-2569 Part A Nov-Dec 2010

(h) Environment

Title	Indian Authors	No. of Indian Authors	Total Author	Ratio I/F*	Institue Name	Source
Technical Aspects of Biodiesel Production by Transesterification - A Review	Meher LC, Sagar DV, Naik SN	3	3	100.00	Ind Technol Inst, Ctr Rural Dev & Technol, New Delhi 110016, India.	Renew Sustain Energy Rev 10 (3): 248-268 Jun 2006
A Review of Imperative Technologies For Wastewater Treatment I: Oxidation Technologies At Ambient Conditions	Gogate PR, Pandit AB	2	2	100.00	Muict, Chem Engn Sect, Bombay 400019, Maharashtra, India.	Adv Environ Res 8 (3-4): 501-551 Mar 2004
The Causes of Land-Use and Land-Cover Change: Moving Beyond the Myths	George PS, Ramakrishnan PS,	2	21	9.52	Ctr Dev Studies, Trivandrum 695011, Kerala, India., Jawaharlal Nehru Univ, Sch Environm Sci, New Delhi 110067, India.	Global Environ Change 11 (4): 261-269 Dec 2001
Salt Tolerance and Salinity Effects on Plants: A Review	Parida AK, Das AB	2	2	100.00	Natl Inst Plant Biodivers Conservat & Res, Bhubaneswar 751015, Orissa, India. Reg Plant Resource Ctr, Bhubaneswar 751015, Orissa, India.	Ecotoxicol Environ Safety 60 (3): 324-349 Mar 2005
Role of Metal-Reducing Bacteria in Arsenic Release From Bengal Delta Sediments	Chatterjee D	1	7	14.29	Univ Kalyani, Dept Chem, Santini Ketan 731235, W Bengal, India.	Nature 430 (6995): 68-71 Jul 1 2004
Single- and Multi-Component Adsorption of Cadmium and Zinc Using Activated Carbon Derived From Bagasse - An Agricultural Waste	Mohan D, Singh KP	2	2	100.00	Ind Toxicol Res Ctr, Environm Chem Div, Lucknow 226001, Uttar Pradesh, India.	Water Res 36 (9): 2304-2318 May 2002

Application of Low-Cost Adsorbents For Dye Removal - A Review	Gupta VK,	1	2	50.00	Indian Inst Technol, Dept Chem, Roorkee 247667, Uttar Pradesh, India.	J Environ Manage 90 (8): 2313-2342 Jun 2009
A Review of Imperative Technologies For Wastewater Treatment Ii: Hybrid Methods	Gogate PR, Pandit AB	2	2	100.00	Uict, Div Chem Engn, Bombay 400019, Maharashtra, India.	Adv Environ Res 8 (3-4): 553-597 Mar 2004
Chromium Toxicity in Plants	Shanker AK, Avudainayagam S	2	4	50.00	Natl Res Ctr Agroforestry, Jhansi, Uttar Pradesh, India. Tamil Nadu Agr Univ, Dept Crop Physiol, Coimbatore 641003, Tamil Nadu, India. Tamil Nadu Agr Univ, Dept Environm Sci, Coimbatore 641003, Tamil Nadu, India.	Environ Int 31 (5): 739-753 Jul 2005
Prospects of Biodiesel Production From Vegetables Oils in India	Barnwal BK, Sharma MP	2	2	100.00	Ind Technol Inst, Alternate Hydro Energy Ctr, Uttaranchal 247667, India.	Renew Sustain Energy Rev 9 (4): 363-378 Aug 2005
Global Biodiversity: Indicators of Recent Declines	Suhel Quader	1	45	2.22	Natl Ctr Biol Sci, Tata Inst Fundamental Res, Bangalore 560065, Karnataka, India.	Science 328 (5982): 1164-1168 May 28 2010
The Status of the World's Land and Marine Mammals: Diversity, Threat, and Knowledge	Syed Ainul Hussain, Sanjay Molur, Chelmala Srinivasulu, Bibhab Kumar Talukdar	4	132	3.03	Aaranyak, Guwhati 781028, Assam, India. Osmania Univ, Dept Zool, Univ Coll Sci, Hyderabad 500007, Andhra Pradesh, India. Zoo Outreach Org, Coimbatore 641004, Tamil Nadu, India.	Science 322 (5899): 225-230 Oct 10 2008
Enhancement of Photocatalytic Activity by Metal Deposition: Characterisation and Photonic Efficiency of Pt, Au and Pd Deposited on Tio2 Catalyst	Shankar MV, Palanichamy M, Arabindoo B, Bahnemann DW, Urugesan V	4	6	66.67	Anna Univ, Dept Chem, Madras 600025, Tamil Nadu, India.	Water Res 38 (13): 3001-3008 Jul 2004

Title	Authors			%	Affiliation	Source
Multivariate Statistical Techniques For the Evaluation of Spatial and Temporal Variations in Water Quality of Gomti River (India) - A Case Study	Singh KP, Malik A, Mohan D, Sinha S	4	4	100.00	Ind Toxicol Res Ctr, Environm Chem Sect, Lucknow 226001, Uttar Pradesh, India. Natl Bot Res Inst, Lucknow 226001, Uttar Pradesh, India.	Water Res 38 (18): 3980-3992 Nov 2004
Asia-Pacific Mussel Watch: Monitoring Contamination of Persistent Organochlorine Compounds in Coastal Waters of Asian Countries	Subramanian A, Karuppiah S,	2	16	12.50	Annamalai Univ, Ctr Adv Study Marine Biol, Parangipettai 608502, Tamil Nadu, India.	Mar Pollut Bull 46 (3): 281-300 Mar 2003
Removal of Cadmium and Nickel From Wastewater Using Bagasse Fly Ash - A Sugar Industry Waste	Gupta VK, Jain CK, Ali I, Sharma M, Saini VK	5	5	100.00	Indian Inst Technol, Dept Chem, Roorkee 247667, Uttar Pradesh, India.Natl Inst Hydrol, Roorkee 247667, Uttar Pradesh, India.	Water Res 37 (16): 4038-4044 Sep 2003
Removal of Congo Red From Aqueous Solution by Bagasse Fly Ash and Activated Carbon: Kinetic Study and Equilibrium Isotherm Analyses	Mall ID, Srivastava VC, Agarwal NK, Mishra IM	4	4	100.00	Indian Inst Technol, Dept Chem Engn, Roorkee 247667, Uttar Pradesh, India.	Chemosphere 61 (4): 492-501 Oct 2005
Biodiesel Production From Jatropha (Jatropha Curcas) With High Free Fatty Acids: An Optimized Process	Tiwari AK, Kumar A, Raheman H	3	3	100.00	Indian Inst Technol, Agr & Food Engn Dept, Kharagpur 721302, W Bengal, India.	Biomass Bioenerg 31 (8): 569-575 Aug 2007
Jatropha Bio-Diesel Production and Use	VP Singh	1	7	14.29	World Agroforestry Ctr Icraf Reg Off S Asia, Natl Agr Sci Ctr, New Delhi 110012, India.	Biomass Bioenerg 32 (12): 1063-1084 Dec 2008
Prospects and Potential of Fatty Acid Methyl Esters of Some Non-Traditional Seed Oils For Use As Biodiesel in India	Azam MM, Waris A, Nahar NM	3	3	100.00	Cent Arid Zone Res Inst, Jodhpur 342003, Rajasthan, India.	Biomass Bioenerg 29 (4): 293-302 2005

Title	Authors				Affiliation	Journal
Review on Thermal Energy Storage With Phase Change Materials and Applications	Tyagi VV, Buddhi D	2	4	50.00	Devi Ahilya Univ, Sch Energy & Environm Studies, Thermal Energy Storage Lab, Indore 452017, Madhya Pradesh, India.	Renew Sustain Energy Rev 13 (2): 318-345 Feb 2009
The Asian Tsunami: A Protective Role For Coastal Vegetation	Vaithilingam Selvam, Vagarappa M Karunagaran	2	12	16.67	Ms Swaminathan Res Fdn, Madras 600113, Tamil Nadu, India.	Science 310 (5748): 643-643 Oct 28 2005
Bacterial Decolorization and Degradation of Azo Dyes	Pandey A, Singh P, Iyengar L	3	3	100.00	Indian Inst Technol, Dept Chem, Biotechnol Lab, Kanpur 208016, Uttar Pradesh, India	Int Biodeterior Biodegrad 59 (2): 73-84 Mar 2007
Biodiesel Production From Mahua (Madhuca Indica) Oil Having High Free Fatty Acids	Ghadge SV, Raheman H	2	2	100.00	Indian Inst Technol, Agr & Food Engn Dept, Kharagpur 721302, W Bengal, India	Biomass Bioenerg 28 (6): 601-605 2005
Stable Colloidal Dispersions of C-60 Fullerenes in Water: Evidence For Genotoxicity			7	0.00	Ind Toxicol Res Ctr, Div Dev Toxicol, Lucknow 226001, Uttar Pradesh, India.	Environ Sci Technol 40 (23): 7394-7401 Dec 1 2006
Fluoride in Drinking Water: A Review on the Status and Stress Effects	Ayoob S, Gupta AK	2	2	100.00	Indian Inst Technol, Dept Civil Engn, Environm Engn Div, Kharagpur 721302, W Bengal, India.	Crit Rev Environ Sci Technol 36 (6): 433-487 Nov-Dec 2006
Low-Cost Adsorbents: Growing Approach to Wastewater Treatmenta Review	Gupta VK,	1	4	25.00	Indian Inst Technol, Dept Chem, Roorkee 247667, Uttar Pradesh, India.	Crit Rev Environ Sci Technol 39 (10): 783-842 2009
Pcm Thermal Storage in Buildings: A State of Art	Tyagi VV, Buddhi D	2	2	100.00	Devi Ahilya Univ, Thermal Energy Storage Lab, Sch Energy & Environm Studies, Fac Engn Sci, Indore 452017	Renew Sustain Energy Rev 11 (6): 1146-1166 Aug 2007

Title	Authors				Affiliation	Journal
Assessing Dangerous Climate Change Through An Update of the Intergovernmental Panel on Climate Change (Ipcc) "Reasons For Concern"	Patwardhan A	1	15	6.67	Indian Inst Technol Bombay Powai, Shailesh J Mehta Sch Management, Mumbai 400076, Maharashtra, India.	Proc Nat Acad Sci USA 106 (11): 4133-4137 Mar 17 2009
Removal of Endosulfan and Methoxychlor From Water on Carbon Slurry	Gupta VK, Ali I	2	2	100.00	Indian Inst Technol, Dept Chem, Roorkee 247667, Uttar Pradesh, India., Jamia Millia Islamia, Cent Univ, Dept Chem, New Delhi 110025, India.	Environ Sci Technol 42 (3): 766-770 Feb 1 2008
Bio-Fuels From Thermochemical Conversion of Renewable Resources: A Review	Goyal HB, Seal D, Saxena RC	3	3	100.00	Indian Inst Petr, Dehra Dun 248005, Uttar Pradesh, India.	Renew Sustain Energy Rev 12 (2): 504-517 Feb 2008
Fungal Dye Decolourization: Recent Advances and Future Potential	Kaushik P, Malik A	2	2	100.00	Indian Inst Technol Delhi, Ctr Rural Dev & Technol, New Delhi 110016, India.	Environ Int 35 (1): 127-141 Jan 2009
Production of First and Second Generation Biofuels: A Comprehensive Review	Naik SN, Goud VV	2	3	66.67	Indian Inst Technol Delhi, Ctr Rural Dev & Technol, Hauz Khas, New Delhi 110016, India.; Indian Inst Technol Guwahati, Dept Chem Engn, N Guwahati, Assam, India.	Renew Sustain Energy Rev 14 (2): 578-597 Feb 2010
A Review on Biodiesel Production, Combustion, Emissions and Performance	Basha SA, Gopal KR, Jebaraj S	3	3	100.00	Gurrala Chavidy, Chilakaluripet 522616, Andhra Pradesh, India.	Renew Sustain Energy Rev 13 (6-7): 1628-1634 Aug-Sep 2009

Title	Authors				Affiliation	Source
The Impact of Conservation on the Status of the World's Vertebrates	Suhel Quader, SD Biju, SK Dutta, Syed Ainul Hussain, Sanjay Molur, Bibhab K Talukdar	6	173	3.47	Tata Inst Fundamental Res, Natl Ctr Biol Sci, Bangalore 560065, Karnataka, India.; Univ Delhi, Sch Environm Studies, Systemat Lab, Delhi 110007, India.; N Orissa Univ, Mayurbhanj 757003, Orissa, India.; Wildlife Inst India, Dehra Dun 248001, Uttar Pradesh, India.; Zoo Outreach Org, Coimbatore 641004, Tamil Nadu, India.; Aaranyak & Int Rhino Fdn, Gauhati 781028, Assam, India.	Science 330 (6010): 1503-1509 Dec 10 2010
Denitrification As the Dominant Nitrogen Loss Process in the Arabian Sea	Naik H, Pratihary A,	2	8	25.00	Natl Inst Oceanog, Panaji 403004, Goa, India.	Nature 461 (7260): 78-U77 Sep 3 2009
Bio-Diesel As An Alternative Fuel For Diesel Engines-A Review	Murugesan A, Umarani C, Subramanian R, Nedunchezhian N	4	4	100.00	Ks Rangasamy Coll Technol, Tiruchengode, Tamil Nadu, India., Inst Rd & Transport Technol, Erode, Tamil Nadu, India.	Renew Sustain Energy Rev 13 (3): 653-662 Apr 2009
Biodiesel Production Through the Use of Different Sources and Characterization of Oils and Their Esters As the Substitute of Diesel: A Review	Singh SP, Singh D	2	2	100.00	Devi Ahilya Univ, Sch Energy & Environm Studies, Khandwa Rd, Takshila Campus, Indore 452011, Madhya Pradesh, India.	Renew Sustain Energy Rev 14 (1): 200-216 Jan 2010
Development of Biodiesel: Current Scenario	Sharma YC, Singh B	2	2	100.00	Banaras Hindu Univ, Inst Technol, Dept Appl Chem, Varanasi 221005, Uttar Pradesh, India.	Renew Sustain Energy Rev 13 (6-7): 1646-1651 Aug-Sep 2009

Title	Authors				Affiliation	Reference
Enhancement of Heat Transfer Using Nanofluids-An Overview	Lazarus Godsona, B Rajab, D Mohan Lala	3	4	75.00	Anna Univ, Coll Engn, Refrigerat & Air Conditioning Div, Dept Mech Engn, Madras 600025, Tamil Nadu, India., Indian Inst Technol, Indian Inst Informat Technol Design & Mfg Kanchee, Madras 600036, Tamil Nadu, India.	Renew Sustain Energy Rev 14 (2): 629-641 Feb 2010
Stability of Biodiesel and Its Blends: A Review	Jain S, Sharma MP	2	2	100.00	Indian Inst Technol, Alternate Hydro Energy Ctr, Roorkee 247667, Uttarakhand, India	Renew Sustain Energy Rev 14 (2): 667-678 Feb 2010
Prospects of Biodiesel From Jatropha in India: A Review	Jain S, Sharma MP	2	2	100.00	Indian Inst Technol, Alternate Hydro Energy Ctr, Roorkee 247667, Uttarakhand, India.	Renew Sustain Energy Rev 14 (2): 763-771 Feb 2010
Interactions Between Protected Areas and Their Surroundings in Human-Dominated Tropical Landscapes	Karanth KK, Pareeth S	2	3	66.67	Ctr Wildlife Studies, Bengaluru 560070, India. Equator Gis, Palarivattom 682025, Kochi, India.	Biol Conserv 143 (12): 2870-2880 Sp. Iss. Si Dec 2010
Role of Renewable Energy Sources in Environmental Protection: A Review	Panwar NL, Kaushik SC, Kothari S	3	3	100.00	Maharana Pratap Univ Agr & Technol, Coll Technol & Engn, Dept Renewable Energy Sources, Udaipur 313001, India. Indian Inst Technol Delhi, Ctr Energy Studies, New Delhi 110016, India.	Renew Sustain Energy Rev 15 (3): 1513-1524 Apr 2011
Dna Barcoding Indian Marine Fishes	Lakra WS, Verma MS, Goswami M, Lal KK, Mohindra V, Punia P, Gopalakrishnan A, Singh KV	8	10	80.00	Natl Bur Fish Genet Resources, Lucknow 226002, Uttar Pradesh, India.	Mol Ecol Resour 11 (1): 60-71 Jan 2011

Title	Indian Authors	No. of Indian Authors	Total Author	Ratio I/F*	Institue Name	Source
A Review of Solar Photovoltaic Technologies	Parida B, Iniyan S,	2	3	66.67	Anna Univ, Coll Engn, Madras 600025, Tamil Nadu, India.	Renew Sustain Energy Rev 15 (3): 1625-1636 Apr 2011
Chromium Removal by Combining the Magnetic Properties of Iron Oxide With Adsorption Properties of Carbon Nanotubes	Gupta VK, Agarwal S	2	3	66.67	Chromium Removal by Combining the Magnetic Properties of Iron Oxide With Adsorption Properties of Carbon Nanotubes	Water Res 45 (6): 2207-2212 Mar 2011
Trends in Reference Evapotranspiration in the Humid Region of Northeast India	Singh VP	1	5	20.00	N Eastern Reg Inst Sci & Technol, Dept Agr Engn, Itanagar 791109, Arunachal Prade	Hydrol Process 26 (3): 421-435 Jan 30 2012
High-Efficiency Dye-Sensitized Solar Cell With A Novel Co-Adsorbent	Chiranjeevi B, Barreddi Chiranjeevi	2	8	25.00	Indian Inst Chem Technol, Inorgan & Phys Chem Div, Hyderabad 500607, Andhra Pradesh	Energy Environ Sci 5 (3): 6057-6060 Mar 2012

(i) Geosciences

Title	Indian Authors	No. of Indian Authors	Total Author	Ratio I/F*	Institue Name	Source
Global Observed Changes in Daily Climate Extremes of Temperature and Precipitation	Kumar KR, Revadekar J	2	24	8.33	Indian Inst Trop Meteorol, Pune 411008, Maharashtra, India.	J Geophys Res-Atmos 111 (D5): Art. No.-D05109 Mar 15 2006
Abrupt Changes in the Asian Southwest Monsoon During the Holocene and Their Links to the North Atlantic Ocean	Gupta AK	1	3	33.33	Indian Inst Technol, Dept Geol & Geophys, Kharagpur 721302, W Bengal, India.	Nature 421 (6921): 354-357 Jan 23 2003

Title	Authors				Affiliation	Reference
Natural Organic Matter in Sedimentary Basins and Its Relation to Arsenic in Anoxic Ground Water: The Example of West Bengal and Its Worldwide Implications	Banerjee DM, Mishra R, Purohit R, Chatterjee A, Talukder T, Chadha DK	6	13	46.15	Cent Ground Water Author, Calcutta 700091, W Bengal, India. Cent Ground Water Author, Delhi 110011, India. Univ Delhi, Dept Geol, Delhi 110007, India.	Appl Geochem 19 (8): 1255-1293 Aug 2004
Adsorption of Methylene Blue on Kaolinite	Ghosh D, Bhattacharyya KG	2	2	100.00	Gauhati Univ, Dept Chem, Gauhati 781014, Assam, India.	Appl Clay Sci 20 (6): 295-300 Feb 2002
Increasing Trend of Extreme Rain Events Over India in A Warming Environment	Goswami BN, Venugopal V, Sengupta D, Madhusoodanan MS, Xavier PK	5	5	100.00	Indian Inst Trop Meteorol, Doctor Homi Bhabha Rd, Pune 411008, Maharashtra, India.	Science 314 (5804): 1442-1445 Dec 1 2006
Intercomparison of the Climatological Variations of Asian Summer Monsoon Precipitation Simulated by 10 Gcms	Satyan V	1	16	6.25	Indian Inst Trop Meteorol, Pune, Maharashtra, India.	Clim Dynam 19 (5-6): 383-395 Aug 2002
The Onset of India-Asia Continental Collision: Early, Steep Subduction Required by the Timing of Uhp Metamorphism in the Western Himalaya	Singh S, Jain AK, Manickavasagam RM	4	5	80.00	Indian Inst Technol, Inst Instrumentat Ctr, Roorkee 247667, Uttar Pradesh, India.	Earth Planet Sci Lett 234 (1-2): 83-97 May 30 2005
Residential Biofuels in South Asia: Carbonaceous Aerosol Emissions and Climate Impacts	Venkataraman C, Habib G	2	5	40.00	Indian Inst Technol, Dept Chem Engn, Bombay 400076, Maharashtra, India.	Science 307 (5714): 1454-1456 Mar 4 2005
Predecessors of the Giant 1960 Chile Earthquake	Malik JK,	1	15	6.67	Ctr Earth Sci Studies, Thiruvananthapuram 69603, Kerala, India. Indian Inst Technol, Kanpur 208016, Uttar Pradesh, India	

Title	Authors			%	Affiliation	Citation
High Resolution Daily Gridded Rainfall Data For the Indian Region: Analysis of Break and Active Monsoon Spells	Rajeevan M, Bhate J, Kale JA, Lal B	4	4	100.00	Indian Meteorol Dept, New Delhi 110003, India.	Curr Sci 91 (3): 296-306 Aug 10 2006
Radiative Effects of Natural Aerosols: A Review	Satheesh SK, Moorthy KK	2	2	100.00	Indian Inst Sci, Ctr Atmospher & Ocean Sci, Bangalore 560012, Karnataka, India. Vikram Sarabhai Space Ctr, Indian Space Res Org, Space Phys Lab, Trivandrum 695022, Kerala	Atmos Environ 39 (11): 2089-2110 Apr 2005
Tropical Cyclones and Climate Change	Srivastava AK	1	10	10.00	Indian Meteorol Dept, Pune 411005, Maharashtra, India.	Nat Geosci 3 (3): 157-163 Mar 2010
Nitrogen and Sulfur Deposition on Regional and Global Scales: A Multimodel Evaluation	U C Kulshrestha	1	37	2.70	Ind Technol Inst, Analyt & Environm Chem Div, Hyderabad, Andhra Pradesh, India.	Global Biogeochem Cycle 20 (4): Art. No.-Gb4003 Oct 28 2006
Variability of Aerosol Parameters Over Kanpur, Northern India	Singh RP, Tripathi SN	2	5	40.00	Indian Inst Technol, Dept Civil Engn, Kanpur 208016, Uttar Pradesh, India.	J Geophys Res-Atmos 109 (D23): Art. No.-D23206 Dec 11 2004
Rise of the Andes	Ghosh P	1	8	12.50	Indian Inst Sci, Ctr Atmospher & Ocean Sci, Bangalore 560012, Karnataka, India.	Science 320 (5881): 1304-1307 Jun 6 2008
Brown Clouds Over South Asia: Biomass or Fossil Fuel Combustion?	Rao PSP	1	10	10.00	Indian Inst Trop Meteorol, Pune 411008, Maharashtra, India.	Science 323 (5913): 495-498 Jan 23 2009
The Impact of Global Warming on the Tropical Pacific Ocean and El Nino	Lengaigne M	1	12	8.33	Natl Inst Oceanog, Phys Oceanog Div, Panaji 403004, Goa, India.	Nat Geosci 3 (6): 391-397 Jun 2010

Title	Authors			%	Affiliation	Source
Rama the Research Moored Array For African-Asian-Australian Monsoon Analysis and Prediction	Murty VSN, Ravichandran M	2	10	20.00	Natl Inst Oceanog, Reg Ctr, Visakhapatnam, Andhra Pradesh, India.; Indian Natl Ctr Ocean Informat Serv, Hyderabad, Andhra Pradesh, India.	Bull Amer Meteorol Soc 90 (4): 459-+ Apr 2009
Dwindling Groundwater Resources in Northern India, From Satellite Gravity Observations	Tiwari VM	1	3	33.33	Csir, Natl Geophys Res Inst, Uppal Rd, Hyderabad 500007, Andhra Pradesh, India.	Geophys Res Lett 36: Art. No.-L18401 Sep 17 2009
Environmental Impact of the 73 Ka Toba Super-Eruption in South Asia	Chattopadhyaya U, Pal J	2	7	28.57	Univ Allahabad, Dept Ancient Hist Culture & Archaeol, Allahabad 211002, Uttar Pradesh, India.	Palaeogeogr Palaeoclimatol 284 (3-4): 295-314 Dec 30 2009
Spatial and Vertical Heterogeneities in Aerosol Properties Over Oceanic Regions Around India: Implications For Radiative Forcing	Moorthy KK, Nair VS, Babu SS, Satheesh SK	4	4	100.00	Vikram Sarabhai Space Ctr, Space Phys Lab, Trivandrum 695022, Kerala, India. Indian Inst Sci, Divecha Ctr Climate Change, Bangalore 560012, Karnataka, India	Quart J Roy Meteorol Soc 135 (645): 2131-2145 Part B Sp. Iss. Si Oct 2009
Mesoproterozoic Ophiolitic Melange From the Se Periphery of the Indian Plate: U-Pb Zircon Ages and Tectonic Implications	Rao CVD	1	3	33.33	Natl Disaster Management Author, New Delhi, India.	Gondwana Res 19 (2): 384-401 Mar 2011
A 1 Year Record of Carbonaceous Aerosols From An Urban Site in the Indo-Gangetic Plain: Characterization, Sources, and Temporal Variability	Ram K, Sarin MM, Tripathi SN	3	3	100.00	Phys Res Lab, Geosci Div, Ahmadabad 380009, Gujarat, India. Indian Inst Technol, Dept Civil Engn, Kanpur 208016, Uttar Pradesh, India.	J Geophys Res-Atmos 115: Art. No.-D24313 Dec 30 2010

Title	Authors				Address	Citation
The Central India Tectonic Zone: A Geophysical Perspective on Continental Amalgamation Along A Mesoproterozoic Suture	Naganjaneyulu K	1	2	50.00	Natl Geophys Res Inst, Council Sci & Ind Res, Hyderabad 500007, Andhra Pradesh, India	Gondwana Res 18 (4): 547-564 Nov 2010
Satellites Measure Recent Rates of Groundwater Depletion in California's Central Valley	Syed TH	1	9	11.11	Indian Sch Mines, Dept Appl Geol, Dhanbad 826004, Bihar, India.	Geophys Res Lett 38: Art. No.-L03403 Feb 5 2011
Seasonal Variation of the Transport of Black Carbon Aerosol From the Asian Continent to the Arctic During the Arctas Aircraft Campaign	Sahu LK	1	12	8.33	Phys Res Lab, Ahmadabad 380009, Gujarat, India.	J Geophys Res-Atmos 116: Art. No.-D05202 Mar 5 2011
Optical and Physical Characteristics of Bay of Bengal Aerosols During W-Icarb: Spatial and Vertical Heterogeneities in the Marine Atmospheric Boundary Layer and in the Vertical Column	Moorthy KK, Beegum SN, Babu SS, Smirnov A, Kumar KR, Narasimhulu K, Dutt CBS, Nair VS	8	9	88.89	Vikram Sarabhai Space Ctr, Space Phys Lab, Trivandrum 695022, Kerala, India. Isro Head Quarters, Bangalore 560094, Karnataka, India. Sri Krishnadevaraya Univ, Dept Phys, Anantapur 515005, Andhra Pradesh, India.	J Geophys Res-Atmos 115: Art. No.-D24213 Dec 22 2010
Geochemical Fingerprinting of the Widespread Toba Tephra Using Biotite Compositions	Korisettar R, Pal JN	2	9	22.22	Karnatak Univ, Dept Hist & Archaeol, Dharwad 580003, Karnataka, India.	Quatern Int 246: 97-104 Dec 20 2011
A Neoarchean Dismembered Ophiolite Complex From Southern India: Geochemical and Geochronological Constraints on Its Suprasubduction Origin	Yellappa T, Chetty TRK, Nagesh P, Mohanty DP, Venkatasivappa V	5	8	62.50	Natl Geophys Res Inst, Csir, Uppal Rd, Hyderabad 500007, Andhra Pradesh, India.	Gondwana Res 21 (1): 246-265 Jan 2012

Title	Authors				Affiliation	Journal
Dhaba: An Initial Report on An Acheulean, Middle Palaeolithic and Microlithic Locality in the Middle Son Valley, North-Central India	Pal JN, Koshy J, Bora J, DI, Ram HP, Dubey A	6	13	46.15	Univ Allahabad, Dept Ancient Hist Culture & Archaeol, Allahabad 211002, Uttar Pradesh, India.	Quatern Int 258: 191-199 May 1 2012
The Influence of A South Asian Dust Storm on Aerosol Radiative Forcing At A High-Altitude Station in Central Himalayas	Srivastava AK, Pant P, Hegde P, Singh S, Dumka UC, Naja M, Singh N, Bhavanikumar Y	8	8	100.00	Natl Phys Lab, Radio & Atmospher Sci Div, Ks Krishnan Rd, New Delhi 110012, India. Aryabhatta Res Inst Observat Sci Aries, Naini Tal, India. Indian Inst Trop Meteorol Branch Iitm, New Delhi, India. Vikram Sarabhai Space Ctr Vssc, Space Phys Lab, Trivandrum, Kerala, India. Natl Atmospher Res Lab, Gadanki, India.	Int J Remote Sens 32 (22): 7827-7845 2011
A Possible Link Between Radon Anomaly and Earthquake	Laskar I, Phukon P, Goswami AK, Chetry G, Roy UC	5	5	100.00	Cotton Coll, Dept Phys, High Energy Phys Lab, Gauhati 781001, Assam, India. Gauhati Univ, Dept Geol Sci, Gauhati 781014, Assam, India.	Geochem J 45 (6): 439-446 Sp. Iss. Si 2011
Accumulation of Aerosols Over the Indo-Gangetic Plains and Southern Slopes of the Himalayas: Distribution, Properties and Radiative Effects During the 2009 Pre-Monsoon Season	Payra S	1	14	7.14	Extens Ctr Jaipur, Birla Inst Technol Mesra, Jaipur, Rajasthan, India.	Atmos Chem Phys 11 (24): 12841-12863 2011

(j) Immunology

Title	Indian Authors	No. of Indian Authors	Total Author	Ratio I/F*	Institue Name	Source
Pathogen Recognition by the Innate Immune System	Kumar H	1	3	33.33	Indian Inst Sci Educ & Res Iiser Bhopal, Dept Biol Sci, Immunol Lab, Bhopal 460023, India.	Int Rev Immunol 30 (1): 16-34 Feb 2011
Molecular Characterization of Nucleotide Binding and Oligomerization Domain (Nod)-2, Analysis of Its Inductive Expression and Down-Stream Signaling Following Ligands Exposure and Bacterial Infection in Rohu (Labeo Rohita)	Swain B, Basu M, Sahoo BR, Maiti NK, Routray P, Eknath AE, Samanta M	6	6	100.00	Cifa, Fish Hlth Management Div, Bhubaneswar 751002, Orissa, India	Develop Comp Immunol 36 (1): 93-103 Jan 2012

(k) Material Sciences

Title	Indian Authors	No. of Indian Authors	Total Author	Ratio I/F*	Institue Name	Source
Controlling the Aspect Ratio of Inorganic Nanorods and Nanowires	Jana NR	1	2	50.00	Rrr Mahavidyalaya, Hooghly 712406, Wb, India.	Advan Mater 14 (1): 80-82 Jan 4 2002
Overview No.144 - Mechanical Behavior of Amorphous Alloys	Ramamurty U	1	3	33.33	Indian Inst Sci, Dept Mat Engn, Bangalore 560012, Karnataka, India.	Acta Mater 55 (12): 4067-4109 Jul 2007

Title	Authors				Affiliation	Reference
Biological Synthesis of Triangular Gold Nanoprisms	Shankar SS, Rai A, Ankamwar B, Singh A, Ahmad A, Sastry M	6	6	100.00	Natl Chem Lab, Div Biochem Sci, Pune 411008, Maharashtra, India.; Abasaheb Garware Coll, Dept Chem, Pune 411004, Maharashtra, India.	Nat Mater 3 (7): 482-488 Jul 2004
Solar Photocatalytic Degradation of Azo Dye: Comparison of Photocatalytic Efficiency of Zno and Tio2	Neppolian B, Shankar MV, Arabindoo B, Palanichamy M, Murugesan V	5	6	83.33	Anna Univ, Dept Chem, Madras 600025, Tamil Nadu, India.	Solar Energ Mater Solar Cells 77 (1): 65-82 Apr 30 2003
Carbon Nanotube Flow Sensors	Ghosh S, Sood AK, Kumar N	3	3	100.00	Indian Inst Sci, Dept Phys, Bangalore 560012, Karnataka, India.Raman Res Inst, Bangalore 560080, Karnataka, India.	Science 299 (5609): 1042-1044 Feb 14 2003
Synthesis, Characterization, Electronic Structure, and Photocatalytic Activity of Nitrogen-Doped Tio2 Nanocatalyst	Sathish M, Viswanathan B, Viswanath RP, Gopinath CS	3	3	100.00	Indian Inst Technol, Dept Chem, Madras 600036, Tamil Nadu, India. Natl Chem Lab, Catalysis Div, Pune 411008, Maharashtra, India.	Chem Mater 17 (25): 6349-6353 Dec 13 2005
Combustion Synthesis: An Update	Patil KC, Aruna ST, Mimani T	3	3	100.00	Indian Inst Technol, Dept Chem, Madras 600036, Tamil Nadu, India. Natl Chem Lab, Catalysis Div, Pune 411008, Maharashtra, India.	Curr Opin Solid State Mat Sci 6 (6): 507-512 Dec 2002
A Review on Fundamentals and Applications of Electrophoretic Deposition (Epd)	Besra L	1	2	50.00	Csir, Colloids & Mat Chem Grp, Reg Res Lab, Bhubaneswar 751013, Orissa, India.	Prog Mater Sci 52 (1): 1-61 Jan 2007
Carrier-Controlled Ferromagnetismin Transparent Oxide Semiconductors	Satpati B	1	10	10.00	Inst Phys, Bhubaneswar 751005, Orissa, India.	Nat Mater 5 (4): 298-304 Apr 2006

Title	Authors				Affiliation	Citation
Carbon Nanotube Filters	Srivastava A, Srivastava ON	2	5	40.00	Banaras Hindu Univ, Dept Phys, Varanasi 221005, Uttar Pradesh, India.	Nat Mater 3 (9): 610-614 Sep 2004
Liquid-Liquid Phase Transition in Supercooled Silicon	Sastry S	1	2	50.00	Jawaharlal Nehru Ctr Adv Sci Res, Jakkur Campus, Bangalore 560064, Karnataka, India.	Nat Mater 2 (11): 739-743 Nov 2003
Photoluminescence and Electron Paramagnetic Resonance of Zno Tetrapod Structure	Rao TKG	1	10	10.00	Indian Inst Technol, Rsic, Bombay 400076, Maharashtra, India.	Adv Funct Mater 14 (9): 856-864 Sep 2004
Ti Based Biomaterials, the Ultimate Choice For Orthopaedic Implants - A Review	Geetha M, Singh AK, Asokamani R, Gogia AK	4	4	100.00	Vellore Inst Technol, Vellore 632014, Tamil Nadu, India. Def Met Res Lab, Hyderabad 500058, Andhra Pradesh, India. Kaveri Engine Program, Project Off Mat, Hyderabad 500058, Andhra Pradesh, India.	Prog Mater Sci 54 (3): 397-425 May 2009
Controlling the Optical Properties of Lemongrass Extract Synthesized Gold Nanotriangles and Potential Application in Infrared-Absorbing Optical Coatings	Shankar SS, Rai A, Ahmad A, Sastry M	4	4	100.00	Nanosci Grp, Mat Chem Div, Pune 411008, Maharashtra, India.Natl Chem Lab, Div Biochem Sci, Pune 411008, Maharashtra, India.	Chem Mater 17 (3): 566-572 Feb 8 2005
Adsorption of Cytochrome C on Mesoporous Molecular Sieves: Influence of Ph, Pore Diameter, and Aluminum Incorporation	Vinu A, Murugesan V,	2	4	50.00	Anna Univ, Dept Chem, Madras 600025, Tamil Nadu, India.	Chem Mater 16 (16): 3056-3065 Aug 10 2004
Absence of Ferromagnetism in Mn- and Co-Doped Zno	Rao CNR, Deepak FL	2	2	100.00	Jawaharlal Nehru Ctr Adv Sci Res, Csir, Bangalore 560064, Karnataka, India.	J Mater Chem 15 (5): 573-578 2005

Title	Authors				Address	Reference
Ultra-Stable Nanoparticles of Cdse Revealed From Mass Spectrometry	Kumar V, Sundararajan V	2	17	11.76	Dr Vijay Kumar Fdn, Madras 600078, Tamil Nadu, India. Ctr Dev Adv Comp, Pune 411007, Maharashtra, India.	
Studies on Mechanical Performance of Biofibre/Glass Reinforced Polyester Hybrid Composites	Mishra S	1	7	14.29	Ravenshaw Coll, Dept Chem, Polymer Composites Lab, Post Grad Dept Chem, Cuttack 753003, Orissa, India. Cent Inst Plast Engn & Technol, Bhubaneswar, Orissa, India.	Composites Sci Technol 63 (10): 1377-1385 Aug 2003
Mechanisms Underlying the Formation of Thick Alumina Coatings Through the Mao Coating Technology	Sundararajan G, Krishna LR	2	2	100.00	Int Adv Res Ctr Powder Met & New Mat, Balapur Po, Hyderabad 500005, Andhra Pradesh, India.	Surf Coat Tech 167 (2-3): 269-277 Apr 22 2003
Recent Advances in Shape Memory Polymers and Composites: A Review	Ratna D	1	2	50.00	Naval Mat Res Lab, Ambernath 421506, Thane, India	J Mater Sci 43 (1): 254-269 Jan 2008
Self-Cross-Linking Biopolymers As Injectable in Situ Forming Biodegradable Scaffolds	Balakrishnan B, Jayakrishnan A	2	2	100.00	Sree Chitra Tirunal Inst Med Sci & Technol, Biomed Technol Wing, Div Polymer Chem, Satelmond Palace Campus, Trivandrum 695012, Kerala, India.	Biomaterials 26 (18): 3941-3951 Jun 2005
Structural and Electrical Studies on Highly Conducting Spray Deposited Fluorine and Antimony Doped Sno2 Thin Films From Sncl2 Precursor	Thangaraju B	1	1	100.00	Indian Inst Sci, Dept Phys, Bangalore 560012, Karnataka, India. Bharathidasan Univ, Dept Phys, Tiruchchirappalli 620024, Tamil Nadu, India	Thin Solid Films 402 (1-2): 71-78 Jan 1 2002

Title	Authors			%	Affiliation	Citation
Understanding the Quantum Size Effects in Zno Nanocrystals	Viswanatha R, Sapra S, Satpati B, Satyam PV, Dev BN, Sarma DD	6	6	100.00	Indian Inst Sci, Solid State & Struct Chem Unit, Bangalore 560012, Karnataka, India. Inst Phys, Bhubaneswar 751005, Orissa, India.	J Mater Chem 14 (4): 661-668 Feb 21 2004
Microstructure and Properties of Ti3alc2 Prepared by the Solid-Liquid Reaction Synthesis and Simultaneous In-Situ Hot Pressing Process	Wang XH, Zhou YC	2	2	100.00	Chinese Acad Sci, Inst Met Res, Shenyang Natl Lab Mat Sci, 72 Wenhua Rd, New Delhi 110016, India	Acta Mater 50 (12): 3141-3149 Jul 17 2002
Synthesis, Structure, and Properties of Boron- and Nitrogen-Doped Graphene	Panchokarla LS, Subrahmanyam KS, Saha SK, Govindaraj A, Krishnamurthy HR, Waghmare UV, Rao CNR	7	7	100.00	Jawaharlal Nehru Ctr Adv Sci Res, Bangalore 560064, Karnataka, India. Indian Inst Sci, Dept Phys, Bangalore 560012, Karnataka, India.	Advan Mater 21 (46): 4726-+ Dec 11 2009
Defect-Induced Ferromagnetism in Co-Doped Zno	Khare N	1	5	20.00	Iit Delhi, Phys Dept, New Delhi 110016, India.	Advan Mater 18 (11): 1449-+ Jun 6 2006
Nanoparticle-Embedded Polymer: in Situ Synthesis, Free-Standing Films With Highly Monodisperse Silver Nanoparticles and Optical Limiting	Porel S, Singh S, Harsha SS, Rao DN, Radhakrishnan TP	5	5	100.00	Univ Hyderabad, Hyderabad 50046, Andhra Pradesh, India. Ctr Cellular & Mol Biol, Hyderabad 500007, Andhra Pradesh, India	Chem Mater 17 (1): 9-12 Jan 11 2005
Pt and Pd Nanoparticles Immobilized on Amine-Functionalized Zeolite: Excellent Catalysts For Hydrogenation and Heck Reactions	Mandal S, Roy D, Chaudhari RV, Sastry M	4	4	100.00	Natl Chem Lab, Pune 411008, Maharashtra, India.	Chem Mater 16 (19): 3714-3724 Sep 21 2004
Grain Refinement of Aluminium and Its Alloys by Heterogeneous Nucleation and Alloying	Murty BS, Kori SA, Chakraborty M	3	3	100.00	Indian Inst Technol, Dept Met & Mat Engn, Kharagpur 721302, W Bengal, India	Int Mater Rev 47 (1): 3-29 2002

Title	Authors				Address	Reference
Novel One-Step Synthesis of Amine-Stabilized Aqueous Colloidal Gold Nanoparticles	Vijayamohanan K	1	5	20.00	Natl Chem Lab, Phys & Mat Chem Div, Pune 411008, Maharashtra, India.	J Mater Chem 14 (12): 1795-1797 2004
Graphene, the New Nanocarbon	Rao CNR, Biswas K, Subrahmanyam KS, Govindaraj A	4	4	100.00	Jawaharlal Nehru Ctr Adv Sci Res, Jakkur Po, Bangalore 560064, Karnataka, India. Indian Inst Sci, Solid State & Struct Chem Unit, Bangalore 560012, Karnataka, India.	Lancet 376 (9735): 112-123 Jul 10 2010
Embrittlement of A Bulk Metallic Glass Due to Sub-T-G Annealing	Murah P, Ramamurty U	2	2	100.00`	Indian Inst Sci, Dept Met, Cv Raman Ave, Bangalore 560012, Karnataka, India	Acta Mater 53 (5): 1467-1478 Mar 2005
A Study of Graphenes Prepared by Different Methods: Characterization, Properties and Solubilization	Subrahmanyam KS, Vivekchand SRC, Govindaraj A, Rao CNR	4	4	100.00	Jawaharlal Nehru Ctr Adv Sci Res, Jakkur Po, Bangalore 560064, Karnataka, India.	J Mater Chem 18 (13): 1517-1523 2008
Hardness and Plastic Deformation in A Bulk Metallic Glass	Ramamurty U, Jana S, Chattopadhyay K	3	4	75.00	Indian Inst Sci, Dept Met, Bangalore 560012, Karnataka, India	Acta Mater 53 (3): 705-717 Feb 2005
Theoretical Prediction of New High-Performance Lead-Free Piezoelectrics	Waghmare UV	1	5	20.00	Jawaharlal Nehru Ctr Adv Sci Res, Theoret Sci Unit, Bangalore 560064, Karnataka, India.	Chem Mater 17 (6): 1376-1380 Mar 22 2005
Ultrafine Single-Crystalline Gold Nanowire Arrays by Oriented Attachment	Halder A, Ravishankar N	2	2	100.00	Indian Inst Sci, Mat Res Ctr, Bangalore 560012, Karnataka, India.	Advan Mater 19 (14): 1854-+ Jul 16 2007
Biosynthesis of Nanoparticles: Technological Concepts and Future Applications	Mohanpuria P, Rana NK, Yadav SK	3	3	100.00	Csir, Inst Himalayan Bioresource, Div Biotechnol, Palampur 176061, Himachal Prades, India.	J Nanopart Res 10 (3): 507-517 Mar 2008

Title	Authors			%	Affiliation	Citation
Biomolecular Immobilization on Conducting Polymers For Biosensing Applications	Ahuja T, Mir IA, Kumar D, Rajesh	4	4	100.00	Csir, Intellectual Property Management Div, 14 Satsang Vihar Marg, New Delhi 110067, India. Univ Delhi, Delhi Coll Engn, Dept Appl Chem, Delhi 110042, India.	Biomaterials 28 (5): 791-805 Feb 2007
Solar Photocatalytic Degradation of A Reactive Azo Dye in Tio2-Suspension	Muruganandham M, Swaminathan M	2	2	100.00	Annamalai Univ, Dept Chem, Annamalainagar 608002, Tamil Nadu, India.	Solar Energ Mater Solar Cells 81 (4): 439-457 Mar 1 2004
Noncovalent and Nonspecific Molecular Interactions of Polymers With Multiwalled Carbon Nanotubes	Baskaran D	1	3	33.33	Natl Chem Lab, Div Polymer Sci & Engn, Pune 411008, Maharashtra, India.	Chem Mater 17 (13): 3389-3397 Jun 28 2005
Strengthening in Carbon Nanotube/Aluminium (Cnt/Al) Composites	George R, Kashyap KT, Raw R, Yamdagni S	4	4	100.00	Ms Ramaiah Inst Technol, Bangalore 560054, Karnataka, India.	Scripta Mater 53 (10): 1159-1163 Nov 2005
Gram-Scale Synthesis of Soluble, Near-Monodisperse Gold Nanorods and Other Anisotropic Nanoparticles	Jana NR	1	1	100.00	Raja Rammohun Roy Mahavidyalaya, Dept Chem, Hooghly 712406, Wb, India	Small 1 (8-9): 875-882 Aug 2005
Properties and Applications of Colloidal Nonspherical Noble Metal Nanoparticles	Tapan K Sau	1	5	20.00	Int Inst Informat Technol, Hyderabad 500032, Andhra Pradesh, India	Advan Mater 22 (16): 1805-1825 Apr 22 2010
Nonspherical Noble Metal Nanoparticles: Colloid-Chemical Synthesis and Morphology Control	Tapan K Sau	1	2	50.00	Int Inst Informat Technol, Hyderabad 500032, Andhra Pradesh, India	Advan Mater 22 (16): 1781-1804 Apr 22 2010

Title	Author			%	Affiliation	Citation
Cdin2s4 Nanotubes and "Marigold" Nanostructures: A Visible-Like Photocatalyst	Kale BB	1	6	16.67	Govt India, Minist Informat & Technol, Ctr Mat Elect Technol, Nanocrystalline Mat Lab, Panchawati Pashran Rd, Pune 411008, Maharashtra, India.Govt India, Minist Informat & Technol, Ctr Mat Elect Technol, Nanocrystalline Mat Lab, Pune 411008, Maharashtra, India.	Adv Funct Mater 16 (10): 1349-1354 Jul 4 2006
Low Loss Dielectric Materials For Ltcc Applications: A Review	Sebastian MT	1	2	50.00	Natl Inst Interdisciplinary Sci & Technol, Mat Minerals Div, Trivandrum 695019, Kerala, India.	Int Mater Rev 53 (2): 57-90 Mar 2008
Supramolecular Gels 'In Action'	Banerjee S, Das RK, Maitra U	3	3	100.00	Indian Inst Sci, Dept Organ Chem, Bangalore 560012, Karnataka, India.	J Mater Chem 19 (37): 6649-6687 2009
Dendritic Polyglycerols For Biomedical Applications	Sharma SK	1	4	25.00	Univ Delhi, Dept Chem, Delhi 110007, India.	Advan Mater 22 (2): 190-218 Jan 12 2010
Natural Products As Corrosion Inhibitor For Metals in Corrosive Media - A Review	Raja PB, Sethuraman MG	2	2	100.00	Deemed Univ, Gandhigram Rural Inst, Dept Chem, Gandhigram 624302, Tamil Nadu, India.	Mater Lett 62 (1): 113-116 Jan 15 2008
A Facile and Novel Synthesis of Ag-Graphene-Based Nanocomposites	Pasricha R, Gupta S, Srivastava AK	3	3	100.00	Natl Phys Lab, Div Mat Characterizat, Dr Krishnan Rd, New Delhi 110012, India.; Natl Phys Lab, Div Mat Characterizat, New Delhi 110012, India.	Small 5 (20): 2253-2259 Oct 16 2009
Hydrothermal Processing of Materials: Past, Present and Future	Byrappa K	1	2	50.00	Univ Mysore, Mysore 570006, Karnataka, India	J Mater Sci 43 (7): 2085-2103 Apr 2008

A Microwell Array System For Stem Cell Culture	Shrivastava S	1	20.00	Indian Inst Technol, Dept Mech Engn, Ctr Nanotechnol, Gauhati, India.	Biomaterials 29 (6): 752-763 Feb 2008
Squaraine Dyes: A Mine of Molecular Materials	Sreejith S, Carol P, Chithra P, Ajayaghosh A	4	100.00	Csir, Nist, Chem Sci & Technol Div, Photosci & Photon Grp, Trivandrum 695019, Kerala, India. Vikram Sarabhai Space Ctr, Battery Dev Div, Trivandrum 695022, Kerala, India	J Mater Chem 18 (3): 264-274 2008
Synthesis of Inorganic Nanotubes	Rao CNR, Govindaraj A	2	100.00	Jawaharlal Nehru Ctr Adv Sci Res, Csir Ctr Excellence Chem, Bangalore 560064, Karnataka, India. Indian Inst Sci, Solid State & Struct Chem Unit, Bangalore 560012, Karnataka, India	Advan Mater 21 (42): 4208-4233 Nov 13 2009
Supramolecular Control of the Magnetic Anisotropy in Two-Dimensional High-Spin Fe Arrays At A Metal Interface	Sen Gupta S, Sarma DD	2	11.76	Indian Inst Sci, Solid State & Struct Chem Unit, Bangalore 560012, Karnataka, India.	Nat Mater 8 (3): 189-193 Mar 2009
Effect of Cefazolin on the Corrosion of Mild Steel in Hcl Solution	Singh AK, Quraishi MA	2	100.00	Banaras Hindu Univ, Inst Technol, Dept Appl Chem, Varanasi 221005, Uttar Pradesh, India.	Corros Sci 52 (1): 152-160 Jan 2010
Solvothermal Synthesis, Cathodoluminescence, and Field-Emission Properties of Pure and N-Doped Zno Nanobullets	LS Panchakarla, Govindaraj, CNR Rao	3	33.33	Jawaharlal Nehru Ctr Adv Sci Res, Chem & Phys Mat Unit, Dst Unit Nanosci, Bangalore 560064, Karnataka, India.	Adv Funct Mater 19 (1): 131-140 Jan 9 2009
Noble Metal Nanoparticles For Water Purification: A Critical Review	Pradeep T, Anshup	2	100.00	Indian Inst Technol, Sophisticated Analyt Instrument Facil, Madras 600036, Tamil Nadu, India.	Thin Solid Films 517 (24): 6441-6478 Oct 30 2009

Title	Authors			%	Affiliation	Citation
Shape-Dependent Electrocatalytic Activity of Platinum Nanostructures	Subhramannia M, Pillai VK	2	2	100.00	Natl Chem Lab, Phys & Mat Chem Div, Pune 411008, Maharashtra, India.	J Mater Chem 18 (48): 5858-5870 2008
Selective Zinc(Ii)-Ion Fluorescence Sensing by A Functionalized Mesoporous Material Covalently Grafted With A Fluorescent Chromophore and Consequent Biological Applications	Sarkar K, Dhara K, Nandi M, Roy P, Bhaumik A, Banerjee P	6	6	100.00	Indian Assoc Cultivat Sci, Dept Inorgan Chem, Calcutta 700032, India.	Adv Funct Mater 19 (2): 223-234 Jan 23 2009
Rgb Emission Through Controlled Donor Self-Assemble and Modulation of Excitation Energy Transfer: A Novel Strategy to White-Light-Emitting Organogels	Vijayakumar C, Praveen VK, Ajayaghosh A	3	3	100.00	Csir, Photosci & Photon Grp, Chem Sci & Technol Div, Niist, Trivandrum 695019, Kerala, India.	Advan Mater 21 (20): 2059-+ May 25 2009
Combustion Synthesis and Nanomaterials	Aruna ST	1	2	50.00	Natl Aerosp Labs, Surface Engn Div, Post Bag 1779, Bangalore 560017, Karnataka, India.	Curr Opin Solid State Mat Sci 12 (3-4): 44-50 Jun-Sep 2008
Nanostructured Biocomposite Substrates by Electrospinning and Electrospraying For the Mineralization of Osteoblasts	Deepika Gupta, S Mitra, VR Giri Dev	3	5	60.00	Amity Univ, Amity Inst Nanotechnol, Noida, India. Anna Univ, Ac Coll Technol, Dept Text Technol, Madras 600025, Tamil Nadu, India.	Biomaterials 30 (11): 2085-2094 Apr 2009
Constitutive Equations to Predict High Temperature Flow Stress in A Ti-Modified Austenitic Stainless Steel	Mandal S, Rakesh V, Sivaprasad PV, Venugopal S, Kasiviswanathan KV	5	5	100.00	Indira Gandhi Ctr Atom Res, Met & Mat Grp, Kalpakkam 603102, Tamil Nadu, India.	Mater Sci Eng A-Struct Mater 500 (1-2): 114-121 Jan 25 2009

Title	Authors				Affiliation	Citation
Pretreatments of Natural Fibers and Their Application As Reinforcing Material in Polymer Composites-A Review	Kalia S, Kaith BS, Kaur I	3	3	100.00	Singhania Univ, Dept Chem, Jhunjhunu 333515, Rajasthan, India.; Deemed Univ, Natl Inst Technol, Dept Chem, Jalandhar 144011, Punjab, India.; Himachal Pradesh Univ, Dept Chem, Shimla 171005, Himachal Prades, India.	Polym Eng Sci 49 (7): 1253-1272 Jul 2009
Dilute Doping, Defects, and Ferromagnetism in Metal Oxide Systems	Ogale SB	1	1	100.00	Natl Chem Lab, Csir, Phys & Mat Chem Div, Pune 411008, Maharashtra, India.; Iiser, Pune, Maharashtra, India.	Advan Mater 22 (29): 3125-3155 Aug 3 2010
Novel Carboxymethyl Derivatives of Chitin and Chitosan Materials and Their Biomedical Applications	Jayakumar R, Prabaharan M, Nair SV, Selvamurugan N	4	6	66.67	Amrita Vishwa Vidyapeetham, Amrita Ctr Nanosci & Mol Med, Amrita Inst Med Sci, Cochin 682041, Kerala, India. Srm Univ, Fac Engn & Technol, Dept Chem, Kattankulathur 603203, India. Srm Univ, Sch Bioengn, Dept Biotechnol, Kattankulathur 603203, India.	Prog Mater Sci 55 (7): 675-709 Sep 2010
Dual Drug Loaded Superparamagnetic Iron Oxide Nanoparticles For Targeted Cancer Therapy	Dilnawaz F, Singh A, Mohanty C, Sahoo SK	4	4	100.00	Inst Life Sci, Lab Nanomed, Bhubaneswar 751023, Orissa, India.	Biomaterials 31 (13): 3694-3706 May 2010
Novel Hydrogen Storage Materials: A Review of Lightweight Complex Hydrides	Jain IP, Jain P, Jain A	3	3	100.00	Univ Rajasthan, Ctr Nonconvent Energy Resources, Jaipur 302004, Rajasthan, India.	J Alloys Compounds 503 (2): 303-339 Aug 6 2010

Title	Authors				Affiliation	Citation
The Intracellular Drug Delivery and Anti Tumor Activity of Doxorubicin Loaded Poly(Gamma-Benzyl L-Glutamate)-B-Hyaluronan Polymersomes	Bhatt AN, Mishra AK, Dwarakanath BS, Jain S, Farooque A, Chandraiah G, Jain AK, Misra A	8	12	66.67	Maharaja Sayajirao Univ Baroda, Dept Pharm, Fac Technol & Engn, Vadodara 390001, Gujarat, India.Drdo, Inmas, Div Radiat Biosci, Delhi 110054, India. Natl Inst Pharmaceut Educ & Res, Dept Pharmaceut, Ctr Pharmaceut Nanotechnol, Sas Nagar 160062, Punjab, India.	Biomaterials 31 (10): 2882-2892 Apr 2010
Solution-Combustion Synthesized Nanocrystalline Li4ti5o12 As High-Rate Performance Li-Ion Battery Anode	Prakash AS, Manikandan P, Ramesha K, Sathiya M, Shukla AK	5	6	83.33	Indian Inst Sci, Solid State & Struct Chem Unit, Bangalore 560012, Karnataka, India. Cent Electrochem Res Inst, Chennai Unit, Madras 600113, Tamil Nadu, India.	Chem Mater 22 (9): 2857-2863 May 11 2010
Facile Synthesis of Water-Soluble Fluorescent Silver Nanoclusters and Hg-Ii Sensing	Adhikari B, Banerjee A	2	2	100.00	Indian Assoc Cultivat Sci, Dept Biol Chem, Calcutta 700032, India.	Chem Mater 22 (15): 4364-4371 Aug 10 2010
Nitrogen Doped Graphene Nanoplatelets As Catalyst Support For Oxygen Reduction Reaction in Proton Exchange Membrane Fuel Cell	Jafri RI, Rajalakshmi N, Ramaprabhu S	3	3	100.00	Indian Inst Technol, Dept Phys, Nfmtc, Alternat Energy & Nanotechnol Lab, Madras 600036, Tamil Nadu, India. Iit, Ctr Fuel Cell Technol, Madras 600113, Tamil Nadu, India	J Mater Chem 20 (34): 7114-7117 2010
Role of Size Scale of Zno Nanoparticles and Microparticles on Toxicity Toward Bacteria and Osteoblast Cancer Cells	Nair S, Sasidharan A, Rani VVD, Menon D, Nair S, Manzoor K, Raina S	7	7	100.00	Amrita Inst Med Sci, Inst Mol Med, Kochi 682026, Kerala, India.	J Mater Sci-Mater Med 20: 235-241 Suppl. 1 Dec 2009

Title	Authors			Affiliation	Reference
Molecular-Receptor-Specific, Non-Toxic, Near-Infrared-Emitting Au Cluster-Protein Nanoconjugates For Targeted Cancer Imaging	Retnakumari A, Setua S, Menon D, Ravindran P, Muhammed H, Pradeep T, Nair S, Koyakutty M	8	8	100.00	Amrita Inst Med Sci, Amrita Ctr Nanosci & Mol Med, Cochin 682041, Kerala, India. Indian Inst Technol, Dst Unit Nanosci, Madras 600036, Tamil Nadu, India	Nanotechnol 21 (5): Art. No.-055103 Feb 5 2010
Thermodynamic, Electrochemical and Quantum Chemical Investigation of Some Schiff Bases As Corrosion Inhibitors For Mild Steel in Hydrochloric Acid Solutions	Ahamad I, Prasad R, Quraishi MA	3	3	100.00	Banaras Hindu Univ, Fac Sci, Dept Chem, Varanasi 221005, Uttar Pradesh, India.	Corros Sci 52 (3): 933-942 Mar 2010
Novel Liquid Precursor-Based Facile Synthesis of Large-Area Continuous, Single, and Few-Layer Graphene Films	Srivastava A, Galande C, Ci L, Song L, Rai C, Jariwala D, Kelly KF, Ajayan PM	8	8	100.00	Banaras Hindu Univ, Dept Phys, Varanasi 221005, Uttar Pradesh, India	Chem Mater 22 (11): 3457-3461 Jun 8 2010
Folate Receptor Targeted, Rare-Earth Oxide Nanocrystals For Bi-Modal Fluorescence and Magnetic Imaging of Cancer Cells	Setua S, Menon D, Asok A, Nair S, Koyakutty M	5	5	100.00	Amrita Vishwa Vidyapeetham Univ, Res Ctr, Cochin 682041, Kerala, India.	Biomaterials 31 (4): 714-729 Feb 2010
A Simple and Effective Modification of Pcbm For Use As An Electron Acceptor in Efficient Bulk Heterojunction Solar Cells	Sharma SS, Sharma GD	5	2	250.00	Jai Narain Vyas Univ, Dept Phys, Mol Elect & Optoelect Device Lab, Jodhpur 342005, Rajasthan, India. Jaipur Engn Coll, R&D Ctr Engn & Sci, Jaipur, Rajasthan, India. Govt Engn Coll Women, Dept Phys, Ajmer, Rajasthan, India. Univ Rajasthan, Thin Film & Membrane Sci Lab, Jaipur 302004, Rajasthan, India.	Adv Funct Mater 21 (4): 746-755 Feb 22 2011

Title	Authors				Affiliation	Citation
A Study of the Synthetic Methods and Properties of Graphenes	Rao CNR, Subrahmanyam KS, Matte HSSR, Abdulhakeem B, Govindaraj A, Das B, Kumar P, Ghosh A, Late DJ	9	9	100.00	Jawaharlal Nehru Ctr Adv Sci Res, Csir, Ctr Bangalore 560064, Karnataka, India. Indian Inst Sci, Solid State & Struct Chem Unit, Bangalore 560012, Karnataka, India.	Sci Technol Adv Mater 11 (5): Art. No.- 054502 Oct 2010
Ultrafast Microwave-Assisted Route to Surfactant-Free Ultrafine Pt Nanoparticles on Graphene: Synergistic Co-Reduction Mechanism and High Catalytic Activity	Kundu P, Nethravathi C, Deshpande PA, Rajamathi M, Madras G, Ravishankar N	6	6	100.00	Indian Inst Sci, Dept Chem Engn, Bangalore 560012, Karnataka, India. St Josephs Coll, Dept Chem, Bangalore, Karnataka, India.	Chem Mater 23 (11): 2772-2780 Jun 14 2011
Fabrication of Silicon Nanowire Arrays Based Solar Cell With Improved Performance	Kumar D, Srivastava SK, Singh PK, Husain M, Kumar V	5	5	100.00	Csir, Natl Phys Lab, New Delhi 110012, India. Jamia Millia Islamia, Dept Phys, New Delhi 110025, India. Indian Inst Technol Delhi, Dept Phys, New Delhi 110016, India.	Solar Energ Mater Solar Cells 95 (1): 215-218 Sp. Iss. Si Jan 2011
Enhanced Magneto-Optical Effects in Magnetoplasmonic Crystals	Kasture S, Vengurlekar AS, Achanta Venu Gopal	3	10	30.00	Tata Inst Fundamental Res, Bombay 400005, Maharashtra, India.	Nat Nanotechnol 6 (6): 370-376 Jun 2011
Inhibitive Effect of Ceftazidime on Corrosion of Mild Steel in Hydrochloric Acid Solution	Singh AK, Shukla SK, Singh M, Quraishi MA	4	4	100.00	Banaras Hindu Univ, Inst Technol, Dept Appl Chem, Varanasi 221005, Uttar Pradesh, India.	Mater Chem Phys 129 (1-2): 68-76 Sep 15 2011
Photo-Catalytic Degradation of Toxic Dye Amaranth on Tio2/ Uv in Aqueous Suspensions	Gupta VK, Jain R, Mittal A, Nayak A, Agarwal S, Sikarwar S	6	7	85.71	Indian Inst Technol Roorkee, Dept Chem, Roorkee 247667, Uttar Pradesh, India.	Mater Sci Eng C 32 (1): 12-17 Jan 1 2012

Title	Indian Authors	No. of Indian Authors	Total Author	Ratio I/F*	Institue Name	Source
Adsorption and Corrosion Inhibition Behaviour of N-(Phenylcarbamothioyl) Benzamide on Mild Steel in Acidic Medium	Gopiraman M, Selvakumaran N, Kesavan D, Karvembu R	4	4	100.00	Natl Inst Technol, Dept Chem, Tiruchchirappalli 620015, Tamil Nadu, India.	Prog Org Coating 73 (1): 104-111 Jan 2012
Infrared Photodetectors Based on Reduced Graphene Oxide and Graphene Nanoribbons	Chitara B, Panchakarla LS, Krupanidhi SB, Rao CNR	4	4	100.00	Indian Inst Sci, Mat Res Ctr, Bangalore 560012, Karnataka, India. Jawaharlal Nehru Ctr Adv Sci Res, Chem & Phys Mat Unit, New Chem Unit, Bangalore 560064, Karnataka, India	Advan Mater 23 (45): 5419-+ Dec 1 2011
Manufacture of Nickel-Ceramic Composite Membranes in Agitated Electroless Plating Baths	Bulasara VK, Uppaluri R, Purkait MK	3	3	100.00	Indian Inst Technol Guwahati, Dept Chem Engn, Gauhati 781039, India.	Mater Manuf Process 26 (6): 862-867 2011
(I) Mathematics						
Title	Indian Authors	No. of Indian Authors	Total Author	Ratio I/F*	Institue Name	Source
Statistical Methods in Assessing Agreement: Models, Issues, and Tools	Sinha B	1	4	25.00	Indian Stat Inst, Calcutta 700035, W Bengal, India.	J Amer Statist Assn 97 (457): 257-270 Mar 2002
Peristaltic Transport of A Newtonian Fluid in An Asymmetric Channel	Mishra M, Rao AR	2	2	100.00	Indian Inst Sci, Dept Math, Bangalore 560012, Karnataka, India.	Z Angew Math Phys 54 (3): 532-550 May 2003
An Iterative Scheme For Equilibrium Problems and Fixed Point Problems of Strict Pseudo-Contraction Mappings	Ansari QH	1	4	25.00	Aligarh Muslim Univ, Dept Math, Aligarh, Uttar Pradesh, India.	J Comput Appl Math 223 (2): 967-974 Jan 15 2009

Title	Authors			%	Institution	Source
Existence and Convergence of Best Proximity Points	Eldred AA, Veeramani P	2	2	100.00	Indian Inst Technol, Dept Math, Madras 600036, Tamil Nadu, India.	J Math Anal Appl 323 (2): 1001-1006 Nov 15 2006
Heat Transfer Enhancement of Copper-Water Nanofluids in A Lid-Driven Enclosure	Muthtamilselvan M, Kandaswamy P	2	3	66.67	Bharathiar Univ, Dept Math, Ugc Drs Ctr Fluid Dynam, Maruthamalai Rd, Coimbatore 641046, Tamil Nadu, India.	Commun Nonlinear Sci Numer Si 15 (6): 1501-1510 Jun 15 2010
A Fuzzy Soft Set Theoretic Approach to Decision Making Problems	Roy AR, Maji PK	2	2	100.00	Indian Inst Technol, Dept Math, Kharagpur 721302, W Bengal, India	J Comput Appl Math 203 (2): 412-418 Jun 15 2007
A Generalisation of Contraction Principle in Metric Spaces	Dutta PN, Choudhury BS	2	2	100.00	Govt Coll Engn & Ceram Technol, Dept Math, Calcutta 700010, W Bengal, India. Bengal Engn & Sci Univ, Dept Math, Howrah 711103, W Bengal, India.	Fixed Point Theory Appl : Art. No.- 406368 2008
Asymptotically Strict Pseudocontractive Mappings in the Intermediate Sense	Sahu DR	1	3	33.33	Banaras Hindu Univ, Dept Math, Varanasi 221005, Uttar Pradesh, India.	Nonlinear Anal-Theor Meth App 70 (10): 3502-3511 May 15 2009
A New Substitution-Diffusion Based Image Cipher Using Chaotic Standard and Logistic Maps	Patidar V, Pareek NK, Sud KK	3	3	100.00	Sir Padampat Singhania Univ, Dept Basic Sci, Udaipur 313601, Rajasthan, India. Mls Univ, Univ Comp Ctr, Udaipur 313002, Rajasthan, India	Commun Nonlinear Sci Numer Si 14 (7): 3056-3075 Jul 2009
Nonlocal Cauchy Problem For Abstract Fractional Semilinear Evolution Equations	Balachandran K	1	2	50.00	Bharathiar Univ, Dept Math, Coimbatore 641046, Tamil Nadu, India.	Nonlinear Anal-Theor Meth App 71 (10): 4471-4475 Nov 15 2009

Title	Authors			%	Affiliation	Source
A Coupled Coincidence Point Result in Partially Ordered Metric Spaces For Compatible Mappings	Choudhury BS, Kundu A	2	2	100.00	Bengal Engn & Sci Univ, Dept Math, Sibpur 711103, Howrah, India. Siliguri Inst Technol, Dept Math, Darjeeling 734009, India.	Nonlinear Anal-Theor Meth App 73 (8): 2524-2531 Oct 15 2010
Fixed Point Results For Mappings Satisfying (Psi, Phi)-Weakly Contractive Condition in Partially Ordered Metric Spaces	Nashine HK	1	2	50.00	Disha Inst Management & Technol, Dept Math, Raipur 492101, Chhattisgarh, India.	Nonlinear Anal-Theor Meth App 74 (6): 2201-2209 Mar 15 2011
Global Exponential Stability Results For Neutral-Type Impulsive Neural Networks	Rakkiyappan R, Balasubramaniam P	2	3	66.67	Gandhigram Rural Univ, Dept Math, Gandhigram 624302, Tamil Nadu, India.	Nonlinear Anal-Real World App 11 (1): 122-130 Feb 2010
The Nonlocal Cauchy Problem For Nonlinear Fractional Integrodifferential Equations in Banach Spaces	Balachandran K	1	2	50.00	Bharathiar Univ, Dept Math, Coimbatore 641046, Tamil Nadu, India.	Nonlinear Anal-Theor Meth App 72 (12): 4587-4593 Jun 15 2010
On Recent Developments in the Theory of Abstract Differential Equations With Fractional Derivatives	Balachandran K	1	3	33.33	Bharathiar Univ, Dept Math, Coimbatore 641046, Tamil Nadu, India.	Nonlinear Anal-Theor Meth App 73 (10): 3462-3471 Nov 15 2010
Mann-Type Steepest-Descent and Modified Hybrid Steepest-Descent Methods For Variational Inequalities in Banach Spaces	Ansari QH	1	3	33.33	Aligarh Muslim Univ, Dept Math, Aligarh, Uttar Pradesh, India.	Numer Func Anal Optimiz 29 (9-10): 987-1033 2008
Stability Analysis of Uncertain Fuzzy Hopfield Neural Networks With Time Delays	Ali MS, Balasubramaniam P	2	2	100.00	Gandhigram Rural Univ, Dept Math, Dindigul 624302, Tamil Nadu, India.	Commun Nonlinear Sci Numer Si 14 (6): 2776-2783 Jun 2009

Title	Authors				Institution	Citation
Heat Transfer in Mhd Viscoelastic Boundary Layer Flow Over A Stretching Sheet With Non-Uniform Heat Source/Sink	Abel MS, Nandeppanavar MM	2	2	100.00	Govt First Grade Coll, Dept Math, Sindhanur 584128, Karnataka, India. Gulbarga Univ, Dept Math, Gulbarga 585106, Karnataka, India.	Commun Nonlinear Sci Numer Si 14 (5): 2120-2131 May 2009
Best Proximity Point Theorems For P-Cyclic Meir-Keeler Contractions	Karpagam S, Agrawal S	2	2	100.00	Univ Madras, Dept Math, Ramanujan Inst Adv Study Math, Madras 600005, Tamil Nadu, India.	Fixed Point Theory Appl : Art. No.- 197308 2009
Synchronization of Different Fractional Order Chaotic Systems Using Active Control	Daftardar-Gejji V	2	2	100.00	Univ Pune, Dept Math, Pune 411007, Maharashtra, India.	Commun Nonlinear Sci Numer Si 15 (11): 3536-3546 Nov 2010
Fixed Points of Weak Contractions in Cone Metric Spaces	Choudhury BS, Metiya N	2	2	100.00	Bengal Engn & Sci Univ, Dept Math, Howrah 711103, W Bengal, India.	Nonlinear Anal-Theor Meth App 72 (3-4): 1589-1593 Feb 1 2010
Optical Solitons With Non-Kerr Law Nonlinearity and Inter-Modal Dispersion With Time-Dependent Coefficients	Sarma AK	1	5	20.00	Indian Inst Technol, Dept Phys, Gauhati 781039, India.	Commun Nonlinear Sci Numer Si 15 (9): 2320-2330 Sep 2010
Existence Results For Fractional Impulsive Integrodifferential Equations in Banach Spaces	Balachandran K, Kiruthika S	2	3	66.67	Bharathiar Univ, Dept Math, Coimbatore 641046, Tamil Nadu, India.	Commun Nonlinear Sci Numer Si 16 (4): 1970-1977 Apr 2011
Likelihood Based Inference For Skew-Normal Independent Linear Mixed Models	Ghosh P	1	3	33.33	Indian Inst Management, Dept Quantitat Methods & Informat Sci, Bangalore 566076, Karnataka, India.	Stat Sinica 20 (1): 303-322 Jan 2010
Property P in G-Metric Spaces	Chugh R, Kadian T, Rani A	3	4	75.00	Maharshi Dayanand Univ, Dept Math, Rohtak 124001, Haryana, India	Fixed Point Theory Appl : Art. No.- 401684 2010

Title	Authors			%	Affiliation	Citation
Random Matrices: Universality of Esds and the Circular Law	Krishnapur M	1	3	33.33	Indian Inst Sci, Dept Math, Bangalore 560012, Karnataka, India.	Ann Probab 38 (5): 2023-2065 Sep 2010
Existence, Uniqueness and Stability Analysis of Recurrent Neural Networks With Time Delay in the Leakage Term Under Impulsive Perturbations	Balasubramaniam P, Rakkiyappan R	2	4	50.00	Gandhigram Rural Univ, Dept Math, Gandhigram 624302, Tamil Nadu, India.	Nonlinear Anal-Real World App 11 (5): 4092-4108 Oct 2010
Peristaltic Transport of A Casson Fluid in A Finite Channel: Application to Flows of Concentrated Fluids in Oesophagus	Pandey SK, Tripathi D	2	2	100.00	Banaras Hindu Univ, Inst Technol, Dept Appl Math, Varanasi 221005, Uttar Pradesh, India	Int J Biomath 3 (4): 453-472 Dec 2010
Passivity Analysis For Neural Networks of Neutral Type With Markovian Jumping Parameters and Time Delay in the Leakage Term	Balasubramaniam P, Nagamani G, Rakkiyappan R	3	3	100.00	Gandhigram Rural Univ, Dept Math, Gandhigram 624302, Tamil Nadu, India	Commun Nonlinear Sci Numer Si 16 (11): 4422-4437 Nov 2011
Unsteady Model of Transportation of Jeffrey-Fluid by Peristalsis	Pandey SK, Tripathi D	2	2	100.00	Banaras Hindu Univ, Inst Technol, Dept Appl Math, Varanasi 221005, Uttar Pradesh, India	Int J Biomath 3 (4): 473-491 Dec 2010
Fixed Point Theorems For Generalized Weakly Contractive Condition in Ordered Metric Spaces	Nashine HK,	1	2	50.00	Disha Inst Management & Technol, Dept Math, Raipur 492101, Chhattisgarh, India.	Fixed Point Theory Appl : Art. No.- 132367 2011
Mittag-Leffler Functions and Their Applications	Saxena RK	1	3	33.33	Ctr Math Sci, Pala 686574, Kerala, India. Jai Narain Vyas Univ, Dept Math & Stat, Jodhpur 342005, Rajasthan, India	J Appl Math : Art. No.-298628 2011

Title	Indian Authors	No. of Indian Authors	Total Author	Ratio I/F*	Institute Name	Source
Fixed Point Results For Contractions Involving Generalized Altering Distances in Ordered Metric Spaces	Nashine HK	1	3	33.33	Disha Inst Management & Technol, Dept Math, Raipur 492101, Chhattisgarh, India.	Fixed Point Theory Appl : Art. No.-5 2011
The Fractional Poisson Process and the Inverse Stable Subordinator	Vellaisamy P	1	3	33.33	Indian Inst Technol, Dept Math, Bombay 400076, Maharashtra, India.	Electron J Probab 16: Art. No.-59 Aug 28 2011
Well-Posed Inhomogeneous Nonlinear Diffusion Scheme For Digital Image Denoising	Prasath VBS, Singh A	2	2	100.00	Indian Inst Technol, Dept Math, Madras 600036, Tamil Nadu, India.	J Appl Math : Art. No.-763847 2010

(m) Microbiology

Title	Indian Authors	No. of Indian Authors	Total Author	Ratio I/F*	Institute Name	Source
The Genome of the Protist Parasite Entamoeba Histolytica	Anuradha Lohia, Chandrama Mukherjee, Alok Bhattacharya, Sudha Bhattacharya	4	52	7.69	Bose Inst, Dept Biochem, Calcutta 700054, India. Jawaharlal Nehru Univ, Sch Environm Sci, New Delhi 110067, India.	Nature 433 (7028): 865-868 Feb 24 2005
Bacterial Alkaline Proteases: Molecular Approaches and Industrial Applications	Gupta R, Beg QK, Lorenz P	1	3	33.33	Univ Delhi, Dept Microbiol, New Delhi 110021, India.	Appl Microbiol Biotechnol 59 (1): 15-32 Jun 2002
Drug Resistance in Leishmaniasis	Sundar S	1	3	33.33	Banaras Hindu Univ, Inst Med Sci, Dept Med, Infect Dis Res Lab, Varanasi 221005, Uttar Pradesh, India.	Clin Microbiol Rev 19 (1): 111-+ Jan 2006

Title	Authors				Affiliation	Reference
Mycobacterium Tuberculosis Complex Genetic Diversity: Mining the Fourth International Spoligotyping Database (Spoidb4) For Classification, Population Genetics and Epidemiology	Joyti Arora, Urvashi B Singh, Gurujaj V Kadival, Savita P Kulkarni	4	65	6.15	Bhabha Atom Res Ctr, Lab Nucl Med Sect, Isotope Grp, Bombay 400012, Maharashtra, India.All India Inst Med Sci, New Delhi, India.	Bmc Microbiol 6: Art. No.-23 Mar 23 2006
Bacterial Lipases: An Overview of Production, Purification and Biochemical Properties	Gupta R, Gupta N, Rathi P	3	3	100.00	Univ Delhi, Dept Microbiol, S Campus, Benito Juarez Rd, New Delhi 110021, India.	Appl Microbiol Biotechnol 64 (6): 763-781 Jun 2004
Antimicrobial Peptides: Premises and Promises	Reddy KVR, Yedery RD, Aranha C	3	3	100.00	Natl Inst Res Reprod Hlth, Dept Immunol, Bombay 400012, Maharashtra, India.	Int J Antimicrobial Agents 24 (6): 536-547 Dec 2004
Environmental Sensing and Signal Transduction Pathways Regulating Morphopathogenic Determinants of Candida Albicans	Subhrajit Biswas	1	3	33.33	Natl Ctr Plant Genome Res, New Delhi 110067, India.	Microbiol Mol Biol Rev 71 (2): 348-+ Jun 2007
Strain Specificity in Antimicrobial Activity of Silver and Copper Nanoparticles	Ruparelia JP, Chatterjee AK, Duttagupta SP, Mukherji S	4	4	100.00	Indian Inst Technol, Ctr Environm Sci & Engn, Bombay 400076, Maharashtra, India.Indian Inst Technol, Dept Elect Engn, Bombay 400076, Maharashtra, India.Innovat Ctr Appl Nanotechnol, Calcutta, India.	Acta Biomater 4 (3): 707-716 May 2008
Vaccine Escape Recombinants Emerge After Pneumococcal Vaccination in the United States	Pai R	1	4	25.00	Christian Med Coll & Hosp, Dept Gastrointestinal Sci, Vellore 632004, Tamil Nadu, India.	Plos Pathog 3 (11): Art. No.-E168 Nov 2007

Early Dissemination of Ndm-1-And Oxa-181-Producing Enterobacteriaceae in Indian Hospitals: Report From the Sentry Antimicrobial Surveillance Program, 2006-2007	Mathai D	1	6	16.67	Christian Med Coll & Hosp, Vellore, Tamil Nadu, India.	Antimicrob Agents Chemother 55 (3): 1274-1278 Mar 2011
Studies on Larvicidal and Pupicidal Activity of Leucas Aspera Willd. (Lamiaceae) and Bacterial Insecticide, Bacillus Sphaericus, Against Malarial Vector, Anopheles Stephensi Liston. (Diptera: Culicidae)	Kovendan K, Murugan K, Vincent S	3	4	75.00	Bharathiar Univ, Div Entomol, Dept Zool, Sch Life Sci, Coimbatore 641046, Tamil Nadu, India.	Parasitol Res 110 (1): 195-203 Jan 2012
Evaluation of Larvicidal Activity of Acalypha Alnifolia Klein Ex Willd. (Euphorbiaceae) Leaf Extract Against the Malarial Vector, Anopheles Stephensi, Dengue Vector, Aedes Aegypti and Bancroftian Filariasis Vector, Culex Quinquefasciatus (Diptera: ...	Kovendan K, Murugan K, Vincent S	3	3	100.00	Bharathiar Univ, Dept Zool, Div Entomol, Sch Life Sci, Coimbatore 641046, Tamil Nadu, India.	Parasitol Res 110 (2): 571-581 Feb 2012
Natural Products For Cancer Chemotherapy	Vaishnav P	1	2	50.00	Gidc, Ankleshwar 393002, Grijarat, India.	Microb Biotechnol 4 (6): 687-699 Nov 2011
Two Pathways For Rnase E Action in Escherichia Coli in Vivo and Bypass of Its Essentiality in Mutants Defective For Rho-Dependent Transcription Termination	Anupama K, Leela JK, Gowrishankar J	3	3	100.00	Ctr Dna Fingerprinting & Diagnost, Lab Bacterial Genet, Hyderabad 500001, Andhra Pradesh, India.	Mol Microbiol 82 (6): 1330-1348 Dec 2011

		(n) Molecular Biology				
Title	**Indian Authors**	**No. of Indian Authors**	**Total Author**	**Ratio I/F***	**Institute Name**	**Source**
Development of Human Protein Reference Database As An Initial Platform For Approaching Systems Biology in Humans	Jonnalagadda CK, Surendranath V, Niranjan V, Muthusamy B, Gandhi TKB, Deshpande N, Shanker K, Shivashankar HN, Rashmi BP, Ramya MA, Deshpande Kschandrika KN, Padma N, Harsha HC, Yatish AJ, Kavitha MP, Menezes M, Choudhury DR, Suresh S, Ghosh N, Saravana R, Chandran S, Krishna S, Joy M, Anand SK, Madavan V, Joseph A,	26	37	70.27	Int Technol Pk Ltd, Inst Bioinformat, Bangalore 560066, Karnataka, India.	Genome Res 13 (10): 2363-2371 Oct 2003
Regulation of Matrix Metalloproteinases: An Overview	Chakraborti S, Mandal M, Das S, Mandal A, Chakraborti T	5	5	100.00	Univ Kalyani, Dept Biochem & Biophys, Kalyani 741235, W Bengal, India.	Mol Cell Biochem 253 (1-2): 269-285 Nov 2003
Nanoscale Organization of Multiple Gpi-Anchored Proteins in Living Cell Membranes	Sharma P, Varma R, Sarasij RC, Ira, Gousset K, Krishnamoorthy G, Rao M, Mayor S	8	8	100.00	Natl Ctr Biol Sci Tifr, Bangalore 560065, Karnataka, India. Raman Res Inst, Bangalore 560080, Karnataka, India. Tata Inst Fundamental Res, Dept Chem Sci, Bombay 400005, Maharashtra, India.	Cell 116 (4): 577-589 Feb 20 2004

Title	Author				Affiliation	Source
Pathways of Clathrin-Independent Endocytosis	Mayor S	1	2	50.00	Uas, Natl Ctr Biol Sci, Bangalore 560065, Karnataka, India.	Nat Rev Mol Cell Biol 8 (8): 603-612 Aug 2007
Heterochromatic Silencing and Hp1 Localization in Drosophila Are Dependent on the Rnai Machinery	Pal-Bhadra M, Gandhi SG, Rao M, Bhadra U	4	7	57.14	Ctr Cellular & Mol Biol, Funct Genom & Gene Silencing Grp, Hyderabad 500007, Andhra Pradesh, India.	Science 303 (5658): 669-672 Jan 30 2004
Global Proteomic Profiling of Phosphopeptides Using Electron Transfer Dissociation Tandem Mass Spectrometry	Mathivanan S	1	5	20.00	Inst Bioinformat, Bangalore 560066, Karnataka, India.	Proc Nat Acad Sci USA 104 (7): 2199-2204 Feb 13 2007
Association Analyses of 249,796 Individuals Reveal 18 New Loci Associated With Body Mass Index	Shah Ebrahim	1	377	0.27	S Asia Network Chron Dis, New Delhi, India.	Nat Genet 42 (11): 937-U53 Nov 2010
Meta-Analysis Identifies 13 New Loci Associated With Waist-Hip Ratio and Reveals Sexual Dimorphism in the Genetic Basis of Fat Distribution	Shah Ebrahim	1	301	0.33	S Asia Network Chron Dis, New Delhi, India.	Nat Genet 42 (11): 949-U160 Nov 2010
Afp-Pred: A Random Forest Approach For Predicting Antifreeze Proteins From Sequence-Derived Properties	Sridharan S	1	7	14.29	Bharathidasan Univ, Tiruchchirappalli 620024, Tamil Nadu, India.	J Theor Biol 270 (1): 56-62 Feb 7 2011
Mapping the Nphp-Jbts-Mks Protein Network Reveals Ciliopathy Disease Genes and Pathways	Devi ARR	1	26	3.85	Rainbow Childrens Hosp, Hyderabad 500034, Andhra Pradesh, India.	Cell 145 (4): 513-528 May 13 2011

Title	Authors				Affiliation	Citation
De Novo Assembly of Chickpea Transcriptome Using Short Reads For Gene Discovery and Marker Identification	Garg R, Patel RK, Tyagi AK, Jain M	4	4	100.00	Nipgr, New Delhi 110067, India.	Dna Res 18 (1): 53-63 Feb 2011
Circadian Modulation of Sodium-Potassium Atpase and Sodium - Proton Exchanger in Human Erythrocytes: in Vitro Effect of Melatonin	Chakravarty S, Rizvi SI	2	2	100.00	Univ Allahabad, Dept Biochem, Allahabad 211002, Uttar Pradesh, India.	Cell Mol Biol 57 (1): 80-86 2011
Role of Aqueous Extract of Cynodon Dactylon in Prevention of Carbofuran-Induced Oxidative Stress and Acetylcholinesterase Inhibition in Rat Brain	Rai DK, Sharma RK, Rai PK, Watal G, Sharma B	5	5	100.00	Univ Allahabad, Dept Chem, Allahabad 211002, Uttar Pradesh, India.	Cell Mol Biol 57 (1): 135-142 2011
Biodegradation of Brassica Haulms by White Rot Fungus Pleurotus Eryngii	Singh MP, Pandey VK, Srivastava AK, Viswakarma SK	4	4	100.00	Vbs Purvanchal Univ, Dept Biotechnol, Jaunpur 222001, Up, India	Cell Mol Biol 57 (1): 47-55 2011
Biochemical and Morphological Perturbations in Rat Erythrocytes Exposed to Ethion: Protective Effect of Vitamin E	Bhatti GK, Bhatti JS, Kiran R, Sandhir R	4	4	100.00	Panjab Univ, Dept Biochem, Chandigarh 160014, India.	Cell Mol Biol 57 (1): 70-79 2011
Studies on in Vitro Antioxidant and Antistaphylococcal Activities of Some Important Medicinal Plants	Mishra A, Kumar S, Bhargava A, Sharma B, Pandey AK	5	5	100.00	Univ Allahabad, Dept Biochem, Allahabad 211002, Uttar Pradesh, India. Mln Med Coll, Dept Microbiol, Allahabad 211002, Uttar Pradesh, India.	Cell Mol Biol 57 (1): 16-25 2011

Title	Authors				Affiliation	Reference
Effect of Trichosanthes Dioica on Oxidative Stress and Cyp450 Gene Expression Levels in Experimentally Induced Diabetic Rats	Rai PK, Rai DK, Sharma B, Watal G	4	6	66.67	Univ Allahabad, Dept Chem, Med Res Lab, Alternat Therapeut Unit, Drug Discovery & Dev Div, Allahabad 211002, Uttar Pradesh, India. Univ Allahabad, Dept Biochem, Allahabad 211002, Uttar Pradesh, India. All India Inst Med Sci, Dept Biotechnol, New Delhi 110029, India.	Cell Mol Biol 57 (1): 31-39 2011
Protective Potential of Bacopa Monniera (Brahmi) Extract on Aluminum Induced Cerebellar Toxicity and Associated Neuromuscular Status in Aged Rats	Tripathi S, Mahdi AA, Hasan M, Mitra K, Mahdi F	5	5	100.00	Chhatrapati Shahuji Maharaj Med Univ, Dept Biochem, Med Elementol & Free Radical Biol Lab, Lucknow 226003, Uttar Pradesh, India. Chhatrapati Shahuji Maharaj Med Univ, Dept Anat, Lucknow 226003, Uttar Pradesh, India. Cent Drug Res Inst, Electron Microscopy Div, Lucknow 226001, Uttar Pradesh, India. Eras Lucknow Med Coll, Dept Biochem, Lucknow, Uttar Pradesh, India.	Cell Mol Biol 57 (1): 3-15 2011
Biochemical Characterization of Dipeptidylcarboxypeptidase of Leishmania Donovani	Baig MS, Gangwar S, Goyal N	3	3	100.00	Cent Drug Res Inst, Div Biochem, Lucknow 226001, Uttar Pradesh, India.	Cell Mol Biol 57 (1): 56-61 2011

Title	Authors				Affiliation	Citation
Sof Gene As A Specific Genetic Marker For Detection of Streptococcus Pyogenes Causing Pharyngitis and Rheumatic Heart Disease	Kumar A, Bhatnagar A, Gupta S, Khare S, Suman	5	5	100.00	Inst Genom & Integrat Biol Csir, Mall Rd, Delhi 110007, India. Natl Ctr Dis Control, Delhi 110054, India. Amity Univ, Inst Adv Res & Studies, Noida 201303, India.	Cell Mol Biol 57 (1): 26-30 2011
Cytotoxic and Genotoxic Effects of Mercury in House Fly Musca Domestica (Diptera: Muscidae)	Mishra N, Tewari RR	2	2	100.00	Univ Allahabad, Dept Zool, Allahabad 211002, Uttar Pradesh, India.	Cell Mol Biol 57 (1): 122-128 2011
Genomic Structure and Sequence Analysis of Lucilia Cuprina Hsp90 Gene	Reddy PVJ, Tiwari PK	2	2	100.00	Jiwaji Univ, Sch Studies Zool, Gwalior 474011, India.	Cell Mol Biol 57 (1): 112-121 2011
In Vitro Induction of the Ubiquitous 60 and 70Kd Heat Shock Proteins by Pesticides Monocrotophos and Endosulphan in Musca Domestica: Potential Biomarkers of Toxicity	Rohilla MS, Reddy PVJ, Sharma S, Tiwari PK	4	4	100.00	Jiwaji Univ, Sch Studies Zool, Mol Cell Biol Lab, Ctr Genom, Gwalior 474011, India.	Cell Mol Biol 57 (1): 100-111 2011
Seasonal Genetic Variation in House Fly Populations, Musca Domestica (Diptera: Muscidae)	Tripathi M, Agrawal UR, Tewari RR	3	3	100.00	Univ Allahabad, Dept Zool, Cmp Degree Coll, Allahabad 211002, Uttar Pradesh, India.	Cell Mol Biol 57 (1): 129-134 2011
Seasonal Genetic Variation in House Fly Populations, Musca Domestica (Diptera: Muscidae)	Tripathi R, Gupta S, Rai S, Mittal PC	4	4	100.00	Univ Allahabad, Dept Biochem, Allahabad 211002, Uttar Pradesh, India.	Cell Mol Biol 57 (1): 62-69 2011
Personalized Oncology Through Integrative High-Throughput Sequencing: A Pilot Study	Kalyana-Sundaram S	1	27	3.70	Bharathidasan Univ, Dept Environm Biotechnol, Tiruchchirappalli 620024, Tamil Nadu, India.	Sci Transl Med 3 (111): Art. No.-111Ra121 Nov 30 2011

Title	Indian Authors	No. of Indian Authors	Total Author	Ratio I/F*	Institute Name	Source
Dense Genotyping Identifies and Localizes Multiple Common and Rare Variant Association Signals in Celiac Disease	Vandana Midah, Akot Sood, Sabyasachi Senapati, B K Thelma	4	62	6.45	Dayanand Med Coll & Hosp, Ludhiana, Punjab, India. Univ Delhi, Dept Genet, New Delhi, India.	Nat Genet 43 (12): 1193-U45 Dec 2011

(o) Pharmacology

Title	Indian Authors	No. of Indian Authors	Total Author	Ratio I/F*	Institute Name	Source
Recent Advances on Chitosan-Based Micro- and Nanoparticles in Drug Delivery	Agnihotri SA, Mallikarjuna NN, Aminabhavi TM	3	3	100.00	Karnatak Univ, Drug Delivery Div, Ctr Excellence Polymer Sci, Dharwad 580003, Karnataka, India.	J Control Release 100 (1): 5-28 Nov 5 2004
Medicinal Plants of India With Anti-Diabetic Potential	Grover JK, Yadav S, Vats V	3	3	100.00	All India Inst Med Sci, Dept Pharmacol, New Delhi 110049, India.	J Ethnopharmacol 81 (1): 81-100 Jun 2002
Hydrogels: From Controlled Release to Ph-Responsive Drug Delivery	Gupta P, Vermani K, Garg S	3	3	100.00	Niper, Sas Nagar 160062, Punjab, India.	Drug Discov Today 7 (10): 569-579 May 15 2002
Paclitaxel and Its Formulations	Singla AK, Garg A, Aggarwal D	3	3	100.00	Panjab Univ, Inst Pharmaceut Sci, Chandigarh 160014, India.	Int J Pharm 235 (1-2): 179-192 Mar 20 2002
Polyionic Hydrocolloids For the Intestinal Delivery of Protein Drugs: Alginate and Chitosan - A Review	George M, Abraham TE	2	2	100.00	Csir, Reg Res Lab, Div Chem Sci, Polymer Sect, Trivandrum 695019, Kerala, India.	J Control Release 114 (1): 1-14 Aug 10 2006
Chitosan Microspheres As A Potential Carrier For Drugs	Sinha VR, Singla AK, Wadhawan S, Kaushik R, Kumria R, Bansal K, Dhawan S	7	7	100.00	Panjab Univ, Univ Inst Pharmaceut Sci, Chandigarh 160014, India.	Int J Pharm 274 (1-2): 1-33 Apr 15 2004

Title	Authors				Affiliation	Citation
Regulation of Inflammation and Redox Signaling by Dietary Polyphenols	Biswas SK,	1	3	33.33	Dr Ambedkar Coll, Dept Biochem, Nagpur, Maharashtra, India.	Biochem Pharmacol 72 (11): 1439-1452 Nov 30 2006
Biodegradable Microspheres For Protein Delivery	Sinha VR, Trehan A	2	2	100.00	Panjab Univ, Univ Inst Pharmaceut Sci, Dept Pharmaceut, Chandigarh 160014, India.	J Control Release 90 (3): 261-280 Jul 31 2003
Oxidant and Antioxidant Balance in the Airways and Airway Diseases	Saibal K Biswas	1	3	33.33	Dr Ambedkar Coll, Dept Biochem, Nagpur, Maharashtra, India.	Eur J Pharmacol 533 (1-3): 222-239 Mar 8 2006
Poly-Epsilon-Caprolactone Microspheres and Nanospheres: An Overview	Sinha VR, Bansal K, Kaushik R, Kumria R, Trehan A	5	5	100.00	Panjab Univ, Inst Pharmaceut Sci, Chandigarh 160014, India.	Int J Pharm 278 (1): 1-23 Jun 18 2004
Nano/Micro Technologies For Delivering Macromolecular Therapeutics Using Poly(D, L-Lactide-Co-Glycolide) and Its Derivatives	Mundargi RC, Babu VR, Rangaswamy V, Patel P, Aminabhavi TM	5	5	100.00	Dhirubhai Ambani Life Sci Ctr, Reliance Life Sci Pvt Ltd, Ind Biotechnol Grp, Thane Belapur Rd, Rabale 400701, Navi Mumbai, India.	J Control Release 125 (3): 193-209 Feb 11 2008
Plga Nanoparticles in Drug Delivery: The State of the Art	Bala I, Hariharan S, Kumar MNVR	3	3	100.00	Niper, Dept Pharmaceut, Sas Nagar 160062, Punjab, India.	Crit Rev Ther Drug Carr Syst 21 (5): 387-422 2004
A Pegylated Dendritic Nanoparticulate Carrier of Fluorouracil	Bhadra D, Bhadra S, Jain S, Jain NK	4	4	100.00	Dr Hs Gour Univ, Pharmaceut Res Labs, Dept Pharmaceut Sci, Sagar 470003, Mp, India.	Int J Pharm 257 (1-2): 111-124 May 12 2003
Role of Antioxidants in Prophylaxis and Therapy: A Pharmaceutical Perspective	Ratnam DV, Ankola DD, Bhardwaj V, Sahana DK, Kumar MNVR	5	5	100.00	Niper, Dept Pharmaceut, Phase X, Sas Nagar, Mohali 160062, Punjab, India.	J Control Release 113 (3): 189-207 Jul 20 2006

Title	Authors				Affiliation	Citation
Butyrylcholinesterase, Paraoxonase, and Albumin Esterase, But Not Carboxylesterase, Are Present in Human Plasma	Indumathi Manoharan, Rathnam Boopathy	2	7	28.57	Bharathiar Univ, Dept Biotechnol, Coimbatore 641046, Tamil Nadu, India.	Biochem Pharmacol 70 (11): 1673-1684 Nov 25 2005
Redox Modifications of Protein-Thiols: Emerging Roles in Cell Signaling	Biswas S	1	3	33.33	Dr Ambedkar Coll, Dept Biochem, Nagpur, Maharashtra, India.	Biochem Pharmacol 71 (5): 551-564 Feb 28 2006
Free Heme Toxicity and Its Detoxification Systems in Human	Kumar S, Bandyopadhyay U	2	2	100.00	Cent Drug Res Inst, Div Drug Target Discovery & Dev, Chatter Manzil Palace, Lucknow 226001, Uttar Pradesh, India.	Toxicol Lett 157 (3): 175-188 Jul 4 2005
Biological Activities of Curcumin and Its Analogues (Congeners) Made by Man and Mother Nature	Thomas SG, Misra K, Priyadarsini IK, Rajasekharan KN	4	11	36.36	Univ Kerala, Dept Chem, Thiruvananthapuram 695034, Kerala, India. Indian Inst Informat Technol, Bioinformat Div, Allahabad, Uttar Pradesh, India. Bhabha Atom Res Ctr, Radiat & Photochem Div, Bombay 400085, Maharashtra, India	Biochem Pharmacol 76 (11): 1590-1611 Sp. Iss. Si Dec 1 2008
Combination of High-Fat Diet-Fed and Low-Dose Streptozotocin-Treated Rat: A Model For Type 2 Diabetes and Pharmacological Screening	Srinivasan K, Viswanad B, Asrat L, Kaul CL, Ramarao P	5	5	100.00	Niper, Dept Pharmacol & Toxicol, Sector 67, Sas Nagar, Phase X, Mohali 160062, Punjab, India	Pharmacol Res 52 (4): 313-320 Oct 2005
Study of the Interaction of An Anticancer Drug With Human and Bovine Serum Albumin: Spectroscopic Approach	Kandagal PB, Ashoka S, Seetharamappa J, Shaikh SMT, Jadegoud Y, Ijare OB	6	6	100.00	Karnatak Univ, Dept Chem, Dharwad 580003, Karnataka, India. Sanjay Gandhi Postgrad Inst Med Sci, Ctr Biomed Magnet Resonance, Lucknow 226014, Uttar Pradesh, India	J Pharmaceut Biomed Anal 41 (2): 393-399 May 3 2006

Title	Authors				Institution	Citation
Curcumin-Phospholipid Complex: Preparation, Therapeutic Evaluation and Pharmacokinetic Study in Rats	Maiti K, Mukherjee K, Gantait A, Saha BP, Mukherjee PK	5	5	100.00	Jadavpur Univ, Sch Nat Prod Studies, Dept Pharmaceut Technol, Calcutta 700032, W Bengal, India	Int J Pharm 330 (1-2): 155-163 Feb 7 2007
Glycogen Synthase Kinase 3: More Than A Namesake	Rayasam GV, Tulasi VK, Sodhi R, Davis JA, Ray A	5	5	100.00	Ranbaxy Res Labs, R&D 3, Dept Pharmacol, Plot 20, Sector 18, Gurgaon, Haryana, India.	Brit J Pharmacol 156 (6): 885-898 Mar 2009
Nanoparticle Encapsulation Improves Oral Bioavailability of Curcumin by At Least 9-Fold When Compared to Curcumin Administered With Piperine As Absorption Enhancer	Shaikh J, Beniwal V, Singh D	3	5	60.00	Niper, Dept Pharmaceut, Sas Nagar 160062, Punjab, India.	Eur J Pharm Sci 37 (3-4): 223-230 Jun 28 2009
Potential of Solid Lipid Nanoparticles in Brain Targeting	Kaur IP, Bhandari R, Bhandari S, Kakkar V	4	4	100.00	Panjab Univ, Univ Inst Pharmaceut Sci, Chandigarh 160014	J Control Release 127 (2): 97-109 Apr 21 2008
Nanoparticles Synthesis Using Supercritical Fluid Technology - Towards Biomedical Applications	Byrappa K	1	3	33.33	Univ Mysore, Mysore 570006, Karnataka, India.	Advan Drug Delivery Rev 60 (3): 299-327 Feb 14 2008
Lipid-Based Systemic Delivery of Sirna	Mozumdar S	1	3	33.33	Univ Delhi, Dept Chem, Delhi 110007, India	Advan Drug Delivery Rev 61 (9): 721-731 Jul 25 2009
Dna Damaging Potential of Zinc Oxide Nanoparticles in Human Epidermal Cells	Sharma V, Shukla RK, Saxena N, Parmar D, Das M, Dhawan A	5	5	100.00	Csir, Indian Inst Toxicol Res, Mahatma Gandhi Marg, Pob 80, Lucknow 226001, Uttar Pradesh, India.	Toxicol Lett 185 (3): 211-218 Mar 28 2009
Oxidative Stress and Neurodegenerative Diseases: A Review of Upstream and Downstream Antioxidant Therapeutic Options	Singh AV, Mahajan RT	2	4	50.00	Mj Coll, Dept Biotechnol, Mj Rd, Jalgaon 425001, India.	Curr Neuropharmacol 7 (1): 65-74 Mar 2009

Title	Authors			%	Affiliation	Citation
Silver Nanoparticles in Therapeutics: Development of An Antimicrobial Gel Formulation For Topical Use	Jain J, Arora S, Rajwade JM, Omray P, Khandelwal S, Paknikar KM	6	6	100.00	Agharkar Res Inst, Ctr Nanobiosci, Gg Agarkar Rd, Pune 411004, Maharashtra, India., Nano Cutting Edge Technol Pvt Ltd, Bombay 400033, Maharashtra, India.	Mol Pharm 6 (5): 1388-1401 Sep-Oct 2009
Targeted Temperature Sensitive Magnetic Liposomes For Thermo-Chemotherapy	Pradhan P, Banerjee R, Bahadur D	3	9	33.33	Iit, Dept Met Engn & Mat Sci, Bombay, Maharashtra, India.	J Control Release 142 (1): 108-121 Feb 25 2010
Cox-2 As A Target For Cancer Chemotherapy	Ghosh N, Chaki R, Mandal V, Mandal SC	4	4	100.00	Dr Bc Roy Coll Pharm & Allied Hlth Sci, Durgapur 713206, India. Jadavpur Univ, Dept Pharmaceut Technol, Calcutta 700032, India	Pharmacol Rep 62 (2): 233-244 Mar-Apr 2010
Intracellular Trafficking of Nuclear Localization Signal Conjugated Nanoparticles For Cancer Therapy	Misra R, Sahoo SK	2	2	100.00	Inst Life Sci, Bhubaneswar, Orissa, India.	Eur J Pharm Sci 39 (1-3): 152-163 Jan 31 2010
Marine Drugs From Sponge-Microbe Association-A Review	Thomas TRA, Kavlekar DP, Lokabharathi PA	3	3	100.00	Natl Inst Oceanog, Panaji 403004, Goa, India.	Mar Drugs 8 (4): 1417-1468 Apr 2010
Nf Kappa B: A Promising Target For Natural Products in Cancer Chemoprevention	Suaib Luqman	1	2	50.00	Cent Inst Med & Aromat Plants, Council Sci & Ind Res, Div Biotechnol, Mol Bioprospect Dept, Lucknow 226015, Uttar Pradesh, India.	Phytother Res 24 (7): 949-963 Jul 2010
Biodegradable Calcium Phosphate Nanoparticle With Lipid Coating For Systemic Sirna Delivery	Mozumdar S	1	5	20.00	Univ Delhi, Dept Chem, Delhi 110007, India.	J Control Release 142 (3): 416-421 Mar 19 2010

Title	Authors				Affiliation	Citation
Plga Nanoparticles Containing Various Anticancer Agents and Tumour Delivery by Epr Effect	Acharya S, Sahoo SK	2	2	100.00	Inst Life Sci, Lab Nanomed, Bhubaneswar 751023, Orissa, India	Advan Drug Delivery Rev 63 (3): 170-183 Mar 18 2011
Ros-Mediated Genotoxicity Induced by Titanium Dioxide Nanoparticles in Human Epidermal Cells	Shukla RK, Sharma V, Pandey AK, Singh S, Sultana S, Dhawan A	6	6	100.00	Csir, Indian Inst Toxicol Res, Lucknow 226001, Uttar Pradesh, India. Csir, Ctr Cellular & Mol Biol, Hyderabad 500007, Andhra Pradesh, India. Jamia Hamdard, Dept Med Elementol & Toxicol, New Delhi 110062, India.	Toxicol Vitro 25 (1): 231-241 Feb 2011
Taurine Suppresses Doxorubicin-Triggered Oxidative Stress and Cardiac Apoptosis in Rat Via Up-Regulation of Pi3-K/Akt and Inhibition of P53, P38-Jnk	Das J, Ghosh J, Manna P, Sil PC	4	4	100.00	Bose Inst, Div Mol Med, Calcutta 700054, W Bengal, India.	Biochem Pharmacol 81 (7): 891-909 Apr 1 2011
Formulation and Evaluation of Quercetin Polycaprolactone Microspheres For the Treatment of Rheumatoid Arthritis	Natarajan V, Krithica N, Madhan B, Sehgal PK	4	4	100.00	Cent Leather Res Inst, Council Sci & Ind Res, Madras 20, Tamil Nadu, India.	J Pharm Sci 100 (1): 195-205 Jan 2011
Self-Emulsifying Drug Delivery Systems: An Approach to Enhance Oral Bioavailability	Kohli K, Chopra S, Dhar D, Arora S, Khar RK	5	5	100.00	Fac Pharm, Dept Pharmaceut, New Delhi 62, India. Ranbaxy Res Labs, Ctr Res & Dev, Gurgaon, India.	Drug Discov Today 15 (21-22): 958-965 Nov 2010
Interplay Between Statins and Ppars in Improving Cardiovascular Outcomes: A Double-Edged Sword?	Balakumar P, Mahadevan N	2	2	100.00	Rits, Inst Pharm, Cardiovasc Pharmacol Div, Dept Pharmacol, Sirsa 125055, Haryana, India	Brit J Pharmacol 165 (2): 373-379 Jan 2012

Title	Indian Authors	No. of Indian Authors	Total Author	Ratio I/F*	Institue Name	Source
Mutagenicity of Industrial Wastewaters Collected From Two Different Stations in Northern India	Tabrez S, Ahmad M	2	2	100.00	Amu, Dept Biochem, Fac Life Sci, Aligarh 202002, Uttar Pradesh, India.	J Appl Toxicol 31 (8): 783-789 Nov 2011
Vitamin E Tpgs Coated Liposomes Enhanced Cellular Uptake and Cytotoxicity of Docetaxel in Brain Cancer Cells	Muthu MS	1	4	25.00	Banaras Hindu Univ, Inst Med Sci, Dept Pharmacol, Varanasi 221005, Uttar Pradesh, India.	Int J Pharm 421 (2): 332-340 Dec 15 2011

(p) Neuroscience

Title	Indian Authors	No. of Indian Authors	Total Author	Ratio I/F*	Institue Name	Source
Chronic Stress Induces Contrasting Patterns of Dendritic Remodeling in Hippocampal and Amygdaloid Neurons	Vyas A, Mitra R, Rao BSS, Chattarji S	4	4	100.00	Tata Inst Fundamental Res, Natl Ctr Biol Sci, Uas-Gkvk Campus, Bangalore 560065, Karnataka, India., Tata Inst Fundamental Res, Natl Ctr Biol Sci, Bangalore 560065, Karnataka, India.	J Neurosci 22 (15): 6810-6818 Aug 1 2002
The Prevalence of Frontotemporal Dementia	Ratnavalli E	1	4	25.00	Natl Inst Mental Hlth & Neurosci, Dept Neurol, Bangalore 560029, Karnataka, India.	Neurology 58 (11): 1615-1621 Jun 11 2002
Loss of Presenilin Function Causes Impairments of Memory and Synaptic Plasticity Followed by Age-Dependent Neurodegeneration	Rao BSS, Chattarji S,	2	12	16.67	Natl Ctr Biol Sci, Bangalore 560065, Karnataka, India.	Neuron 42 (1): 23-36 Apr 8 2004

Title	Authors				Affiliation	Citation
Correction of Fragile X Syndrome in Mice	Rao BSS, Chattarji S	2	7	28.57	Natl Inst Mental Hlth & Neurosci, Dept Neurophysiol, Bangalore 560002, Karnataka, India. Tata Inst Fundamental Res, Natl Ctr Biol Sci, Bangalore 560002, Karnataka, India.	Neuron 56 (6): 955-962 Dec 20 2007
Stress, Memory and the Amygdala	Chattarji S	1	3	33.33	Tata Inst Fundamental Res, Natl Ctr Biol Sci, Bangalore 560065, Karnataka, India.	Nat Rev Neurosci 10 (6): 423-433 Jun 2009
Molecular Targets in Cerebral Ischemia For Developing Novel Therapeutics	Mehta SL, Manhas N, Rahubir R	3	3	100.00	Cent Drug Res Inst, Div Pharmacol, Chatter Manzil Palace, Pob 173, Lucknow 226001, Uttar Pradesh, India.	Brain Res Rev 54 (1): 34-66 Apr 2007
Melatonin-A Pleiotropic, Orchestrating Regulator Molecule	Venkatramanujam Srinivasan	1	6	16.67	Sri Sathya Sai Med Educ & Res Fdn, Coimbatore 641014, Tamil Nadu, India. Karpagam Univ, Fac Med, Dept Physiol, Coimbatore 641021, Tamil Nadu, India.	Prog Neurobiol 93 (3): 350-384 Mar 2011
Long Term Exposure to Cypermethrin Induces Nigrostriatal Dopaminergic Neurodegeneration in Adult Rats: Postnatal Exposure Enhances the Susceptibility During Adulthood	Singh AK, Tiwari Mn, Upadhyay G, Patel DK, Singh D, Prakash O, Singh MP	7	7	100.00	Csir, Iitr, Lucknow -2; Banaras Hindu Univ, Varanasi	Neurobiol Aging 33 (2): 404-415 Feb 2012

(q) Physics						
Title	Indian Authors	No. of Indian Authors	Total Author	Ratio I/F*	Institute Name	Source
Review of Particle Physics	Gurtu, A	1	169	0.59	Tata Inst Fundamental Res, Bombay	J Phys G-Nucl Particle Phys 33 (1): 1-+ Sp. Iss. Si Jul 2006
Review of Particle Physics	Gurtu, A	1	173	0.58	Tata Inst Fundamental Res, Bombay	Phys Lett B 667 (1-5): 1-+ Sep 18 2008
Review of Particle Physics	Gurtu, A	1	155	0.65	Tata Inst Fundamental Res, Bombay	Phys Lett B 592 (1-4): 1-1109 Jul 15 2004
Review of Particle Physics	Gurtu, A	1	140	0.71	Tata Inst Fundamental Res, Bombay	Phys Rev D 66 (1): Art. No.-010001 Part 1 Jul 1 2002
Review of Particle Physics	Gurtu, A	1	179	0.56	Tata Inst Fundamental Res, Bombay	J Phys G-Nucl Particle Phys 37 (7A): Art. No.-075021 Jul 2010
Epitaxial Bifeo3 Multiferroic Thin Film Heterostructures	Waghmare UV	1	14	7.14	J Nehru Centre For Advanced Scientific Research, Jakkur, Bangalore	Science 299 (5613): 1719-1722 Mar 14 2003
De Sitter Vacua in String Theory	Trivedi, Sandip P	1	4	25	Tata Inst Fundamental Res, Bombay	Phys Rev D 68 (4): Art. No.-046005 Aug 15 2003
Cosmological Constant - the Weight of the Vacuum	Padmanabhan T	1	1	100	Iucaa, Pune	Phys Rep-Rev Sect Phys Lett 380 (5-6): 235-320 Jul 2003

Experimental and Theoretical Challenges in the Search For the Quark-Gluon Plasma: The Star Collaboration's Critical Assessment of the Evidence From Rhic Collisions	MM Aggarwal, Z Ahammed, SK Badyal, S Bharadwaj, A Bhasin, AK Bhati, VS Bhatia, Chattopadhyay, D Das, S Das, SM Dogra, AK Dubey, MR Dutta Mazumdar, MS Ganti, P Ghosh, A Gupta, A Kumar, S Mahajan, DP Mahapatra, LK Mangotra, DK Mishra, B Mohanty, BK Nandi, SK Nayak, TK Nayak, PK Netrakanti, SK Pal, SC Phatak, BVKS Potukuchi, R Raniwala, S Raniwala, R Sahoo, M Sharma, RN Singaraju, YP Viyogi	35	371	9.43	Inst Phys, Bhubaneswar 751005, Orissa, India. Univ Jammu, Jammu 180001, India.Indian Inst Technol, Bombay 400076, Maharashtra, India. Panjab Univ, Chandigarh 160014, India. Univ Rajasthan, Jaipur 302004, Rajasthan, India. Ctr Variable Energy Cyclotron, Calcutta 700064, W Bengal, India.	Nucl Phys A 757 (1-2): 102-183 Aug 8 2005
Formation of Dense Partonic Matter in Relativistic Nucleus-Nucleus Collisions At Rhic: Experimental Evaluation by the Phenix Collaboration	SS Kapoor, SK Gupta, D Dutta, JD Ojha, CP Singh, V Singh, SK Tuli, P Chand, RK Choudhary, BV Disnesh	10	512	1.95	Banaras Hindu Univ, Dept Phys, Varanasi 221005, Uttar Pradesh, India. Bhabha Atom Res Ctr, Bombay 400085, Maharashtra, India.	Nucl Phys A 757 (1-2): 184-283 Aug 8 2005
Suppression of Hadrons With Large Transverse Momentum in Central Au+Au Collisions At Root S(Nn)=130 Gev	P Chand, RK Choudhury, BV Dinesh, D Dutta, SK Gupta, SS Kapoor, AK Mohanty, ID Ojha, CP Singh, V Singh, SK Tuli	11	307	3.58	Banaras Hindu Univ, Dept Phys, Varanasi 221005, Uttar Pradesh, India. Bhabha Atom Res Ctr, Bombay 400085, Maharashtra, India.	Phys Rev Lett 88 (2): Art. No.-022301 Jan 14 2002
Rolling Tachyon	Sen A	1	1	100	Harish Chandra Res Inst, Chhatnag Rd, Allahabad 211019, Uttar Pradesh, India.	J High Energy Phys (4): Art. No.-048 Apr 2002

Tachyon Matter	Sen A	1	1	100	Harish Chandra Res Inst, Allahabad 211019, Uttar Pradesh, India.; Penn State Univ, Dept Phys, University Pk, Pa USA.	J High Energy Phys (7): Art. No.-065 Jul 2002
Elliptic Flow of Identified Hadrons in Au+Au Collisions At Root S(Nn)=200 Gev	Chand P, Choudhury RK, Dutta D, Kapoor SS, Mohanty AK, Ojha ID, Singh CP, Singh V, Tuli SK	9	325	2.77	Banaras Hindu Univ, Dept Phys, Varanasi 221005, Uttar Pradesh, India. Bhabha Atom Res Ctr, Bombay 400085, Maharashtra, India.	Phys Rev Lett 91 (18): Art. No.-182301 Oct 31 2003
Accelerated Expansion of the Universe Driven by Tachyonic Matter	Padmanabhan T	1	1	100	Iucaa, Pune 411007, Maharashtra, India	Phys Rev D 66 (2): Art. No.-021301 Jul 15 2002
Identified Charged Particle Spectra and Yields in Au Plus Au Collisions At Root(Snn)=200 Gev	Ojha ID, Singh CP, Singh V, Tuli SK, Chand P, Choudhury RK, Dutta D, Kapoor SS, Mohanty AK	9	326	2.76	Banaras Hindu Univ, Dept Phys, Varanasi 221005, Uttar Pradesh, India. Bhabha Atom Res Ctr, Bombay 400085, Maharashtra, India.	Phys Rev C 69 (3): Art. No.-034909 Mar 2004
Observation of A Narrow Charmoniumlike State in Exclusive B+/--> K+/-Pi(+)Pi(-) J/Psi Decays	Kumar S, Soni N, Banerjee S, Majumder G, Sarangi TR, Satapathy M	6	174	3.45	Panjab Univ, Chandigarh 160014, India.Tata Inst Fundamental Res, Bombay 400005, Maharashtra, India. Utkal Univ, Bhubaneswar 751004, Orissa, India.	Phys Rev Lett 91 (26): Art. No.-262001 Dec 31 2003

Title	Authors		Year		Affiliations	Journal
Cms Physics Technical Design Report, Volume Ii: Physics Performance	A Bhardwaj, S Bhattacharya, S Chatterji, S Chauhan, BC Choudhary, P Gupta, M Jha, K Ranjan, RK Shivpuri, AK Srivastava, HS Bawa, SB Beri, V Bhandari, V Bhatnagar, M Kaur, R Kaur, JM Kohli, A Kumar, JB Singh, S Borkar, M Dixit, M Ghodgaonkar, SK Kataria, SK Lalwani, V Mishra, AK Mohanty, A Topkar, T Aziz, S Banerjee, S Bose, N Cheere, S Chendvankar, PV Deshpande, M Guchait, A Gurtu, M Maity, G Majumder, K Mazumdar, A Nayak, MR Patil, S Sharma, K Sudhakar, SC Tonwar, Bannanje Sripath Acharya, S Banerjee, S Bheesette, S Dugad, SD Kalmani, VR Lakkireddi, NK Mondal, N Panyam, P Verma,	52	2010	2.59	Panjab Univ, Chandigarh 160014, India. Univ Delhi, Delhi 110007, India. Bhabha Atom Res Ctr, Bombay 400085, Maharashtra, India. Tata Inst Fundamental Res, Ehep, Bombay 400005, Maharashtra, India. Hecr, Bombay 400005, Maharashtra, India. Visva Bharati Univ, Santini Ketan, W Bengal, India. Inst Studies Theoret Phys & Math, Tehran, Iran.	J Phys G-Nucl Particle Phys 34 (6): 995-1579 Jun 2007
Ferromagnetism As A Universal Feature of Nanoparticles of the Otherwise Nonmagnetic Oxides	Sundaresan A, Bhargavi R, Rangarajan N, Siddesh U, Rao CNR	5	5	100	Jawaharlal Nehru Ctr Adv Sci Res, Jakkur Po, Bangalore 560064, Karnataka, India. Jawaharlal Nehru Ctr Adv Sci Res, Bangalore 560064, Karnataka, India.	Phys Rev B 74 (16): Art. No.-161306 Oct 2006

Title	Authors				Affiliation	Citation
First-Principles Study of Spontaneous Polarization in Multiferroic Bifeo3	Waghmare UV	1	5	20	Jawaharlal Nehru Ctr Adv Sci Res, Bangalore 560064, Karnataka, India.	Phys Rev B 71 (1): Art. No.-014113 Jan 2005
Astrophysical Magnetic Fields and Nonlinear Dynamo Theory	Subramanian K	1	2	50	Punjabi Univ, Inter Univ Ctr Astron & Astrophys, Pune 411007, Maharashtra, India.	Phys Rep-Rev Sect Phys Lett 417 (1-4): 1-209 Oct 2005
Transverse-Momentum and Collision-Energy Dependence of High-P(T) Hadron Suppression in Au+Au Collisions At Ultrarelativistic Energies	Dubey AK, Mahapatra DP, Mishra D, Phatak SC, Sahoo R, Badyal SK, Gupta A, Mahajan S, Mangotra LK, Nayak SK, Potukuchi BVKS, Sood G, Sharma M, Kumar A, Bhati AK, Aggarwal MM, Bhardwaj S, Raniwala R, Raniwala S, Voloshin SA, Trivedi MD, Singaraju RN, Pal SK, Nayak TK, Mohanty B, Nandi BK, Ghosh P, Ganti MS, Dutta Majumdar MR, Das D, Das S, Chattopadhyay S, Bhaskar P	34	364	9.34	Inst Phys, Bhubaneswar 751005, Orissa, India. Univ Jammu, Jammu 180001, India.Panjab Univ, Chandigarh 160014, India. Univ Rajasthan, Jaipur 302004, Rajasthan, India. Bhabha Atom Res Ctr, Ctr Variable Energy Cyclotron, Calcutta 700064, W Bengal, India.	Phys Rev Lett 91 (17): Art. No.-172302 Oct 24 2003
Physics of Negative Refractive Index Materials	Ramakrishna SA	1	1	100	Indian Inst Technol, Dept Phys, Kanpur 208016, Uttar Pradesh, India.	Rep Progr Phys 68 (2): 449-521 Feb 2005

Title	Authors				Affiliation	Citation
Particle-Type Dependence of Azimuthal Anisotropy and Nuclear Modification of Particle Production in Au Plus Au Collisions At Root S(Nn)=200 Gev	Dubey AK, Mahapatra DP, Mishra D, Phatak SC, Sahoo R, Potukuchi BVKS, Nayak SK, Mangotra LK, Mahajan S, Gupta A, Badyal SK, Aggarwal MM, Bhati AK, Kumar A, Sharma M, Das D, Das S, Dutta MR Majumdar, Ganti MS, Ghosh P, Mohanty B, Nandi K, Nayak TK, Pal SK, Singaraju RN, Trivedi MD, Viyogi YP	17	161	16.77	Inst Phys, Bhubaneswar 751005, Orissa, India. Univ Jammu, Jammu 180001, India. Panjab Univ, Chandigarh 160014, India. Univ Rajasthan, Jaipur 302004, Rajasthan, India. Bhabha Atom Res Ctr, Ctr Variable Energy Cyclotron, Calcutta 700064, W Bengal, India.	Phys Rev Lett 92 (5): Art. No.-052302 Feb 6 2004
Field Theory of Tachyon Matter	Sen A	1	1	100	Harish Chandra Res Inst, Allahabad 211019, Uttar Pradesh, India.	Mod Phys Lett A 17 (27): 1797-1804 Sep 7 2002
Suppressed Pi(0) Production At Large Transverse Momentum in Central Au Plus Au Collisions At Root S(Nn)=200 Gev	Chand P, Choudhury RK, Dutta D, Kapoor SS, Mohanty AK, Ojha ID, Singh CP, Singh V, Tuli SK	9	326	2.76	Banaras Hindu Univ, Dept Phys, Varanasi 221005, Uttar Pradesh, India. Bhabha Atom Res Ctr, Bombay 400085, Maharashtra, India.	Phys Rev Lett 91 (7): Art. No.-072301 Aug 15 2003
Search For Neutral Mssm Higgs Bosons At Lep	Aziz T, Banerjee S, Ganguli SN, Gurtu A, Jindal P, Kaur M, Mazumdar K, Mohanty GB, Raja N, Rahaman MA, Sudhakar K, Tonwar SC	12	1211	0.99	Panjab Univ, Chandigarh 160014, India. Tata Inst Fundamental Res, Bombay 400005, Maharashtra, India. Utkal Univ, Bhubaneswar 751004, Orissa, India.	Eur Phys J C 47 (3): 547-587 Sep 2006
Cosmological Dynamics of A Phantom Field	Singh P, Sami A, Dadhich N	3	3	100	Iucaa, Post Bag 4, Pune 411007, Maharashtra, India.	Phys Rev D 68 (2): Art. No.-023522 Jul 15 2003

Title	Authors				Affiliation	Citation
The Physics Behind High-Temperature Superconducting Cuprates: The 'Plain Vanilla' Version of Rvb	Randeria M, Trivedi N	2	6	33.33	Tata Inst Fundamental Res, Bombay 400005, Maharashtra, India.	J Phys-Condens Matter 16 (24): R755-R769 Jun 23 2004
Azimuthal Anisotropy in Au Plus Au Collisions At Root S-Nn=200 Gev	Aggarwal MM, Ahammed Z, Badyal SK, Bharadwaj S, Bhasin A, Bhati AK, Bhatia VS, Chattopadhyay S, Das D, Das S, Dogra SM, Dubey AK, Dutta Mazumdar MR, Ganti MS, Ghosh P, Gupta A, Mahajan S, Mahapatra DP, Mangotra LK, Mishra DK, Mohanty B, Nandi BK, Nayak SK, Nayak TK, Netrakanti PK, Pal SK, Potukuchi BVKS, Sahoo R, Sharma M, Singaraju N, Varma R, Viyogi YP	32	377	8.49	Inst Phys, Bhubaneswar 751005, Orissa, India. Indian Inst Technol, Bombay 400076, Maharashtra, India. Univ Jammu, Jammu 180001, India. Panjab Univ, Chandigarh 160014, India. Ctr Variable Energy Cyclotron, Calcutta 700064, W Bengal, India.	Phys Rev C 72 (1): Art. No.-014904 Jul 2005
Absence of Suppression in Particle Production At Large Transverse Momentum in Root S(Nn)=200 Gev D+Au Collisions	Chand P, Dutta D, Mohanty A	3	368	0.82	Bhabha Atom Res Ctr, Bombay 400085, Maharashtra, India.	Phys Rev Lett 91 (7): Art. No.-072303 Aug 15 2003
Strong-Driving-Assisted Multipartite Entanglement in Cavity Qed	Agarwal GS	1	3	33.33	Phys Res Lab, Ahmedabad 380009, Gujarat, India.	Phys Rev Lett 90 (2): Art. No.-027903 Jan 17 2003
Cosmology With Tachyon Field As Dark Energy	Bagla JS, Jassal HK, Padmanabhan T	3	3	100	Harish Chandra Res Inst, Allahabad 211019, Uttar Pradesh, India. Interuniv Ctr Astron & Astrophys, Pune 411007, Maharashtra, India.	Phys Rev D 67 (6): Art. No.-063504 Mar 2003

Title	Authors				Affiliation	Citation
Statefinder - A New Geometrical Diagnostic of Dark Energy	Varun Sahnia, Tarun Deep Sainia, Ujjaini Alam	3	4	75	Interuniv Ctr Astron & Astrophys, Pune 411007, Maharashtra, India.	Jetp Lett-Engl Tr 77 (5): 201-206 2003
Oxide Materials For Development of Integrated Gas Sensors - A Comprehensive Review	Eranna G, Joshi BC, Runthala DP, Gupta RP	4	4	100	Cent Elect Engn Res Inst, Pilani 333031, Rajasthan, India.	Crit Rev Solid State Mat Sci 29 (3-4): 111-188 2004
Precision Electroweak Measurements on the Z Resonance	Aziz T, Banerjee S, Bhattacharya S, Dutta S, Ganguli SN, Gurtu A, Jindal P, Kaur M, Mazumdar K, Moulik T, Mohanty GB, Rahaman MA, Raja N, Sudhakar K, Tonwar SC	15	2512	0.6	Panjab Univ, Chandigarh 160014, India. Tata Inst Fundamental Res, Ehep, Bombay 400005, Maharashtra, India.	Phys Rep-Rev Sect Phys Lett 427 (5-6): 257-454 May 2006
Thermal Conductivities of Naked and Monolayer Protected Metal Nanoparticle Based Nanofluids: Manifestation of Anomalous Enhancement and Chemical Effects	Patel HE, Das SK, Sundararajan T, Sreekumaran Nair A, George B, Pradeep T	6	6	100	Indian Inst Technol, Reg Sophisticated Instrumentat Ctr, Madras 600036, Tamil Nadu, India	Appl Phys Lett 83 (14): 2931-2933 Oct 6 2003
First Principles Based Design and Experimental Evidence For A Zno-Based Ferromagnet At Room Temperature	Raju AR, Rout C, Waghmare UV	3	7	42.86	Jawaharlal Nehru Ctr Adv Sci Res, Bangalore 560064, Karnataka, India.	Phys Rev Lett 94 (18): Art. No.-187204 May 13 2005
Nnlo Corrections to the Total Cross Section For Higgs Boson Production in Hadron-Hadron Collisions	Ravindran V	1	3	33.33	Harish Chandra Res Inst, Chhatnag Rd, Allahabad 211019, Uttar Pradesh, India.	Nucl Phys B 665 (1-3): 325-366 Aug 18 2003
Electronic Model For Coo2 Layer Based Systems: Chiral Resonating Valence Bond Metal and Superconductivity	Baskaran G	1	1	100	Inst Math Sci, Cit Campus, Madras 600113, Tamil Nadu, India.	Phys Rev Lett 91 (9): Art. No.-097003 Aug 29 2003

Physics Interplay of the Lhc and the Ilc	Godbole R	1	50	2	Indian Inst Sci, Ctr Theoret Studies, Bangalore 560012, Karnataka, India	Phys Rep-Rev Sect Phys Lett 426 (2-6): 47-358 Apr 2006
Qcd Equation of State With Almost Physical Quark Masses	Datta S	1	15	6.67	Tata Inst Fundamental Res, Dept Theoret Phys, Bombay 400005, Maharashtra, India	Phys Rev D 77 (1): Art. No.-014511 Jan 2008
Energy Loss and Flow of Heavy Quarks in Au+Au Collisions At Root S(Nn) = 200 Gev	Mishra M, Singh CP, Singh V, Tuli SK	4	422	0.95	Banaras Hindu Univ, Dept Phys, Varanasi 221005, Uttar Pradesh, India.	Phys Rev Lett 98 (17): Art. No.-172301 Apr 27 2007
Nonlinear Fluid Dynamics From Gravity	Sayantani Bhattacharyya, Shiraz Minwalla	2	4	50	Tata Inst Fundamental Res, Dept Theoret Phys, Bombay 400005, Maharashtra, India.	J High Energy Phys (2): Art. No.-045 Feb 2008
Multistrange Baryon Elliptic Flow in Au Plus Au Collisions At Root(Nn)-N-S=200 Gev	Dubey AK, Mahapatra DP, Mishra DK, Phatak SC, Sahoo R, Varma R, Potukuchi BVKS, Nayak SK, Mangotra LK, Mahajan S, Gupta N, Gupta A, Dogra SM, Bhasin A, Badyal SK, Aggarwal MM, Bhati AK, Bhatia VS, Kumar A, Sharma M, Raniwala R, Raniwala S, Bharadwaj S, Ahammed Z, Chattopadhyay S, Das D, Das S, Dutta Mazumdar MR, Ganti MS, Ghosh P, Mohanty B, Nandi BK, Nayak TK, Netrakanti PK, Pal SK, Singaraju RN, Viyogi YP	37	388	9.54	Inst Phys, Bhubaneswar 751005, Orissa, India. Indian Inst Technol, Bombay 400076, Maharashtra, India. Univ Jammu, Jammu 180001, India. Univ Rajasthan, Jaipur 302004, Rajasthan, India. Panjab Univ, Chandigarh 160014, India. Variable Energy Cyclotron Ctr, Calcutta 700064, W Bengal, India	Phys Rev Lett 95 (12): Art. No.-122301 Sep 16 2005

Title	Authors				Affiliation	Publication
Black Hole Entropy Function and the Attractor Mechanism in Higher Derivative Gravity	Sen A	1	1	100	Harish Chandra Res Inst, Chhatnag Rd, Allahabad 211019, Uttar Pradesh, India.	J High Energy Phys (9): Art. No.-038 Sep 2005
Tachyon Dynamics in Open String Theory	Sen A	1	1	100	Harish Chandra Res Inst, Allahabad 211019, Uttar Pradesh, India.	Int J Mod Phys A 20 (24): 5513-5656 Sep 30 2005
Dense-Medium Modifications to Jet-Induced Hadron Pair Distributions in Au+Au Collisions At Root(Nn)-N-S=200 Gev	Joha ID, Singh CP, Singh V, Tuli SK, Chand P, Kapoor SS, Mohanty AK, Dutta D	8	370	2.16	Banaras Hindu Univ, Dept Phys, Varanasi 221005, Uttar Pradesh, India. Bhabha Atom Res Ctr, Bombay 400085, Maharashtra, India.	Phys Rev Lett 97 (5): Art. No.-052301 Aug 4 2006
Observation of Double C(C) Over-Bar Production in E(+) E(-) Annihilation At Root S Approximate to 10.6 Gev	Mojumder G, Kumar S, Singh JB, Behera PK, Satapathy M	5	208	2.4	Panjab Univ, Chandigarh 160014, India. Tata Inst Fundamental Res, Bombay 400005, Maharashtra, India. Utkal Univ, Bhubaneswar 751004, Orissa, India.	Phys Rev Lett 89 (14): Art. No.-142001 Sep 30 2002
Observation of Saturated Polarization and Dielectric Anomaly in Magnetoelectric Bifeo3 Thin Films	Palkar VR, John J, Pinto R	3	3	100	Tata Inst Fundamental Res, Dept Condensed Matter Phys & Mat Sci, Mumbai 400005, India.	Appl Phys Lett 80 (9): 1628-1630 Mar 4 2002

Title	Authors				Institution	Reference
Identified Particle Distributions in Pp and Au+Au Collisions At Root S(Nn)=200 Gev	Viyogi YP, Rahiwala R, Raniwala S, Agarwal MM, Kumar A, Sharma M, Dubey K, Mahapatra DP, Mishra DK, Phatak SC, Sahoo R, Varma R, Badyal SK, Bharadwaj S, Bhati AK, Bhasin A, Dogra SM, Gupta A, Mahajan S, Mangotra LK, Nayak SK, Potukuchi BVKS, Ahammed Z, Chattopadhyay S, Das D, Das S, Dutta Mazumdar MR, Ganti MS, Ghosh P, Mohanty B, Nandi BK, Nayak TK, Netrakanti PK, Pal SK, Singaraju RN	35	370	9.46	Inst Phys, Bhubaneswar 751005, Orissa, India.Univ Jammu, Jammu 180001, India. Univ Rajasthan, Jaipur 302004, Rajasthan, India. Panjab Univ, Chandigarh 160014, India.Bhabha Atom Res Ctr, Ctr Variable Energy Cyclotron, Calcutta 700064, W Bengal, India.	Phys Rev Lett 92 (11): Art. No.-112301 Mar 19 2004
Diffusion in Metallic Glasses and Supercooled Melts	Suman K Sharma	1	9	11.11	Malaviya Natl Inst Technol, Dept Phys, Jaipur 302017, Rajasthan, India.	Rev Mod Phys 75 (1): 237–280 Jan 2003
Ligo: The Laser Interferometer Gravitational-Wave Observatory	Dhurandhar S, Mukhopadhyay H	2	511	0.39	Interuniv Ctr Astron & Astrophys, Pune 411007, Maharashtra, India.	Rep Progr Phys 72 (7): Art. No.-076901 Jul 2009
Midrapidity Neutral-Pion Production in Proton-Proton Collisions At Root S 200 Gev	Ojha ID, Singh CP, Singh V, Tuli SK, Chand P, Chowdhary RK, Dutta D, Kapoor SS, Mohanty AK	9	376	2.39	Banaras Hindu Univ, Dept Phys, Varanasi 221005, Uttar Pradesh, India. Bhabha Atom Res Ctr, Bombay 400085, Maharashtra, India.	Phys Rev Lett 91 (24): Art. No.-241803 Dec 12 2003
J/Psi Production Versus Centrality, Transverse Momentum, and Rapidity in Au+Au Collisions At Root S-Nn=200 Gev	Mishra M, Singh CP, Singh V, Tuli SK	4	420	0.95	Banaras Hindu Univ, Dept Phys, Varanasi 221005, Uttar Pradesh, India.	Phys Rev Lett 98 (23): Art. No.-232301 Jun 8 2007

Title	Authors				Affiliation	Reference
Limits on the Time Variation of the Electromagnetic Fine-Structure Constant in the Low Energy Limit From Absorption Lines in the Spectra of Distant Quasars	Srianand R, Chand H,	2	4	50	Iucaa, Post Bag 4, Pune 411007, Maharashtra, India.	Phys Rev Lett 92 (12): Art. No.-121302 Mar 26 2004
Centrality Dependence of Pi(+/-), K+/-, P, and (P)Over-Bar Production From Root(Nn)-N-S = 130 Gev Au+Au Collisions At Rhic		11	307	0.59	Banaras Hindu Univ, Dept Phys, Varanasi 221005, Uttar Pradesh, India. Bhabha Atom Res Ctr, Bombay 400085, Maharashtra, India.	Phys Rev Lett 88 (24): Art. No.-242301 Jun 17 2002
Hydrodynamic Fluctuations and Instabilities in Ordered Suspensions of Self-Propelled Particles	Simha RA, Ramaswamy S	2	2	100	Indian Inst Sci, Dept Phys, Ctr Condensed Matter Theory, Bangalore 560012, Karnataka, India.	Phys Rev Lett 89 (5): Art. No.-058101 Jul 29 2002
Wmap Constraints, Supersymmetric Dark Matter, and Implications For the Direct Detection of Supersymmetry	Chattopadhyay U,	1	3	33.33	Indian Assoc Cultivat Sci, Dept Theoret Phys, Jadavpur 700032, Kolkata, India.	Phys Rev D 68 (3): Art. No.-035005 Aug 1 2003
Quantum Nature of the Big Bang: An Analytical and Numerical Investigation	Ashtekar A, Singh P	2	3	66.67	Interuniv Ctr Astron & Astrophys, Pune 411017, Maharashtra, India.	Phys Rev D 73 (12): Art. No.-124038 Jun 2006
Aspects of Semiclassical Strings in Ads(5)	Mandal G, Wadia SR	2	3	66.67	Tata Inst Fundamental Res, Homi Bhabha Rd, Mumbai 400005, India.	Phys Lett B 543 (1-2): 81-88 Sep 5 2002
High-P(T) Charged Hadron Suppression in Au+Au Collisions At Root(Snn)=200 Gev	Ojha ID, Singh CP, Singh V, Chand P, Choudhury RK, Dutta D, Kapoor SS, Mohanty AK	8	371	2.16	Banaras Hindu Univ, Dept Phys, Varanasi 221005, Uttar Pradesh, India. Bhabha Atom Res Ctr, Bombay 400085, Maharashtra, India.	Phys Rev C 69 (3): Art. No.-034910 Mar 2004

Supersymmetry Parameter Analysis: Spa Convention and Project	Dutta S, Godbole Pm, Mukhopadhyaya B, Roy P	4	125	3.2	Harish Chandra Res Inst, Allahabad, Uttar Pradesh, India. Indian Inst Sci, Ctr High Energy Phys, Bangalore 560012, Karnataka, India. Univ Delhi, Delhi 110007, India. Tata Inst Fundamental Res, Bombay 400005, Maharashtra, India.	Eur Phys J C 46 (1): 43-60 Apr 2006
Gravity and the Thermodynamics of Horizons	Padmanabhan T	1	1	100	Iucaa, Post Bag 4, Ganeshkhind, Pune 411007, Maharashtra, India.	Phys Rep-Rev Sect Phys Lett 406 (2): 49-125 Jan 2005
Distributions of Charged Hadrons Associated With High Transverse Momentum Particles in Pp and Au Plus Au Collisions At Root(S)(Nn)=200 Gev	Varma R, Dubey AK, Mahapatra DP, Mishra D, Phatak SC, Sahoo R, Badya SK, Gupta A, Mahajan S, Mangotra LK, Nayak SK, Potukuchi BVKS, Aggarwal MM, Ahammed Z, Bhati AK, Kumar A, Sharma M, Sood G, Bhardwaj S, Raniwala R, Raniwala S, Chattopadhyay S, Das D, Das Ś, Dutta Majumdar MR, Ganti MS, Ghosh P, Mohanty B, Nandi BK, Nayak TK, Netrakanti PK, Pal SK, Singaraju RN, Viyogi YP	32	372	8.6	Indian Inst Technol, Bombay 400076, Maharashtra, India. Inst Phys, Bhubaneswar 751005, Orissa, India. Univ Jammu, Jammu 180001, India. Panjab Univ, Chandigarh 160014, India. Univ Rajasthan, Jaipur 302004, Rajasthan, India. Bhabha Atom Res Ctr, Ctr Variable Energy Cyclotron, Calcutta 700064, W Bengal, India.	Phys Rev Lett 95 (15): Art. No.-152301 Oct 7 2005

Title	Authors				Affiliation	Reference
First Measurement of the Transverse Spin Asymmetries of the Deuteron in Semi-Inclusive Deep Inelastic Scattering	Dasgupta SS, Dhara L, Sinha L	3	214	1.4	Univ Burdwan, Burdwan 713104, W Bengal, India. Matrivani Inst Expt Res & Educ, Calcutta 700030, W Bengal, India.	Phys Rev Lett 94 (20): Art. No.-202002 May 27 2005
Improved Measurement of Mixing-Induced Cp Violation in the Neutral B Meson System	Kumar S, Singh JB, Soni N, Majumder G, Behera PK, Satapathy M	6	230	2.61	Panjab Univ, Chandigarh 160014, India.Tata Inst Fundamental Res, Bombay 400005, Maharashtra, India. Utkal Univ, Bhubaneswar 751004, Orissa, India.	Phys Rev D 66 (7): Art. No.-071102 Oct 1 2002
Azimuthal Anisotropy At the Relativistic Heavy Ion Collider: The First and Fourth Harmonics	Dubey AK, Mahapatra DP, Mishra D, Phatak SC, Sahoo R, Badyal SK, Gupta A, Mahajan S, Mangotra LK, Nayak SK, Potukuchi BVKS, Aggarwal MM, Bhati AK, Ahammed Z, Kumar A, Sharma M, Sood G, Bhardwaj S, Bhaskar P, Raniwala R, Raniwala S, Chattopadhyay S, Das D, Das S, Dutta Majumdar MR, Ganti MS, Ghosh P, Mohanty B, Nandi BK, Nayak TK, Pal SK, Singaraju RN, Trivedi MD, Viyogi YP	34	371	9.16	Inst Phys, Bhubaneswar 751005, Orissa, India. Univ Jammu, Jammu 180001, India.Panjab Univ, Chandigarh 160014, India. Univ Rajasthan, Jaipur 302004, Rajasthan, India. Bhabha Atom Res Ctr, Ctr Variable Energy Cyclotron, Kolkata 700064, India.	Phys Rev Lett 92 (6): Art. No.-062301 Feb 13 2004

Title	Authors				Affiliation	Publication
Pion Interferometry in Au+Au Collisions At Root S(Nn)=200 Gev	Dubey AK, Mahapatra DP, Mishra DK, Phatak SC, Sahoo R, Varma R, Badyal SK, Bhasin A, Bhati AK, Bhatia VS, Dogra SM, Gupta A, Mahajan S, Mangotra LK, Nayak SK, Nayak TK, Potukuchi BVKS, Aggarwal MM, Ahammed Z, Kumar A, Sharma M, Bharadwaj S, Raniwala R, Raniwala S, Chattopadhyay S, Das D, Das S, Dutta Mazumdar MR, Ganti MS, Ghosh P, Mohanty B, Nandi BK, Netrakanti PK, Pal SK, Singaraju RN, Viyogi YP	35	367	9.54.	Inst Phys, Bhubaneswar 751005, Orissa, India. Indian Inst Technol, Bombay 400076, Maharashtra, India. Univ Jammu, Jammu 180001, India. Panjab Univ, Chandigarh 160014, India. Univ Rajasthan, Jaipur 302004, Rajasthan, India. Bhabha Atom Res Ctr, Ctr Variable Energy Cyclotron, Calcutta 700064, W Bengal, India.	Phys Rev C 71 (4): Art. No.-044906 Apr 2005
Dark Energy Cosmology From Higher-Order, String-Inspired Gravity, and Its Reconstruction	Sami M	1	3	33.33	Jamia Millia Islamia, Ctr Theoret Phys, New Delhi 110025, India.	Phys Rev D 74 (4): Art. No.-046004 Aug 2006
Classical and Quantum Thermodynamics of Horizons in Spherically Symmetric Spacetimes	Padmanabhan T	1	1	100	Interuniv, Ctr Astron & Astrophys, Post Bag 4, Pune 411007, Maharashtra, India.	Class Quantum Gravity 19 (21): 5387-5408 Nov 7 2002
Scaling Properties of Azimuthal Anisotropy in Au Plus Au and Cu Plus Cu Collisions At Root S(Nn)=200 Gev	Tuli SK, Singh CP, Singh V	3	488	0.61	Banaras Hindu Univ, Dept Phys, Varanasi 221005, Uttar Pradesh, India.	Phys Rev Lett 98 (16): Art. No.-162301 Apr 20 2007
Evidence For M-B and M-C Phases in the Morphotropic Phase Boundary Region of (1-X)[Pb(Mg1/3Nb2/3)O-3]-Xpbtio(3): A Rietveld Study	Singh AK, Pandey D	2	2	100	Banaras Hindu Univ, Sch Mat Sci & Technol, Varanasi 221005, Uttar Pradesh, India.	Phys Rev B 67 (6): Art. No.-064102 Feb 1 2003

Title	Authors				Affiliation	Citation
Cosmological Consequences of A Chaplygin Gas Dark Energy	Dev A, Jain D	2	3	66.67	Univ Delhi, Dept Phys & Astrophys, Delhi 110007, India.Univ Delhi, Deen Dayal Upadhyaya Coll, Delhi 110015, India.	Phys Rev D 67 (2): Art. No.-023515 Jan 15 2003
New Supersymmetric String Compactifications	Tripathy PK, Trivedi SP	2	4	50	Tata Inst Fundamental Res, Bombay 400005, Maharashtra, India.	J High Energy Phys (3): Art. No.-061 Mar 2003
Evidence For the Likely Occurrence of Magnetoferroelectricity in the Simple Perovskite, Bimno3	Parashar S, Raju AR, Rao CNR	3	6	50	Jawaharlal Nehru Ctr Adv Sci Res, Chem & Phys Mat Unit, Jakkur Po, Bangalore 560064, Karnataka, India.	Solid State Commun 122 (1-2): 49-52 2002
Observation and Properties of the X(3872) Decaying to J/Psi Pi(+)Pi(-) in P(P)Over-Bar Collisions At Root S=1.96 Tev	SB Beri, V Bhatnaga, JM Kohli, S Chakrabarti, BS Acharya, S Banerjee, A Chandra, SR Dugad, MR Krishnaswamy, PK Mal, NK Mondal, VS Narasimham, KJ Rani	13	585	2.22	Panjab Univ, Chandigarh 160014, India. Tata Inst Fundamental Res, Bombay 400005, Maharashtra, India.	Phys Rev Lett 93 (16): Art. No.-162002 Oct 15 2004
Inclusive Measurement of the Photon Energy Spectrum in B -> S Gamma Decays	S Kumar, N Soni, G Gokhroo, G Majumder, TR Sarangi	5	174	2.87	Panjab Univ, Chandigarh 160014, India.Tata Inst Fundamental Res, Bombay 400005, Maharashtra, India. Utkal Univ, Bhubaneswar 751004, Orissa, India.	Phys Rev Lett 93 (6): Art. No.-061803 Aug 6 2004
Effect of Substrate-Induced Strain on the Structural, Electrical, and Optical Properties of Polycrystalline Zno Thin Films	Ghosh R, Basak D,	2	3	66.67	Indian Assoc Cultivat Sci, Dept Solid State Phys, Calcutta 700032, W Bengal, India.	J Appl Phys 96 (5): 2689-2692 Sep 1 2004

Title	Authors				Affiliation	Citation
Observation of the Dsj(2317) and Dsj(2457) in B Decays	Singh JB, Soni N, Majumder G, Behera PK, Sarangi TR	5	178	2.81	Panjab Univ, Chandigarh 160014, India.Tata Inst Fundamental Res, Bombay 400005, Maharashtra, India. Utkal Univ, Bhubaneswar 751004, Orissa, India.	Phys Rev Rev Lett 91 (26): Art. No.-262002 Dec 31 2003
Observation of A Power-Law Memory Kernel For Fluctuations Within A Single Protein Molecule	Cherayil BJ	1	5	20	Indian Inst Sci, Dept Inorgan & Phys Chem, Bangalore 560012, Karnataka, India.	Phys Rev Lett 94 (19): Art. No.-198302 May 20 2005
Moduli Stabilization From Fluxes in A Simple lib Orientifold	Trivedi SP	1	3	33.33	Tata Inst Fundamental Res, Bombay 400005, Maharashtra, India.	J High Energy Phys (10): Art. No.-007 Oct 2003
General Logarithmic Corrections to Black-Hole Entropy	Majumdar P	1	3	33.33	Inst Math Sci, Chennai 600113, India.	Class Quantum Gravity 19 (9): 2355-2367 May 7 2002
Search For W' Bosons Decaying to An Electron and A Neutrino With the D0 Detector	Beri SB, Bhatnagar V, Kaur R, Kohli JM, Choudhary B, Ranjan K, Shivpuri RK, Acharya BS, Banerjee S, Banerjee P, Dugad SR, Mondal NK	12	556	2.16	Panjab Univ, Chandigarh 160014, India. Univ Delhi, Delhi 110007, India. Tata Inst Fundamental Res, Bombay 400005, Maharashtra, India.	Phys Rev Lett 1 (3): Art. No.-031804 Jan 25 2008
Atomistic Calculations of Elastic Properties of Metallic Fcc Crystal Surfaces	Shenoy VB	1	1	100	Indian Inst Sci, Mat Res Ctr, Bangalore 560012, Karnataka, India.	Phys Rev B 71 (9): Art. No.-094104 Mar 2005
Centrality Dependence of Direct Photon Production in Root S-Nn=200 Gevau+Aucollisions	R K Choudhury, P Chand, D Dutta, SS Kapoor, AK Mohanty, ID Ojha, CP Singh, V Singh, SK Tuli	9	330	2.73	Banaras Hindu Univ, Dept Phys, Varanasi 221005, Uttar Pradesh, India. Bhabha Atom Res Ctr, Bombay 400085, Maharashtra, India.	Phys Rev Lett 94 (23): Art. No.-232301 Jun 17 2005

Title	Authors				Affiliation	Citation
Field Emission From Open Ended Aluminum Nitride Nanotubes	Tondare VN, Balasubramanian C, Shende SV, Joag DS, Godbole VP, Bhoraskar SV, Bhadbhade M	7	7	100	Univ Poona, Dept Phys, Ctr Adv Studies Mat Sci & Solid State Phys, Pune 411007, Maharashtra, India.Natl Chem Lab, Pune 411008, Maharashtra, India.	Appl Phys Lett 80 (25): 4813-4815 Jun 24 2002
Cross Sections and Transverse Single-Spin Asymmetries in Forward Neutral-Pion Production From Proton Collisions At Root S=200 Gev	Dubey AK, Mahapatra DP, Mishra D, Phatak SC, Sahoo R, Varma R, Potukuchi BVKS, Nayak SK, Mangotra LK, Mahajan S, Gupta N, Gupta A, Dogra SM, Bhasin A, Badyal SK, Aggarwal MM, Bhati AK, Kumar A, Sharma M, Sood G, Raniwala R, Raniwala S, Bhardwaj S, Ahammed Z, Bhaskar P, Chattopadhyay S, D Das, Das S, Dutta Majumdar MR, Ganti MS, Ghosh P, Mohanty B, Nandi BK, Nayak TK, Pal SK, Singaraju RN, Trivedi MD, Viyogi YP	38	367	10.35	Inst Phys, Bhubaneswar 751005, Orissa, India. Univ Jammu, Jammu 180001, India.Panjab Univ, Chandigarh 160014, India. Univ Rajasthan, Jaipur 302004, Rajasthan, India. Bhabha Atom Res Ctr, Ctr Variable Energy Cyclotron, Calcutta 700064, India.	Phys Rev Lett 92 (17): Art. No.-171801 Apr 30 2004
Model For Heat Conduction in Nanofluids	Kumar DH, Patel HE, Kumar VRR, Sundararajan T, Pradeep T, Das SK	6	6	100	Indian Inst Technol, Madras 600036, Tamil Nadu, India.	Phys Rev Lett 93 (14): Art. No.-144301 Oct 1 2004
Synthesis and Size-Selective Catalysis by Supported Gold Nanoparticles: Study on Heterogeneous and Homogeneous Catalytic Process	Panigrahi S, Basu S, Praharaj S, Pande S, Jana S, Pal A, Ghosh SK, Pal T	8	8	100	Indian Inst Technol, Dept Chem, Kharagpur 721302, W Bengal, India	J Phys Chem C 111 (12): 4596-4605 Mar 29 2007

Title	Authors				Affiliation	Publication
Perfect Teleportation and Superdense Coding With W States	Agrawal P, Pati A	2	2	100	Inst Phys, Bhubaneswar 751005, Orissa, India	Phys Rev A 74 (6): Art. No.-062320 Dec 2006
Mode-Coupling Theory and the Glass Transition in Supercooled Liquids	Das SP	1	1	100	Jawaharlal Nehru Univ, Sch Phys Sci, New Delhi 110067, India	Rev Mod Phys 76 (3): 785-851 Part 1 Jul 2004
Superconductivity in Coo2 Layers and the Resonating Valence Bond Mean-Field Theory of the Triangular Lattice T-J Model	Kumar B, Shastry BS	2	2	100	Indian Inst Sci, Dept Phys, Bangalore 560012, Karnataka, India	Phys Rev B 68 (10): Art. No.-104508 Sep 1 2003
Magnetoelectricity At Room Temperature in the Bi0.9-Xtbxla0.1feo3 System	Palkar VR, Malik SK, Bhattacharya S, Darshan C Kundaliya	4	4	100	Indian Institute of Technology Bombay, Mumbai, India; Indian Institute of Technology Madras, Chennai, State of Tamil Nadu, India; Tata Inst Fundamental Res, Dept Condensed Matter Phys & Mat Sci, Homi Bhabha Rd, Bombay 400005, Maharashtra, India	Phys Rev B 69 (21): Art. No.-212102 Jun 2004
Gravitational Radiation From Inspiralling Compact Binaries Completed At the Third Post-Newtonian Order	Iyer BR	4	1	25	Raman Res Inst, Bangalore 560080, Karnataka, India.	Phys Rev Lett 93 (9): Art. No.-091101 Aug 27 2004
Theory of Neutrinos: A White Paper	Joshipura A, Mohanty S	30	2	6.67	Phys Res Lab, Ahmadabad 380009, Gujarat, India.	Rep Progr Phys 70 (11): 1757-1867 Nov 2007

Title	Authors				Affiliation	Reference
Identified Hadron Spectra At Large Transverse Momentum in P+P and D+Au Collisions At, Root(Nn)-N-S=200 Gev	Aggarwal MM, Ahammed Z, Badyal SK, Bhasin A, Bhati AK, Das D, Das S, Chattopadhyay S, Dutta Mazumdar MR, Ganti MS, Ghosh P, Gupta A, Gupta N, Kumar A, Mahajan S, Mahapat DP, Mangotra LK, Mishra DK, Mohanty B, Nandi BK, Nayak SK, Nayak TK, Netrakanti PK, Pal SK, Phatak SC, Potukuchi BVKS, Raniwala R, Raniwala S, Sahoo R, Sharma M, Singaraju RN, Viyogi YP	32	384	8.33	Bhabha Atom Res Ctr, Ctr Variable Energy Cyclotron, Calcutta 700064, W Bengal, India. Inst Phys, Bhubaneswar 751005, Orissa, India. Indian Inst Technol, Bombay 400076, Maharashtra, India. Univ Jammu, Jammu 180001, India. Panjab Univ, Chandigarh 160014, India. Univ Rajasthan, Jaipur 302004, Rajasthan, India.	Phys Lett B 637 (3): 161-169 Jun 8 2006
Exact Counting of Supersymmetric Black Hole Microstates	Dabholkar A	1	1	100	Tata Inst Fundamental Res, Dept Theoret Phys, Bombay 400005, Maharashtra, India.	Phys Rev Lett 94 (24): Art. No.-241301 Jun 24 2005
Hydrodynamics and Phases of Flocks	Ramaswamy S	3	1	33.33	Indian Inst Sci, Dept Phys, Ctr Condensed Matter Theory, Bangalore 560012, Karnataka, India.	Ann Phys N Y 318 (1): 170-244 Jul 2005
Nonsupersymmetric Attractors	Goldstein K, Iizuka N, Jena RP, Trivedi SP	4	4	100	Tata Inst Fundamental Res, Homi Bhabha Rd, Bombay 400005, Maharashtra, India	Phys Rev D 72 (12): Art. No.-124021 Dec 2005
Nuclear Modification of Electron Spectra and Implications For Heavy Quark Energy Loss in Au + Au Collisions At Root S(Nn)=200 Gev	Mishra M, Singh CP, Singh V, Tuli SK	4	423	0.95	Banaras Hindu Univ, Dept Phys, Varanasi 221005, Uttar Pradesh, India. Bhabha Atom Res Ctr, Bombay 400085, Maharashtra, India.	Phys Rev Lett 96 (3): Art. No.-032301 Jan 27 2006

Title	Authors				Affiliation	Source
Evolution of the Electronic Structure With Size in Ii-Vi Semiconductor Nanocrystals	Sapra S, Sarma DD	2	2	100	Indian Inst Sci, Solid State & Struct Chem Unit, Bangalore 560012, Karnataka, India. Jawaharlal Nehru Ctr Adv Sci Res, Bangalore, Karnataka, India. Iisc, Ctr Condensed Matter Theory, Bangalore, Karnataka, India	Phys Rev B 69 (12): Art. No.-125304 Mar 2004
Quantum Tunneling and Back Reaction	Banerjee R, Majhi BR	2	2	100	Sn Bose Natl Ctr Basic Sci, Jd Block, Sector 3, Calcutta 700098, India	Phys Lett B 662 (1): 62-65 Apr 10 2008
Open Charm Yields in D+Au Collisions At Root S(Nn)=200 Gev	Dubey AK, Mahapatra DP, Mishra DK, Phatak SC, Sahoo R, Varma R, Potukuchi BVKS, Nayak SK, Mangotra LK, Mahajan S, Gupta N, Gupta A, Dogra SM, Bhasin A, Badyal SK, Aggarwal MM, Sharma M, Kumar A, Bhatia VS, Bhati AK, Bharadwaj S, Raniwala R, Raniwala S, Viyogi YP	24	367	6.54	Inst Phys, Bhubaneswar 751005, Orissa, India. Indian Inst Technol, Bombay 400076, Maharashtra, India. Univ Jammu, Jammu 180001, India. Panjab Univ, Chandigarh 160014, India. Univ Rajasthan, Jaipur 302004, Rajasthan, India. Ctr Variable Energy Cyclotron, Calcutta 700064, W Bengal, India.	Phys Rev Lett 94 (6): Art. No.-062301 Feb 18 2005
Study of B--> D-**0 Pi(-)(D-**0 -> D(*)+Pi(-)) Decays	Singh JB, Soni N, Banerjee S, Sarangi TR, Satapathy M, Behera PK	6	181	3.31	Panjab Univ, Chandigarh 160014, India. Tata Inst Fundamental Res, Bombay 400005, Maharashtra, India. Utkal Univ, Bhubaneswar 751004, Orissa, India.	Phys Rev D 69 (11): Art. No.-112002 Jun 2004

Systematic Measurements of Identified Particle Spectra in Pp, D Plus Au, and Au Plus Au Collisions At the Star Detector	Dash S, Mahapatra DP, Phatak SC, Viyogi YP, Varma R, Nandi BK, Bhasin A, Dogra SM, Gupta A, Gupta N, Potukuchi BVKS, Mangotra LK, Aggarwal MM, Bhati AK, Kumar L, Pruthi NK, Bhardwaj S, Raniwala R, Raniwala S, Ahammed Z, Chattopadhyay S, Dutta Mazumdar MR, Ganti MS, Ghosh P, Mohanty B, Nayak TK, Pal SK, Singaraju RN	28	374	7.49	Univ Rajasthan, Jaipur 302004, Rajasthan, India. Panjab Univ, Chandigarh 160014, India. Inst Phys, Bhubaneswar 751005, Orissa, India. Indian Inst Technol, Bombay 400076, Maharashtra, India. Univ Jammu, Jammu 180001, India.	Phys Rev C 79 (3): Art. No.-034909 Mar 2009
Effect of Crystallographic Structure of Mno2 on Its Electrochemical Capacitance Properties	Devaraj S, Munichandraiah N	2	2	100	Indian Inst Sci, Dept Inorgan & Phys Chem, Bangalore 560012, Karnataka, India.	J Phys Chem C 112 (11): 4406-4417 Mar 20 2008
Evidence of the Purely Leptonic Decay B- -> Tau(-)(V)Over-Bar(Tau)	Kumar R, Gokhroo, T Aziz	3	184	1.63	Panjab Univ, Chandigarh 160014, India.	Phys Rev Lett 97 (25): Art. No.-251802 Dec 22 2006
Measurement of High-P(T) Single Electrons From Heavy-Flavor Decays in P+P Collisions At Root S=200 Gev	Adare A	1	1	100	Banaras Hindu Univ, Dept Phys, Varanasi 221005, Uttar Pradesh, India.	Phys Rev Lett 97 (25): Art. No.-252002 Dec 22 2006
What Is Needed of A Tachyon If It Is to Be the Dark Energy?	Sami M	4	1	25	Iucaa, Pune 411007, Maharashtra, India.	Phys Rev D 71 (4): Art. No.-043003 Feb 2005
M2-Branes on M-Folds	Mukhi S, Papageorgakis C	4	2	50	Tata Inst Fundamental Res, Bombay 400005, Maharashtra, India.	J High Energy Phys (5): Art. No.-38 May 2008

Title	Authors				Affiliation	Journal
M2 to D2	Mukhi S, Papageorgakis C	2	2	100	Tata Inst Fundamental Res, Bombay 400005, Maharashtra, India.	J High Energy Phys (5): Art. No.-085 May 2008
Direct Limits on the B-S(0) Oscillation Frequency	Choudhary B, Mondal NK, Ravi KJ, Ranjan K, Shivpuri RK	5	5	100	Panjab Univ, Chandigarh 160014, India. Univ Delhi, Delhi 110007, India. Tata Inst Fundamental Res, Bombay 400005, Maharashtra, India.	Phys Rev Lett 97 (2): Art. No.-021802 Jul 14 2006
Elliptic Flow and Incomplete Equilibration At Rhic	Bhalerao RS	4	1	25	Tata Inst Fundamental Res, Dept Theoret Phys, Homi Bhabha Rd, Bombay 400005, Maharashtra, India.	Phys Lett B 627: 49-54 Oct 27 2005
The Fate of (Phantom) Dark Energy Universe With String Curvature Corrections	Sami M	4	1	25	Iucaa, Post Bag 4, Pune 411007, Maharashtra, India.	Phys Lett B 619 (3-4): 193-200 Jul 21 2005
Persistent Supersolid Phase of Hard-Core Bosons on the Triangular Lattice	Heidarian D, Damle K	2	2	100	Tata Inst Fundamental Res, Dept Theoret Phys, Homi Bhabha Rd, Bombay 400005, Maharashtra, India	Phys Rev Lett 95 (12): Art. No.-127206 Sep 16 2005
Ferromagnetism in Fe-Doped Zno Nanocrystals: Experiment and Theory	Karmakar D, Mandal SK, Kadam RM, Paulose PL, Rajarajan AK, Nath TK, Das AK, Dasgupta I, Das GP	9	9	100	Bhabha Atom Res Ctr, Tech Phys & Prototype Engn Div, Bombay 400085, Maharashtra, India. Indian Inst Technol, Dept Phys & Meteorol, Kharagpur 721302, W Bengal, India.	Phys Rev B 75 (14): Art. No.-144404 Apr 2007
Observation of A Near-Threshold Omega J/Psi Mass Enhancement in Exclusive B -> K Omega J/Psi Decays	Soni N, Gokhroo G	2	150	1.33	Panjab Univ, Chandigarh 160014, India. Tata Inst Fundamental Res, Bombay 400005, Maharashtra, India.	Phys Rev Lett 94 (18): Art. No.-182002 May 13 2005

Title	Authors				Affiliation	Source
Minijet Deformation and Charge-Independent Angular Correlations on Momentum Subspace (Eta, Phi) in Au-Au Collisions At Root S-Nn=130 Gev	Ahammed Z, Chattopadhyay S, Das D, Das S, Dutta Mazumdar MR, Mohanty B, Singaraju, Gupta A, Sahoo R, Varma R, Aggarwal MM, Bharadwaj S	12	130	9.23	Inst Phys, Bhubaneswar 751005, Orissa, India. Indian Inst Technol, Bombay 400076, Maharashtra, India. Univ Jammu, Jammu 180001, India. Panjab Univ, Chandigarh 160014, India. Univ Rajasthan, Jaipur 302004, Rajasthan, India. Bhabha Atom Res Ctr, Ctr Variable Energy Cyclotron, Calcutta 700064, W Bengal, India.	Phys Rev C 73 (6): Art. No.-064907 Jun 2006
The Muon G-2	Andreas Nyffeler	1	3	33.33	Harish Chandra Res Inst, Reg Ctr Accelerator Based Particle Phys, Allahabad 211019, Uttar Pradesh, India.	Phys Rep-Rev Sect Phys Lett 477 (1-3): 1-110 Jun 2009
Hybrid Inorganic-Organic Materials: A New Family in Condensed Matter Physics	Rao CNR, Thirumurugan A	3	2	66.67	Jawaharlal Nehru Ctr Adv Sci Res, Chem & Phys Mat Unit, Bangalore 560064, Karnataka, India.	J Phys-Condens Matter 20 (8): Art. No.-083202 Feb 27 2008
Measurement of the Inclusive Jet Cross Section in P(P)Over-Bar Collisions At Root S=1.96 Tev	Beri SB, Bhatnagar V, Kaur R, Kohli JM, Shivpuri RK, Ranjan K, Choudhary B, Acharya BS, Banerjee P, Banerjee S, Dugad SR, Mondal NK	12	539	2.23	Panjab Univ, Chandigarh 160014, India. Univ Delhi, Delhi 110007, India. Tata Inst Fundamental Res, Bombay 400005, Maharashtra, India.	Phys Rev Lett 101 (6): Art. No.-062001 Aug 8 2008

Title	Authors				Affiliation	Source
Direct Observation of Dijets in Central Au Plus Au Collisions At Root S(Nn)=200 Gev	Sahoo R, Phatak SC, Mishra DK, Mahapatra DP, Varma R, Nandi BK, Potukuchi BVKS, Mangotra LK, Gupta N, Dogra SM, Bhasin A, Sharma M, Kumar A, Bhati AK, Aggarwal MM, Bhardwaj S, Raniwala R, Raniwala S, Viyogi YP, Singaraju RN, Pal SK, Netrakanti PK, Nayak TK, Mohanty B, Ghosh P, Ganti MS, Dutta Mazumdar MR, Das D, Das S, Chattopadhyay S, Ahammed Z	31	232	13.36	Inst Phys, Bhubaneswar 751005, Orissa, India. Indian Inst Technol, Bombay 400076, Maharashtra, India. Univ Jammu, Jammu 180001, India. Panjab Univ, Chandigarh 160014, India. Univ Rajasthan, Jaipur 302004, Rajasthan, India. Bhabha Atom Res Ctr, Ctr Variable Energy Cyclotron, Calcutta 700064, W Bengal, India.	Phys Rev Lett 97 (16): Art. No.-162301 Oct 20 2006
Glass Transition in Biomolecules and the Liquid-Liquid Critical Point of Water	Sastry S	1	8	12.5	Javaharlal Nehru Ctr Adv Sci Res, Bangalore 560061, Karnataka, India.	Phys Rev Lett 97 (17): Art. No.-177802 Oct 27 2006
Intrinsic Half-Metallicity in Modified Graphene Nanoribbons	Dutta S, Manna AK, Pati SK	3	3	100	Jawaharlal Nehru Ctr Adv Sci Res, Theoret Sci Unit, Jakkur Campus, Bangalore 560064, Karnataka, India.	Phys Rev Lett 102 (9): Art. No.-096601 Mar 6 2009
A New Measurement of the Collins and Sivers Asymmetries on A Transversely Polarised Deuteron Target	Dasgupta SS, Das S, Dhara L, Sinha L	4	242	1.65	Univ Burdwan, Burdwan 713104, W Bengal, India. Matrivani Inst Expt Res & Educ, Calcutta 700030, W Bengal, India.	Nucl Phys B 765 (1-2): 31-70 Mar 12 2007
Nanocrystalline Metal Oxides Dispersed Multiwalled Carbon Nanotubes As Supercapacitor Electrodes	Reddy ALM, Ramaprabhu S	2	2	100	Indian Inst Technol, Dept Phys, Alternat Energy Technol Lab, Madras 600036, Tamil Nadu	J Phys Chem C 111 (21): 7727-7734 May 31 2007

Title	Authors				Affiliation	Citation
A Benchmark Study on the Thermal Conductivity of Nanofluids	Sarit K Das, Hrishikesh E Patel, Pawan K Singh, Thirumalachari Sundararajan,	4	47	8.51	Indian Inst Technol, Dept Mech Engn, Madras 600036, Tamil Nadu, India.	J Appl Phys 106 (9): Art. No.-094312 Nov 1 2009
Susceptibilities and Speed of Sound From the Polyakov-Nambu-Jona-Lasinio Model	Ghosh SK, Mukherjee TK, Mustafa MG, Ray R	4	4	100	Bose Inst, Dept Phys, Calcutta 700009, W Bengal, India. Saha Inst Nucl Phys, Div Theory, Calcutta 700064, W Bengal, India	Phys Rev D 73 (11): Art. No.-114007 2006
Evidence For D-0-(D)Over-Bar(0) Mixing	Kumar R, Aziz T	2	185	1.08	Panjab Univ, Chandigarh 160014, India.Tata Inst Fundamental Res, Bombay 400005, Maharashtra, India.	Phys Rev Lett 98 (21): Art. No.-211803 May 25 2007
Thermodynamical Aspects of Gravity: New Insights	Padmanabhan T	1	1	100	Iucaa, Pune Univ Campus, Pune 411007, Maharashtra, India.Iucaa, Pune 411007, Maharashtra, India.	Rep Progr Phys 73 (4): Art. No.-046901 Apr 2010
J/Psi Production Versus Transverse Momentum and Rapidity in P+P Collisions At Root S=200 Gev	Mishra M, Singh CP, Singh V	3	374	0.8	Banaras Hindu Univ, Dept Phys, Varanasi 221005, Uttar Pradesh, India.	Phys Rev Lett 98 (23): Art. No.-232002 Jun 8 2007
Thermodynamic Route to Field Equations in Lanczos-Lovelock Gravity	Aseem Paranjape, Sudipta Sarkar, Padmanabhan T	3	3	100	Tata Inst Fundamental Res, Homi Bhabha Rd, Bombay 400005, Maharashtra, India.; Iucaa, Pune 411007, Maharashtra, India.	Phys Rev D 74 (10): Art. No.-104015 Nov 2006

Title	Authors				Affiliation	Reference
Search For Supersymmetry in Pp Collisions At 7 Tev in Events With Jets and Missing Transverse Energy	Bansal S, Beri SB, Bhatnagar V, Dhingra N, Gupta R, Jindal M, Kaur M, Kohli JM, Mehta MZ, Nishu N, Saini LK, Sharma A, Singh AP, Singh JB, Singh SP, Ahuja S, Bhattacharya S, Choudhary BC, Gupta P, Jain S, Jain S, Kumar A, Shivpuri RK, Choudhury RK, Dutta D, Kailas S, Kataria SK, Mohanty AK, Pant LM, Shukla P, Aziz T, Guchait M, Gurtu A, Maity M, Majumder D, Majumdar G, Mazumdar K, Mohanty GB, Saha A, Sudhakar K, Wickramage N, Banerjee S, Dugad S, Mondal NK	44	627	7.02	Panjab Univ, Chandigarh 160014, India.; Univ Delhi, Delhi 110007, India.; Bhabha Atom Res Ctr, Bombay 400085, Maharashtra, India.; Tata Inst Fundamental Res, Ehep, Bombay 400005, Maharashtra, India.;	Phys Lett B 698 (3): 196-218 Apr 11 2011
An Index For 4 Dimensional Super Conformal Theories	Shiraz Minwalla	1	4	25	Tata Inst Fundamental Res, Bombay 400005, Maharashtra, India.	Commun Math Phys 275 (1): 209-254 Oct 2007
Upper Limits on Gravitational Wave Emission From 78 Radio Pulsars	Dhurandhar S, Mitra S	2	457	0.44	Interuniv Ctr Astron & Astrophys, Pune 411007, Maharashtra, India.	Phys Rev D 76 (4): Art. No.-042001 Aug 2007

Title	Authors				Affiliation	Citation
Suppression of Charged Particle Production At Large Transverse Momentum in Central Pb-Pb Collisions At Root S(Nn)=2.76 Tev	Viyogi YP, Varma R, Singaraju R, Singh R, Sinha BC, Sinha T, Sharma N, Sharma S, Samanta T, Sambyal S, Sahu PK, Raniwala R, Raniwala S, Potukuchi B, Pal SK, Nyatha A, Nayak TK, Nandi BK, Muhuri S, Mohanty B, Mangotra L, Khan MM, Irfan M, Gupta A, Gupta R, Ghosh P, Dutta Majumdar AK, Dutta Majumdar MR, Dubey AK, Das D, Das I, Dash A, S De, Chattopadhyay S, Chattopadhyay S, Bhasin A, Bhati AK, Azmi MD, Ahammed Z, Ahmad N, Ahmad Masoodi A, Aggarwal MM	42	917	4.58	Panjab Univ, Dept Phys, Chandigarh 160014, India., Ctr Variable Energy Cyclotron, Calcutta, India., Aligarh Muslim Univ, Dept Phys, Aligarh 202002, Uttar Pradesh, India.; Univ Jammu, Dept Phys, Jammu 180004, India. Saha Inst Nucl Phys, Calcutta, India. Indian Inst Technol, Bombay 400076, Maharashtra, India.; Univ Rajasthan, Dept Phys, Jaipur 302004, Rajasthan, India.	Phys Lett B 696 (1-2): 30-39 Jan 24 2011
Simple Method of Preparing Graphene Flakes by An Arc-Discharge Method	Subrahmanyam KS, Panchakarla LS, Govindaraj A, Rao CNR	4	4	100	Jawaharlal Nehru Ctr Adv Sci Res, Chem & Phys Mat Unit, Jakkur Po, Bangalore 560064, Karnataka, India; Indian Inst Sci, Solid State & Struct Chem Unit, Bangalore 560012, Karnataka, India.	J Phys Chem C 113 (11): 4257-4259 Mar 19 2009
Dark Energy and Gravity	Padmanabhan T	1	1	100	Iucaa, Post Bag 4, Ganeshkhind, Pune 411007, Maharashtra, India.	Gen Relativ Gravit 40 (2-3): 529-564 Feb 2008

Title	Authors				Affiliation	Publication
Centrality Dependence of Charged Hadron and Strange Hadron Elliptic Flow From Root S(Nn)=200 Gevau+Au Collisions	Dash S, Mahapatra DP, Phatak SC, Viyogi YP, Varma R, Nandi BK, Bhasin A, Dogra SM, Gupta A, Gupta N, Potukuchi BVKS, Mangotra LK, Aggarwal MM, Bhati AK, Kumar A, Kumar L, Pruthi NK, Raniwala R, Raniwala S, Bhardwaj S	20	368	5.43	Inst Phys, Bhubaneswar 751005, Orissa, India.; Indian Inst Technol, Mumbai 400076, Maharashtra, India; Univ Jammu, Jammu 180001, India. Panjab Univ, Chandigarh 160014,; India.; Univ Rajasthan, Jaipur 302004, Rajasthan, India.; Ctr Variable Energy Cyclotron, Calcutta 700064, India.	Phys Rev C 77 (5): Art. No.-054901 May 2008
Suppression Pattern of Neutral Pions At High Transverse Momentum in Au Plus Au Collisions At Root S-Nn=200 Gev and Constraints on Medium Transport Coefficients	Mishra M, Singh CP, Singh V, Tuli SK	4	422	0.95	Banaras Hindu Univ, Dept Phys, Varanasi 221005, Uttar Pradesh, India.	Phys Rev Lett 101 (23): Art. No.-232301 Dec 5 2008
Measurement of the E(+)E(-) -> Pi(+)Pi(-)J/Psi Cross Section Via Initial-State Radiation At Belle	Kumar R, Singh JB	2	142	1.41		
Heat Transport in Low-Dimensional Systems	Dhar A	1	1	100	Raman Res Inst, Bangalore 560080, Karnataka, India.	Advan Phys 57 (5): 457-537 2008
Measurement of B-S(0) Mixing Parameters From the Flavor-Tagged Decay B-S(0)-> J/Psi Phi	Beri SB, Beri SB, Kaur R, Kohli JM, Ranjan K, Choudhary B, Shivpuri RK, Banerjee S, Acharya BS, Banerjee P, Dugad SR, Dugad SR, Mondal NK	13	537	2.42	Panjab Univ, Chandigarh 160014, India., Univ Delhi, Delhi 110007, India.; Tata Inst Fundamental Res, Bombay 400005, Maharashtra, India.	Phys Rev Lett 101 (24): Art. No.-241801 Dec 12 2008
High-Temperature Superconductivity in Eu0.5k0.5fe2as2	Hossain Z	1	6	16.67	Indian Inst Technol, Dept Phys, Kanpur 208016, Uttar Pradesh, India.	Phys Rev B 78 (9): Art. No.-092406 Sep 2008

Title	Authors				Affiliation	Citation
Long-Lived Giant Number Fluctuations in A Swarming Granular Nematic	Narayan V, Ramaswamy S,	2	3	66.67	Jawaharlal Nehru Ctr Adv Sci Res, Condensed Matter Theory Unit, Bangalore 560064, Karnataka, India.	Science 317 (5834): 105-108 Jul 6 2007
Azimuthal Charged-Particle Correlations and Possible Local Strong Parity Violation	Phatak SC, Mahapatra DP, Jena C, Dash S, Varma R, Pujahari PR, Nandi BK, Bhasin A, Dogra SM, Gupta A, Gupta N, Mangotra LK, Potukuchi BVKS, Pruthi NK, Kumar L, Bhati AK, Aggarwal MM, Raniwala R, Raniwala S, Viyogi YP, Singaraju RN, Pal SK, Nayak TK, Mohanty B, Ghosh P, Ganti MS, Dutta Mazumdar MR, Chattopadhyay S, Ahammed Z	29	376	7.71	Indian Inst Technol, Bombay 400076, Maharashtra, India. Inst Phys, Bhubaneswar 751005, Orissa, India. Univ Jammu, Jammu 180001, India.Panjab Univ, Chandigarh 160014, India. Univ Rajasthan, Jaipur 302004, Rajasthan, India.Ctr Variable Energy Cyclotron, Calcutta 700064, W Bengal, India.	Phys Rev Lett 103 (25): Art. No.-251601 Dec 18 2009
In Situ Synthesis of Metal Nanoparticles and Selective Naked-Eye Detection of Lead Ions From Aqueous Media	Yoosaf K, Ipe BI, Suresh CH, Thomas KG	4	4	100	Csir, Photosci & Photon Grp, Trivandrum 695019, Kerala, India. Csir, Reg Res Lab, Computat Modeling & Simulat Sect, Trivandrum 695019, Kerala, India.	J Phys Chem C 111 (34): 12839-12847 Aug 30 2007

Enhancement of Ferromagnetism Upon Thermal Annealing in Pure Zno	Banerjee S, Mandal M, Gayathri N, Sardar M	4	4	100	Bhabha Atom Res Ctr, Ctr Variable Energy Cyclotron, Mat Sci Sect, 1-Af Bidhannagar, Calcutta 700064, W Bengal, India. Saha Inst Nucl Phys, Div Chem Sci, Calcutta 700064, W Bengal, India.Indira Gandhi Ctr Atom Res, Div Mat Sci, Kalpakkam 603102, Tamil Nadu, India.	Appl Phys Lett 91 (18): Art. No.-182501 Oct 29 2007
An Upper Limit on the Stochastic Gravitational-Wave Background of Cosmological Origin	H Mukhopadhyay, S Dhurandhar	2	661	0.3	Inter Univ Ctr Astron & Astrophys, Pune 411007, Maharashtra, India.	Nature 460 (7258): 990-994 Aug 20 2009
Rotating Attractors	Astefanesei D, Goldstein K, Jena RP, Sen A, Trivedi SP	5	5	100	Harish Chandra Res Inst, Allahabad 211019, Uttar Pradesh, India. Tata Inst Fundamental Res, Bombay 400005, Maharashtra, India.	J High Energy Phys (10): Art. No.-058 Oct 2006
Susy Les Houches Accord 2	Godbole R, Choudhury D, Guchait M	3	34	8.52	Tata Inst Fundamental Res, Bombay 400005, Maharashtra, India. Indian Inst Sci, Ctr Theoret Studies, Bangalore 560012, Karnataka, India. Harish Chandra Res Inst, Allahabad 211019, Uttar Pradesh, India.	Comput Phys Commun 180 (1): 8-25 Jan 1 2009

Title	Authors				Affiliation	Citation
Charged-Particle Multiplicity Measurement in Proton-Proton Collisions At Root S=0.9 and 2.36 Tev With Alice At Lhc	Bhasin A, Raniwala S, Verma S, Sinha BC, Singh R, Gupta R, Lal C, Mangotra M, Potukuchi B, Sambyal S, Sharma S, Tripady P, Sharma N	13	1064	1.22	Saha Inst Nucl Phys, Calcutta, India. Ctr Variable Energy Cyclotron, Calcutta, India. Univ Rajasthan, Dept Phys, Jaipur 302004, Rajasthan, India. Indian Inst Technol, Mumbai 400076, Maharashtra, India. Univ Jammu, Dept Phys, Jammu 180004, India.	Eur Phys J C 68 (1-2): 89-108 Jul 2010
N-Doped Tio2 Nanoparticle Based Visible Light Photocatalyst by Modified Peroxide Sol-Gel Method	Jagadale TC, Takale SP, Sonawane RS, Joshi HM, Patil SI, Kale BB, Ogale SB	6	6	100	Natl Chem Lab, Pune 411008, Maharashtra, India. Univ Poona, Dept Phys, Pune 411008, Maharashtra, India. Ctr Mat Elect Technol, Pune 411008, Maharashtra, India.	J Phys Chem C 112 (37): 14595-14602 Sep 18 2008
Evidence For An Anomalous Like-Sign Dimuon Charge Asymmetry	Beri SB, Bhatnagar V, Dutt S, Kholi JM, Choudhary B, Dubey A, Naimuddin M, Rnayyar, Ranjan K, Shivpuri RK, Acharya BS, Banerjee S, Mandal NK	13	368	3.53	Panjab Univ, Chandigarh 160014, India. Univ Delhi, Delhi 110007, India. Tata Inst Fundamental Res, Bombay 400005, Maharashtra, India.	Phys Rev D 82 (3): Art. No.-032001 Aug 16 2010
Corrected Entropy of Btz Black Hole in Tunneling Approach	Modak SK	1	1	100	Sn Bose Natl Ctr Basic Sci, Calcutta 700098, India.	Phys Lett B 671 (1): 167-173 Jan 12 2009
Electrical Resistivity and Specific Heat of Single-Crystalline Eufe2as2: A Magnetic Homologue of Srfe2as2	Hossain Z	1	6	16.67	Indian Inst Technol, Dept Phys, Kanpur 208016, Uttar Pradesh, India.	Phys Rev B 78 (5): Art. No.-052502 Aug 2008
Two New Diagnostics of Dark Energy	Sahni V, Shafieloo A	2	3	66.67	Interuniv Ctr Astron & Astrophys, Pune 411007, Maharashtra, India.	Phys Rev D 78 (10): Art. No.-103502 Nov 2008

Title	Authors				Affiliation	Publication
The Fourth Family: A Simple Explanation For the Observed Pattern of Anomalies in B-Cp Asymmetries	Alok AK, Giri A, Mohanta R, Nandi S	4	5	80	Tata Inst Fundamental Res, Bombay 400005, Maharashtra, India. Panjab Univ, Dept Phys, Patiala 147002, Punjab, India. Iit Hyderabad, Dept Phys, Hyderabad 502205, Andhra Pradesh, India. Univ Hyderabad, Sch Phys, Hyderabad 500046, Andhra Pradesh, India. Harish Chandra Res Inst, Allahabad 211019, Uttar Pradesh, India.	Phys Lett B 683 (4-5): 302-305 Jan 25 2010
Search For Cp Violation in B-S(0) -> Mu(+) D-S(-) X Decays in P(P)Over-Bar Collisions At Root S=1.96 Tev	Beri SB, Bhatnagar V, Dutt S, Kohli JM, Shivpuri RK, Ranjan K, Dubey A, Choudhary B, Acharya BS, Banerjee S, Mondal NK	11	502	2.19	Panjab Univ, Chandigarh 160014, India. Univ Delhi, Delhi 110007, India. Tata Inst Fundamental Res, Mumbai 400005, Maharashtra, India.	Phys Rev D 82 (1): Art. No.-012003 Jul 26 2010
Voros Product, Noncommutative Schwarzschild Black Hole and Corrected Area Law	Banerjee R, Gangopadhyay S, Modak SK	3	3	100	W Bengal State Univ, Barasat, W Bengal, India. Sn Bose Natl Ctr Basic Sci, Calcutta 700098, India.	Phys Lett B 686 (2-3): 181-187 Mar 22 2010
Helioseismology and Solar Abundances	Antia HM	1	2	50	Tata Inst Fundamental Res, Bombay 400005, Maharashtra, India.	Phys Rep-Rev Sect Phys Lett 457 (5-6): 217-283 Mar 2008
Black Hole Entropy Function, Attractors and Precision Counting of Microstates	Sen A	1	1	100	Harish Chandra Res Inst, Chhatnag Rd, Allahabad 211019, Uttar Pradesh, India.	Gen Relativ Gravit 40 (11): 2249-2431 Nov 2008

Title	Authors				Affiliation	Reference
Flavor Physics of Leptons and Dipole Moments	DK Ghosh, SK Vempati	2	91	2.2	Phys Res Lab, Div Theoret Phys, Ahmadabad 380009, Gujarat, India.Indian Inst Sci, Ctr High Energy Phys, Bangalore 560012, Karnataka, India.	Eur Phys J C 57 (1-2): 13-182 Sep 2008
Predictions For the Rates of Compact Binary Coalescences Observable by Ground-Based Gravitational-Wave Detectors	S Dhurandhar	1	710	0.14	Inter Univ, Ctr Astron & Astrophys, Pune 411007, Maharashtra, India.	Class Quantum Gravity 27 (17): Art. No.-173001 Sep 7 2010
Quantum Tunneling Beyond Semiclassical Approximation	Banerjee R, Majhi BR	2	2	100	Sn Bose Natl Ctr Basic Sci, Jd Block, Sector 3, Calcutta 700098, India.	J High Energy Phys (6): Art. No.-095 Jun 2008
The Deuteron Spin-Dependent Structure Function G(1)(D) and Its First Moment	Das S, Dasgupta SS, Dhara L, Sinha L	4	230	1.74	Matrivani Inst Expt Res & Educ, Calcutta 700030, W Bengal, India.Univ Burdwan, Burdwan 713104, W Bengal, India.	Phys Lett B 647 (1): 8-17 Mar 29 2007
Role of Surface Energy Coefficients and Nuclear Surface Diffuseness in the Fusion of Heavy-Ions	Dutt I, Puri RK	2	2	100	Panjab Univ, Dept Phys, Chandigarh 160014, India.	Phys Rev C 81 (4): Art. No.-047601 Apr 2010
Intrinsic Rotation and Electric Field Shear	Singh R	1	4	25	Inst Plasma Res, Bhat 382428, Gandhinagar, India.	Phys Plasmas 14 (4): Art. No.-042306 Apr 2007

Elliptic Flow of Charged Particles in Pb-Pb Collisions At Root S(Nn)=2.76 Tev	Aggarwal MM, Bhati AK, Sharma N, Viyogi YP, Sinha BC, Singhal V, Singaraju R, Samanta T, Saini J, Pal SK, Nayak TK, Muhuri S, Mohanty B, Ghosh P, Ganti MS, Dutta Majumdar MR, Dubey AK, De S, Chattopadhyay S, Khan MM, Khan SA, Irfan M, Azmi MD, Ahmad N, Ahammed Z, Ahmad Masoodi A, Sharma S, Singh R, Sambyal S, Potukuchi B, Mangotra L, Gupta A, Gupta R, Prasad SK, Sharma S, Sambyal S, Bose S, Das D, Das I, Das K, Dutta Majumdar AK, Varma R, Roy P, Sinha T, Sahu PK, Mahapatra DP, Roy P, Bhasin A, Chattopadhyay S, Dash A, Jena S, Koyithatta G Meethaleveedu, Nandi BK, Nyatha A, Raniwala R, Raniwala S	56	967	5.79	Univ Rajasthan, Dept Phys, Jaipur 302004, Rajasthan, India. Indian Inst Technol, Mumbai 400076, Maharashtra, India. Inst Phys, Bhubaneswar 751007, Orissa, India. Saha Inst Nucl Phys, Calcutta, India. Univ Jammu, Dept Phys, Jammu 180004Aligarh Muslim Univ, Dept Phys, Aligarh 202002, Uttar Pradesh, India. Ctr Variable Energy Cyclotron, Calcutta, India. Panjab Univ, Dept Phys, Chandigarh 160014, India.	Phys Rev Lett 105 (25): Art. No.-252302 Dec 13 2010

Title	Authors			Affiliation	Reference
Transverse-Momentum and Pseudorapidity Distributions of Charged Hadrons in Pp Collisions At Root S=7 Tev	Bansal S, Beri SB, Bhatnagar V, Jindal M, Kaur M, Kohli JM, Mehta MZ, Nishu N, Saini LK, Sharma A, Sharma R, Singh AP, Singh JB, Singh SP, Ahuja S, Bhattacharya S, Chauhan S, Choudhary BC, Gupta P, Jain S, Jain S, Kumar A, Ranjan K, Shivpuri RK, Choudhury RK, Dutta D, Kailas S, Kataria SK, Mohanty AK, Pant LM, Shukla P, Suggisetti P, Aziz T, Guchait M, Gurtu A, Maity12 M, Majumder D, Majumder G, Mazumdar K, Banerjee S, Dugad S, Mondal NK, Mohanty GB, Saha A, Sudhakar K, Wickramage N	46	2071	2.22	Univ Delhi, Delhi 110007, India. Bhabha Atom Res Ctr, Mumbai 400085, Maharashtra, India. Tata Inst Fundamental Res Ehep, Mumbai, Maharashtra, India. Inst Studies Theoret Phys & Math Ipm, Tehran, Iran.	Phys Rev Lett 105 (2): Art. No.-022002 Jul 6 2010
Fermion Tunneling Beyond Semiclassical Approximation	Majhi BR	1	1	100	Sn Bose Natl Ctr Basic Sci, Calcutta 700098, India.	Phys Rev D 79 (4): Art. No.-044005 Feb 2009
Search For Squarks and Gluinos in Events With Jets and Missing Transverse Energy Using 2.1 Fb(-1) of P(P)Over-Bar Collision Data At Root S=1.96 Tev - D Phi Collaboration	Beri SB, Bhatnagar V, Kaur R, Kohli JM, Ranjan K, Shivpuri RK, Mondal NK, Dugad SR, Banerjee S, Banerjee P, Acharya BS	11	541	2.03	Panjab Univ, Chandigarh 160014, India. Univ Delhi, Delhi 110007, India. Tata Inst Fundamental Res, Bombay 400005, Maharashtra, India.	Phys Lett B 660 (5): 449-457 Mar 6 2008
Einstein@Home Search For Periodic Gravitational Waves in Early S5 Ligo Data	Djiramdjar S, Mukhopadhyay H	2	513	0.39	Inter Univ, Ctr Astron & Astrophys, Pune 411007, Maharashtra, India.	Phys Rev D 80 (4): Art. No.-042003 Aug 2009

Title	Authors				Affiliation	Citation
Fluorescent Carbon Nanoparticles: Synthesis, Characterization, and Bioimaging Application	Ray SC, Arindam Saha, Nikhil R Jana, Rupa Sarkar	4	4	100	Indian Assoc Cultivat Sci, Ctr Adv Mat, Calcutta 700032, India.	J Phys Chem C 113 (43): 18546-18551 Oct 29 2009
Cold Nuclear Matter Effects on J/Psi Production As Constrained by Deuteron-Gold Measurements At Root S-Nn=200 Gev	Mishra M, Ojha ID, Singh CP, Singh V	4	453	0.88	Banaras Hindu Univ, Dept Phys, Varanasi 221005, Uttar Pradesh, India.; Bhabha Atom Res Ctr, Bombay 400085, Maharashtra, India.	Phys Rev C 77 (2): Art. No.-024912 Feb 2008
Buckling Analysis of A Single-Walled Carbon Nanotube Embedded in An Elastic Medium Based on Nonlocal Elasticity and Timoshenko Beam Theory and Using Dqm	Murmu T, Pradhan SC	2	2	100	Indian Inst Technol, Dept Aerosp Engn, Kharagpur 721302, W Bengal, India.	Physica E 41 (7): 1232-1239 Jun 2009
Hawking Radiation and Covariant Anomalies	Banerjee R, Kulkarni S	2	2	100	Satyendra Nath Bose Natl Ctr Basic Sci, Block Jd Sect Iii Salt Lake, Calcutta 700098, W Bengal, India.	Phys Rev D 77 (2): Art. No.-024018 Jan 2008
Exact Results For Quench Dynamics and Defect Production in A Two-Dimensional Model	Sengupta K, Sen D, Mondal S	3	3	100	Saha Inst Nucl Phys, Tcmp Div, 1-Af Bidhannagar, Calcutta 700064, India.; Indian Inst Sci, High Energy Phys Res Ctr, Bangalore 560012, Karnataka, India.	Phys Rev Lett 1 (7): Art. No.-077204 Feb 22 2008
Ligand Exchange of Au(25)Sg(18) Leading to Functionalized Gold Clusters: Spectroscopy, Kinetics, and Luminescence	Shibu ES, Muhammed MAH, Pradeep T	3	4	75	Indian Inst Technol, Sophisticated Analyt Instrument Facil, Madras 600036, Tamil Nadu, India.	J Phys Chem C 112 (32): 12168-12176 Aug 14 2008

Quantum Tunneling and Trace Anomaly	Banerjee R, Majhi BR	2	2	100	Sn Bose Natl Ctr Basic Sci, Jd Block, Sector 3, Calcutta 700098, India.	Phys Lett B 674 (3): 218-222 Apr 20 2009
Long Range Rapidity Correlations and Jet Production in High Energy Nuclear Collisions	Dash S, Jena C, Mahapatra DP, Phatak SC, Viyogi YP, Varma R, Pujahari PR, Nandi BK, Bhasin A, Dogra SM, Gupta A, Gupta N, Mangotra LK, Potukuchi BVKS, Aggarwal MM, Bhati AK, Kumar L, Pruthi NK, Raniwala R, Raniwala S, Singaraju RN, Pal SK, Nayak TK, Mohanty B, Ghosh P, Ganti MS, Dutta Mazumdar MR, Chattopadhyay S, Ahammed Z	29	387	7.49	Inst Phys, Bhubaneswar 751005, Orissa, India.; Indian Inst Technol, Bombay 400076, Maharashtra, India.; Univ Jammu, Jammu 180001, India.; Panjab Univ, Chandigarh 160014, India.; Univ Rajasthan, Jaipur 302004, Rajasthan, India.; Ctr Variable Energy Cyclotron, Calcutta 700064, India.	Phys Rev C 80 (6): Art. No.-064912 Dec 2009

Title	Authors			Affiliation		Journal
Charged-Particle Multiplicity Measurement in Proton-Proton Collisions At Root S=7 Tev With Alice At Lhc	Ahmad A, Ahmad N, Azmi MD, Irfan M, Khan MM, Rath S, Mahapatra DP, Jena C, Dash A, Dash S, Aggarwal MM, Bhati AK, Kumar L, Kumar N, Sharma N, Raniwala R, Raniwala S, Singh R, Sharma S, Sambyal S, Potukuchi B, Mangotra L, Lal C, Gupta A, Gupta R, Bhasin A, Bose S, Chattopadhyay S, Das I, Das S, Dutta Majumdar AK, Pal S, Roy P, Sinha T, Viyogi YP, Tribedy P, Singhal V, Sinha BC, Singaraju R, Samanta T, Saini J, Prasad SK, Pal SK, Nayak TK, Muhuri S, Mondal MM, Mohanty B, Khan SA, Ghosh P, Ganti MS, Dutta Majumdar MR, Dubey AK, Chattopadhyay S, Ahammed Z, Jena S, Nandi BK, Nyatha A, Pujahari P, Varma R	59	1038	Aligarh Muslim Univ, Dept Phys, Aligarh, Uttar Pradesh, India.; Inst Phys, Bhubaneswar 751007, Orissa, India.; Panjab Univ, Dept Phys, Chandigarh 160014, India.; Univ Rajasthan, Dept Phys, Jaipur 302004, Rajasthan, India.; Univ Jammu, Dept Phys, Jammu 180004, India.; Saha Inst Nucl Phys, Calcutta, India.; Variable Energy Cyclotron Ctr, Calcutta, India.; Indian Inst Technol, Bombay 400076, Maharashtra, India.	5.68	Eur Phys J C 68 (3-4): 345-354 Aug 2010
Colloquium: Nonequilibrium Dynamics of Closed Interacting Quantum Systems	Sengupta K	4	1	Indian Assoc Cultivat Sci, Dept Theoret Phys, Calcutta 700032, India.	25	Rev Mod Phys 83 (3): 863-883 Aug 15 2011

Title	Authors				Affiliation	Reference
Standard Model With Four Generations: Selected Implications For Rare B and K Decays	Giri A, Mohanta R	2	5	40	Iit Hyderabad, Dept Phys, Hyderabad 502205, Andhra Pradesh, India. Univ Hyderabad, Sch Phys, Hyderabad 500046, Andhra Pradesh, India.	Phys Rev D 82 (3): Art. No.-033009 Aug 13 2010
Physics of Non-Diffusive Turbulent Transport of Momentum and the Origins of Spontaneous Rotation in Tokamaks	Singh R	1	10	10	Inst Plasma Res, Bhat 382428, Gandhinagar, India.	Nucl Fusion 49 (4): Art. No.-045002 Apr 2009
Physics At A Future Neutrino Factory and Super-Beam Facility	Bandyopadhyay A, Choubey S, Gandhi R, Goswami S, Umasankar S, Mondal N	6	96	6.25	Harish Chandra Res Inst, Chhatnag Rd, Allahabad 211019, Uttar Pradesh, India, Inst Math Sci, Chennai 600113, Tamil Nadu, India, Tata Inst Fundamental Res, Sch Nat Sci, Mumbai 400005, Maharashtra, India,	Rep Progr Phys 72 (10): Art. No.-106201 Oct 2009
Phonon Renormalization in Doped Bilayer Graphene	Das A, Chakraborty B, Sood AK	3	6	50	Indian Inst Sci, Dept Phys, Bangalore 560012, Karnataka, India.	Phys Rev B 79 (15): Art. No.-155417 Apr 2009
Enhanced Production of Direct Photons in Au Plus Au Collisions At Root S(Nn)=200 Gev and Implications For the Initial Temperature	Mishra M, Singh CP, Singh V, Tuli SK	4	421	0.95	Banaras Hindu Univ, Dept Phys, Varanasi 221005, Uttar Pradesh, India.	Phys Rev Lett 104 (13): Art. No.-132301 Apr 2 2010
Detailed Measurement of the E(+)E(-) Pair Continuum in P Plus P and Au Plus Au Collisions At Root S(Nn)=200 Gev and Implications For Direct Photon Production	Mishra M, Singh CP, Singh V, Tuli SK	4	423	0.95		

Title	Authors				Affiliation	Reference
Equipartition of Energy in the Horizon Degrees of Freedom and the Emergence of Gravity	Padmanabhan T	1	1	100	Iucaa, Pune Univ Campus, Pune 411007, Maharashtra, India	Mod Phys Lett A 25 (14): 1129-1136 May 10 2010
First Results From 2+1 Dynamical Quark Flavors on An Anisotropic Lattice: Light-Hadron Spectroscopy and Setting the Strange-Quark Mass	Nilmani Mathur	1	15	6.67	Tata Inst Fundamental Res, Dept Theoret Phys, Mumbai 400005, Maharashtra, India.	Phys Rev D 79 (3): Art. No.-034502 Feb 2009
Connecting Anomaly and Tunneling Methods For the Hawking Effect Through Chirality	Banerjee R, Majhi BR	2	2	100	Sn Bose Natl Ctr Basic Sci, Jd Block, Sector 3, Calcutta 700098, India.	Phys Rev D 79 (6): Art. No.-064024 Mar 2009
Electrochemical Reduction of Oriented Graphene Oxide Films: An in Situ Raman Spectroelectrochemical Study	Ramesha GK, Sampath S	2	2	100	Indian Inst Sci, Dept Inorgan & Phys Chem, Bangalore 560012, Karnataka, India.	J Phys Chem C 113 (19): 7985-7989 May 14 2009
Holography of Charged Dilaton Black Holes	Shiroman Prakash, Sandip P Trivedi	2	4	50	Tata Inst Fundamental Res, Bombay 400005, Maharashtra, India.	J High Energy Phys (8): Art. No.-078 Aug 2010
Antiferromagnetic Ordering and Structural Phase Transition in Ba2fe2as2 With Sn Incorporated From the Growth Flux	Mittal R	1	12	8.33	Bhabha Atom Res Ctr, Div Solid State Phys, Bombay 400085, Maharashtra, India.	Phys Rev B 79 (6): Art. No.-064504 Feb 2009

Title	Authors				Affiliation	Reference
First Proton-Proton Collisions At the Lhc As Observed With the Alice Detector: Measurement of the Charged-Particle Pseudorapidity Density At Root S=900 Gev	Khan MM, Kamal A, Irfan M, Dash A, Dash S, Jena C, Mahapatra DP, Rath S, Viyogi YP, Sharma N, Kumar L, Kumar N, Bhati AK, Aggarwal MM, Raniwala R, Raniwala S, Gupta A, Gupta R, Lal C, Mahajan A, Mangotra L, Potukuchi B, Sambyal S, Sharma S, Singh R, Sinha T, Roy P, Pal S, Dutta Majumdar AK, Das I, Das S, Chattopadhyay S, Chattopadhyay S, Bose S, Ahammed Z, Dubey AK, Dutta Majumdar MR, Ganti MS, Ghosh P, Khan SA, Mohanty B, Mondal MM, Muhuri S, Nayak TK, Pal SK, Prasad SK, Saini J, Samanta T, Singaraju R, Singhal V, Sinha BC, Tribedy P, Varma R, Pujahari P, Nyatha A, Nandi BK, Jena S	56	1055	5.31	Aligarh Muslim Univ, Dept Phys, Aligarh 202002, Uttar Pradesh, India., Inst Phys, Bhubaneswar 751007, Orissa, India.; Panjab Univ, Dept Phys, Chandigarh 160014, India.; Univ Rajasthan, Dept Phys, Jaipur 302004, Rajasthan, India.; Univ Jammu, Dept Phys, Jammu 180004, India.; Saha Inst Nucl Phys, Calcutta, India.; Ctr Variable Energy Cyclotron, Calcutta, India.; Indian Inst Technol, Bombay 400076, Maharashtra, India.	Eur Phys J C 65 (1-2): 111-125 Jan 2010
Effect of the Symmetry Energy on Nuclear Stopping and Its Relation to the Production of Light Charged Fragments	Kumar S, Kumar S, Puri RK	3	3	100	Thapar Univ, Sch Phys & Mat Sci, Patiala 147004, Punjab, India. Panjab Univ, Dept Phys, Chandigarh 160014, India.	Phys Rev C 81 (1): Art. No.-014601 Jan 2010

Title	Authors				Affiliation	Citation
Charged-Particle Multiplicity Density At Midrapidity in Central Pb-Pb Collisions At Root S(Nn)=2.76 Tev	Sharma N, Sharma S, Bhati AK, Aggarwal MM, Viyogi YP, Varma R, Singhal V, Sinha BC, Singaraju R, Singh R, Samanta T, Sambyal S, Saini J, Pal SK, Ahammed Z, Chattopadhyay S, De S, Dubey AK, Dutta Majumdar MR, Ganti MS, Ghosh P, Khan SA, Mohanty B, Muhuri S, Nayak TK, Ahmad Masoodi A, Ahmad N, Azmi MD, Irfan M, Khan MM, Potukuchi B, Mangotra L	34	964	3.53	Panjab Univ, Dept Phys, Chandigarh 160014, India. Ctr Variable Energy Cyclotron, Calcutta, India. Aligarh Muslim Univ, Dept Phys, Aligarh 202002, Uttar Pradesh, India. Univ Jammu, Dept Phys, Jammu 180004, India. Saha Inst Nucl Phys, Calcutta, India. Inst Phys, Bhubaneswar 751007, Orissa, India. Indian Inst Technol, Mumbai 400076, Maharashtra, India. Univ Rajasthan, Dept Phys, Jaipur 302004, Rajasthan, India.	Phys Rev Lett 105 (25): Art. No.-252301 Dec 13 2010
Combination of Tevatron Searches For the Standard Model Higgs Boson in the W+W- Decay Mode	Kohli JM, Dutt S, Bhatnagar V, SB Beri, B Choudhary, A Dubey, R Nayyar, M Naimuddin, K Ranjan, RK Shivpuri, NK Mondal, S Banerjee, BS Acharya, A Gupta, R Gupta, A Bhasin, S Bose, S Chattopadhyay, D Das, I Das, K Das, AK Dutta Majumdar, T Sinha, P K Sahu, D P Mahapatra, A Dash, S Jena, G Koyithatta Meethaleveedu, BK Nandi, A Nyatha, R Varma, R Raniwala, S Raniwala	13	1044	1.25	Panjab Univ, Chandigarh 160014, India. Univ Delhi, Delhi 110007, India. Tata Inst Fundamental Res, Mumbai 400005, Maharashtra, India.	Phys Rev Lett 104 (6): Art. No.-061802 Feb 12 2010

Title	Authors				Affiliation	Reference
Hawking Radiation and Black Hole Spectroscopy in Horava-Lifshitz Gravity	Majhi Br	1	1	100	Sn Bose Natl Ctr Basic Sci, Jd Block, Sector 3, Calcutta 700098, India.	Phys Lett B 686 (1): 49-54 Mar 15 2010
Observation of Charge-Dependent Azimuthal Correlations and Possible Local Strong Parity Violation in Heavy-Ion Collisions	Dash S, Jena C, Mahapatra DP, Phatak SC, Viyogi YP, Varma R, Pujahari PR, Nandi BK, Bhasin A, Dogra SM, Gupta A, Gupta N, Mangotra LK, Potukuchi BVKS, Aggarwal MM, Bhati AK, Kumar L, Raniwala R, Raniwala S	19	378	5.03	Inst Phys, Bhubaneswar 751005, Orissa, India.Indian Inst Technol, Bombay 400076, Maharashtra, India. Univ Jammu, Jammu 180001, India. Panjab Univ, Chandigarh 160014, India. Univ Rajasthan, Jaipur 302004, Rajasthan, India. Bhabha Atom Res Ctr, Ctr Variable Energy Cyclotron, Calcutta 700064, India.	Phys Rev C 81 (5): Art. No.-054908 May 2010
Indications of Conical Emission of Charged Hadrons At the Bnl Relativistic Heavy Ion Collider	Dash S, Mahapatra DP, Phatak SC, Viyogi YP, Potukuchi BVKS, Mangotra LK, Gupta A, Gupta N, Dogra SM, Bhasin A, Bhati AK, Kumar L, Pruthi NK, Bhardwaj S, Raniwala R, Raniwala S, Singaraju RN, Pal SK, Nayak TK, Mohanty B, Ghosh P, Ganti MS, Dutta Mazumdar MR, Chattopadhyay S, Ahammed Z	25	373	6.7	Inst Phys, Bhubaneswar 751005, Orissa, India.Indian Inst Technol, Mumbai 400076, Maharashtra, India. Univ Jammu, Jammu 180001, India. Panjab Univ, Chandigarh 160014, India. Univ Rajasthan, Jaipur 302004, Rajasthan, India. Ctr Variable Energy Cyclotron, Calcutta 700064, India.	Phys Rev Lett 102 (5): Art. No.-052302 Feb 6 2009
Is Cosmic Acceleration Slowing Down?	Varun Sahni	1	3	33.33	Inter Univ Ctr Astron & Astrophys, Pune 411007, Maharashtra, India.	Phys Rev D 80 (10): Art. No.-101301 Nov 2009

Title	Authors				Affiliations	Reference
Centrality Dependence of the Charged-Particle Multiplicity Density At Midrapidity in Pb-Pb Collisions At Root S(Nn)=2.76 Tev	Bhati AK, Sharma N, Viyogi YP, Sinha BC, Singaraju R, Samanta T, Pal SK, Muhuri S, Mohanty B, Ghosh P, Dutta Majumdar MR, Dubey AK, De S, Chattopadhyay S, Ahammed Z, Ahmad N, Ahmad Masoodi A, Azmi MD, Irfan M, Khan MM, Singh R, Sharma S, Sambyal S, Potukuch B, Mangotra L, Gupta A, Gupta R, Bhasin A, Bose S, Chattopadhyay S, Das D, Das I, Dutta Majumdar AK, Roy P, Sinha T, Sahu PK, Mahapatra DP, Dash A, Varma R, Nyatha A, Nandi BK, Koyithatta Meethaleveedu G, Jena S, Raniwala R, Raniwala S	45	917	4.91	Panjab Univ, Dept Phys, Chandigarh 160014, India.; Ctr Variable Energy Cyclotron, Calcutta, India.; Aligarh Muslim Univ, Dept Phys, Aligarh 202002, Uttar Pradesh, India.; Univ Jammu, Dept Phys, Jammu 180004, India.; Saha Inst Nucl Phys, Calcutta, India.; Inst Phys, Bhubaneswar 751007, Orissa, India.; Indian Inst Technol, Mumbai 400076, Maharashtra, India.; Univ Rajasthan, Dept Phys, Jaipur 302004, Rajasthan, India.	Phys Rev Lett 106 (3): Art. No.-032301 Jan 20 2011
Evidence For An Anomalous Like-Sign Dimuon Charge Asymmetry	Beri SB, Bhatnagar V, Dutt S, Kohli JM, Shivpuri RK, Ranjan K, Nayyar R, Naimuddin M, Dubey A, Choudhary B, Acharya BS, Banerjee S, Mondal NK	13	449	2.9	Panjab Univ, Chandigarh 160014, India. Univ Delhi, Delhi 110007, India. Tata Inst Fundamental Res, Mumbai 400005, Maharashtra, India.	Phys Rev Lett 105 (8): Art. No.-081801 Aug 16 2010
The Influence of Slip Conditions, Wall Properties and Heat Transfer on Mhd Peristaltic Transport	Srinivas S, Gayathri R, Kothandapani M	3	3	100	Vit Univ, Div Appl Math, Vellore 632014, Tamil Nadu, India.	Comput Phys Commun 180 (11): 2115-2122 Nov 2009

Transverse-Momentum and Pseudorapidity Distributions of Charged Hadrons in Pp Collisions At Root S=0.9 and 2.36 Te	Bansal S, Beri SB, Bhatnagar V, Jindal M, Kaur M, Kohli JM, Mehta MZ, Nishu N, Saini LK, Sharma A, Sharma R, Singh AP, Singh JB, Singh SP, Ahuja S, Bhattacharya S, Chauhan S, Choudhary BC, Gupta P, Jain S, Jain S, Kumar A, Ranjan K, Shivpuri RK, Choudhury RK, Dutta D, Kailas S, Kataria SK, Mohanty AK, Pant LM, Shukla P, Suggisetti P, Guchait M, Banerjee S, Dugad S, Mondal NK, Maity M	37	1967	1.88	Panjab Univ, Chandigarh 160014, India.Univ Delhi, Delhi 110007, India.Bhabha Atom Res Ctr, Bombay 400085, Maharashtra, India. Tata Inst Fundamental Res Ehep, Bombay, Maharashtra, India. Visva Bharati Univ, Santini Ketan, W Bengal, India.	J High Energy Phys (2): Art. No.-041 Feb 2010
Effects of Nematic Fluctuations on the Elastic Properties of Iron Arsenide Superconductors	Bhattacharya S	1	10	10	Tata Inst Fundamental Res, Bombay 400005, Maharashtra, India.	Phys Rev Lett 105 (15): Art. No.-157003 Oct 4 2010
Production and Decay of Element 114: High Cross Sections and the New Nucleus (277)Hs	Lahiri, Maiti M	2	45	4.44	Saha Inst Nucl Phys, Calcutta 700064, W Bengal, India.	Phys Rev Lett 104 (25): Art. No.-252701 Jun 21 2010

Title	Authors				Affiliations	Journal
Higher Harmonic Anisotropic Flow Measurements of Charged Particles in Pb-Pb Collisions At Root S(Nn)=2.76 Tev	Sharma N, Rathee D, Bhati AK, Aggarwal MM, Ahammed Z, Chattopadhyay S, De S, Dubey AK, Dutta Majumdar MR, Ghosh P, Mohanty B, Muhuri S, Nayak TK, Pal SK, Singaraju R, Singha S, Sinha BC, Viyogi YP, Khan MM, Irfan M, Azmi MD, Ahmad N, Ahmad Masoodi A, Baral RC, Dash A, Mahapatra DP, Sahu PK, Singh R, Bhasin A, Gupta R, Gupta A, Mangotra L, Potukuchi B, Sambyal S, Sharma S, Bose S, Chattopadhyay S, Das I, Das D, Dutta Majumdar AK, Khan P, Roy P, Sinha T, Varma R, Nyatha A, Jena S, Koyithatta Meethaleveedu G, Nandi BK, Raniwala R, Raniwala S	51	951	5.36	Panjab Univ, Dept Phys, Chandigarh 160014, India. Ctr Variable Energy Cyclotron, Calcutta, India. Aligarh Muslim Univ, Dept Phys, Aligarh 202002, Uttar Pradesh, India. Inst Phys, Bhubaneswar 751007, Orissa, India. Univ Jammu, Dept Phys, Jammu 180004, India. Saha Inst Nucl Phys, Calcutta, India. Indian Inst Technol, Bombay 400076, Maharashtra, India. Univ Rajasthan, Dept Phys, Jaipur 302004, Rajasthan, India.	Phys Rev Lett 107 (3): Art. No.-032301 Jul 11 2011
Combined Measurement and Qcd Analysis of the Inclusive E(+/-)P Scattering Cross Sections At Hera	Kaur M, Kaur P, Singh I	3	542	0.55	Panjab Univ, Dept Phys, Chandigarh 160014, India.	J High Energy Phys (1): Art. No.-109 Jan 2010

Title	Authors				Affiliations	Citation
Search For Supersymmetry At the Lhc in Events With Jets and Missing Transverse Energy	Beri SB, Bhatnagar V, Dhingra N, Gupta R, Jindal M, Kaur M, Kohli JM, Mehta MZ, Nishu N, Saini LK, Sharma A, Singh AP, Singh J, Singh SP, Ahuja S, Choudhary BC, Gupta P, Kumar A, Kumar A, Malhotra S, Naimuddin M, Ranjan K, Shivpuri RK, Banerjee S, Bhattacharya S, Dutta S, Gomber B, Jain S, Jain S, Khurana R, Sarkar S, Choudhury RK, Dutta D, Kailas S, Kumar V, Mehta P, Mohanty AK, Pant LM, Shukla P, Aziz T, Guchait M, Gurtu A, Maity M, Majumder D, Majumder G, Mathew T, Mazumdar K, Mohanty GB, Parida B, Saha A, Sudhakar K, Wickramage N, Banerjee S, Dugad S, Mondal NK	55	2255	2.44	Panjab Univ, Chandigarh 160014, India. Univ Delhi, Delhi 110007, India. Saha Inst Nucl Phys, Calcutta, India. Bhabha Atom Res Ctr, Mumbai 400085, Maharashtra, India. Tata Inst Fundamental Res, Ehep, Mumbai 400005, Maharashtra, India. Tata Inst Fundamental Res, Hecr, Mumbai 400005, Maharashtra, India. Visva Bharati Univ, Santini Ketan, W Bengal, India.	Phys Rev Lett 107 (22): Art. No.-221804 Nov 21 2011
Superconductivity in Ru-Substituted Polycrystalline Bafe2-Xruxas2	Shilpam Sharma, Bharathi A, Sharat Chandra, Raghavendra Reddy, Paulraj S, Satya AT, Sastry VS, Ajay Gupta, Sundar CS	9	9	100	Indira Gandhi Ctr Atom Res, Div Mat Phys, Mat Sci Grp, Kalpakkam 603102, Tamil Nadu, India. Ugc Dae Consortium Sci Res, Indore 452017, Madhya Pradesh, India. Periyar Univ, Dept Phys, Salem 636011, India	Phys Rev B 81 (17): Art. No.-174512 May 1 2010

Title	Authors				Affiliation	Publication
Enhanced B-S-(B)Over-Bar(S) Lifetime Difference and Anomalous Like-Sign Dimuon Charge Asymmetry From New Physics in B-S -> Tau(+)Tau(-)	Amol Dighe, Anirban Kundu	2	3	66.67	Tata Inst Fundamental Res, Bombay 400005, Maharashtra, India. Univ Calcutta, Dept Phys, Calcutta 700009, India.	Phys Rev D 82 (3): Art. No.-031502 Aug 18 2010
The Mechanics and Statistics of Active Matter	Paulraj S	1	1	100	Indian Inst Sci, Dept Phys, Ctr Condensed Matter Theory, Bangalore 560012, Karnataka, India. Jncasr, Cmtu, Bangalore 560064, Karnataka, India.	Annu Rev Condens Matter Phys 1: 323-345 2010
Holographic C-Theorems in Arbitrary Dimensions	Aninda Sinha	1	2	50	Indian Inst Sci, Ctr High Energy Phys, Bangalore 560012, Karnataka, India.	J High Energy Phys (1): Art. No.-125 Jan 2011
Measurement of W+W- Production and Search For the Higgs Boson in Pp Collisions At Root S=7 Tev	Bansal S, Beri SB, Bhatnagar V, Dhingra N, Gupta R, Jindal M, Kaur M, Kohli JM, Mehta MZ, Nishu N, Saini LK, Sharma A, Singh AP, Singh JB, Singh SP, Ahuja S, Bhattacharya S, Choudhary BC, Gupta P, Jain S, Jain S, Kumar A, Ranjan K, Shivpuri RK, Choudhury RK, Dutta D, Kailas S, Kumar V, Mohanty1 AK, Pant LM, Shukla P, Aziz T, Guchait14 M, Gurtu A, Maity15 M, Majumder D, Majumder G, Mazumdar K, Mohanty GB, Saha A, Sudhakar K, Wickramage N, Banerjee S, Dugad S, Mondal NK	45	2151	2.09	Panjab Univ, Chandigarh 160014, India. Univ Delhi, Delhi 110007, India. Bhabha Atom Res Ctr, Mumbai 400085, Maharashtra, India. Tata Inst Fundamental Res, Mumbai 400005, Maharashtra, India.	Phys Lett B 699 (1-2): 25-47 May 2 2011

Chemical Reduction of Graphene Oxide to Graphene by Sulfur-Containing Compounds	Sundar CS	1	3	33.33	Indian Inst Chem Technol, Inorgan & Phys Chem Div, Hyderabad 500607, Andhra Pradesh, India.	J Phys Chem C 114 (47): 19885-19890 Dec 2 2010
Galileon Gravity and Its Relevance to Late Time Cosmic Acceleration	Gannouji R, Sami M	2	2	100	Iucaa, Pune 411007, Maharashtra, India. Jamia Millia Islamia, Ctr Theoret Phys, New Delhi 110025, India.	Phys Rev D 82 (2): Art. No.-024011 Jul 14 2010
Observation of Long-Range, Near-Side Angular Correlations in Proton-Proton Collisions At the Lhc	Beri SB, Bhatnagar V, Dhingra N, Gupta R, Kaur M, Mehta MZ, Nishu N, Saini LK, Sharma A, Singh JB, Ashok Kumar, Arun Kumar, Ahuja S, Bhardwaj A, Choudhary BC, Malhotra S, Naimuddin M, Ranjan K, Sharma V, Shivpuri RK, Banerjee, Bhattacharya S, Dutta S, Gomber B, Jain S, Jain S, Khurana R, Sarkar S, Sharan M, Abdulsalam A, Dutta D, Kailas S, Kumar V, Mohanty2 AK, Pant LM, Shukla P, Aziz, Ganguly S, Guchait22 M, Gurtu23 A, Maity24 M, Majumder G, Mazumdar K, Mohanty GB, Parida B, Sudhakar K, Wickramage N, Banerjee S, Dugad S	50	2148	2.33	Panjab Univ, Chandigarh 160014, India. Univ Delhi, Delhi 110007, India. Bhabha Atom Res Ctr, Bombay 400085, Maharashtra, India. Tata Inst Fundamental Res, Bombay, Maharashtra, India.	J High Energy Phys (9): Art. No.-091 Sep 2010

Title	Authors				Affiliation	Journal
First Measurement of the Cross Section For Top-Quark Pair Production in Proton-Proton Collisions At Root S=7 Tev	Bansal S, Beri SB, Bhatnagar V, Jindal M, Kaur M, Kohli JM, Mehta MZ, Nishu N, Saini LK, Sharma A, Sharma R, Singh AP, Singh JB, Singh SP, Ahuja S, Bhattacharya S, Chauhan S, Choudhary BC, Gupta P, Jain S, Jain S, Kumar A, Shivpuri RK, Choudhury RK, Dutta D, Kailas S, Kataria SK, Mohanty1 AK, Pant LM, Shukla P, Suggisetti P, Aziz T, Guchait14 M, Gurtu A, Maity15 M, Majumder D, Majumder G, Mazumdar K, Mohanty GB, Saha A, Sudhakar K, Wickramage N, Banerjee S, Dugad S, Mondal NK	45	1983	2.27	Panjab Univ, Chandigarh 160014, India. Univ Delhi, Delhi 110007, India. Bhabha Atom Res Ctr, Mumbai 400085, Maharashtra, India. Tata Inst Fundamental Res Ehep, Mumbai, Maharashtra, India. Visva Bharati Univ, Santini Ketan, W Bengal, India.	Phys Lett B 695 (5): 424-443 Jan 17 2011
Biomolecules Under Mechanical Force	Kumar S	1	2	50	Banaras Hindu Univ, Dept Phys, Varanasi 221005, Uttar Pradesh, India	Phys Rep-Rev Sect Phys Lett 486 (1-2): 1-74 Jan 2010
Forward-Backward Asymmetry in Top Quark Production From Light Colored Scalars in So(10) Model	Patel KM, Sharma P	2	2	100	Phys Res Lab, Ahmadabad 380009, Gujarat, India	J High Energy Phys (4): Art. No.-085 Apr 2011
Damping of Exciton Rabi Rotations by Acoustic Phonons in Optically Excited Ingaas/ Gaas Quantum Dots	Gopal A	1	7	14.29	Tata Inst Fundamental Res, Dcmp & Ms, Bombay 400005, Maharashtra, India.	Phys Rev Lett 104 (1): Art.No.-017402 Jan 8 2010

Title	Authors				Affiliation	Reference
Bounds on An Anomalous Dijet Resonance in W Plus Jets Production in P(P)Over-Bar Collisions At Root S=1.96 Tev	Banerjee S, Acharya, BS	2	321	0.62	Panjab Univ, Chandigarh 160014, India. Univ Delhi, Delhi 110007, India. Tata Inst Fundamental Res, Mumbai 400005, Maharashtra, India.	Phys Rev Lett 107 (1): Art. No.-011804 Jun 30 2011
Flavor Physics in the Quark Sector	Libby J, Mohanty G	2	140	1.43	Indian Inst Technol, Madras 600032, Tamil Nadu, India. Tata Inst Fundamental Res, Bombay 400005, Maharashtra, India.	Phys Rep-Rev Sect Phys Lett 494 (3-4): 197-414 Sep 2010
An Ads(3) Dual For Minimal Model Cfts	Gopakumar R	1	2	50	Harish Chandra Res Inst, Allahabad 211019, Uttar Pradesh, India.	Phys Rev D 83 (6): Art. No.-066007 Mar 8 2011
Observation and Studies of Jet Quenching in Pbpb Collisions At Root S(Nn)=2.76 Tev	Bansal S, Beri SB, Bhatnagar V, Dhingra N, Gupta R, Jindal M, Kaur M, Kohli JM, Mehta MZ, Nishu N, Saini LK, Sharma A, Singh AP, Singh JB, Singh SP, Ahuja S, Bhattacharya S, Choudhary BC, Gupta P, Jain S, Jain S, Kumar A, Ranjan K, Shivpuri RK, Choudhury RK, Dutta D, Kailas S, Mohanty AK, Pant LM, Shukla P, Aziz T, Guchait M, Gurtu A, Maity M, Majumder D, Majumder G, Mazumdar K, Mohanty GB, Saha A, Sudhakar K, Wickramage N, Banerjee S, Dugad S, Mondal NK	44	2145	2.05	Panjab Univ, Chandigarh 160014, India. Univ Delhi, Delhi 110007, India. Bhabha Atom Res Ctr, Mumbai 400085, Maharashtra, India. Tata Inst Fundamental Res Ehep, Mumbai, Maharashtra, India. Tata Inst Fundamental Res Hecr, Mumbai, Maharashtra, India.	Phys Rev C 84 (2): Art. No.-024906 Aug 12 2011

Title	Authors				Affiliation	Reference
Surface Density of Spacetime Degrees of Freedom From Equipartition Law in Theories of Gravity	Padmanabhan T	1	1	100	Iucaa, Pune 411007, Maharashtra, India.	Phys Rev D 81 (12): Art. No.-124040 Jun 22 2010
Top Quark Forward-Backward Asymmetry At the Tevatron and Its Implications At the Lhc	Bhattacherjee B, Biswal SS, Ghosh D	3	3	100	Tata Inst Fundamental Res, Mumbai 400005, Maharashtra, India.	Phys Rev D 83 (9): Art. No.-091501 May 4 2011
Measurements of Inclusive W and Z Cross Sections in Pp Collisions At Root S=7 Tev	Beri SB, Bhatnagar V, Dhingra N, Gupta R, Jindal M, Kaur M, Kohli JM, Mehta MZ, Nishu N, Saini LK, Sharma A, Singh AP, Singh J, Singh SP, Ahuja S, Choudhary BC, Gupta P, Jain S, Kumar A, Kumar A, Naimuddin M, Ranjan K, Shivpuri RK, Banerjee S, Bhattacharya S, Dutta S, Gomber B, Jain S, Khurana R, Sarkar S, Choudhury RK, Dutta D, Kailas S, Kumar V, Mehta P, Mohanty1 AK, Pant LM, Shukla P, Aziz T, Guchait14 M, Gurtu A, Maity15 M, Majumder D, Majumder G, Mazumdar K, Mohanty GB, Saha A, Sudhakar K, Wickramage N, Banerjee S, Dugad S, Mondal NK	52	2248	2.31	Panjab Univ, Chandigarh 160014, India. Univ Delhi, Delhi 110007, India. Bhabha Atom Res Ctr, Bombay 400085, Maharashtra, India. Tata Inst Fundamental Res, Ehep, Mumbai 400005, Maharashtra, India. Tata Inst Fundamental Res, Hecr, Bombay 400005, Maharashtra, India.	J High Energy Phys (1): Art. No.-080 Jan 2011

Title	Authors				Affiliation	Reference
Search For A Heavy Neutral Gauge Boson in the Dielectron Channel With 5.4 Fb(-1) of P(P) Over-Bar Collisions At Root S=1.96 Tev	Beri SB, Bhatnagar V, Dutt S, Joshi J, Kohli JM, Choudhary B, Dubey A, Naimuddin M, Nayyar R, Ranjan K, Shivpuri RK, Mondal NK, Banerjee S, Acharya BS	14	437	3.2	Panjab Univ, Chandigarh 160014, India. Univ Delhi, Delhi 110007, India. Tata Inst Fundamental Res, Bombay 400005, Maharashtra, India.	Phys Lett B 695 (1-4): 88-94 Jan 10 2011
Top Polarization, Forward-Backward Asymmetry, and New Physics	Choudhury D, Rindani SD, Saha P	3	4	75	Univ Delhi, Dept Phys & Astrophys, Delhi 110007, India. Phys Res Lab, Div Theoret Phys, Ahmadabad 380009, Gujarat, India. Indian Inst Sci, Ctr High Energy Phys, Bangalore 560012, Karnataka, India.	Phys Rev D 84 (1): Art. No.-014023 Jul 19 2011
Two-Pion Bose-Einstein Correlations in Central Pb-Pb Collisions At Root(Nn)-N-S=2.76 Tev	Aggarwal MM, Ahammed Z, Ahmad N, Ahmad Masoodi A, Azmi MD, Bhasin A, Bhati AK, Bose S, Chattopadhyay S, Chattopadhyay S, Das D, Das I, Dash A, De S, Dubey AK, Ghosh P, Gupta A, Gupta R, Khan MM, Koyithatta Meethaleveedu G, Mahapatra DP, Mangotra L, Mohanty B, Muhuri S, Nandi BK, Nayak TK, Nyatha A, Pal SK, Potukuchi B, Raniwala R, Raniwala S, Roy P, Sahu PK, Samanta T, Sambyal S, Sharma N, Sharma S, Singaraju R, Singha R, Sinha BC, Sinha T, Varma R, Viyogi YP	43	921	4.67	Panjab Univ, Dept Phys, Chandigarh 160014, India. Ctr Variable Energy Cyclotron, Calcutta, India. Aligarh Muslim Univ, Dept Phys, Aligarh 202002, Uttar Pradesh, India. Univ Jammu, Dept Phys, Jammu 180004, India. Saha Inst Nucl Phys, Calcutta, India. Inst Phys, Bhubaneswar 751007, Orissa, India. Indian Inst Technol, Bombay 400076, Maharashtra, India. Univ Rajasthan, Dept Phys, Jaipur 302004, Rajasthan, India.	Phys Lett B 696 (4): 328-337 Feb 7 2011

Title	Authors				Affiliations	Reference
Search For Resonances in the Dijet Mass Spectrum From 7 Tev Pp Collisions At Cms	Beri SB, Bhatnagar V, Dhingra N, Gupta R, Jindal M, Kaur M, Kohli JM, Mehta MZ, Nishu N, Saini LK, Sharma A, Singh AP, Singh J, Singh SP, Ahuja S, Choudhary BC, Gupta P, Kumar A, Kumar A, Malhotra S, Naimuddin M, Ranjan K, Shivpuri RK, Banerjee S, Bhattacharya S, Dutta S, Gomber B, Jain S, Jain S, Khurana R, Sarkar S, Choudhury RK, Dutta D, Kailas S, Kumar V, Mehta P, Mohanty1 AK, Pant LM, Shukla P, Aziz T, Guchait16 M, Gurtu A, Maity17 M, Majumder D, Majumder G, Mazumdar K, Mohanty GB, Saha A, Sudhakar K, Wickramage N, Banerjee S, Dugad S, Mondal NK	53	2301	2.3	Panjab Univ, Chandigarh 160014, India. Univ Delhi, Delhi 110007, India. Saha Inst Nucl Phys, Calcutta, India. Bhabha Atom Res Ctr, Mumbai 400085, Maharashtra, India. Tata Inst Fundamental Res Ehep, Mumbai, Maharashtra, India. Tata Inst Fundamental Res Hecr, Mumbai, Maharashtra, India. Isva Bharati Univ, Santini Ketan, W Bengal, India.	Phys Lett B 704 (3): 123-142 Oct 13 2011
Forward-Backward Asymmetry in Top Quark-Antiquark Production	Shivpuri RK, Ranjan K, 5 Nayyar R, Naimuddin M, Dubey A, Choudhary B, Beri SB, Bhatnagar V, Dutt S, Joshi J, Kohli JM, Mondal NK, Acharya BS, Banerjee S	14	421	3.33	Panjab Univ, Chandigarh 160014, India. Univ Delhi, Delhi 110007, India. Tata Inst Fundamental Res, Bombay 400005, Maharashtra, India.	Phys Rev D 84 (11): Art. No.-112005 Dec 12 2011
Hydrodynamics From Charged Black Branes	Banerjee N, Bhattacharya J, Bhattacharyya S, Dutta S, Loganayagam R	5	6	83.33	Harish Chandra Res Inst, Allahabad 211019, Uttar Pradesh, India. Tata Inst Fundamental Res, Dept Theoret Phys, Bombay 400005, Maharashtra, India.	J High Energy Phys (1): Art. No.-094 Jan 2011

Title	Authors			Affiliation	Reference
Search For Neutral Minimal Supersymmetric Standard Model Higgs Bosons Decaying to Tau Pairs in Pp Collisions At Root S=7 Tev	Bansal S, Beri SB, Bhatnagar V, Dhingra N, Gupta R, Jindal M, Kaur M, Kohli JM, Mehta MZ, Nishu N, Saini LK, Sharma A, Singh AP, Singh JB, Singh SP, Ahuja S, Bhattacharya S, Choudhary BC, Gupta P, Jain S, Jain S, Kumar A, Ranjan K, Shivpuri RK, Choudhury RK, Dutta D, Kailas S, Kumar V, Mohanty AK, Pant LM, Shukla P, Aziz T, Guchaiti M, Gurtu A, Maity M, Majumder D, Majumder G, Mazumdar K, Mohanty GB, Saha A, Sudhakar K, Wickramage N, Banerjee S, Dugad S, Mondal NK	46	2183	2.11	Univ Delhi, Delhi 110007, India. Bhabha Atom Res Ctr, Bombay 400085, Maharashtra, India. Tata Inst Fundamental Res Ehep, Bombay, Maharashtra, India. Tata Inst Fundamental Res Hecr, Bombay, Maharashtra, India.	Phys Rev Lett 106 (23): Art. No.-231801 Jun 8 2011
BCS-BEC Crossover Induced by A Synthetic Non-Abelian Gauge Field	Vyasanakere JP, Shenoy VB	2	3	66.67	Indian Inst Sci, Dept Phys, Ctr Condensed Matter Theory, Bangalore 560012, Karnataka, India	Phys Rev B 84 (1): Art. No.-014512 Jul 25 2011
Measurement of the Anomalous Like-Sign Dimuon Charge Asymmetry With 9 Fb(-1) of P(P)Over-Bar Collisions	Acharya BS, Banerjee S, Beri SB, Bhatnagar V, Dutt S, Joshi J, Kohli JM, Choudhary B, Dubey A, Naimuddin M, Nayyar R, Ranjan K, Shivpuri RK, Mondal NK	14	432	3.24	Panjab Univ, Chandigarh 160014, India. Univ Delhi, Delhi 110007, India. Tata Inst Fundamental Res, Mumbai 400005, Maharashtra, India.	Phys Rev D 84 (5): Art. No.-052007 Sep 16 2011

Title	Authors				Affiliation	Journal
Holography of Dyonic Dilaton Black Branes	Prakash S, Trivedi SP	2	6	33.33	Tata Inst Fundamental Res, Bombay 400005, Maharashtra, India.	J High Energy Phys (10): Art. No.-027 Oct 2010
Modified Gravity A La Galileon: Late Time Cosmic Acceleration and Observational Constraints	Ali A, Gannouji R, Sami M	3	3	100	Jamia Millia Islamia, Ctr Theoret Phys, New Delhi 110025, India. Iucaa, Pune 411007, Maharashtra, India.	Phys Rev D 82 (10): Art. No.-103015 Nov 24 2010
Bound States of Two Spin-1/2 Fermions in A Synthetic Non-Abelian Gauge Field	Vyasanakere JP, Shenoy VB	2	2	100	Indian Inst Sci, Dept Phys, Ctr Condensed Matter Theory, Bangalore 560012, Karnataka, India.	Phys Rev B 83 (9): Art. No.-094515 Mar 11 2011
Search For Resonances in the Dilepton Mass Distribution in Pp Collisions At Root S=7 Tev	Bansal S, Beri SB, Bhatnagar V, Dhingra N, Gupta R, Jindal M, Kaur M, Kohli JM, Bansal S, Beri SB, Bhatnagar V, Dhingra N, Gupta R, Jindal M, Kaur M, Kohli JM, Mehta MZ, Nishu N, Saini LK, Sharma A, Singh AP, Singh JB, Singh SP, Ahuja S, Bhattacharya S, Choudhary BC, Gupta P, Jain S, Jain S, Kumar A, Ranjan K, Shivpuri RK, Choudhury RK, Dutta D, Kailas S, Kumar V, Mohanty1 AK, Pant LM, Shukla P, Aziz T, Guchait14 M, Gurtu A, Maity15 M, Majumder D, Majumder G, Mazumdar K, Mohanty GB, Saha A, Sudhakar K, Wickramage N, Banerjee S, Dugad S, Mondal NK	53	2179	2.43	Panjab Univ, Chandigarh 160014, India. Univ Delhi, Delhi 110007, India. Bhabha Atom Res Ctr, Bombay 400085, Maharashtra, India. Tata Inst Fundamental Res, Bombay 400005, Maharashtra, India.	J High Energy Phys (5): Art. No.-093 May 2011

| Search For A Heavy Bottom-Like Quark in Pp Collisions At Root S=7 Tev | Bansal S, Beri SB, Bhatnagar V, Dhingra N, Gupta R, Jindal M, Kaur M, Kohli JM, Mehta MZ, Nishu N, Saini LK, Sharma A, Singh AP, Singh JB, Singh SP, Ahuja S, Bhattacharya S, Choudhary BC, Gupta P, Jain S, Jain S, Kumar A, Ranjan K, Shivpuri RK, Choudhury RK, Dutta D, Kailas S, Kumar V, Mohanty1 AK, Pant LM, Shukla P, Aziz T, Guchait13 M, Gurtu A, Maity14 M, Majumder D, Majumder G, Mazumdar K, Mohanty GB, Saha A, Sudhakar K, Wickramage N, Banerjee S, Dugad S, Mondal NK | 45 | 2164 | 2.08 | Panjab Univ, Chandigarh 160014, India. Univ Delhi, Delhi 110007, India. Bhabha Atom Res Ctr, Mumbai 400085, Maharashtra, India. Tata Inst Fundamental Res Ehep, Mumbai, Maharashtra, India. | Phys Lett B 701 (2): 204-223 Jul 4 2011 |

Title	Authors				Affiliation	Journal
Search For Stopped Gluinos in Pp Collisions At Root S=7 Tev	Bansal S, Beri SB, Bhatnagar V, Dhingra N, Jindal M, Kaur M, Kohli JM, Mehta MZ, Nishu N, Saini LK, Sharma A, Singh AP, Singh JB, Singh SP, Ahuja S, Bhattacharya S, Choudhary BC, Gupta P, Jain S, Jain S, Kumar A, Shivpuri RK, Choudhury RK, Dutta D, Kailas S, Kataria SK, Mohanty1 AK, Pant LM, Shukla P, Suggisetti P, Aziz T, Guchait13 M, Gurtu A, Maity14 M, Majumder D, Majumder G, Mazumdar K, Mohanty GB, Saha A, Sudhakar K, Wickramage N, Banerjee S, Dugad S, Mondal NK	44	2173	2.02	Panjab Univ, Chandigarh 160014, India. Univ Delhi, Delhi 110007, India. Bhabha Atom Res Ctr, Mumbai 400085, Maharashtra, India. Tata Inst Fundamental Res Ehep, Mumbai, Maharashtra, India.	Phys Rev Lett 106 (1): Art. No.-011801 Jan 7 2011
Constraints on the Four-Generation Quark Mixing Matrix From A Fit to Flavor-Physics Data	Dighe A	1	3	33.33	Tata Inst Fundamental Res, Bombay 400005, Maharashtra, India.	Phys Rev D 83 (7): Art. No.-073008 Apr 29 2011

Measurement of Dijet Angular Distributions and Search For Quark Compositeness in Pp Collisions At Root S=7 Tev	Bansal S, Beri SB, Bhatnagar V, Dhingra N, Gupta R, Jindal M, Kaur M, Kohli JM, Mehta MZ, Nishu N, Saini LK, Sharma A, Singh AP, Singh JB, Singh SP, Ahuja S, Bhattacharya S, Choudhary BC, Gupta P, Jain S, Jain S, Kumar A, Shivpuri RK, Choudhury RK, Dutta D, Kailas S, Kataria SK, Mohanty AK, Pant LM, Shukla P, Aziz T, Guchait M, Gurtu HA, Maity M, Majumder D, Majumder G, Mazumdar K, Mohanty GB, Saha A, Sudhakar K, Wickramage N, Banerjee S, Dugad S, Mondal NK	44	2396	1.84	Panjab Univ, Chandigarh 160014, India. Univ Delhi, Delhi 110007, India. Bhabha Atom Res Ctr, Mumbai 400085, Maharashtra, India. Tata Inst Fundamental Res Ehep, Mumbai, Maharashtra, India.	Phys Rev Lett 106 (20): Art. No.-201804 May 18 2011
Electron Acoustic Solitary Waves and Double Layers With Superthermal Hot Electrons	Sahu B	1	1	100	W Bengal State Univ, Dept Math, Calcutta 700126, India.	Phys Plasmas 17 (12): Art. No.-122305 Dec 2010
A New Route to Explosive Percolation	Manna SS	1	2	50	Satyendra Nath Bose Natl Ctr Basic Sci, Calcutta 700098, India.	Physica A 390 (2): 177-182 Jan 15 2011
Metal Oxide Thin Film Based Supercapacitors	Lokhande CD	1	3	33.33	Shivaji Univ, Dept Phys, Thin Film Phys Lab, Kolhapur 416004, Ms, India.	Curr Appl Phys 11 (3): 255-270 May 2011

Search For Same-Sign Top-Quark Pair Production At Root S=7 Tev and Limits on Flavour Changing Neutral Currents in the Top Sector	Beri SB, Bhatnagar V, Dhingra N, Gupta R, Jindal M, Kaur M, Kohli JM, Mehta MZ, Nishu N, Saini LK, Sharma, Singh AP, Singh J, Singh SP, Ahuja S, Choudhary BC, Gupta P, Jain S, Kumar A, Kumar A, Naimuddin M, Ranjan K, Shivpuri RK, Banerjee S, Bhattacharya S, Dutta S, Gomber B, Jain S, Khurana R, Sarkar S, Choudhury RK, Dutta D, Kailas S, Kumar V, Mehta P, Mohanty AK, Pant, Shukla P, Aziz T, Guchait M, Gurtu A, Maity M, Majumder D, Majumder G, Mazumdar K, Mohanty GB, Saha A, Sudhakar K, Wickramage N, Banerjee S, Dugad S, Mondal NK	52	2209	2.35	Panjab Univ, Chandigarh 160014, India. Univ Delhi, Delhi 110007, India. Bhabha Atom Res Ctr, Mumbai 400085, Maharashtra, India. Tata Inst Fundamental Res Ehep, Mumbai, Maharashtra, India.	J High Energy Phys (8): Art. No.-005 Aug 2011
Optimal Pulse Spacing For Dynamical Decoupling in the Presence of A Purely Dephasing Spin Bath	Ajoy A	1	3	33.33	Birla Inst Technol & Sci, Zuarinagar 403726, Goa, India. Indian Inst Sci, Nmr Res Ctr, Bangalore 560012, Karnataka, India.	Phys Rev A 83 (3): Art. No.-032303 Mar 8 2011

Search For Microscopic Black Hole Signatures At the Large Hadron Collider	Bansal S, Beri SB, Bhatnagar V, Dhingra N, Jindal M, Kaur M, Kohli JM, Mehta MZ, Nishu N, Saini LK, Sharma A, Singh AP, Singh JB, Singh SP, Ahuja S, Bhattacharya S, Choudhary BC, Gupta P, Jain S, Jain S, Kumar A, Shivpuri RK, Choudhury RK, Dutta D, Kailas S, Kataria SK, Mohanty1 AK, Pant LM, Shukla P, Aziz T, Guchait M, Gurtu A, Maity M, Majumder D, Majumder G, Mazumdar K, Mohanty GB, Saha A, Sudhakar K, Wickramage N, Banerjee S, Dugad S, Mondal NK	43	2179	1.97	Panjab Univ, Chandigarh 160014, India. Univ Delhi, Delhi 110007, India. Bhabha Atom Res Ctr, Mumbai 400085, Maharashtra, India. Tata Inst Fundamental Res Ehep, Mumbai, Maharashtra, India.	Phys Lett B 697 (5): 434-453 Mar 21 2011
Scale For the Phase Diagram of Quantum Chromodynamic	Gupta S, Mohanty B	2	5	40	Ctr Variable Energy Cyclotron, Expt High Energy Phys & Applicat Grp, Calutta 700064, India. Tata Inst Fundamental Res, Dept Theoret Phys, Mumbai 400005, Maharashtra, India.	Science 332 (6037): 1525-1528 Jun 24 2011

Title	Authors				Affiliation	Source
Search For B-S(0) -> Mu(+)Mu(-) and B-0 -> Mu(+)Mu(-) Decays in Pp Collisions At Root S=7 Tev	Beri SB, Bhatnagar V, Dhingra N, Gupta R, Jindal M, Kaur M, Kohli JM, Mehta MZ, Nishu N, Saini LK, Sharma A, Singh AP, Singh J, Singh SP, Ahuja S, Choudhary BC, Gupta P, Kumar A, Kumar A, Malhotra S, Naimuddin M, Ranjan K, Shivpuri RK, Banerjee S, Bhattacharya S, Dutta S, Gomber B, Jain S, Jain S, Khurana R, Sarkar S, Choudhury RK, Dutta D, Kailas S, Kumar V, Mehta P, Mohanty AK, Pant LM, Shukla P, Aziz T, Guchait M, Gurtu A, Maity M, Majumder D, Majumder G, Mazumdar K, Mohanty GB, Saha A, Sudhakar K, Wickramage N, Banerjee S, Dugad S, Mondal NK	53	2258	2.35	Panjab Univ, Chandigarh 160014, India. Univ Delhi, Delhi 110007, India. Bhabha Atom Res Ctr, Mumbai 400085, Maharashtra, India. Tata Inst Fundamental Res Ehep, Mumbai, Maharashtra, India.	Phys Rev Lett 107 (19): Art. No.-191802 Nov 1 2011
Ambient- and Low-Temperature Synchrotron X-Ray Diffraction Study of Bafe2as2 and Cafe2as2 At High Pressures Up to 56 Gpa	Mittal R, Mishra SK, Chaplot SL	3	12	25	Bhabha Atom Res Ctr, Div Solid State Phys, Bombay 400085, Maharashtra, India.	Phys Rev B 83 (5): Art. No.-054503 Feb 9 2011
Saturation Models of Hera Dis Data and Inclusive Hadron Distributions in P Plus P Collisions At the Lhc	Prithwish Tribedy	1	2	50	Ctr Variable Energy Cyclotron, Calcutta 700064, W Bengal, India.	Nucl Phys A 850 (1): 136-156 Jan 15 2011

Title	Authors				Affiliation	Citation
Lattice Qcd Predictions For Shapes of Event Distributions Along the Freezeout Curve in Heavy-Ion Collisions	Gavai RV, Gupta S	2	2	100	Tata Inst Fundamental Res, Dept Theoret Phys, Mumbai 400005, Maharashtra, India.	Phys Lett B 696 (5): 459-463 Feb 14 2011
Quantum W-Symmetry in Ads(3)	Rajesh Gopakumar, Arunabha Saha	2	3	66.67	Harish Chandra Res Inst, Allahabad 211019, Uttar Pradesh, India.	J High Energy Phys (2): Art. No.-004 Feb 2011
Prompt and Non-Prompt J/Psi Production in Pp Collisions At Root S=7 Tev	Bansal S, Beri SB, Bhatnagar V, Dhingra N, Jindal M, Kaur M, Kohli JM, Mehta MZ, Nishu N, Saini LK, Sharma A, Singh AP, Singh JB, Singh SP, Ahuja S, Bhattacharya S, Choudhary BC, Gupta P, Jain S, Jain S, Kumar A, Shivpuri RK, Choudhury RK, Dutta D, Kailas S, Kataria SK, Mohanty1 AK, Pant LM, Shukla P, Suggisetti P, Aziz T, Guchait13 M, Gurtu A, Maity14 M, Majumder D, Majumder G, Mazumdar K, Mohanty GB, Saha A, Sudhakar K, Wickramage N, Banerjee S, Dugad S, Mondal NK	44	2155	2.04	Panjab Univ, Chandigarh 160014, India. Univ Delhi, Delhi 110007, India. Bhabha Atom Res Ctr, Mumbai 400085, Maharashtra, India. Tata Inst Fundamental Res Ehep, Mumbai, Maharashtra, India.	Eur Phys J C 71 (3): Art. No.-1575 Mar 2011

Title	Authors				Affiliation	Publication
Search For Supersymmetry in Pp Collisions At Root S=7 Tev in Events With Two Photons and Missing Transverse Energy	Bansal S, Beri SB, Bhatnagar V, Dhingra N, Gupta R, Jindal M, Kaur M, Kohli JM, Mehta MZ, Nishu N, Saini LK, Sharma A, Singh AP, Singh JB, Singh SP, Ahuja S, Bhattacharya S, Choudhary BC, Gupta P, Jain S, Jain S, Kumar A, Ranjan K, Shivpuri RK, Choudhury RK, Dutta D, Kailas S, Kumar V, Pant LM, Shukla P, Aziz T, Gurtu A, Majumder D, Majumder, Mazumdar K, Mohanty GB, Saha A, Sudhakar K, Wickramage N	40	2175	1.84	Panjab Univ, Chandigarh 160014, India. Univ Delhi, Delhi 110007, India. Bhabha Atom Res Ctr, Mumbai 400085, Maharashtra, India. Tata Inst Fundamental Res Ehep, Mumbai, Maharashtra, India.	Phys Rev Lett 106 (21): Art. No.-211802 May 24 2011
Model For T2k Indication With Maximal Theta(23) and Trimaximal Theta(12)	Patel KM	1	3	33.33	Phys Res Lab, Ahmadabad 380009, Gujarat, India.	Phys Rev D 84 (5): Art. No.-053002 Sep 6 2011
Chiral Couplings of W ' and Top Quark Polarization At the Lhc	Gopalakrishna S	1	5	20	Inst Math Sci, Madras 600113, Tamil Nadu, India.	Phys Rev D 82 (11): Art. No.-115020 Dec 21 2010
Superluminal Neutrinos At Opera Confront Pion Decay Kinematics	Sarkar U	1	3	33.33	Phys Res Lab, Ahmadabad 380009, Gujarat, India.	Phys Rev Lett 107 (25): Art. No.-251801 Dec 16 2011
Measurements of Higher Order Flow Harmonics in Au Plus Au Collisions At Root S(Nn)=200 Gev	Chowdhary RK, Dutta D, Mohanty AK, Shukla P, Mishra P, Singh BK, Singh V	7	387	1.81	Banaras Hindu Univ, Dept Phys, Varanasi 221005, Uttar Pradesh, India. Bhabha Atom Res Ctr, Bombay 400085, Maharashtra, India.	Phys Rev Lett 107 (25): Art. No.-252301 Dec 14 2011

Model-Independent Measurement of T-Channel Single Top Quark Production in P(P)Over-Bar Collisions At, Root S=1.96 Tev	Acharya BS, Beri SB, Bhatnagar V, Dutt S, Joshi J, Kohli JM, Choudhary B, Dubey A, Naimuddin M, Nayyar R, Ranjan K, Shivpuri RK, Banerjee S, Mondal NK	14	428	3.27	Panjab Univ, Chandigarh; Univ Delhi, Delhi; Tata Inst Fundamental Res, Mumbai;	Phys Lett B 705 (4): 313-319 Nov 17 2011
Trapped Two-Dimensional Condensates With Synthetic Spin-Orbit Coupling	Sinha S	1	3	33.33	Indian Inst Sci Educ & Res Kolkata, Mohanpur	Phys Rev Lett 107 (27): Art. No.-270401 Dec 27 2011
Relevance of the Heisenberg-Kitaev Model For the Honeycomb Lattice Iridates A(2)Iro(3)	Singh Y	1	8	12.5	Indian Inst Sci Educ & Res Mohali, Sect 81, Mohali - 2	Phys Rev Lett 108 (12): Art. No.-127203 Mar 20 2012
Positron Scattering From Argon: Total Cross Sections and the Scattering Length	Sarkar A, Chattopadhyay S	2	9	22.22	Bangabasi Morning Coll, Dept Phys, Calcutta 700009, India. Maulana Azad Coll, Dept Phys, Calcutta 700013, India.	J Phys-B-At Mol Opt Phys 45 (1): Art. No.-015203 Jan 14 2012

Title	Authors				Affiliation	Journal
Inclusive Search For Squarks and Gluinos in Pp Collisions At Root S=7 Tev	Beri SB, Bhatnagar V, Dhingra N, Gupta R, Jindal M, Kaur M, Kohli JM, Mehta MZ, Nishu N, Saini LK, Sharma A, Singh AP, Singh J, Singh SP, Ahuja S, Choudhary BC, Gupta P, Jain S, Kumar A, Kumar A, Naimuddin M, Ranjan K, Shivpuri RK, Banerjee S, Bhattacharya S, Dutta S, Gomber B, Jain S, Khurana R, Sarkar S, Choudhury RK, Dutta D, Kailas S, Kumar V, Mehta P, Mohanty AK, Pant LM, Shukla P, Aziz T, Guchait M, Gurtu A, Maity M, Majumder D, Majumder G, Mazumdar K, Mohanty GB, Saha A, Sudhakar K, Wickramage N, Banerjee S, Dugad S, Mondal NK	52	2217	2.35	Panjab Univ, Chandigarh 160014, India. Univ Delhi, Delhi 110007, India. Saha Inst Nucl Phys, Calcutta, India. Bhabha Atom Res Ctr, Bombay 400085, Maharashtra, India. Tata Inst Fundamental Res Ehep, Bombay, Maharashtra, India. Tata Inst Fundamental Res Hecr, Bombay, Maharashtra, India. Isva Bharati Univ, Santini Ketan, W Bengal, India.	Phys Rev D 85 (1): Art. No.-012004 Jan 11 2012
Correlation Functions in Holographic Minimal Models	Suvrat Raju	1	2	50	Harish Chandra Res Inst, Chatnag Marg, Allahabad 211019, Uttar Pradesh, India.	Nucl Phys B 856 (2): 607-646 Mar 11 2012

Title	Authors				Affiliation	Journal
Combined Results of Searches For the Standard Model Higgs Boson in Pp Collisions At Root S=7 Tev	Beri SB, Bhatnagar V, Dhingra N, Gupta R, Jindal M, Kaur M, Kohli JM, Mehta MZ, Nishu N, Saini LK, Sharma A, Singh AP, Singh J, Singh SP, Ahuja S, Choudhary BC, Kumar A, Kumar A, Malhotra S, Naimuddin M, Ranjan K, Sharma V, Shivpuri RK, Banerjee S, Bhattacharya S, Dutta S, Gomber B, Jain S, Jain S, Khurana R, Sarkar S, Choudhury RK, Dutta D, Kailas S, Kumar V, Mohanty1 AK, Pant LM, Shukla P, Aziz T, Ganguly S, Guchait17 M, Gurtu18 A, Maity19 M, Majumder G, Mazumdar K, Mohanty GB, Parida B, Saha A, Sudhakar K, Wickramage N, Banerjee S, Dugad S, Mondal NK	53	2256	2.35	Panjab Univ, Chandigarh 160014, India. Univ Delhi, Delhi 110007, India. Saha Inst Nucl Phys, Calcutta, India. Bhabha Atom Res Ctr, Bombay 400085, Maharashtra, India. Tata Inst Fundamental Res Ehep, Bombay, Maharashtra, India. Tata Inst Fundamental Res Hecr, Bombay, Maharashtra, India. Visva Bharati Univ, Santini Ketan, W Bengal, India.	Phys Lett B 710 (1): 26-48 Mar 29 2012
Kerr Naked Singularities As Particle Accelerators	Patil M, Joshi P	2	2	100	Tata Inst Fundamental Res, Homi Bhabha Rd, Bombay 400005, Maharashtra, India	Class Quantum Gravity 28 (23): Art. No.-235012 Dec 7 2011
Near Maximal Atmospheric Mixing in Neutrino Mass Matrices With Two Vanishing Minors	Dev S, Gupta S, Gautam RR, Singh L	2	4	50	Himachal Pradesh Univ, Dept Phys, Shimla	Phys Lett B 706 (2-3): 168-176 Dec 6 2011

Title	Authors				Affiliation	Reference
Systematic Study of the Decay of Ba-118, Ba-122* Formed in Kr-78, Kr-82-Induced Reactions At E-Lab=5.5 Mev/Nucleon	Kaur M, Kumar R, Sharma MK	3	3	100	Thapar Univ, Sch Phys & Mat Sci, Patiala 147004, Punjab, India	Phys Rev C 85 (1): Art. No.-014609 Jan 19 2012
Fusion-Evaporation Residue As A Dynamical Decay Process in the Ca-48+Bk-249 -> (297)117*Reaction	Sandhu K, Sharma MK, Gupta RK	3	3	100	Thapar Univ, Sch Phys & Mat Sci, Patiala 147004, Punjab, India. Panjab Univ, Dept Phys, Chandigarh 160014, India	Phys Rev C 85 (2): Art. No.-024604 Feb 6 2012
Harmonic Decomposition of Two Particle Angular Correlations in Pb-Pb Collisions At Root S(Nn)=2.76 Tev	Jena S, Meethaleveedu G Koyithatta, Nandi BK, Nyatha A, Varma R, Bose S, Das I, Das K, Majumdar AK Dutta Khan, Roy P, Sinha T Bhasin A, Gupta A, Gupta R, Potukuchi B, Sambyal S, Sharma S, Singh R, Baral RC, Mahapatra DP, Sahu PK, Ahmad N, Masoodi A Ahmad, Ahn SU, Azmi MD, Irfan M, Khan MM, Chattopadhyay S, Chattopadhyay S, De S, Ahammed Z, Dubey AK, Dutta MR, Ghosh P, Muhuri S, Khan SA, Mohanty B, Muhuri S, Nayak TK, Pal S, Pal SK, Saini J, Singaraju R, Singha S, Sinha BC, Viyogi YP, Aggarwal MM, Sharma N	49	943	5.2	Ctr Variable Energy Cyclotron, Calcutta, India. Aligarh Muslim Univ, Dept Phys, Aligarh 202002, Uttar Pradesh, India. Panjab Univ, Dept Phys, Chandigarh 160014, India. Inst Phys, Bhubaneswar 751007, Orissa, India. Univ Jammu, Dept Phys, Jammu 180004, India. Saha Inst Nucl Phys, Calcutta, India. Univ Rajasthan, Dept Phys, Jaipur 302004, Rajasthan, India. Indian Inst Technol, Mumbai 400076, Maharashtra, India.	Phys Lett B 708 (3-5): 249-264 Feb 28 2012

Title	Authors				Affiliation	Reference
Skyrme Forces and the Fusion-Fission Dynamics of the Sn-132+Ni-64 -> Pt-196* Reaction	Jain D, Kumar R, Sharma MK, Gupta RK	4	4	100	Thapar Univ, Sch Phys & Mat Sci, Patiala 147004, Punjab, India. Panjab Univ, Dept Phys, Chandigarh 160014, India.	Phys Rev C 85 (2): Art. No.-024615 Feb 22 2012
Equilibrium Configurations From Gravitational Collapse	Joshi PS, Malafarina D	2	3	66.67	Tata Inst Fundamental Res, Bombay - 2; Harvard Smithsonian Ctr Astrophys, Cambridge	Class Quantum Gravity 28 (23): Art. No.-235018 Dec 7 2011
Action Principle For the Fluid-Gravity Correspondence and Emergent Gravity	Kolekar S, Padmanabhan T	2	2	100	Iucaa, Pune Univ Campus, Pune 411007, Maharashtra, India	Phys Rev D 85 (2): Art. No.-024004 Jan 4 2012
First Observation of the P-Wave Spin-Singlet Bottomonium States H(B)(1P) and H(B)(2P)	Bhuyan B, Libby J, Bhardwaj V, Aziz T, Gaur V, Mohanty GB	6	167	3.59	Indian Inst Technol Guwahati, Gauhati, India. Indian Inst Technol, Madras 600036, Tamil Nadu, India. Panjab Univ, Chandigarh 160014, India. Tata Inst Fundamental Res, Bombay 400005, Maharashtra, India.	Phys Rev Lett 108 (3): Art. No.-032001 Jan 18 2012
Holographic Fermi and Non-Fermi Liquids With Transitions in Dilaton Gravity	Kundu N, Narayan P, Trivedi SP	3	4	75	Tata Inst Fundamental Res, Bombay 400005, Maharashtra, India.	J High Energy Phys (1): Art. No.-094 Jan 2012
Spin Waves and Revised Crystal Structure of Honeycomb Iridate Na2iro3	Singh Y	1	14	7.14	Indian Inst Sci Educ & Res Mohali, Sect 81, Mohali, India.	Phys Rev Lett 108 (12): Art. No.-127204 Mar 20 2012

Title	Authors				Affiliation	Source
Search For the Standard Model Higgs Boson Decaying Into Two Photons in Pp Collisions At Root S=7 Tev	Beri SB, Bhatnagar V, Dhingra N, Gupta R, Jindal M, Kaur M, Kohli JM, Mehta MZ, Nishu N, Saini LK, Sharma A, Singh AP, Singh J, Singh SP, Ahuja S, Choudhary BC, Kumar A, Kumar A, Malhotra S, Naimuddin M, Ranjan K, Sharma V, Shivpuri RK, Banerjee S, Bhattacharya S, Dutta S, Gomber B, Jain S, Jain S, Khurana R, Sarkar S, Choudhury RK, Dutta D, Kailas S, Kumar V, Mohanty 1 AK, Pant LM, Shukla P, Aziz T, Ganguly S, Guchait 17 M, Gurtu18 A, Maity19 M, Majumder G, Mazumdar K, Mohanty GB, Parida B, Saha A, Sudhakar K, Wickramage N, Banerjee S, Dugad S, Mondal NK	53	2259	2.35	Visva Bharati Univ, Santini Ketan, W Bengal, India. Panjab Univ, Chandigarh 160014, India. Univ Delhi, Delhi 110007, India. Saha Inst Nucl Phys, Calcutta, India. Bhabha Atom Res Ctr, Mumbai 400085, Maharashtra, India. Tata Inst Fundamental Res Ehep, Mumbai, Maharashtra, India.	Phys Lett B 710 (3): 403-425 Apr 12 2012
New Physics in B -> S Mu(+)Mu(-): Cp-Violating Observables	Dighe A, Ghosh D,	2	6	33.33	Tata Inst Fundamental Res, Bombay 400005, Maharashtra, India.	J High Energy Phys (11): Art. No.-122 Nov 2011
New Physics in B -> S Mu(-): Cp-Conserving Observables	Dighe A, Ghosh D,	2	6	33.33	Tata Inst Fundamental Res, Bombay 400005, Maharashtra, India.	J High Energy Phys (11): Art. No.-121 Nov 2011

Title	Authors				Affiliation	Citation
Effect of Solvent Volume on the Physical Properties of Undoped and Fluorine Doped Tin Oxide Films Deposited Using A Low-Cost Spray Technique	Muruganantham G, Ravichandran K, Saravanakumar K, Ravichandran AT, Sakthivel B	5	5	100	Avvm Sri Pushpam Coll Autonomous, Pg & Res Dept Phys, Thanjavur 613503, Tamil Nadu, India. Natl Coll Autonomous, Dept Phys, Tiruchchirappalli 620001, Tamil Nadu, India.	Superlattice Microstruct 50 (6): 722-733 Dec 2011
Search For A Vectorlike Quark With Charge 2/3 in T Plus Z Events From Pp Collisions At Root S 7 Tev	Beri SB, Bhatnagar V, Dhingra N, Gupta R, Jindal M, Kaur M, Kohli JM, Mehta MZ, Nishu N, Saini LK, Sharma A, Singh AP, Singh J, Singh SP, Ahuja S, Choudhary BC, Gupta P, Kumar A, Kumar A, Malhotra S, Naimuddin M, Ranjan K, Shivpuri RK, Banerjee S, Bhattacharya S, Dutta S, Gomber B, Jain S, Jain S, Khurana R, Sarkar S, Choudhury RK, Dutta D, Kailas S, Kumar V, Mehta P, Mohanty AK, Pant LM, Shukla P, Aziz T, Guchait M, Gurtu A, Maity M, Majumder D, Majumder G, Mathew T, Mazumdar K, Mohanty GB, Parida B, Saha A, Sudhakar K, Wickramage N, Banerjee S, Dugad S, Mondal NK	55	2254	2.44	Visva Bharati Univ, Santini Ketan, W Bengal, India. Univ Delhi, Delhi 110007, India. Saha Inst Nucl Phys, Calcutta, India. Bhabha Atom Res Ctr, Bombay 400085, Maharashtra, India. Tata Inst Fundamental Res Ehep, Bombay, Maharashtra, India. Inst Res & Fundamental Sci Ipm, Tehran, Iran.	Phys Rev Lett 107 (27): Art. No.-271802 Dec 29 2011

(r) Plant Animal Sciences

Title	Indian Authors	No. of Indian Authors	Total Author	Ratio I/F*	Institue Name	Source
The Map-Based Sequence of the Rice Genome		2	263	0.76	University of Delhi, New Delhi, Indian Agr Res Inst, New Delhi	Nature 436 (7052): 793-800 Aug 11 2005
The Sorghum Bicolor Genome and the Diversification of Grasses	Hash CT	1	21	4.76	Int Crops Res Inst Semi Arid Trop, Patancheru 502324, Andhra Pradesh, India.	Nature 457 (7229): 551-556 Jan 29 2009
Drought-Induced Responses of Photosynthesis and Antioxidant Metabolism in Higher Plants	Reddy AR, Chaitanya KV, Vivekanandan M	3	3	100.00	Pondicherry Univ, Sch Life Sci, Pondicherry 605014, India. Bharathidasan Univ, Sch Life Sci, Dept Biotechnol, Tiruchchirappalli 620024, India.	J Plant Physiol 161 (11): 1189-1202 Nov 2004
Effect of Cadmium on Lipid Peroxidation, Superoxide Anion Generation and Activities of Antioxidant Enzymes in Growing Rice Seedlings	Shah K, Kumar RG, Verma S, Dubey RS	4	4	100.00	Banaras Hindu Univ, Fac Sci, Dept Biochem, Varanasi 221005, Uttar Pradesh, India.	Plant Sci 161 (6): 1135-1144 Nov 2001
An Integrated Physical and Genetic Map of the Rice Genome	Trilochan Mohapatra, Nagendra K Singh, Akhilesh K Tyagi	3	39	7.69	Natl Res Ctr Plant Biotechnol, Indian Agr Res Inst, Indian Initiat Rice Genome Sequencing, New Delhi 110112, India. Indian Initiat Rice Genome Sequencing, New Delhi 110021, India.	Plant Cell 14 (3): 537-545 Mar 2002

Title	Authors				Affiliation	Source
Lead Toxicity Induces Lipid Peroxidation and Alters the Activities of Antioxidant Enzymes in Growing Rice Plants	Verma S, Dubey RS	2	2	100.00	Banaras Hindu Univ, Fac Sci, Dept Biochem, Varanasi 221005, Uttar Pradesh, India.	Plant Sci 164 (4): 645-655 Apr 2003
Small Rnas As Big Players in Plant Abiotic Stress Responses and Nutrient Deprivation	Chinnusamy V	1	4	25.00	Indian Agr Res Inst, Water Technol Ctr, New Delhi 110012, India.	Trends Plant Sci 12 (7): 301-309 Jul 2007
Genetic Mapping of 66 New Microsatellite (Ssr) Loci in Bread Wheat	Gupta PK, Balyan HS	2	20	10.00	Ccs Univ, Meerut, Uttar Pradesh, India.	Theor Appl Genet 105 (2-3): 413-422 Aug 2002
Physiology and Molecular Biology of Salinity Stress Tolerance in Plants	Sairam RK, Tyagi A	2	2	100.00	Indian Agr Res Inst, Div Plant Physiol, New Delhi 110012, India.	Curr Sci 86 (3): 407-421 Feb 10 2004
Toxicological Effects of Malachite Green	Srivastava S, Sinha R, Roy D	3	3	100.00	Smm Town Postgrad Coll, Dept Zool, Ballia 277001, India. Cent Inland Capture Fisheries Res Inst, Barakpur 743101, W Bengal, India.	Aquat Toxicol 66 (3): 319-329 Feb 25 2004
Cold Stress Regulation of Gene Expression in Plants	Chinnusamy V	1	3	33.33	Indian Agr Res Inst, Water Technol Ctr, New Delhi 110012, India.	Trends Plant Sci 12 (10): 444-451 Oct 2007
The Significance of Amino Acids and Amino Acid-Derived Molecules in Plant Responses and Adaptation to Heavy Metal Stress	Sharma SS	1	2	50.00	Himachal Pradesh Univ, Dept Biosci, Shimla 171005, Himachal Prades, India.	J Exp Bot 57 (4): 711-726 Mar 2006
Beneficial Interactions of Mitochondrial Metabolism With Photosynthetic Carbon Assimilation	Raghavendra AS, Padmasree K	2	2	100.00	Univ Hyderabad, Sch Life Sci, Dept Plant Sci, Hyderabad 500046, Andhra Pradesh, India.	Trends Plant Sci 8 (11): 546-553 Nov 2003

Title	Authors				Affiliation	Source
Differential Response of Wheat Genotypes to Long Term Salinity Stress in Relation to Oxidative Stress, Antioxidant Activity and Osmolyte Concentration	Raghavendra AS, Padmasree K	2	2	100.00	Univ Hyderabad, Sch Life Sci, Dept Plant Sci, Hyderabad 500046, Andhra Pradesh, India.	Trends Plant Sci 8 (11): 546-553 Nov 2003
Molecular Mechanisms of Quenching of Reactive Oxygen Species by Proline Under Stress in Plants	Bhalu B, Mohanty P	2	4	50.00	Jawaharlal Nehru Univ, New Delhi 110067, India.	Curr Sci 82 (5): 525-532 Mar 10 2002
Role of Dreb Transcription Factors in Abiotic and Biotic Stress Tolerance in Plants	Agarwal PK, Agarwal P, Reddy MK, Sopory SK	4	4	100.00	Int Ctr Genet Engn & Biotechnol, Aruna Asaf Ali Rd, New Delhi 110067, India. Bhavnagar Univ, Dept Life Sci, Bhavnagar 364002, Gujarat, India	Plant Cell Rep 25 (12): 1263-1274 Dec 2006
Oxidative Stress and Antioxidative System in Plants	Arora A, Sairam RK, Srivastava GC	3	3	100.00	Indian Agr Res Inst, Div Plant Physiol, New Delhi 110012, India	Curr Sci 82 (10): 1227-1238 May 25 2002
Metabolite Fingerprinting and Profiling in Plants Using Nmr	Krishnan P	1	3	33.33	Cent Rice Res Inst, Div Biochem Plant Physiol & Environm Sci, Cuttack 753006, Orissa, India.	J Exp Bot 56 (410): 255-265 Jan 2005
F-Box Proteins in Rice. Genome-Wide Analysis, Classification, Temporal and Spatial Gene Expression During Panicle and Seed Development, and Regulation by Light and Abiotic Stress	Jain M, Nijhawan A, Arora R, Agarwal P, Ray S, Sharma P, Kapoor S, Tyagi AK, Khurana JP	9	9	100.00	Univ Delhi, Interdisciplinary Ctr Plant Genom, S Campus, New Delhi 110021, India	Plant Physiol 143 (4): 1467-1483 Apr 2007

Title	Authors			%	Affiliation	Citation
Regulation of Proline Biosynthesis, Degradation, Uptake and Transport in Higher Plants: Its Implications in Plant Growth and Abiotic Stress Tolerance	Kishor PBK, Sangam S, Amrutha RN, Laxmi PS, Naidu KR, Rao S,	6	10	60.00	Osmania Univ, Dept Bot, Hyderabad 500007, Andhra Pradesh, India. Nagarjuna Univ, Nagarjuna Nagar 522510, Andhra Pradesh, India. Gulbarga Univ, Dept Bot, Gulbarga 585106, India.	Curr Sci 88 (3): 424-438 Feb 10 2005
Guar Seed Beta-Mannan Synthase Is A Member of the Cellulose Synthase Super Gene Family	Gursharn S Randhawa	1	11	9.09	Indian Inst Technol, Dept Biosci & Biotechnol, Roorkee 247667, Uttar Pradesh, India.	Science 303 (5656): 363-366 Jan 16 2004
Comparison of Mercury, Lead and Arsenic With Respect to Genotoxic Effects on Plant Systems and the Development of Genetic Tolerance	Patra M, Bhowmik N, Bandopadhyay B, Sharma A	4	4	100.00	Univ Calcutta, Dept Biochem, 35 Ballygunge Circular Rd, Calcutta 700019, W Bengal, India. Dept Genet & Plant Breeding, Cooch Behar, India. Bijoygarh Coll, Dept Bot, Calcutta, W Bengal, India. Univ Calcutta, Ctr Adv Study, Dept Bot, Calcutta 700019, W Bengal, India.	Environ Exp Bot 52 (3): 199-223 Dec 2004
Linkage Disequilibrium and Association Studies in Higher Plants: Present Status and Future Prospects	Gupta PK, Rustgi S, Kulwal PL	3	3	100.00	Ch Charan Singh Univ, Dept Genet & Plant Breeding, Mol Biol Lab, Meerut 250004, Uttar Pradesh, India. Indian Natl Sci Acad, New Delhi, India.	Plant Mol Biol 57 (4): 461-485 Mar 2005
Overexpression of A Zinc-Finger Protein Gene From Rice Confers Tolerance to Cold, Dehydration, and Salt Stress in Transgenic Tobacco	Mukhopadhyay A, Vij S, Tyagi AK	3	3	100.00	Univ Delhi, Dept Plant Mol Biol, S Campus, Benito Juarez Rd, New Delhi 110021, India.	Proc Nat Acad Sci USA 101 (16): 6309-6314 Apr 20 2004

Title	Authors				Affiliation	Citation
Dinosaur Coprolites and the Early Evolution of Grasses and Grazers	Prasad V, Alimohammadian H, Sahni A	3	4	75.00	Birbal Sahni Inst Paleobot, Lucknow 226007, Uttar Pradesh, India. Panjab Univ, Ctr Adv Study Geol, Chandigarh 160014, India.	Science 310 (5751): 1177-1180 Nov 18 2005
Genomics-Assisted Breeding For Crop Improvement	Rajeev K Varshneya	1	3	33.33	Int Crops Res Inst Semi Arid Trop, Patancheru 502324, Andhra Pradesh, India	Trends Plant Sci 10 (12): 621-630 Dec 2005
Phytochelatin Synthesis and Response of Antioxidants During Cadmium Stress in Bacopa Monnieri L	Mishra S, Srivastava S, Tripathi RD, Govindarajan R, Kuriakose SV, Prasad MNV	6	6	100.00	Natl Bot Res Inst, Ecotoxicol & Bioremediat Grp, Lucknow 226001, Uttar Pradesh, India.Univ Hyderabad, Sch Life Sci, Dept Plant Sci, Hyderabad 500046, Andhra Pradesh, India.	Plant Physiol Biochem 44 (1): 25-37 Jan 2006
Coastal Mangrove Forests Mitigated Tsunami	Kathiresan K, Rajendran N	2	2	100.00	Annamalai Univ, Ctr Adv Study Marine Biol, Parangipettai 608502, Tamil Nadu, India.	Estuar Coast Shelf Sci 65 (3): 601-606 Nov 2005
Epigenetic Regulation of Stress Responses in Plants	Chinnusamy V,	1	2	50.00	Indian Agr Res Inst, Water Technol Ctr, New Delhi 110012, India.	Curr Opin Plant Biol 12 (2): 133-139 Apr 2009
Plant Products As Fumigants For Stored-Product Insect Control	Rajendran S, Sriranjini V	2	2	100.00	Cent Food Technol Res Inst, Food Protectants & Infestat Control Dept, Mysore 570020, Karnataka, India.	J Stored Prod Res 44 (2): 126-135 2008
An Assessment of the Use of Sediment Traps For Estimating Upper Ocean Particle Fluxes	Sarin M	1	12	8.33	Phys Res Lab, Ahmadabad 380009, Gujarat, India.	J Mar Res 65 (3): 345-416 May 2007
A High Density Barley Microsatellite Consensus Map With 775 Ssr Loci	Varshney RK	1	10	10.00	Nt Crops Res Inst Semi Arid Trop, Patancheru 502324, Andhra Pradesh, India.	Theor Appl Genet 114 (6): 1091-1103 Apr 2007

Title	Authors				Affiliation	Citation
Reactive Oxygen Species and Antioxidant Machinery in Abiotic Stress Tolerance in Crop Plants	Gill SS, Tuteja N	2	2	100.00	Int Ctr Genet Engn & Biotechnol, Plant Mol Biol Grp, Aruna Asaf Ali Marg, New Delhi 110067, India.	Plant Physiol Biochem 48 (12): 909-930 Dec 2010
Transgenic Approaches For Abiotic Stress Tolerance in Plants: Retrospect and Prospects	Bhatnagar-Mathur P, Vadez V, Sharma KK	3	3	100.00	Int Crops Res Inst Semi Arid Trop, Patancheru 502324, Andhra Pradesh, India.	Plant Cell Rep 27 (3): 411-424 Mar 2008
Improvement of Water Use Efficiency in Rice by Expression of Hardy, An Arabidopsis Drought and Salt Tolerance Gene	Karaba N Nataraja, Makarla Udayakumar, Andy Pereira, Aarati Karaba	4	10	40.00	Univ Agr Sci Bangalore, Dept Crop Physiol, Bangalore 560065, Karnataka, India.	Proc Nat Acad Sci USA 104 (39): 15270-15275 Sep 25 2007
Wide Geographic Distribution of Cryptosporidium Bovis and the Deer-Like Genotype in Bovines	Pradeep Das	1	10	10.00	Rajendra Mem Res Inst Med Sci, Patna, Bihar, India.	Vet Parasitol 144 (1-2): 1-9 Mar 15 2007
Responses of Antioxidant Defense System of Catharanthus Roseus (L.) G Don. to Paclobutrazol Treatment Under Salinity	Jaleel CA, Gopi R, Manivannan P, Panneerselvam R	4	4	100.00	Annamalai Univ, Dept Bot, Div Plant Physiol, Annamalainagar 608002, Tamil Nadu, India	Acta Physiol Plant 29 (3): 205-209 Jun 2007
Studies on Germination, Seedling Vigour, Lipid Peroxidation and Proline Metabolism in Catharanthus Roseus Seedlings Under Salt Stress	Jaleel CA, Gopi R, Sankar B, Manivannan P, Kishorekumar A, Sridharan R, Panneerselvam R	7	7	100.00	Annamalai Univ, Dept Bot, Div Plant Physiol, Annamalainagar 608002, Tamil Nadu, India	S Afr J Bot 73 (2): 190-195 Apr 2007
Advances in Molecular Marker Techniques and Their Applications in Plant Sciences	Agarwal M, Shrivastava N, Padh H	3	3	100.00	Bv Patel Pharmaceut Educ Res & Dev Ctr, Thaltej Gandhinagar Highway, Ahmadabad 380054, India.	Plant Cell Rep 27 (4): 617-631 Apr 2008

Title	Authors			%	Institution	Source
Ethnobiology, Socio-Economics and Management of Mangrove Forests: A Review	Syed Ainul Hussain, Ruchi Badola	2	9	22.22	Wildlife Inst India, Dehra Dun 248001, Uttar Pradesh, India.	Aquat Bot 89 (2): 220-236 Aug 2008
Inter and Intra-Population Variability of Jatropha Curcas (L.) Characterized by Rapd and Issr Markers and Development of Population-Specific Scar Markers	Basha SD, Sujatha M	2	2	100.00	Directorate Oilseeds Res, Crop Improvement Sect, Hyderabad 500030, Andhra Pradesh, India.	Euphytica 156 (3): 375-386 Sep 2007
Plant Physiology Meets Phytopathology: Plant Primary Metabolism and Plant-Pathogen Interactions	Sinha AK,	1	3	33.33	Natl Inst Plant Genome Res, New Delhi 110067, India.	J Exp Bot 58 (15-16): 4019-4026 2007
Aba Perception and Signalling	Raghavendra AS, Gonugunta VK, Christmann A,	3	4	75.00	Univ Hyderabad, Dept Plant Sci, Sch Life Sci, Hyderabad 500134, Andhra Pradesh, India.	Trends Plant Sci 15 (7): 395-401 Jul 2010
Gamete Formation Without Meiosis in Arabidopsis	Ravi M, Marimuthu MPA, Siddiqi I	3	3	100.00	Ctr Cellular & Mol Biol, Uppal Rd, Hyderabad 500007, Andhra Pradesh, India.	Nature 451 (7182): 1121-U10 Feb 28 2008
Genome Sequence and Analysis of the Tuber Crop Potato	Chakrabarti SK, Virupaksh U Patil	2	96	2.08	Cent Potato Res Inst, Shimla 171001, Himachal Prades, India.	Nature 475 (7355): 189-U94 Jul 14 2011
Lysozyme: An Important Defence Molecule of Fish Innate Immune System	Saurabh S, Sahoo PK	2	2	100.00	Central Inst Freshwater Aquaculture, Fish Hlth Management Div, Bhubaneswar 751002, Orissa, India.	Aquac Res 39 (3): 223-239 Feb 14 2008

Title	Authors			%	Affiliation	Citation
Abscisic Acid-Mediated Epigenetic Processes in Plant Development and Stress Responses	Viswanathan Chinnusamy	3	1	33.33	Indian Agr Res Inst, Water Technol Ctr, New Delhi 110012, India.	J Integr Plant Biol 50 (10): 1187-1195 Oct 2008
The Relationship Between Metal Toxicity and Cellular Redox Imbalance	Sharma SS,	2	1	50.00	Himachal Pradesh Univ, Dept Biosci, Shimla 171005, Himachal Prades, India.	Trends Plant Sci 14 (1): 43-50 Jan 2009
Biological Control of Postharvest Diseases of Fruits and Vegetables by Microbial Antagonists: A Review	Sharma RR, Singh D, Singh R	3	3	100.00	Indian Agr Res Inst, New Delhi 110012, India. Cent Inst Postharvest Engn & Technol, Div Hort Crops Proc, Abohar 152116, India.	Biol Control 50 (3): 205-221 Sep 2009
Heat Stress: An Overview of Molecular Responses in Photosynthesis	Mohanty, P	6	1	16.67	Reg Plant Resource Ctr, Bhubaneswar, Orissa, India., Jawaharlal Nehru Univ, New Delhi 110067, India.	Photosynth Res 98 (1-3): 541-550 Oct 2008
Arbuscular Mycorrhizal Fungi in Alleviation of Salt Stress: A Review	Evelin H, Kapoor R, Giri B	3	3	100.00	Univ Delhi, Swami Shraddhanand Coll, Dept Bot, Delhi 110036, India.	Ann Bot 104 (7): 1263-1280 Dec 2009
Effects of Brassinosteroids on the Plant Responses to Environmental Stresses	Hayat S	2	1	50.00	Aligarh Muslim Univ, Dept Bot, Plant Physiol & Biochem Div, Aligarh 202002, Uttar Pradesh, India.	Plant Physiol Biochem 47 (1): 1-8 Jan 2009
Antioxidant Defense Responses: Physiological Plasticity in Higher Plants Under Abiotic Constraints	Jaleel CA, Gopi R, Manivannan P, Panneerselvam R	9	4	44.44	Annamalai Univ, Dept Bot, Stress Physiol Lab, Annamalainagar 608002, Tamil Nadu, India.	Acta Physiol Plant 31 (3): 427-436 May 2009
Orphan Legume Crops Enter the Genomics Era!	Rajeev K Varshney, Nagendra K Singh, David A Hoisington	5	3	60.00	Int Crops Res Inst Semi Arid Trop, Patancheru 502324, Andhra Pradesh, India. Nrcpb, New Delhi 110012, India.	Curr Opin Plant Biol 12 (2): 202-210 Apr 2009

Title	Authors				Affiliation	Source
Effect of Exogenous Salicylic Acid Under Changing Environment: A Review	Hayat Q, Hayat S, Irfan M	3	4	75.00	Aligarh Muslim Univ, Dept Chem, Plant Physiol Sect, Aligarh 202002, Uttar Pradesh, India	Environ Exp Bot 68 (1): 14-25 Mar 2010
Integration of Novel Ssr and Gene-Based Snp Marker Loci in the Chickpea Genetic Map and Establishment of New Anchor Points With Medicago Truncatula Genome	Nayak SN, Varghese N, Singh J, Hoisington DA, Varshney RK, Kavikishor PB, Srivaramakrishnan S, Dutta S	8	15	53.33	Int Crops Res Inst Semi Arid Trop, Ctr Excellence Genom, Patancheru 502324, Andhra Pradesh, India. Osmania Univ, Dept Genet, Hyderabad 500007, Andhra Pradesh, India. Indian Inst Pulses Res, Kanpur 208024, Uttar Pradesh, India. Acharya Ng Ranga Agr Univ, Dept Agr Biotechnol, Hyderabad 500030, Andhra Pradesh, India.	Theor Appl Genet 120 (7): 1415-1441 May 2010
Heavy Metals Toxicity in Plants: An Overview on the Role of Glutathione and Phytochelatins in Heavy Metal Stress Tolerance of Plants	Yadav SK	1	1	100.00	Csir, Inst Himalayan Bioresource Technol, Div Biotechnol, Palampur 176061, Himachal Prades, India.	S Afr J Bot 76 (2): 167-179 Apr 2010
Rna Interference in Lepidoptera: An Overview of Successful and Unsuccessful Studies and Implications For Experimental Design	Sriramana Kanginakudru, Archana S Gandhe, Nirotpal Mrinal, Piotr Bebas, Joanna Kotwica, Teruyuki Niimi, Javaregowda Nagaraju	7	73	9.59	Ctr Dna Fingerprinting & Diagnost, Mol Genet Lab, Hyderabad 500001, Andhra Pradesh, India.Univ Delhi S Campus, Dept Genet, New Delhi 110021, India.	J Insect Physiol 57 (2): 231-245 Feb 2011
Chlorophyll A Fluorescence Study Revealing Effects of High Salt Stress on Photosystem Ii in Wheat Leaves	Mehta P, Jajoo A, Mathur S, Bharti S	4	4	100.00	Devi Ahilya Univ, Sch Life Sci, Indore 452017, Madhya Pradesh, India.	Plant Physiol Biochem 48 (1): 16-20 Jan 2010

Title	Author				Affiliation	Citation
One Hundred New Species of Lichenized Fungi: A Signature of Undiscovered Global Diversity	Upreti DK	1	103	0.97	Natl Bot Res Inst, Lichenol Lab, Lucknow 226001, Uttar Pradesh, India.	Phytotaxa 18: 1-127 Feb 18 2011
Phylogenetic Generic Classification of Parmelioid Lichens (Parmeliaceae, Ascomycota) Based on Molecular, Morphological and Chemical Evidence	Upreti DK	1	43	2.33	Csir, Natl Bot Res Inst, Lucknow 226001, Uttar Pradesh, India.	Taxon 59 (6): 1735-1753 Dec 2010
Microsatellite Markers: An Overview of the Recent Progress in Plants	Kalia RK, Rai MK, Kalia S, Singh R, Dhawan AK	5	5	100.00	Ctr Plant Biotechnol, Plant Mol Biol Div, Hisar 125004, Haryana, India. Nrc Plant Biotechnol, New Delhi 110012, India.	Euphytica 177 (3): 309-334 Feb 2011
Impact of Plant Products on Innate and Adaptive Immune System of Cultured Finfish and Shellfish	Balasundaram C	1	3	33.33	Bharathidasan Univ, Dept Anim Sci, Tiruchchirappalli 620024, India.	Aquaculture 317 (1-4): 1-15 Jul 4 2011

Title	Authors				Affiliation	Reference
Development of Genic-Ssr Markers by Deep Transcriptome Sequencing in Pigeonpea [Cajanus Cajan (L.) Millspaugh]	Dutta S, Kumawat G, Singh BP, Gupta DK, Singh S, Dogra V, Gaikwad K, Sharma TR, Raje RS, Bandhopadhya TK, Datta S, Singh MN, Bashasab F, Kulwal P, Wanjari KB, Varshney RK, Singh NK	17	18	94.44	Indian Agr Res Inst, Natl Res Ctr Plant Biotechnol, New Delhi 110012, India. Univ Kalyani, Dept Mol Biol & Biotechnol, Kalyani 741235, W Bengal, India. Indian Agr Res Inst, Div Genet, New Delhi 110012, India. Indian Inst Pulses Res, Kanpur 208024, Uttar Pradesh, India. Banaras Hindu Univ, Inst Agr Sci, Varanasi 221005, Uttar Pradesh, India. Univ Agr Sci, Dharwad 580005, Karnataka, India. Panjabrao Deshmukh Krishi Vidyapeeth, Akola 444104, Maharasthra, India. Int Crops Res Inst Semi Arid Trop, Patancheru 502324, Ap, India.	Bmc Plant Biol 11: Art. No.-17 Jan 20 2011
Genome-Wide Association Mapping: A Case Study in Bread Wheat (Triticum Aestivum L.)	Varshney RK	1	5	20.00	Icrisat, Patancheru 502324, Greater Hyderab, India.	Mol Breeding 27 (1): 37-58 Jan 2011
Identification of Water Deficit Stress Upregulated Genes in Sugarcane	Prabu G, Kawar PG, Pagariya MC, Prasad DT	4	4	100.00	Vasantdada Sugar Inst, Mol Biol & Genet Engn Div, Pune 412307, Maharashtra, India.Shivaji Univ, Dept Biotechnol, Kolhapur 416004, Maharashtra, India.	Pl Mol Biol Rep 29 (2): 291-304 Jun 2011

Title	Authors			%	Affiliation	Citation
Agricultural Biotechnology For Crop Improvement in A Variable Climate: Hope or Hype?	Varshney RK, Bansal KC, Aggarwal PK, Datta SK, Craufurd PQ	5	5	100.00	Icrisat, Patancheru 502324, Andhra Pradesh, India. Nrcpb, New Delhi 110012, India.Indian Agr Res Inst, Div Environm Sci, New Delhi 110012, India. Iwmi, Cgiar Challenge Program Climate Change Agr & Food, New Delhi 110012, India. Icar, Div Crop Sci, New Delhi 110114, India.	Trends Plant Sci 16 (7): 363-371 Jul 2011
Taxonomic Reassessment of the Genus Mychonastes (Chlorophyceae, Chlorophyta) Including the Description of Eight New Species	Dadheech PK	1	4	25.00	Govt Coll, Dept Bot, Ajmer 305001, Rajasthan, India.	Phycologia 50 (1): 89-106 Jan 2011
Synthetic Clonal Reproduction Through Seeds	Marimuthu MPA, Davda JN	2	11	18.18	Csir, Ctr Cellular & Mol Biol, Hyderabad 500007, Andhra Pradesh, India.	Science 331 (6019): 876-876 Feb 18 2011
Overexpression of Brassica Juncea Wild-Type and Mutant Hmg-Coa Synthase 1 in Arabidopsis Up-Regulates Genes in Sterol Biosynthesis and Enhances Sterol Production and Stress Tolerance	Nagegowda DA	1	7	14.29	Cent Inst Med & Aromat Plants, Lucknow 226016, Uttar Pradesh, India.	Plant Biotechnol J 10 (1): 31-42 Jan 2012
Chemical Modification of Simul Wood With Styrene-Acrylonitrile Copolymer and Organically Modified Nanocla	Devi RR, Maji TK	2	2	100.00	Tezpur Univ, Dept Chem Sci, Napaam 784028, Assam, India.	Wood Sci Technol 46 (1-3): 299-315 Jan 2012

Title	Indian Authors	No. of Indian Authors	Total Author	Ratio I/F*	Institue Name	Source
Effects of Nacl and Iso-Osmotic Peg Stress on Growth, Osmolytes Accumulation and Antioxidant Defense in Cultured Sugarcane Cells	Patade VY, Bhargava S, Suprasanna P	3	3	100.00	Bhabha Atom Res Ctr, Funct Plant Biol Sect, Nucl Agr & Biotechnol Div, Bombay 400085, Maharashtra, India. Univ Pune, Dept Bot, Pune 411007, Maharashtra, India	Plant Cell Tissue Organ Cult 108 (2): 279-286 Feb 2012
Effect of Coriolus Versicolor Supplemented Diet on Innate Immune Response and Disease Resistance in Kelp Grouper Epinephelus Bruneus Against Listonella Anguillarum	Balasundaram C	1	5	20.00	Bharathidasan Univ, Dept Anim Sci, Tiruchchirappalli 620024, India.	Fish Shellfish Immunol 32 (2): 339-344 Feb 2012
(s) Psychology						
Prevalence and Correlates of Bipolar Spectrum Disorder in the World Mental Health Survey Initiative	Sagar Rajesh	1	17	5.88	All India Inst Med Sci, Dept Psychiat, Delhi, India.	Arch Gen Psychiat 68 (3): 241-251 Mar 2011
Wpa Guidance on Steps, Obstacles and Mistakes to Avoid in the Implementation of Community Mental Health Care	Thara R	1	18	5.56	Schizophrenia Res Fdn Scarf, Chennai, Tamil Nadu, India.	World Psychiatry 9 (2): 67-77 Jun 2010
Childhood Adversities and Adult Psychopathology in the Who World Mental Health Surveys	Sagar Rajesh	1	33	3.03	All India Inst Med Sci, Dept Psychiat, New Delhi, India.	Brit J Psychiat 197 (5): 378-385 Nov 2010

Title	Indian Authors	No. of Indian Authors	Total Author	Ratio I/F*	Institue Name	Source
Screening For Serious Mental Illness in the General Population With the K6 Screening Scale: Results From the Who World Mental Health (Wmh) Survey Initiative	Rajesh Sagar	1	19	5.26	All India Inst Med Sci, Dept Psychiat, New Delhi, India.	Int J Meth Psychiatr Res 19: 4-22 Suppl. 1 Jun 2010
Development of Lifetime Comorbidity in the World Health Organization World Mental Health Surveys	Rajesh Sagar	1	27	3.70	All India Inst Med Sci, New Delhi, India.	Arch Gen Psychiat 68 (1): 90-100 Jan 2011
Physical Illness in Patients With Severe Mental Disorders. I. Prevalence, Impact of Medications and Disparities in Health Care	Shiv Gautam	1	13	7.69	Coll Med, Ctr Psychiat, Jaipur, Rajasthan, India	World Psychiatry 10 (1): 52-77 Feb 2011
Serotonin Reuptake Inhibitor Antidepressants and Abnormal Bleeding: A Review For Clinicians and A Reconsideration of Mechanisms	Andrade C, Sandarsh S, Chethan KB, Nagesh KS	4	4	100.00	Natl Inst Mental Hlth & Neurosci, Dept Psychopharmacol, Bangalore 560029, Karnataka, India.	J Clin Psychiat 71 (12): 1565-1575 Dec 2010
Twelve-Month Prevalence of and Risk Factors For Suicide Attempts in the World Health Organization World Mental Health Surveys	Sagar R	1	27	3.70	All India Inst Med Sci, New Delhi, India.	J Clin Psychiat 71 (12): 1617-1628 Dec 2010
-	-	-	-	-	-	-

(t) Multidisciplinary

(u) Social Sciences

Title	Indian Authors	No. of Indian Authors	Total Author	Ratio I/F*	Institue Name	Source
Diet, Nutrition and the Prevention of Hypertension and Cardiovascular Diseases	Reddy KS	1	2	50.00	All India Inst Med Sci, Ctr Cardiothorac, Dept Cardiol, New Delhi 110029, India.	Public Health Nutr 7 (1A): 167-186 Sp. Iss. Si Feb 2004
The Sonagachi Project: A Sustainable Community Intervention Program	Smarajit Jana, Ishika Basu	2	4	50.00	Std Hiv Intervent Programme, Calcutta, W Bengal, India.	Aids Educ Prev 16 (5): 405-414 Oct 2004
Effect of Iron Supplementation on Mental and Motor Development in Children: Systematic Review of Randomised Controlled Trials	Sachdev HPS, Gera T,	2	3	66.67	E-6-12 Vasant Vihar, New Delhi 110057, India. Maulana Azad Med Coll, Dept Pediat, Div Clin Epidemiol, New Delhi 110002, India. Sl Jain Hosp, New Delhi 110052, India.	Public Health Nutr 8 (2): 117-132 Apr 2005
Arsenic Contamination in Groundwater: A Global Perspective With Emphasis on the Asian Scenario	Mukherjee A, Sengupta MK, Hossain MA, Ahamed S, Das B, Nayak B, Lodh D	7	9	77.78	Jadavpur Univ, Sch Environm Studies, Calcutta 700032, W Bengal, India	J Health Popul Nutr 24 (2): 142-163 Jun 2006
The Impact of Traumatic Brain Injuries: A Global Perspectiv	Gururaj G, Puvanachandra P	2	5	40.00	Natl Inst Mental Hlth & Neurosci, Bangalore, Karnataka, India.	Neurorehabilitation 22 (5): 341-353 2007
The Protocols For the 10/66 Dementia Research Group Population-Based Research Programme	Jacob KS, Krishnamoorthy ES	2	18	11.11	Christian Med Coll & Hosp, Vellore, Tamil Nadu, India., Srinivasan Ctr Clin Neurosci, Inst Neurol Sci, Voluntary Hlth Serv, Madras, Tamil Nadu, India.	Bmc Public Health 7: Art. No.-165 Jul 20 2007

Title	Authors				Institution	Source
Application of Passive Solar Architecture For Intelligent Building Construction: A Review	Ralegaonkar RV, Gupta R	2	2	100.00	Vnit, Dept Civil Engn, Nagpur 440010, Maharashtra, India. Bits, Dean Ehd, Civil Engn Grp, Pilani 333031, Rajasthan, India.	Energy Educ Sci Technol-Pt A 26 (1): 75–85 Oct 2010
Research Approaches to Mobile Use in the Developing World: A Review of the Literature	Donner J	1	1	100.00	Microsoft Res India, Bangalore 560080, Karnataka, India.	Inform Soc 24 (3): 140–159 2008
Energy and Economic Evaluation of Fixed Focus Type Solar Parabolic Concentrator For Community Cooking Applications	Rajamohan P, Rajasekhar RVJ, Shanmugan S, Ramanathan K	4	4	100.00	Thiagarajar Coll Engn, Dept Phys, Madurai 625015, Tamil Nadu, India. Madurai Kamaraj Univ, Reg Test Ctr Solar Thermal, Madurai 625021, Tamil Nadu, India	Energy Educ Sci Technol-Pt A 26 (1): 49–59 Oct 2010
Poverty, Gender Inequities, and Women's Risk of Human Immunodeficiency Virus/Aids	Suneeta Krishnan	1	6	16.67	Indian Inst Management, Ctr Publ Policy, Bangalore, Karnataka, India.	Ann N Y Acad Sci 1136: 101–110 2008
Global Retail Chains and Poor Farmers: Evidence From Madagascar	Minten B, Randrianarison L, Swinnen JFM	3	3	100.00	Int Food Policy Res Inst, New Delhi, India.	World Develop 37 (11): 1728–1741 Sp. Iss. Si Nov 2009
Performance Evaluation of A Smooth Flat Plate Solar Air Heater	Kumar TS, Mittal V, Thakur NS, Kumar A	4	4	100.00	Brcm Coll Engn & Technol, Med, Bahal, Bhiwani, India. Nit Hamirpur, Dept Mech Engn, Hamirpur, India.	Energy Educ Sci Technol-Pt A 23 (1-2): 105–117 Apr-Jul 2009
Is There A Place For A Mock H-Index?	Prathap G	1	1	100.00	Natl Inst Sci Commun & Informat Resources, New Delhi 110012, India.	Scientometrics 84 (1): 153–165 Jul 2010

Title	Author			%	Affiliation	Journal
Do Changes in Spousal Employment Status Lead to Domestic Violence? Insights From A Prospective Study in Bangalore, India	Krishnan S, Subbiah K	2	6	33.33	Samata Hlth Study, Bangalore, Karnataka, India. Indian Inst Management, Ctr Publ Policy, Bangalore, Karnataka, India.	Soc Sci Med 70 (1): 136-143 Sp. Iss. Si Jan 2010
Cyanobacteria and Ultraviolet Radiation (Uvr) Stress: Mitigation Strategies	Sinha RP	1	3	33.33	Banaras Hindu Univ, Lab Photobiol & Mol Microbiol, Ctr Adv Study Bot, Varanasi 221005, Uttar Pradesh, India.	Ageing Res Rev 9 (2): 79-90 Sp. Iss. Si Apr 2010
Biodiesel From Jatropha: Can India Meet the 20% Blending Target?	Biswas PK, Pohit S, Kumar R	3	3	100.00	Natl Council Appl Econ Res, New Delhi 110002, India.	Energ Policy 38 (3): 1477-1484 Mar 2010
The Political Ecology of Jatropha Plantations For Biodiesel in Tamil Nadu, India	Lele S,	1	5	20.00	Ashoka Trust Res Ecol & Environm, Ctr Environm & Dev, Bangalore, Karnataka, India.	J Peasant Stud 37 (4): 875-897 2010
The 74 Ka Toba Super-Eruption and Southern Indian Hominins: Archaeology, Lithic Technology and Environments At Jwalapuram Locality 3	Korisettar R	1	8	12.50	Karnatak Univ, Dept Hist & Archaeol, Dharwad 580003, Karnataka, India.	J Archaeol Sci 37 (12): 3370-3384 Dec 2010
The Energy-Exergy-Entropy (Or Eee) Sequences in Bibliometric Assessment	Prathap G	1	1	100.00	Natl Inst Sci Commun & Informat Resources, New Delhi 110012, India.	Scientometrics 87 (3): 515-524 Jun 2011

Title	Authors			%	Affiliation	Journal
Population Health Metrics Research Consortium Gold Standard Verbal Autopsy Validation Study: Design, Implementation, and Development of Analysis Datasets	Ramesh Ahuja, Lalit Dandona, Vinita Das, Aarti Kumar, Vishwajeet Kumar, Rajendra Prasad, Devarsetty Praveen,	7	35	20.00	Community Empowerment Lab, Shivgarh, India. Inclen Trust Int, New Delhi, India. Publ Hlth Fdn India, New Delhi, India. Csm Med Univ, Lucknow, Uttar Pradesh, India. George Inst Global Hlth, Hyderabad, Andhra Pradesh, India.	Popul Health Metr 9: Art. No.-27 Aug 4 2011
Prevalence of Intellectual Disability: A Meta-Analysis of Population-Based Studies	Maulik PK	1	5	20.00	George Inst Global Hlth, Hyderabad - 2	Res Develop Disabil 32 (2): 419-436 Mar-Apr 2011
Early Pleistocene Presence of Acheulian Hominins in South India	Pappu S, Akhilesh K	2	7	28.57	Sharma Ctr Heritage Educ, Madras - 2	Science 331 (6024): 1596-1599 Mar 25 2011
Milwaukee Police Department Retirees Cardiovascular Disease Risk and Morbidity Among Aging Law Enforcement Officers	Franke WD	1	3	33.33	Coll Nursing, Trivandrum	Aaohn J 57 (11): 448-453 Nov 2009
Product Yields and Kinetics of Pyrolysis of Sawdust and Bagasse Particles	Kapoor L, Chaurasia AS	2	2	100.00	Juet, Dept Chem Engn, Guna 473226, Madhya Pradesh, India.	Energy Educ Sci Technol-Pt A 29 (1): 419-426 Apr 2012
Medical Tourism: A Review of the Literature and Analysis of A Role For Bi-Lateral Trade	Rupa Chanda	1	3	33.33	Econ Indian Inst Management Bangalore, Bangalore 560076, Karnataka, India.	Health Policy 103 (2-3): 276-282 Dec 2011
Model Projections For Household Energy Use in India	Balachandra P	1	7	14.29	Indian Inst Sci, Dept Management Studies, Bangalore 560012, Karnataka, India.	Energ Policy 39 (12): 7747-7761 Dec 2011

(v) Space Sciences

Title	Indian Authors	No. of Indian Authors	Total Author	Ratio I/F*	Institue Name	Source
Dynamics of Dark Energy	Sami M	1	3	33.33	Jamia Millia Islamia, Ctr Theoret Phys, New Delhi	Int J Mod Phys D 15 (11): 1753–1935 Nov 2006
The Seventh Data Release of the Sloan Digital Sky Survey	Yogesh Wadadekar	1	239	0.42	Tata Inst Fundamental Res, Natl Ctr Radio Astrophys, Pune 411007, Maharashtra, India.	Astrophys J Suppl Ser 182 (2): 543–558 Jun 2009
Is There Supernova Evidence For Dark Energy Metamorphosis?	Alam U, Sahni V	2	4	50.00	Inter Univ Ctr Astron & Astrophys, Pune, Maharashtra, India.	Mon Notic Roy Astron Soc 354 (1): 275–291 Oct 11 2004
An Increased Estimate of the Merger Rate of Double Neutron Stars From Observations of A Highly Relativistic System	Joshi BC	1	13	7.69	Natl Ctr Radio Astrophys, Pune 411007, Maharashtra, India.	Nature 426 (6966): 531–533 Dec 4 2003
A Double-Pulsar System: A Rare Laboratory For Relativistic Gravity and Plasma Physics	Joshi BC	1	12	8.33	Natl Ctr Astrophys, Pune 411007, Maharashtra, India.	Science 303 (5661): 1153–1157 Feb 20 2004
Exploring the Expanding Universe and Dark Energy Using the Statefinder Diagnostic	Alam U, Sahni V	2	4	50.00	Inter Univ Ctr Astron & Astrophys, Pune, Maharashtra, India	Mon Notic Roy Astron Soc 344 (4): 1057–1074 Oct 1 2003
Reconstructing Dark Energy	Sahni V	1	2	50.00	Interuniv Ctr Astron & Astrophys, Post Bag 4, Pune 411007, Maharashtra, India.	Int J Mod Phys D 15 (12): 2105–2132 Dec 2006

Title	Author				Institution	Reference
Multiwavelength Study of Massive Galaxies At Z Similar to 2. Ii. Widespread Compton-Thick Active Galactic Nuclei and the Concurrent Growth of Black Holes and Bulges	Ravindranath S	1	19	5.26	Inter Univ Ctr Astron & Astrophys, Pune 411007, Maharashtra, India.	Astrophys J 670 (1): 173-189 Part 1 Nov 20 2007
Broadband Observations of the Naked-Eye Gamma-Ray Burst Grb 080319B	Pandey SB	1	94	1.06	Aryabhatta Res Inst Observ Sci Aries, Naini Tal 263129, India.	Nature 455 (7210): 183-188 Sep 11 2008
Elemental Abundance Survey of the Galactic Thick Disc	Reddy BE	1	2	50.00	Indian Inst Astrophys, Bangalore 560034, Karnataka, India.	Mon Notic Roy Astron Soc 367 (4): 1329-1366 Apr 21 2006
Segue: A Spectroscopic Survey of 240, 000 Stars With G=14-20	Yogesh Wadadekar	1	55	1.82	Natl Ctr Radio Astrophys, Pune 411007, Maharashtra, India.	Astron J 137 (5): 4377-4399 May 2009
The Eighth Data Release of the Sloan Digital Sky Survey: First Data From Sdss-Ii	Thirupathi Sivarani	1	180	0.56	Indian Inst Astrophys, Bangalore 560034, Karnataka, India.	Astrophys J Suppl Ser 193 (2): Art. No.-29 Apr 2011
Character and Spatial Distribution of Oh/H2o on the Surface of the Moon Seen by M-3 on Chandrayaan-1	Goswami JN, Annadurai M, Kumar S	3	28	10.71	Phys Res Lab, Ahmadabad 380009, Gujarat, India.; Indian Space Res Org, Bangalore 562140, Karnataka, India.; Natl Remote Sensing Agcy, Hyderabad, Andhra Pradesh, India..	Science 326 (5952): 568-572 Oct 23 2009
Solar-Like Oscillations in Low-Luminosity Red Giants: First Results From Kepler	Mathur S	1	50	2.00	Indian Inst Astrophys, Bangalore 560034, Karnataka, India.	Astrophys J Lett 713 (2): L176-L181 Apr 20 2010

Title	Author		Count	Ratio I/F*	Affiliation	Journal
Sdss-lii: Massive Spectroscopic Surveys of the Distant Universe, the Milky Way, and Extra-Solar Planetary Systems	Thirupathi Sivarani	1	243	0.41	Indian Inst Astrophys, Bangalore 560034, Karnataka, India.	Astron J 142 (3): Art. No.-72 Sep 2011
Candels: The Cosmic Assembly Near-Infrared Deep Extragalactic Legacy Survey	Ravindranath S	1	1	100.00	Interuniv Ctr Astron & Astrophys, Pune, Maharashtra, India.	Astrophys J Suppl Ser 197 (2): Art. No.-35 Dec 2011
Candels: The Cosmic Assembly Near-Infrared Deep Extragalactic Legacy Survey- The Hubble Space Telescope Observations, Imaging Data Products, and Mosaics	Swara Ravindranth	1	125	0.80	Interuniv Ctr Astron & Astrophys, Pune 411007, Maharashtra, India.	Astrophys J Suppl Ser 197 (2): Art. No.-36 Dec 2011
Ensemble Asteroseismology of Solar-Type Stars With the Nasa Kepler Mission	Mazumdar A	1	59	1.69	Homi Bhabha Ctr Sci Educ, Bombay 400088, Maharashtra, India.	Science 332 (6026): 213-216 Apr 8 2011

Ratio I/F* : Ratio of Indian papers and International collaborating authored papers

APPENDIX - II: Highly Cited Papers: Subject-wise (a-v)

	(a) Agricultural Sciences				
Sr. No.	Journal	CPP	Rank	Papers	Impact Factor
1	Advances In Agronomy	20.86	160	2	5.204
2	Agricultural Water Management	7.87	30	2	1.998
3	American Journal Of Potato Research	4.14	115	1	1.234
4	British Journal Of Nutrition	12.09	115	1	3.013
5	Compr Rev Food Sci Food Saf	10.62	163	1	3.724
6	Crit Rev Food Sci Nutr	21.97	123	5	4.789
7	Crop Science	8.08	7	2	1.641
8	Field Crop Res	9.68	35	3	2.474
9	Food And Chemical Toxicol	9.82	9	5	2.999
10	Food Bioprocess Technol	3.81	119	1	3.703
11	Food Chem	11.15	2	9	3.655
12	Food Microbiology	10.72	42	1	3.283
13	Food Nutr Bull	5.05	131	1	1.922
14	Food Res Int	9.15	23	4	3.15
15	Food Technol. Biotechnol.	5.88	93	1	1.195
16	Ind Crops Products	6.04	39	3	2.469
17	Int J Food Microbiol	15.11	12	3	3.327
18	International Journal Of Agricultural Sustainability	1.33	10	1	1.696
19	J Agr Food Chem	14.04	1	2	2.823
20	J Food Compos Anal	10.10	51	1	2.079
21	J Nutr.	20.46	4	1	3.916
22	Journal Of Plant Nutrition And Soil Science	7.39	55	1	1.596
23	Lwt-Food Sci Technol	6.23	27	1	2.545
24	Molecular Nutrition & Food Research	12.56	45	1	4.301
25	Nutrition	3.42	107	1	3.025
26	Pest Manag Science	8.46	29	1	2.251

Sr. No.	Journal	CPP	Rank	Papers	Impact Factor
27	Postharvest Biol Technol	11.03	36	1	2.411
28	Science	131.62	165	1	31.201
29	Trends In Food Science & Technology	20.12	95	2	3.672
(b) Bio and Biochemistry					
Sr. No.	Journal	CPP	Rank	Papers	Impact Factor
1	J Biosciences	13.87	27	1	1.648
2	Arch Biochem Biophys	13.87	27	1	2.935
3	Biochem Biophys Res Commun	12.91	2	2	2.484
4	Biomacromolecules	20.50	18	1	5.479
5	Bioresource Technol	12.30	7	7	4.98
6	Biosens Bioelectron	17.13	20	3	5.602
7	Biotechnol Adv	25.46	212	8	9.646
8	Biotechnol Progr	11.30	54	1	2.34
9	Colloid Surface B	8.69	33	4	3.456
10	J Biol Chem	28.97	1	1	4.773
11	J Biomed Nanotechnol	9.37	1056	353	4.216
12	J Biomol Struct Dyn	7.98	183	2	0*
13	J Bone Miner Res	25.96	46	1	6.373
14	J Ind Microbiol Biotechnol	8.67	110	1	2.735
15	Life Sci	13.42	16	1	2.527
16	Nat Biotechnol	93.51	3	3	23.268
17	Nat Chem Biol	40.40	198	1	14.69
18	Nat Protoc	24.82	100	1	9.924
19	Nucl Acid Res	29.36	4	5	8.026
20	Proc Nat Acad Sci USA	46.15	8	1	9.681
21	Process Biochem	13.16	36	1	2.627
22	Protein-Struct Funct Genet	36.88	210	1	3.337
23	Science	137.26	173	1	31.201
24	Trends In Biotechnology	33.58	178	1	9.148
* Impact Factor score for the years: 2013/2014 - 2.983; 2012 -0; 2011 -0; 2010 - 4.986; 2009 - 1.124; 2008 - 1.289					

(c) Chemistry				
Sr. No.	Journal	CPP	Rank	Papers	Impact Factor
1	Account Chem Res	77.57	218	8	21.64
2	Acs Nano	14.58	102	1	10.774
3	Acta Crystallogr B-Struct Sci	18.86	260	1	2.286
4	Advan Colloid Interface Sci	27.06	280	6	8.120
5	Anal Bioanal Chem	11.34	33	1	3.778
6	Anal Chim Acta	15.55	29	4	4.555
7	Analyst	11.11	99	2	4.23
8	Angew Chem Int Ed	38.69	10	14	13.455
9	Appl Catal A-Gen	16.38	52	1	3.903
10	Biomed Chromatogr	6.16	198	1	1.966
11	Bioorgan Med Chem	11.51	32	1	2.921
12	Carbohyd Polym	9.63	73	3	3.628
13	Carbon	17.38	71	2	5.378
14	Chem Commun	18.13	8	8	6.169
15	Chem Eng J	7.89	59	2	3.461
16	Chem Phys Lett	11.72	14	3	2.337
17	Chem Rev	150.54	179	12	40.197
18	Chem Soc Rev	58.76	195	9	28.76
19	Chem-Eur J	18.01	20	2	5.925
20	Chemphyschem	13.01	95	1	3.412
21	Colloid Surface A	8.94	44	1	2.236
22	Coord Chem Rev	50.25	211	4	12.110
23	Cryst Growth Des	13.53	63	1	4.720
24	Crystengcomm	8.41	93	2	3.842
25	Dalton Trans	10.36	27	2	3.838
26	Desalination	7.05	36	1	2.590
27	Dye Pigment	10.71	162	5	3.126
28	Eur J Med Chem	8.71	94	1	3.346
29	Eur J Org Chem	11.20	40	1	3.329
30	Eur Polym J	11.49	90	1	2.739

31	Fuel	10.14	78	3	3.248
33	Inorg Chem	18.19	13	3	4.601
34	J Am Chem Soc	38.51	2	8	9.907
35	J Appl Cryst	17.01	188	1	5.152
36	J Biol Inorg Chem	14.14	246	1	3.289
37	J Chem Phys	13.32	3	3	3.333
38	J Chem Sci	1.06	410	1	1.177
39	J Cheminformatics	11.54	175	1	3.419
40	J Colloid Interface Sci	11.98	25	6	3.07
41	J Fluoresc	6.96	226	1	2.107
42	J Instrum	1.48	15	2	1.869
43	J Membrane Sci	14.93	47	2	3.85
44	J Nanophotonics	1.52	37	1	1.57
45	J Org Chem	18.13	11	3	4.45
46	J Photochem Photobiol C-Photo	43.95	303	1	10.36
47	J Phys Chem A	12.76	9	9	2.946
48	J Phys Chem Lett	4.85	245	1	6.213
49	J Therm Anal Calorim	5.50	54	1	1.604
50	Langmuir	18.08	6	4	4.186
51	Molecules	4.21	105	1	2.386
52	Nano Lett	44.64	35	4	13.198
53	Nano Today	30.43	300	1	15.355
54	Nanoscale	4.13	200	1	5.914
55	Org Biomol Chem	11.15	45	2	3.696
56	Org Lett	20.94	12	10	5.862
57	Polym Degrad Stabil	11.74	117	1	2.769
58	Polymer	16.89	28	1	3.438
59	Proc Nat Acad Sci USA	52.38	145	1	9.681
60	Prog Cryst Growth Charact	1.64	12	1	5.75
61	Prog Polym Sci	91.17	291	12	24.1
62	Prog Solid State Chem	2.67	2	1	4.188
63	Pure Appl Chem	12.46	180	2	2.789

64	Qsar Comb Sci	1.94	29	1	1.55
65	React Funct Polym	10.06	220	1	2.479
66	Science	184.83	255	1	31.201
67	Sep Purif Technol	10.90	125	1	2.921
68	Surf Sci Rep	83.21	302	1	11.696
69	Synthesis-Stuttgart	10.25	56	1	2.466
70	Talanta	12.61	37	1	3.794
71	Tetrahedron	12.84	17	11	3.025
(d) Clinical Medicine					
Sr. No.	Journal	CPP	Rank	Papers	Impact Factor
1	Am J Transplant	22.37	2	127	6.394
2	Amer J Gastroenterol 106 (2): 199-212 Feb 2011	26.28	1	111	7.282
3	Amer J Infect Control 38 (2): 95-U31 Mar 2010	10.12	1	544	2.396
4	Amer J Respir Crit Care Med 184 (1): 132-140 Jul 1 2011	40.58	1	94	11.08
5	Amer J Trop Med Hyg 83 (6): 1178-1182 Dec 2010	11.94	1	81	2.592
6	Ann Intern Med 154 (8): 523-W177 Apr 19 2011	66.39	1	354	16.733
7	Ann N Y Acad Sci 1215: 1-8 2011	14.26	1	67	3.155
8	Ann Rheum Dis 65 (7): 936-941 Jul 2006	22.01	1	92	8.727
9	Bba-Rev Cancer 1785 (2): 182-206 Apr 2008	38.31	1	1067	9.38
10	Bju Int 101 (1): 83-88 Jan 2008	10.34	1	37	2.844
11	Blood 119 (5): 1123-1129 Feb 2 2012	39.34	1	2	9.898
12	Cancer Res 69 (9): 3892-3900 May 1 2009	39.69	3	3	7.856
13	Circulation 113 (11): 1434-1441 Mar 21 2006	64.12	3	9	14.739
14	Clin Infect Dis 45 (7): 883-893 Oct 1 2007	28.96	1	35	9.154

15	Clin J Am Soc Nephrol 5 (12): 2207-2212 Dec 2010	11.20	1	434	5.227
16	Diabetes Care 32 (1): 84-90 Jan 2009	30.72	2	39	8.087
17	Diabetologia 49 (2): 289-297 Feb 2006	23.49	4	148	6.814
18	Eur Heart J 31 (6): 703-711C Mar 2010	35.52	1	152	10.478
19	Eur Resp J 38 (3): 516-528 Sep 2011	22.32	1	106	5.895
20	Hepatol Int 3 (1): 269-282 Mar 2009	2.68	1	767	2.645
21	Hepatology 54 (1): 91-100 Jul 2011	40.86	1	88	11.665
22	Int Orthop 36 (2): 261-269 Sp. Iss. Si Feb 2012	4.79	1	395	2.025
23	J Amer Acad Dermatol 60 (5): S1-S50 Suppl. S May 2009	14.26	1	115	3.991
24	J Amer Coll Cardiol 43 (7): 1149-1153 Apr 7 2004	47.78	1	36	14.156
25	J Clin Oncol 21 (11): 2101-2109 Jun 1 2003	54.12	10	12	18.372
26	J Clin Pharmacol 52 (1): 6-17 Jan 2012	12.36	1	438	2.911
27	J Endourol 25 (1): 11-17 Jan 2011	6.29	2	187	1.847
28	J Exp Med 206 (11): 2407-2416 Oct 26 2009	72.92	1	149	13.853
29	J Intern Med 259 (3): 247-258 Mar 2006	23.29	1	572	5.483
30	J Nat Cancer Inst 95 (23): 1772-1783 Dec 3 2003	63.04	1	458	13.757
31	J Pharm Pharm Sci 6 (1): 33-66 Jan-Apr 2003	9.59	1	1038	1.646
32	J Thromb Haemost 4 (10): 2103-2114 Oct 2006	19.15	1	173	5.731
33	J Vasc Interven Radiol 22 (3): 265-278 Mar 2011	9.01	1	285	2.075
34	Jama-J Am Med Assn 293 (4): 437-446 Jan 26 2005	124.24	5	151	30.026

35	Lancet 378 (9790): 487-497 Aug 6 2011	107.90	38	69	38.278
36	Lancet Infect Dis 10 (9): 597-602 Sep 2010	51.63	3	916	17.391
37	Lancet Neurol 9 (12): 1157-1163 Dec 2010	57.12	1	784	23.462
38	Lancet Oncol 11 (11): 1048-1056 Nov 2010	43.93	2	644	22.589
39	N Engl J Med 347 (22): 1739-1746 Nov 28 2002	192.67	31	102	53.298
40	Nanomedicine 2 (1): 23-39 Feb 2007	12.10	1	1023	5.055
41	Nanomed-Nanotechnol Biol Med 6 (1): 9-24 Feb 2010	9.37	4	1056	6.692
42	Nature 430 (6998): 419-421 Jul 22 2004	248.65	5	747	36.28
43	Neoplasia 6 (1): 1-6 Jan-Feb 2004	20.07	1	631	5.946
44	Neuro-Oncology 13 (1): 132-142 Jan 2011	14.22	1	901	5.723
45	Nmr Biomed 22 (1): 104-113 Jan 2009	16.16	1	716	3.214
46	Oncogene 30 (14): 1615-1630 Apr 2011	33.91	1	11	6.373
47	Osteoporosis Int 20 (11): 1807-1820 Nov 2009	16.54	2	239	4.58
48	Periton Dialysis Int 25 (2): 107-131 Mar-Apr 2005	7.58	2	592	2.097
49	Plos Med 8 (1): Art. No.-E1000406 Jan 2011	27.72	5	668	16.269
50	Proc Nat Acad Sci USA 101 (25): 9309-9314 Jun 22 2004	64.65	2	42	9.681
51	Radiother Oncol 89 (2): 180-191 Nov 2008	14.93	2	256	2.321
52	Stem Cells 26 (2): 300-311 Feb 2008	29.92	2	248	7.781
53	Thorax 64 (6): 476-483 Jun 2009	26.39	1	360	6.84
54	Urology 72 (2): 260-263 Aug 2008	10.83	1	23	2.428

(e) Computer Sciences					
Sr. No.	Journal	CPP	RANK	PAPERS	Impact Factor
1	Appl Soft Comput	3.31	35	1	2.612
2	Bioinformatics	26.48	2	6	5.468
3	Bmc Bioinformatics	10.75	4	5	2.751
4	Brief Bioinform	40.06	143	1	5.202
5	Commun Acm	6.04	24	1	1.919
6	Comput Biol Chem	6.72	117	1	1.551
7	Comput Chem Eng	7.99	19	2	2.320
8	Comput Ind Eng	4.89	27	2	1.589
9	Comput Math Appl	4.02	6	6	1.747
10	Comput Sci Eng	3.27	108	1	1.422
11	Comput Struct	6.83	20	1	1.874
12	Curr Bioinform	0.75	19	2	0.898
13	Environ Modell Softw	9.27	31	1	3.114
14	Evol Comput	8.02	182	1	1.061
15	Ieee Trans Commun	6.90	10	4	1.677
16	Inform Sciences	7.02	12	1	2.833
17	J Netw Comput Appl	1.44	84	1	1.065
(f) Economics and Business					
Sr. No.	Journal	CPP	Rank	Papers	Impact Factor
1	Ecol Econ	20.86	160	2	2.713
2	Int J Manag Rev	7.87	30	1	3.33
3	J Econometrics	4.14	115	1	1.349
(g) Engineering					
Sr. No.	Journal	CPP	RANK	Papers	Impact Factor
1	Prog Electromagn Res	7.29	114	1	5.298
2	Acta Astronaut	1.48	68	1	0.569
3	Algorithmica	2.62	273	1	0.604
4	Annu Rev Fluid Mech	48.15	498	2	12.767
5	Appl Energ	6.13	80	8	5.106
6	Appl Math Comput	4.39	6	4	1.317

7	Appl Therm Eng	6.13	48	1	2.127
8	Chemometr Intell Lab Syst	9.28	218	1	1.92
9	Cold Reg Sci Technol	5.08	302	1	1.429
10	Desalin Water Treat	1.14	83	1	0.614
11	Elec Power Syst Res	3.63	117	1	1.478
12	Electroanal	11.33	42	2	2.872
13	Energ Conv Manage	8.31	45	3	2.216
14	Energ Fuel	8.52	18	3	2.721
15	Energy	7.04	55	4	3.487
16	Eur J Oper Res	7.28	14	4	1.815
17	Exp Therm Fluid Sci	5.31	164	1	1.414
18	Expert Syst Appl	3.82	11	1	2.203
19	Heat Mass Transfer	3.24	173	2	0.896
20	Ieee Electron Dev Lett	8.50	37	1	2.849
21	Ieee J Solid-State Circuits	10.28	63	1	3.226
22	Ieee Sens J	3.75	74	1	1.52
23	Ieee Signal Process Mag	14.85	394	1	4.066
24	Ieee Trans Dielect Electr In	4.44	118	1	1.094
25	Ieee Trans Evol Computat	22.26	407	6	3.341
26	Ieee Trans Fuzzy Syst	12.14	274	1	4.26
27	Ieee Trans Geosci Remot Sen	10.20	36	1	2.895
28	Ieee Trans Image Processing	9.68	77	1	3.042
29	Ieee Trans Ind Electron	9.47	39	4	5.16
30	Ieee Trans Microwave Theory	8.53	31	1	1.853
31	Ieee Trans Patt Anal Mach Int	18.26	107	2	4.908
32	Ieee Trans Power Syst	8.04	76	3	2.678
33	Image Vision Comput	5.66	197	1	1.723
34	Ind Eng Chem Res	3.70	12	2	2.237
35	Int Commun Heat Mass Trans	4.75	131	1	1.892
36	Int J Adv Manuf Technol	3.17	20	2	1.103
37	Int J Heat Mass Transfer	9.35	16	9	2.407
38	Int J Hydrogen Energ	9.69	9	19	4.054
39	Int J Mech Sci	6.02	190	1	1.231

40	Int J Non-Linear Mech	6.18	196	2	1.209
41	Int J Therm Sci	5.85	124	2	2.142
42	Int Rev Electr Eng-Iree	1.74	220	1	1.36
43	J Franklin Inst-Eng Appl Math	3.15	298	1	2.724
44	J Global Optim	4.90	232	1	1.196
45	J Hazard Mater	11.28	7	97	4.173
46	J Heat Transfer	5.52	99	1	1.83
47	J Hydrol	11.85	28	3	2.656
48	J Irrig Drain Eng-Asce	5.34	310	1	0.941
49	J Power Sources	15.66	4	7	4.951
50	Macromol Mater Eng	9.10	200	1	1.986
51	Math Comput Modelling	3.42	58	1	1.346
52	Mech Mach Theor	4.45	212	1	1.366
53	Med Biol Eng Comput	5.62	185	1	1.878
54	Microelectron Rel	3.50	54	1	1.167
55	Nav Res Log	4.29	381	1	1.038
56	Neurocomputing	4.07	41	1	1.58
57	Nucl Data Sheets	8.34	499	1	3.821
58	Nucl Eng Des	2.83	53	1	0.765
59	Nucl Instrum Meth Phys Res A	4.68	1	8	1.207
60	Nucl Instrum Meth Phys Res B	5.03	3	1	1.266
61	Org Electron	8.67	144	1	4.047
62	Pattern Recognition Lett	5.61	181	1	2.292
63	Proc Ieee 97	16.27	177	1	6.810
64	Prog Energ Combust Sci	42.66	505	2	14.22
65	Prog Photovoltaics	12.48	361	1	5.789
66	Renewable Energy	6.93	65	8	2.978
67	Chemometrics And Intelligent Laboratory Systems	1.77	18	1	2.291
68	Sensor Actuator B-Chem	10.00	12	15	3.898
69	Signal Process	3.84	72	1	1.503
70	Spectrochim Acta Pt A-Mol Bio	6.20	15	16	2.098
71	Thermochim Acta	7.91	40	1	2.162

| 72 | Vib Spectrosc | 7.60 | 222 | 1 | 1.65 |
| 73 | Waste Management | 8.53 | 92 | 4 | 2.428 |

(h) Environment					
Sr. No.	**Journal**	**CPP**	**RANK**	**Papers**	**Impact Factor**
1	Adv Environ Res	24.88	183	2	-
2	Biol Conserv	16.70	17	1	4.115
3	Biomass Bioenerg	10.80	41	4	3.646
4	Chemosphere	14.70	2	1	3.206
5	Crit Rev Environ Sci Technol	16.89	182	2	4.841
6	Ecotoxicol Environ Safety	9.21	37	1	2.294
7	Energy Environ Sci	10.29	68	1	9.610
8	Environ Int	19.13	57	2	5.297
9	Environ Sci Technol	21.31	1	2	5.228
10	Global Environ Change	7.10	93	1	6.868
11	Hydrol Process	9.14	23	1	2.488
12	Int Biodeterior Biodegrad	7.22	74	1	2.074
13	J Environ Manage	8.29	28	1	3.245
14	Mar Pollut Bull	11.05	25	1	2.503
15	Mol Ecol Resour	10.75	24	1	3.062
16	Nature	127.67	174	2	36.28
17	Proc Nat Acad Sci USA	51.02	103	1	9.681
18	Renew Sustain Energy Rev	9.64	50	15	6.018
19	Science	124.82	163	4	31.201
20	Water Res	19.15	7	5	4.865

(i) Geography Sciences					
Sr. No.	**Journal**	**CPP**	**Rank**	**Papers**	**Impact Factor**
1	Appl Clay Sci	8.14	39	1	2.474
2	Appl Geochem	10.70	34	2	2.176
3	Atmos Chem Phys	15.44	7	1	5.52
4	Atmos Environ	14.86	3	1	3.465
5	Bull Amer Meteorol Soc	22.87	108	1	6.026
6	Clim Dynam	15.62	47	1	4.602

7	Curr Sci	4.73	114	1	0.935
8	Earth Planet Sci Lett	19.77	4	1	4.180
9	Geochem J	4.84	164	1	0.711
10	Geophys Res Lett	12.90	1	2	3.792
11	Global Biogeochem Cycle	20.12	67	1	4.785
12	Gondwana Res	11.27	101	3	6.659
13	Int J Remote Sens	6.17	8	1	1.117
14	J Geophys Res-Atmos	15.79	2	4	3.021
15	Nat Geosci	18.69	148	2	11.754
16	Nature	67.00	77	1	36.28
17	Palaeogeogr Palaeoclimatol	12.41	13	1	2.392
18	Quart J Roy Meteorol Soc	13.23	40	1	2.907
19	Quatern Int	6.78	33	3	1.874
20	Science	76.58	79	4	31.201

(j) Immunology

S. No.	Journal	CPP	RANK	Papers	Impact Factor
1	Develop Comp Immunol 36 (1): 93-103 Jan 2012	12.39	30	1	3.268
2	Int Rev Immunol 30 (1): 16-34 Feb 2011	17.70	26	1	3.426

(k) Material Sciences

Sr. No.	Journal	CPP	RANK	Papers	Impact Factor
1	Acta Mater	16.82	20	4	3.755
2	Adv Funct Mater	29.19	36	5	10.179
3	Advan Mater	42.36	17	11	13.877
4	Biomaterials	28.91	16	7	7.404
5	Chem Mater	28.54	11	11	7.286
6	Composites Sci Technol	14.21	44	1	3.144
7	Corros Sci	10.98	38	2	0.692
8	Curr Opin Solid State Mat Sci	26.96	172	2	4.233
9	Int Mater Rev	30.68	184	2	6.962
10	J Alloys Compounds	6.55	1	1	2.289
11	J Mater Chem	14.25	5	8	5.968

12	J Mater Sci	5.73	8	2	2.015
13	J Mater Sci-Mater Med	8.25	53	1	2.316
14	J Nanopart Res	6.90	67	1	3.287
15	Lancet	170.9	69	1	38.278
16	Mater Chem Phys	7.35	19	1	2.234
17	Mater Lett	7.58	6	1	2.307
18	Mater Manuf Process	2.62	86	1	1.058
19	Mater Sci Eng A-Struct Mater	8.74	4	1	2.003
20	Mater Sci Eng C	2.18	126	1	2.404
21	Nanotechnol	9.37	9	1	3.979
22	Nat Mater	112.99	93	5	32.841
23	Nat Nanotechnol	26.27	171	1	27.27
24	Polym Eng Sci	6.28	56	1	1.302
25	Prog Mater Sci	63.44	180	3	18.216
26	Prog Org Coating	8.05	81	1	1.977
27	Sci Technol Adv Mater	7.30	25	1	3.513
28	Science	142.59	185	1	31.201
29	Scripta Mater	11.65	22	1	2.699
30	Small	20.07	57	2	8.349
31	Solar Energ Mater Solar Cells	11.37	39	3	4.542
32	Thin Solid Films	7.76	3	2	1.89
33	Nature Materials	176.70	188	1	35.749
34	Surf Coat Tech 167 (2-3): 269-277 Apr 22 2003	8.90	10	1	1.941

(I) Mathematics

Sr No.	Journal	CPP	RANK	Papers	Impact Factor
1	Ann Probab	7.00	87	1	1.789
2	Commun Nonlinear Sci Numer Si	4.03	23	8	2.806
3	Electron J Probab	2.63	157	1	0.713
4	Fixed Point Theory Appl	2.47	131	5	1.634
5	Int J Biomath	0.69	43	2	0.364
6	J Amer Statist Assn	16.15	45	1	1.992
7	J Appl Math	5.31	232	2	0.656

8	J Comput Appl Math	4.34	3	2	1.112
9	J Math Anal Appl	4.75	1	1	1.001
10	Nonlinear Anal-Real World App	4.49	24	9	2.043
11	Numer Func Anal Optimiz	2.81	134	1	0.711
12	Stat Sinica	6.49	110	1	1.017
13	Z Angew Math Phys	4.11	123	1	0.951

(m) Microbiology

Sr. No.	Journal	CPP	RANK	Papers	Impact Factor
1	Acta Biomater	9.61	27	1	4.865
2	Antimicrob Agents Chemother	19.43	4	1	4.841
3	Appl Microbiol Biotechnol	12.47	6	2	3.425
4	Bmc Microbiol	7.82	37	1	3.044
5	Clin Microbiol Rev	82.92	72	1	16.129
6	Int J Antimicrobial Agents	10.83	24	1	4.128
7	Microb Biotechnol	4.44	18	2	2.534
8	Microbiol Mol Biol Rev	92.14	75	1	13.018
9	Mol Microbiol	27.39	7	1	5.01
10	Nature	146.80	74	1	36.28
11	Parasitol Res	6.27	12	1	2.149
12	Plos Pathog	16.07	23	1	9.127

(n) Molecular Biology

Sr. No.	Journal	CPP	RANK	Papers	Impact Factor
1	Cell	133.18	21	2	32.403
2	Cell Mol Biol	7.38	113	14	0.975
3	Dna Res	20.09	150	1	5.164
4	Genome Res	50.48	32	1	13.608
5	J Theor Biol	9.45	18	1	2.208
6	Mol Cell Biochem	8.74	20	1	2.057
7	Nat Genet	120.08	34	3	35.532
8	Nat Rev Mol Cell Biol	165.56	107	1	39.123
9	Proc Nat Acad Sci USA	38.20	3	1	9.681
10	Sci Transl Med	13.63	17	1	7.804
11	Science	138.57	71	1	31.201

(o) Pharmcology					
Sr. No.	Journal	CPP	RANK	Papers	Impact Factor
1	Advan Drug Delivery Rev	52.70	47	3	11.502
2	Biochem Pharmacol	17.77	6	5	4.705
3	Brit J Pharmacol	17.28	5	2	4.409
4	Crit Rev Ther Drug Carr Syst	22.50	125	1	2.609
5	Curr Neuropharmacol	1.70	149	1	2.847
6	Drug Discov Today	26.09	43	2	6.828
7	Eur J Pharm Sci	13.06	37	2	3.212
8	Eur J Pharmacol	11.70	1	1	2.516
9	Int J Pharm	13.50	2	6	3.35
10	J Appl Toxicol	7.89	76	1	2.478
11	J Ethnopharmacol	11.35	6	1	3.014
12	J Pharm Sci	11.72	10	1	3.055
13	J Pharmaceut Biomed Anal	10.78	4	1	2.967
14	Mar Drugs	4.70	100	2	3.854
15	Pharmacol Rep	6.68	68	2	2.445
16	Phytother Res	8.20	20	1	2.086
17	Toxicol Lett	13.49	21	2	3.23
18	Toxicol Vitro	9.28	32	1	2.775
(p) Neuro Sciences					
Sr. No.	Journal	CPP	RANK	Papers	Impact Factor
1	Brain Res Rev	0	19	1	10.342
2	J Neurosci	35.10	1	1	7.115
3	Nat Rev Neurosci	135.65	104	1	30.445
4	Neurobiol Aging	18.83	37	1	6.189
5	Neurology	31.34	6	1	8.312
6	Neuron	71.25	20	2	14.736
7	Prog Neurobiol	48.06	127	1	8.874
(q) Physics					
Sr. No.	Journal	CPP	RANK	Papers	Impact Factor
1	Advan Phys	78.74	185	1	37
2	Ann Phys N Y	11.21	148	1	2.857

3	Annu Rev Condens Matter Phys	10.54	6	1	12.389
4	Appl Phys Lett	13.98	2	4	3.844
5	Class Quantum Gravity	10.22	46	7	3.32
6	Commun Math Phys	9.19	107	1	1.941
7	Comput Phys Commun	8.20	96	2	3.268
8	Crit Rev Solid State Mat Sci	2.939	34	1	9.467
9	Curr Appl Phys	4.12	98	1	1.900
10	Eur Phys J C	11.27	77	6	3.631
11	Gen Relativ Gravit	5.96	131	1	2.069
12	Int J Mod Phys A	3.66	52	1	1.053
13	J Appl Phys	8.18	3	2	2.168
14	J High Energy Phys	14.10	12	24	5.831
15	J Phys Chem C	10.20	9	10	4.805
16	J Phys G-Nucl Particle Phys	7.27	78	3	4.178
17	J Phys-B-At Mol Opt Phys	6.96	44	1	1.875
18	J Phys-Condens Matter	8.07	10	2	2.546
19	Jetp Lett-Engl Tr	4.42	84	1	1.352
20	Mod Phys Lett A	4.73	83	1	1.083
21	Nature	167.37	153	1	36.28
22	Nucl Fusion	10.90	115	1	4.09
23	Nucl Phys A	7.72	37	3	1.54
24	Nucl Phys B	20.63	59	3	4.661
25	Phys Lett B	16.64	18	27	3.955
26	Phys Plasmas	8.09	26	2	2.147
27	Phys Rep-Rev Sect Phys Lett	82.51	172	11	20.394
28	Phys Rev A	9.66	7	2	2.878
29	Phys Rev B	12.36	1	16	3.691
30	Phys Rev C	10.73	20	17	3.308
31	Phys Rev D	15.83	5	43	4.558
32	Phys Rev Lett	27.56	4	77	7.37
33	Physica A	6.50	31	1	1.373
34	Physica E	4.42	43	1	1.532

35	Rep Progr Phys	46.69	176	5	14.72
36	Rev Mod Phys	168.48	179	3	43.933
37	Science	138.47	160	3	31.201
38	Superlattice Microstruct	4.20	133	1	1.487
39	Solid State Commun	7.44	41	1	1.534
(r) Plant Animal Sciences					
Sr. No.	Journal	CPP	RANK	Papers	Impact Factor
1	Acta Physiol Plant	2.78	136	1	1.639
2	Ann Bot	15.35	36	1	4.03
3	Aquac Res	5.70	39	1	1.203
4	Aquaculture	10.98	2	1	2.041
5	Aquat Bot	8.97	195	1	1.516
6	Aquat Toxicol	15.04	52	1	3.761
7	Biol Control	8.21	65	1	2.003
8	Bmc Plant Biol	7.41	198	1	3.447
9	Curr Opin Plant Biol	41.55	182	2	9.272
10	Curr Sci	6.13	289	4	0.935
11	Environ Exp Bot	10.52	131	2	2.985
12	Estuar Coast Shelf Sci	10.24	22	1	2.247
13	Euphytica	6.69	30	2	1.554
14	Fish Shellfish Immunol	10.24	62	1	3.322
15	J Exp Bot	19.64	8	3	5.364
16	J Insect Physiol	8.73	73	1	2.236
17	J Integr Plant Biol	5.10	154	1	2.534
18	J Mar Res	8.33	400	1	0.766
19	J Plant Physiol	8.99	55	1	2.791
20	J Stored Prod Res	6.97	328	1	1.414
21	Mol Breeding	9.15	181	1	2.852
22	Nature	120.62	344	4	36.28
23	Photosynth Res	12.67	153	1	3.243
24	Phycologia 50	8.07	291	1	1.647
25	Phytotaxa	1.21	98	1	1.797

26	Pl Mol Biol Rep	5.07	345	1	2.453
27	Plant Biotechnol J	15.70	280	1	5.442
28	Plant Cell	49.01	28	1	8.987
29	Plant Cell Rep	9.73	66	3	2.274
30	Plant Cell Tissue Organ Cult	6.63	74	1	3.09
31	Plant Mol Biol	20.48	60	1	4.15
32	Plant Physiol	31.27	3	1	6.535
33	Plant Physiol Biochem	10.20	95	4	2.838
34	Plant Sci	11.56	35	2	2.945
35	Proc Nat Acad Sci USA	48.86	56	2	9.681
36	S Afr J Bot	3.82	185	2	1.659
37	Science	116.88	335	3	31.201
38	Taxon	8.59	210	1	2.703
39	Theor Appl Genet	18.12	20	3	3.297
40	Trends Plant Sci	43.72	202	7	11.047
41	Vet Parasitol	9.06	7	1	2.579
42	Wood Sci Technol	6.06	322	1	1.727
43	Trends Plant Sci	43.72	202	1	11.808

(s) Psychiatry / Psychology					
S.No.	Journal	CPP	RANK	Papers	Impact Factor
1	Arch Gen Psychiat	69.15	32	2	12.016
2	Brit J Psychiat	25.79	15	1	6.619
3	Int J Meth Psychiatr Res	17.13	274	1	2.462
4	J Clin Psychiat	25.93	7	2	5.799
5	World Psychiatry	8.90	5	2	6.233

(t) Multidisciplinary					
S. No.	Journal	CPP	RANK	Papers	Impact Factor
-	-	-	-	-	-

(u) Social Sciences					
S. No.	Journal	CPP	RANK	Papers	Impact Factor
1	Aaohn J	0	103	1	1.468
2	Ageing Res Rev	23.59	393	2	6.174

3	Aids Educ Prev	11.35	158	3	1.484
4	Ann N Y Acad Sci	7.15	924	4	3.155
5	Bmc Public Health	5.17	3	5	1.997
6	Energ Policy	7.10	1	6	2.723
7	Energy Educ Sci Technol-Pt A	11.91	409	7	31.677
8	Health Policy	5.93	20	8	1.506
9	Inform Soc	6.28	583	9	0.532
10	J Archaeol Sci	5.84	10	10	1.914
11	J Health Popul Nutr	4.88	145	11	0.954
12	J Peasant Stud	3.40	533	12	5.805
13	Neurorehabilitation	6.11	149	13	1.635
14	Popul Health Metr	2.67	13	14	1.024
15	Public Health Nutr	11.14	12	15	2.169
16	Res Develop Disabil	6.36	48	16	0
17	Science	35.17	933	17	31.201
18	Scientometrics	7.33	19	18	1.966
19	Soc Sci Med	15.17	2	19	2.699
20	World Develop	9.07	23	20	0

(v) Space Sciences

Sr. No.	Journal	CPP	Rank	Papers	Impact Factor
1	Astron J	22.89	5	2	4.035
2	Astrophys J	23.51	1	1	6.024
3	Astrophys J Lett	9.62	8	3	6.345
4	Astrophys J Suppl Ser	50.64	17	4	16.238
5	Int J Mod Phys D	6.13	12	2	1.03
6	Mon Notic Roy Astron Soc 3	17.25	3	3	4.90
7	Nature	63.34	29	2	36.28
8	Science	59.12	34	3	31.201

CPP : Citations Per Paper Rank: Rank of the Journal in the given Subject Category Papers : Total Number of Papers Published in the given journal Imapct Factor : Impact Factor of the Journal
NB: '-' details unavailable from the source

Appendix III: Unique Institutions Contributing to the Indian Highly Cited Papers (HCP) (based on Top 1% of ESI database Jan 2002-June 2012)

Name of Institutions	Occurrence Frequency*	Subject**	Total No. of HCPs Across Fields
Tata Inst Fundamental Res EHEP, Bombay, Maharashtra, India.	96	Physics	110
Panjab Univ, Chandigarh 160014, India.	84	Physics	100
Bhabha Atom Res Ctr, Backend Technol Dev Div, Bombay 400085, Maharashtra, India	54	Physics	69
Indian Inst Sci, Bangalore 560012, Karnataka, India.	20 each	Chemistry / Physics	73
Univ Delhi, Deen Dayal Upadhyaya Coll, Delhi 110015, India.	42	Physics	72
Banaras Hindu Univ, Dept Microbiol, Inst Med Sci, Varanasi 221005, Uttar Pradesh, India.	22	Physics	58
Indian Inst Technol Bombay Powai, Shailesh J Mehta Sch Management, Mumbai 400076, Maharashtra, India.	25	Physics	26
Jawaharlal Nehru Ctr Adv Sci Res, Bangalore 560061, Karnataka, India.	13	Chemistry	35
Anna Univ Tirunelveli, Tirunelveli 627007, TN, India	10	Engineering	21
Inter Univ Ctr Astron & Astrophys, Pune 411007, Maharashtra, India.	12	Physics	19
Univ Rajasthan, Ctr Nonconvent Energy Resources, 14 Vigyan Bhavan, Jaipur 302004, Rajasthan, India.	29	Physics	37
Indian Inst Technol Madras, Dept Phys, Nano Funct Mat Technol Ctr, Alternat Energy & Nanotechnol Lab, Madras 600036, Tamil Nadu, India.	16	Engineering	32
Indian Inst Technol, Agr & Food Engn Dept, Kharagpur 721302, W Bengal, India	13	Engineering	31
Indian Inst Technol, Dept Biol Sci & Bioengn, Kanpur 208016, Uttar Pradesh, India.	12	Engineering	33
Inst Phys, Bhubaneswar 751005, Orissa, India.	29	Physics	32
Univ Jammu, Dept Phys, Jammu 180004	10	Physics	30
Indian Inst Technol Roorkee, Dept Chem Engn, Roorkee 247667, Uttar Pradesh, India	13	Engineering	31

Indian Inst Technol Delhi, Appl Microbiol Lab, Ctr Rural Dev & Technol, New Delhi 110016, India	2	Agricultural Sciences	29
Natl Chem Lab, Biochem Sci Div, Mat Chem Div, Nanosci Grp, Pune 411008, Maharashtra, India.	3	Biology & Biochemistry	28
Univ Hyderabad, Dept Comp & Informat Sci, Hyderabad 500046, Andhra Pradesh, India	1	Computer Science	26
All India Inst Med Sci, Ctr Cardiothorac, Dept Cardiol, New Delhi 110029, India.	1	Social Sciences	21
Saha Inst Nucl Phys, Calcutta 700064, W Bengal, India.	20	Physics	21
Harish Chandra Res Inst, Allahabad 211019, Uttar Pradesh, India.	18	Physics	18
Jadavpur Univ, Calcutta 700032, India.	10	Engineering	18
CSIR, Div Biotechnol, NIIST, Thiruvananthapuram 695019, Kerala, India	1 each	Biology & Biochemistry / Engineering	2
St Johns Med Coll & Res Inst, Bangalore, Karnataka, India	13	Clinical Medicine	13
Indian Assoc Cultivat Sci, Ctr Adv Mat, Calcutta 700032, India.	10	Chemistry	17
Indian Inst Chem Technol, Biochem & Environm Engn Grp, Hyderabad 500007, Andhra Pradesh, India.	12	Chemistry	17
Indian Inst Technol Guwahati, Dept Chem Engn, Gauhati 781039, Assam, India.	6	Engineering	14
Indian Agr Res Inst, New Delhi 110012, India.	11	Plant and Animal Science	15
Bharathiar Univ, Bioinformat Ctr, Coimbatore 641046, Tamil Nadu, India.	5	Mathematics	14
Aligarh Muslim Univ, Dept Bot, Plant Physiol & Biochem Div, Aligarh 202002, Uttar Pradesh, India.	9	Physics	14
Satyendra Nath Bose Natl Ctr Basic Sci, Calcutta 700098, India.	2	Physics	3
Ind Toxicol Res Ctr, Aquat Toxicol Div, POB 80,MG Marg, Lucknow 226001, Uttar Pradesh, India.	8	Engineering	12
Univ Allahabad, Dept Ancient Hist Culture & Archaeol, Allahabad 211002, Uttar Pradesh, India.	7	Molicular Biology	11

Visva Bharati Univ, Dept Stat, Santini Ketan, W Bengal, India.	7	Physics	10
Phys Res Lab, Ahmadabad 380009, Gujarat, India	7	Physics	11
Cent Food Technol Res Inst, Food Protectants & Infestat Control Dept, Mysore 570020, Karnataka, India.	4	Agricultural Sciences	5
Int Crops Res Inst Semi Arid Trop, Ctr Excellence Genom, Patancheru 502324, Andhra Pradesh, India.	6	Plant and Animal Science	9
Indian Stat Inst, Econ Res Unit, Calcutta 108, India.	4	Engineering	9
Jamia Millia Islamia Univ, Dept Biosci, New Delhi 110025, India.	3	Physics	9
Tata Mem Hosp, Adv Ctr Treatment, Res & Educ Canc, Navi Mumbai 41021, Maharashtra, India.	9	Clinical Medicine	9
Gandhigram Rural Univ, Dept Chem, Gandhigram 624302, Tamil Nadu, India.	5	Engineering	10
Natl Phys Lab, Biomol Elect & Conduct Polymer Res Grp, Dr KS Krishnan Marg, New Delhi 110012, India.	3	Biology & Biochemistry	9
Utkal Univ, Bhubaneswar 751004, Orissa, India.	7	Physics	8
VIT Univ, Div Appl Math, Vellore 632014, Tamil Nadu, India.	6	Engineering	7
Bharathidasan Univ, Dept Anim Sci, Tiruchchirappalli 620024, India.	3	Plant and Animal Science	7
Christian Med Coll & Hosp, Vellore, Tamil Nadu, India.	5	Clinical Medicine	8
Guru Jambheshwar Univ, Dept Environm Sci & Engn, Hisar 125001, Haryana, India.	6	Engineering	7
NIPER, Dept Med Chem, Sector 67, SAS Nagar 160062, Punjab, India.	5	Pharmcology & Toxicology	7
Publ Hlth Fdn India, New Delhi, India.	7	Clinical Medicine	8
Raman Res Inst, Bangalore 560080, Karnataka, India.	2 each	Chemistry/ Physics	7
CSIR, Indian Inst Toxicol Res, Lucknow 226001, Uttar Pradesh, India.	2 each	Pharmcology & Toxicology / Clinical Medicine	7
Inst Bioinformat, Bangalore 560066, Karnataka, India.	3	Biology & Biochemistry	8

Natl Bot Res Inst, Council Sci & Ind Res, Ecoauditing Grp, Rana Pratap Marg, Lucknow 226001, Uttar Pradesh, India.	3 each	Engineering / Plant Animal Science	7
Univ Burdwan, Burdwan 713104, W Bengal, India.	3 each	Physics / Engineering	6
Univ Calcutta, Ctr Adv Study, Dept Bot, Calcutta 700019, W Bengal, India.	2	Engineering	5
Bose Inst, Dept Biochem, Calcutta 700054, W Bengal, India.	1 each	Biology & Biochemistry / Microbiology	2
Cent Drug Res Inst, Div Drug Target Discovery & Dev, Chatter Manzil Palace, Lucknow 226001, Uttar Pradesh, India.	2	Molecular Biology	5
Cent Electro Chem Res Inst, ElectroOrgan Div, Karaikkudi 630006, Tamil Nadu, India.	5	Engineering	5
Cent Salt & Marine Chem Res Inst CSIR, Bhavnagar 364002, Gujarat, India	5	Chemistry	6
Guru Nanak Dev Univ, Dept Chem, Amritsar 143005, Punjab, India	4	Chemistry	5
Himachal Pradesh Univ, Dept Biosci, Shimla 171005, Himachal Prades, India.	2	Plant and Animal Science	5
Inst Microbial Technol, Bioinformat Ctr, Chandigarh 160036, India.	6	Computer Science	6
Jawaharlal Nehru Univ, New Delhi 110067, India	2	Plant and Animal Science	5
Jiwaji Univ, Sch Studies Chem, Gwalior 474011, MP, India.	3	Engineering	5
Karnatak Univ, Dept Chem, Dharwad 580003, Karnataka, India.	2	Pharmcology & Toxicology	5
Mar Ivanios Coll, Dept Phys, Ctr Mol & Biophys Res, Thiruvananthapuram 695015, Kerala, India.	5	Engineering	5
Natl Ctr Biol Sci TIFR, Bangalore 560065, Karnataka, India.	4	Neuroscience & Behaviour	10
Natl Inst Mental Hlth & Neurosci, Bangalore 560029, Karnataka, India.	2	Neuroscience & Behaviour	5
Shivaji Univ, Dept Biotechnol, Kolhapur 416004, Maharashtra, India.	3	Engineering	5
Sri Venkateswara Univ, Dept Chem, Analyt & Environm Chem Div, Tirupati 517502, Andhra Pradesh, India.	5	Engineering	5

Tata Inst Fundamental Res, Natl Ctr Biol Sci, Bangalore 560002, Karnataka, India.	3	Neuroscience & Behaviour	8
Thapar Inst Engn & Technol, Dept Biotechnol & Environm Sci, Patiala 147004, Punjab, India.	1	Biology & Biochemistry	1
Amrita Vishwa Vidhyapeetham Univ, Amrita Inst Med Sci & Res Ctr, Amrita Ctr Nanosci & Mol Med, Cochin 682041, Kerala, India.	2 each	Biology & Biochemistry / Materials Science	5
Bengal Engn & Sci Univ, Dept Math, Howrah 711103, W Bengal, India.	3	Mathematics	4
CSIR, Ctr Cellular & Mol Biol, Hyderabad 500007, Andhra Pradesh, India.	1 each	Plant and Animal Science / Pharmcology & Toxicology	2
Dayanand Med Coll & Hosp, Ludhiana, Punjab, India.	3	Clinical Medicine	4
Devi Ahilya Univ, Sch Energy & Environm Studies, Khandwa Rd,Takshila Campus, Indore 452011, Madhya Pradesh, India.	3	Environment / Ecology	4
Disha Inst Management & Technol, Chhatisgarh-2	3	Mathematics	4
Dr Ambedkar Coll, Dept Biochem & Biotechnol, Nagpur, Maharashtra, India.	3	Pharmcology & Toxicology	4
Gauhati Univ, Dept Chem, Assam, India.	2 each	Chemistry / Geosciences	4
Gulbarga Univ, Dept Bot, Gulbarga 585106, India.	2	Engineering	4
Indian Council Med Res, Natl Inst Malaria Res, New Delhi 110077, India.	4	Clinical Medicine	4
Indian Inst Astrophys, Bangalore 560034, Karnataka, India.	4	Space Science	4
Indian Inst Trop Meteorol, Doctor Homi Bhabha Rd, Pune 411008, Maharashtra, India.	4	Geosciences	4
Inst Life Sci, Bhubaneswar, Orissa, India.	2	Pharmcology & Toxicology	4
Inst Math Sci, Chennai 600113, Tamil Nadu, India,	4	Physcis	4
Int Ctr Genet Engn & Biotechnol, Aruna Asaf Ali Marg, New Delhi 110067, India.	2	Plant Animal Science	4
Karunya Deemed Univ, Dept Elect Sci, Coimbatore 641114, Tamil Nadu, India.	2	Engineering	2

Matrivani Inst Expt Res & Educ, Calcutta 700030, W Bengal, India.	3	Physics	4
Maulana Azad Natl Inst Technol, Dept Appl Chem, Bhopal 462007, MP, India.	3	Engineering	4
Natl Inst Hydrol, Deltaic Reg Ctr, Kakinada 533003, India.	2	Engineering	4
Natl Inst Technol Calicut, Calicut 673601, Kerala, India.	3	Engineering	3
Natl Inst Technol, Dept Chem, Tiruchchirappalli 620015, Tamil Nadu, India.	2	Engineering	3
Post Grad Inst Med Educ & Res, Dept Hepatol, Chandigarh, India	1	Clinical Medicine	1
PSG Coll, Coimbatore 641014, Tamil Nadu, India	2	Engineering	2
Sanjay Gandhi Postgrad Inst Med Sci, Ctr Biomed Magnet Resonance, Lucknow 226014, Uttar Pradesh, India	3	Clinical Medicine	4
Sitaram Bhartia Inst Sci & Res, Dept Paediat & Clin Epidemiol, New Delhi, India.	4	Clinical Medicine	4
Sree Chitra Tirunal Inst Med Sci & Technol, Biomed Technol Wing, Div Biosurface Technol, Thiruvananthapuram 695012, Kerala, India	2	Chemistry	4
St Longowal Inst Engn & Technol, Dept Chem Technol, Sangrur 148106, Punjab, India.	3	Engineering	4
Univ Kalyani, Dept Biochem & Biophys, Kalyani 741235, W Bengal, India.	1	Molicular Biology	4
Univ Madras, Dept Math, Ramanujan Inst Adv Study Math, Madras 600005, Tamil Nadu, India.	2	Clinical Medicine	4
Univ Mysore, Dept Food Sci & Nutr, Mysore 570006, Karnataka, India.	1	Agricultural Sciences	4
Univ Poona, Dept Phys, Ctr Adv Studies Mat Sci & Solid State Phys, Pune 411007, Maharashtra, India	2	Physics	3
Univ Pune, Dept Math, Pune 411007, Maharashtra, India.	1	Mathematics	4
Vikram Sarabhai Space Ctr, Battery Dev Div, Trivandrum 695022,	4	Geosciences	5
W Bengal State Univ, Barasat, W Bengal, India.	2	Physics	2
Amity Inst Nanotechnol, Noida 201301, India.	1 each	Engineering / Materials Science	2

CARE Hosp, Dept Cardiol, Exhibit Rd,Nampally, Hyderabad 500001, Andhra Pradesh, India.	3	Clinical Medicine	3
CSIR, Inst Himalayan Bioresource Technol, Div Biotechnol, Palampur 176061, Himachal Prades, India	1 each	Biology & Biochemistry / Materials Science / Plant Animal Science	3
Ctr DNA Fingerprinting & Diagnost, Immunol Lab, Hyderabad 500076, Andhra Pradesh, India.	1 each	Chemistry / Microbiology / Plant & Animal Science	3
Ctr Fuel Cell Technol ARCI, Madras 600113, Tamil Nadu, India.	2	Engineering	2
Def R&D Org, Ctr Fire Explosives & Environm Safety, Brig SK Majumdar Marg, Delhi 110054, India.	2	Engineering	2
Diabet Unit, 6th Floor,Banoo Coyaji Bldg, Pune 411011, Maharashtra, India.	3	Clinical Medicine	3
Dr Hari Singh Gour Univ, Pharmaceut Res Projects Lab, Dept Pharmaceut Sci, Sagar, MP, India.	1	Clinical Medicine	1
High Energy Mat Res Lab, Pune 41102, Maharashtra, India.	3	Engineering	3
ICRISAT, Patancheru 502324, Andhra Pradesh, India.	2 each	Plant and Animal Science / Agricultural Sciences	5
Ind Inst Technol Roorkee, Dept Chem, Roorkee 247667, Uttar Pradesh, India.	3	Chemistry	3
Indian Inst Sci Educ & Res Mohali, Sect 81, Mohali - 2	2	Physcis	2
Indian Institute of Science, Bangalore	1 each	Agricultural Sciences / Chemistry / Clinical Medicine	3
Indira Gandhi Ctr Atom Res, Kalpakkam 603102, Tamil Nadu, India.	2	Physics	3
Inst Res & Fundamental Sci IPM, Tehran, Iran	1	Physics	1
Karaikal Polytech Coll, Dept Sci & Humanities, Karaikal 609609, Puducherry, India.	2	Engineering	2

Karpagam Arts & Sci Coll, Dept Biotechnol, Coimbatore 641021, Tamil Nadu, India.	1 each	Agricultural Sciences / engineering / neuro	3
Kidwai Mem Inst Oncol, Bangalore, Karnataka, India.	3	Clinical Medicine	3
King Edward Mem Hosp & Res Ctr, Maharashtra	2	Clinical Medicine	3
KS Rangasamy Coll Technol, Dept Elect & Elect Engn, Tiruchengode, Tamilnadu, India	2	Engineering	3
Kuvempu Univ, Dept Biotechnol, Shankaraghatta, Karnataka, India.	1	Biology & Biochemistry	2
Madurai Kamaraj Univ, Reg Test Ctr Solar Thermal, Madurai 625021, Tamil Nadu, India	1 each	Social Science / Chemistry	2
Maulana Azad Med Coll, Bahadur Shah Zafar Marg, Delhi 110002, India.	1 each	Clinical / Social Science	2
Muljibhai Patel Urol Hosp, Nadiad, India.	3	Clinical Medicine	3
Natl Ctr Astrophys, Pune 411007, Maharashtra, India.	1	Space Science	1
Natl Environm Engn Res Inst, Environm Imapct & Risk Assesment Div, Nagpur 440020, Maharashtra, India.	2	Engineering	3
Natl Inst Oceanog, Panaji 403004, Goa, India.	1 each	Environment/ Geosciences/ Pharmacology & Toxicology	3
Natl Inst Technol Silchar, Dept Chem, Silchar 788010, Assam, India.	1 each	Chemistry / Engineering	2
Nizam Inst Med Sci, Hyderabad, Andhra Pradesh, India.	1	Clinical Medicine	1
NTPC Ltd Kahalgaon, Kahalgaon STPP, Bhagalpur 813214, India.	2	Engineering	2
Osmania Univ, Dept Bot, Hyderabad 500007, Andhra Pradesh, India.	2	Plant and Animal Science	3
Periyar Maniammai Univ, Dept Biotechnol, Thanjavur 613403, Tamil Nadu, India	1	Engineering	1
Reg Canc Ctr, Div Radiat Oncol, Trivandrum 695011, Kerala, India.	3	Clinical Medicine	3
S Asia Network Chron Dis, New Delhi, India.	2	Molicular Biology	3

Sangath Ctr, Alto Porvorim 403521, Goa, India.	3	Clinical Medicine	3
St Josephs Coll Engn, Dept Chem, Madras 600119, Tamil Nadu, India.	1	Engineering	1
Tamil Nadu Agr Univ, Coimbatore 641003, Tamil Nadu, India.	1 each	Biology & Biochemistry / Environment/ Ecology	2
Univ Agr Sci Bangalore, Dept Biotechnol, Coll Agr, GKVK, Bangalore 560065, Karnataka, India.	1 each	Agriculture / Plant Animal Science	2
Univ Kerala, Ctr Arthropod Bioresources & Biotechnol, Trivandrum 695001, Kerala, India.	1 each	Chemistry / Engineering / Pharmocology & Toxicology	3
Wildlife Inst India, Dehra Dun 248001, Uttar Pradesh, India.	1 each	Environment/ Ecology/Clinical Medicine/ Plant & Animal Science	3
YR Gaitonde Ctr AIDS Res & Educ, Voluntary Hlth Serv, Madras, Tamil Nadu, India	3	Clinical Medicine	3
Aaranyak & Int Rhino Fdn, Gauhati 781028, Assam, India.	1	Environment/ Ecology / Clinical Medicine/ Plant & Animal Science	1
Affiliated Univ Delhi, GB Pant Hosp, Dept Gastroenterol, Jawahar Lal Nehru Rd, New Delhi 110002, India;	2	Clinical Medicine	2
Agharkar Res Inst, Ctr Nanobiosci, GG Agarkar Rd, Pune 411004, Maharashtra, India.	1 each	Pharmcology & Toxicology / Clinical Medicine	2
Amrita Inst Med Sci, Amrita Ctr Nanosci & Mol Med, Cochin 682041, Kerala, India.	2	Materials Science	2
Apollo Hosp, Dept Microbiol, Madras, Tamil Nadu, India.	2	Clinical Medicine	2
Aryabhatta Res Inst Observ Sci ARIES, Naini Tal 263129, India.	1 each	Space Science / Geosciences	2
AVC Coll Autonomous, Dept Phys, Mappadugai 609305, Mayiladuthurai, India.	2	Engineering	2

Birla Inst Technol & Sci Pilani, Dept Chem, Zuarinagar 403726, Goa, India.	1 each	Chemistry / Physics	2
Birla Inst Technol & Sci Pilani, Dept Math, Hyderabad 500078, Andhra Pradesh, India.	1 each	Computer Science / engineering	2
BITS, Dean EHD, Civil Engn Grp, Pilani 333031, Rajasthan, India.	1	Social Science	1
Canc Epigenet Res, Kalyani B-7-183, Nadia 741235, W Bengal, India.	1	Clinical Medicine	1
Cent Electrochem Res Inst, Chennai Unit, Madras 600113, Tamil Nadu, India.	1	Materials Science	1
Cent Inst Med & Aromat Plants, Council Sci & Ind Res, Div Biotechnol, Mol Bioprospect Dept, Lucknow 226015, Uttar Pradesh, India.	1 each	Pharmcology & Toxicology / Plant & Animal Science	2
Cent Leather Res Inst, Chem Lab, Madras 600020, Tamil Nadu, India	2	Engineering	2
Cent Mech Engn Res Inst, Durgapur 713209, W Bengal, India.	1 each	Engineering / Chemistry	2
Chhatrapati Shahuji Maharaj Med Univ, Dept Anat, Lucknow 226003, Uttar Pradesh, India.	2	Molicular Biology	2
Christian Fellowship Community Hlth Area, Ambilikai, Tamil Nadu, India.	2	Clinical Medicine	2
CSIR, Chem & Phys Mat Unit, Ctr excellence Chem, Bangalore 560064, Karnataka, India.	2	Chemistry	2
CSIR, CSTD, NIIST, Trivandrum 695019, Kerala, India.	1	Engineering	1
CSIR, IITR, Lucknow -2	1	Neuroscience & Behaviour	1
CSIR, Intellectual Property Management Div, 14 Satsang Vihar Marg, New Delhi 110067, India.	1	Materials Science	1
CSIR, Nat Prod Chem Div, Reg Res Lab, Jorhat 785006, Assam, India.	1	Chemistry	1
CSIR, Natl Phys Lab, Biomed Instrumentat Sect, Mat Phys & Engn Div,Ctr Biomol Elect,Dept Sci & T, New Delhi 110012, India.	1 each	Biology & Biochemistry / Materials science	2
Ctr Cellular & Mol Biol, Funct Genom & Gene Silencing Grp, Hyderabad 500007, Andhra Pradesh, India.	1 each	Molicular Biology / Materials Science	2

Ctr Fire Explos & Environm Safety, Brig SK Mazumdar Rd, Delhi 110054, India.	2	Engineering	2
Deemed Univ, Gandhigram Rural Inst, Dept Chem, Gandhigram 624302, India.	1 each	Engineering / Materials Science	2
Dr. S K Sachdev, Gynaecologists and Obstetricians, E-6-12 Vasant Vihar, New Delhi 110057, India. (91-11-26149787)	1	Clinical Medicine	2
Escorts Heart Inst & Res Ctr, Dept Cardiol, New Delhi, India.	1	Clinical Medicine	1
George Inst Global Hlth, Hyderabad - 2	2	Social Science	2
GIDC, Ankleshwar 393002, Grijarat, India.	1 each	Microbiology / Biology & Biochemistry	2
Govind Ballabh Pant Univ Agr & Technol, Coll Basic Sci & Humanities, Dept Math Stat & Comp Sci, Pantnagar 263145, Uttar Pradesh, India.	1 each	Computer Science / Agriculture	2
Govt India, Minist Informat & Technol, Ctr Mat Elect Technol, Nanocrystalline Mat Lab, Panchawati Pashran Rd, Pune 411008, Maharashtra, India.	1	Materials Science	1
IIT Bombay, Dept Aerosp Engn, Bombay, Maharashtra, India.	1	Computer Science	1
ILBS, Dept Hepatol, New Delhi 110070, India	2	Clinical Medicine	2
INCLEN Trust Int, New Delhi, India.	1	Social Science	1
India Diabet Res Fdn, Chennai, Tamil Nadu, India.	2	Clinical Medicine	2
India Inst Management, Ctr Publ Policy, Bangalore, Karnataka, India.	2	Social Science	2
Indian Inst Integrat Med, Dept Canc Pharmacol, Jammu 180001	2	Chemistry	2
Indian Inst Management, Dept Quantitat Methods & Informat Sci, Bangalore 566076, Karnataka, India.	1	Mathematics	1
Indian Inst Pulses Res, Kanpur 208024, Uttar Pradesh, India.	2	Plant and Animal Science	2
Indian Inst Sci Educ & Res Bhopal, Bhopal 462023, India.	1	Chemistry	1
Indian Inst Technol, Hyderabad, Dept Phys, Hyderabad 502205, Andhra Pradesh, India.	2	Physcis	2

Indian Meteorol Dept, Pune 411005, Maharashtra, India.	2	Geosciences	2
Indian Oil Corp Ltd, R&D Ctr, Faridabad 121007, India.	1	Engineering	2
Indraprastha Apollo Hosp, New Delhi 110044, India.	2	Clinical Medicine	2
Inst Chem Technol, Food Engn & Technol Dept, Bombay 400019, Maharashtra, India.	1 each	Biology & Biochemistry / Engineering	2
Inst Genom & Integrat Biol CSIR, Mall Rd, Delhi 110007, India.	1 each	Biology & Biochemistry / Molicular Biology	2
Inst Plasma Res, Bhat 382428, Gandhinagar, India.	2	Physics	2
Int Inst Informat Technol, Hyderabad 500032, Andhra Pradesh, India	2	Materials Science	2
Int Tech Pk, Inst Bioinformat, Bangalore 560066, Karnataka, India.	1	Biology & Biochemistry	1
International Crops Research Institute for Semi-Arid Tropics (ICRISAT), Patancheru 502324, India	2	Agricultural Sciences	2
Jai Narain Vyas Univ, Dept Math & Stat, Jodhpur 342005, Rajasthan, India	1 each	Mathematics / Engineering	2
Jaipur Engn Coll, Jaipur, Rajasthan, India	1 each	Engineering / Materials Science	2
Jaslok Hosp & Res Ctr, Bombay, Maharashtra, India.	2	Clinical Medicine	2
Jehangir Hosp & Med Ctr, Pune, Maharashtra, India.	1	Clinical Medicine	1
KEM Hosp & Res Ctr, Diabet Unit, Pune, Maharashtra, India	3	Clinical Medicine	3
KLDAV PG Coll, Dept Chem, Roorkee, Uttar Pradesh, India.	1	Engineering	1
Med Res Ctr Hinduja, Bombay, Maharashtra, India.	1	Clinical Medicine	1
Mepco Schlenk Engn Coll, Dept Biotechnol, Sivakasi 608502, Tamil Nadu, India.	1	Clinical Medicine	1
Minist Hlth & Family Welf, Natl Vector Borne Dis Control Programme, New Delhi, India.	2	Clinical Medicine	2

MLS Univ, Ctr Comp, Udaipur 313002, Rajasthan, India.	1 each	Engineering / Mathematics	2
MS Ramaiah Inst Technol, Bangalore 560054, Karnataka, India.	1	Materials Science	1
N Maharashtra Univ, Sch Chem Sci, Jalgaon 425001, Maharashtra, India.	1	Engineering	1
Natl AIDS Res Inst, Pune, Maharashtra, India.	2	Clinical Medicine	2
Natl Dairy Res Inst, Dairy Technol Div, Karnal 132001, Haryana, India.	2	Agricultural Sciences	2
Natl Geophys Res Inst, Council Sci & Ind Res, Hyderabad 500007, Andhra Pradesh, India	2	Geosciences	2
Natl Inst Interdisciplinary Sci & Technol, Mat Minerals Div, Trivandrum 695019, Kerala, India.	1 each	Materials Science / Biology & Biochemistry / Chemistry / Engineering	4
Natl Inst Pharmaceut Educ & Res, Dept Pharmaceut, Ctr Pharmaceut Nanotechnol, Sas Nagar 160062, Punjab, India.	1 each	Materials Science / Biology / Chemistry	3
Natl Inst Sci Commun & Informat Resources, New Delhi 110012, India.	2	Social Science	2
Natl Inst Technol Durgapur, Dept Biotechnol, Durgapur 713209, WB, India.	2	Chemistry	2
Natl Inst Technol Warangal, Dept Chem, Warangal 506004, Andhra Pradesh, India.	1	Engineering	1
Natl Inst Technol, Dept Civil Engn, Kurukshetra 136119, Haryana, India.	2	Engineering	2
Natl Inst Virol, Pune 411001, Maharashtra, India.	1	Chemistry	1
NIT , orrisa	1	Engineering	1
Pondicherry Univ, Dept Biotechnol, Kalapet 605014, Puducherry, India.	1 each	Chemistry / Plant & Animal Science	2
Pratap Coll, Dept Phys, Mat Res Lab, Amalner 425401, Maharashtra, India.	2	Engineering	2
PRIST Univ, Ctr Res & Dev, Thanjavur, India.	3	Engineering	3
Rajiv Gandhi Canc Inst & Res Ctr, New Delhi, India.	2	Clinical Medicine	2

Ranbaxy Res Labs, Ctr Res & Dev, Gurgaon, India.	2	Pharmcology & Toxicology	2
Reg Plant Resource Ctr, Bhubaneswar 751015, Orissa, India.	1 each	Environment/ Plant Animal Science	2
Sambalpur Univ, Dept Chem, Ctr Studies Surface Sci & Technol, Jyoti Vihar 768019, India.	2	Chemistry	2
Sant Longowal Inst Engn & Technol, Dept Chem Technol, Sangrur 148106, Punjab, India.	2	Engineering	2
Sardar Patel Univ, BRD Sch Biosci, Vallabh Vidyanagar 388120, Gujarat, India	1 each	Engineering / Computer Science	2
Schizophrenia Res Fdn SCARF, Chennai, Tamil Nadu, India.	1	Psychiatry/ Psychology	1
SL Jain Hosp, Dept Paediat, Delhi, India.	1 each	Clinical Medicine/ Social Science	2
SRM Univ, Fac Engn & Technol, Dept Chem, Kattankulathur 603203, India.	1 each	Materials Science / Biology & Biochemistry	2
World Hlth Org Collaborating Ctr Noncommunicable, Dr Mohans Diabet Special Ctr, Madras, Tamil Nadu, India.	2	Clinical Medicine	2
Zoo Outreach Org, Coimbatore 641004, Tamil Nadu, India.	2	Environment/ Ecology	2
Amrita Center for Nanosciences, Amrita Institute, Amrita Vishwa Vidyapeetham University, Kochi 682041, India	1	Chemistry	1
Assam University, Department of Chemistry, Silchar, India	1	Biology & Biochemistry	1
Aartral Energy Res Org, 42-9 Avvai Nagar,Kannagi St, Madras 600094, Tamil Nadu, India.	1	Engineering	1
Abasaheb Garware Coll, Dept Chem, Pune 411004, Maharashtra, India.	1	Materials Science	1
Abeda Inamdar Senior Coll, Interdisciplinary Sci & Technol Res Acad, Pune, Maharashtra, India.	1	Clinical Medicine	1
Acharya NG Ranga Agr Univ, Dept Agr Biotechnol, Hyderabad 500030, Andhra Pradesh, India.	1	Plant and Animal Science	1

AG Hosp Bikaner, Bikaner, Rajasthan, India.	1	Clinical Medicine	1
AIST, Computat Biol Res Ctr, CBRC, Koto Ku, Tokyo 1350064, Japan.	1	Computer Science	1
Allahabad Agricultural Institute	1	Agriculture	1
AMA Coll Engn, Dept Math, Vadamavandal 604410, Tamil Nadu, India.	1	Engineering	1
Amravati Univ, Dept Biotechnol, Amravati 444602, Maharashtra, India.	1	Biology & Biochemistry	1
AMRI Hosp, Calcutta, India.	1	Clinical Medicine	1
Amritsar Coll Engn & Technol, Dept Appl Sci, Amritsar 143001, Punjab, India.	1	Engineering	1
AMU, Dept Biochem, Fac Life Sci, Aligarh 202002, Uttar Pradesh, India.	1	Pharmcology & Toxicology	1
Apollo Gleneagles Hosp, Dept Microbiol, Calcutta, India.	1	Clinical Medicine	1
Army Inst Technol, Dept Mech Engn, Pune 411015, Maharashtra, India.	1	Engineering	1
Arneja Heart Inst, Nagpur, Maharashtra, India.	1	Clinical Medicine	1
Ashoka Trust Res Ecol & Environm, Ctr Environm & Dev, Bangalore, Karnataka, India.	1	Social Science	1
Associates Clin Endocrinol Educ & Res, Madras, Tamil Nadu, India.	1	Clinical Medicine	1
AstraZeneca India, Bangalore, Karnataka, India.	1	Chemistry	1
ATREE, Ctr Environm & Dev, Bangalore, Karnataka, India.	1	Economics & Business	1
AVVM Sri Pushpam Coll Autonomous, PG & Res Dept Phys, Thanjavur 613503, Tamil Nadu, India.	1	Physcis	1
Bharathiar Univ, Dept Chem, Coimbatore 641046, Tamil Nadu, India.	1	Chemistry	1
Balaji Utthan Sanastan, Patna, Bihar, India.	1	Clinical Medicine	1
Bangabasi Morning Coll, Dept Phys, Calcutta 700009, India.	1	Physcis	1
Bangalore Inst Oncol, Bangalore, Karnataka, India.	1	Clinical Medicine	1
Bangalore Univ, Dept Post Grad Studies Chem, Cent Coll City Campus,Dr BR Ambedkar Veedi, Bangalore 560001, Karnataka, India.	1	Engineering	1

Bhavnagar Univ, Dept Life Sci, Bhavnagar 364002, Gujarat, India	1	Plant and Animal Science	1
Bijoygarh Coll, Dept Bot, Calcutta, W Bengal, India.	1	Plant and Animal Science	1
Bioeth & Global Hlth, Pune, Maharashtra, India.	1	Clinical Medicine	1
Bioinformat Ctr, Inst Microbial Technol, Chandigarh, India.	1	Computer Science	1
Birbal Sahni Inst Paleobot, Lucknow 226007, Uttar Pradesh, India.	1	Plant and Animal Science	1
Birla Inst Technol & Sci, Dept Chem Engn, Pilani 333031, Rajasthan, India	1	Computer Science	1
Birla Inst Technol, Dept Mech Engn, Ranchi 835215, Bihar, India.	1	Engineering	1
BJ Med Coll, Pune, Maharashtra, India.	1	Clinical Medicine	1
BN Coll, Dept Chem, Dhubri 783324, Assam, India	1	Chemistry	1
Bombay Hosp & Med Res Ctr, Inst Med Sci, Bombay, Maharashtra, India.	1	Clinical Medicine	1
BRCM Coll Engn & Technol, MED, Bahal, Bhiwani, India.	1	Social Science	1
BV Patel Pharmaceut Educ Res & Dev Ctr, Thaltej Gandhinagar Highway, Ahmadabad 380054, India.	1	Plant and Animal Science	1
BVB Coll Engn & Technol, Dept Mech Engn, Hubli 580031, India.	1	Engineering	1
Canc Inst WIA, Div Epidemiol, Madras, Tamil Nadu, India.	1	Clinical Medicine	1
Canc Prevent & Relief Soc, Raipur, Madhya Pradesh, India.	1	Clinical Medicine	1
CCS Univ, Meerut, Uttar Pradesh, India.	1	Plant and Animal Science	1
Cent Arid Zone Res Inst, Jodhpur 342003, Rajasthan, India.	1	Environment	1
Cent Bldg Res Inst, Roorkee 247667, Uttar Pradesh, India.	1	Chemistry	1
Cent Elect Engn Res Inst, Pilani 333031, Rajasthan, India.	1	Physics	1
Cent Ground Water Author, Calcutta 700091, W Bengal, India.	1	Geosciences	1

Cent Ground Water Author, Delhi 110011, India.	1	Geosciences	1
Cent Inland Capture Fisheries Res Inst, Barakpur 743101, W Bengal, India.	1	Plant and Animal Science	1
Cent Inst Brackishwater Aquaculture, Madras 600028, Tamil Nadu, India.	1	Engineering	1
Cent Inst Plast Engn & Technol, Bhubaneswar, Orissa, India.	1	Materials Science	1
Cent Inst Postharvest Engn & Technol, Div Hort Crops Proc, Abohar 152116, India.	1	Plant and Animal Science	1
Cent Pollut Control Board, New Delhi 110092, India.	1	Engineering	1
Cent Potato Res Inst, Shimla 171001, Himachal Prades, India.	1	Plant and Animal Science	1
Cent Rice Res Inst, Div Biochem Plant Physiol & Environm Sci, Cuttack 753006, Orissa, India.	1	Plant and Animal Science	1
Center for Arthropod Bioresources and Biotechnology, Kerala University, Trivandrum, India	1	Biology & Biochemistry	1
Center for Chronic Disease Control, New Delhi, India	1	Clinical Medicine	1
Center for Plant Molecular Biology, Tamil Nadu Agricultural University, Coimbatore	1	Agricultural Sciences	1
Central Inst Freshwater Aquaculture, Fish Hlth Management Div, Bhubaneswar 751002, Orissa, India.	1	Plant and Animal Science	1
Centre for Cellular and Molecular Biology (CCMB), Council of Scientific and Industrial Research (CSIR), Uppal Road, Hyderabad 500 007, India;	1	Clinical Medicine	1
Ch Charan Singh Univ, Dept Genet & Plant Breeding, Mol Biol Lab, Meerut 250004, Uttar Pradesh, India.	1	Plant and Animal Science	1
Chem Sci & Technol, Reg Res Lab, Trivandrum 695019, Kerala, India.	1	Engineering	1
Chennai Cancer Inst, Madras, India.	1	Clinical Medicine	1
Chest Res Fdn, Pune 411014, Maharashtra, India	1	Clinical Medicine	1
Chinese Acad Sci, Inst Met Res, Shenyang Natl Lab Mat Sci, 72 Wenhua Rd, New Delhi 110016, India	1	Materials Science	1

Christian Coll, Dept Phys, Kattakada, Kerala, India.	1	Engineering	1
CHROMIUM REMOVAL BY COMBINING THE MAGNETIC PROPERTIES OF IRON OXIDE WITH ADSORPTION PROPERTIES OF CARBON NANOTUBES	1	Environment/ Ecology	1
CIFA, Fish Hlth Management Div, Bhubaneswar 751002, Orissa, India	1	Immunology	1
Coll Engn, Dept Civil Engn, Badnera 444701, Maharashtra, India.	1	Engineering	1
Coll Med, Ctr Psychiat, Jaipur, Rajasthan, India	1	Psychiatry/ Psychology	1
Coll Nursing, Trivandrum	1	Social Science	1
Community Empowerment Lab, Shivgarh, India.	1	Social Science	1
Cotton Coll, Dept Phys, High Energy Phys Lab, Gauhati 781001, Assam, India.	1	Geosciences	1
CSIR, Cent Leather Res Inst, Chem Lab, Madras 600020, Tamil Nadu, India.	1	Engineering	1
CSIR, Colloids & Mat Chem Grp, Reg Res Lab, Bhubaneswar 751013, Orissa, India.	1	Materials Science	1
CSIR, Salt & Marine Chem Div, Cent Salt & Marine Chem Res Inst, Bhavnagar 364002, Gujarat, India.	1	Engineering	1
CSIR, Special Inst Area, Delhi 110067, India.	1	Engineering	1
CSM Med Univ, Lucknow, Uttar Pradesh, India.	1	Social Science	1
Ctr Adv Technol, Laser Mat Div, Indore 452013, India.	1	Chemistry	1
Ctr Dev Adv Comp, Pune 411007, Maharashtra, India.	1	Materials Science	1
Ctr Dev Studies, Trivandrum 695011, Kerala, India.	1	Environment	1
Ctr Earth Sci Studies, Thiruvananthapuram 69603, Kerala, India.	1	Geosciences	1
Ctr Interdisciplinary Studies, Calcutta 700123, India.	1	Clinical Medicine	1
Ctr Math Sci, Pala 686574, Kerala, India.	1	Mathematics	1
Ctr Plant Biotechnol, Plant Mol Biol Div, Hisar 125004, Haryana, India.	1	Plant and Animal Science	1
Ctr Variable Energy Cyclotron, Calcutta, India.	1	Chemistry	1

Ctr Wildlife Studies, Bengaluru 560070, India. Equator GIS, Palarivattom 682025, Kochi, India.	1	Environment	1
Curie Manavata Canc Ctr, Nasik, Maharashtra, India.	1	Clinical Medicine	1
DDU Gorakhpur Univ, Chem Dept, Gorakhpur 273009, Uttar Pradesh, India.	1	Agricultural Sciences	1
Deemed Univ, Allahabad Agr Inst, Dept Chem, Allahabad, Uttar Pradesh, India.	1	Biology & Biochemistry	1
Deemed Univ, Natl Inst Technol, Dept Chem, Jalandhar 144011, Punjab, India.	1	Materials Science	1
Deemed Univ, Sathysbama Inst Sci & Technol, Dept Mech & Prod Engn, Madras 119, Tamil Nadu, India	1	Engineering	1
Deenanath Mangeshkar Hosp, Pune, Maharashtra, India.	1	Clinical Medicine	1
Def Met Res Lab, Hyderabad 500058, Andhra Pradesh, India.	1	Materials Science	1
Defence Research and Development Organisation India, Biotechnology Division, Dehradun, India	1	Biology & Biochemistry	1
Delhi Dermatol Grp, New Delhi, India.	1	Clinical Medicine	1
Delhi Sch Econ, Delhi, India.	1	Agricultural Sciences	1
Delta Reg Ctr, Natl Inst Hydrol, Siddartha Nagar, Kakinada, India.	1	Engineering	1
Department of Biochemistry and Nutrition, Central Food Technological Research Institute, Mysore-570 013, India.	1	Agricultural Sciences	1
Department of Chemistry, Motilal Nehru National Institute of Technology, Allahabad 211004, India;	1	Agricultural Sciences	1
Department of Food & Agricultural Process Engineering, Tamil Nadu Agricultural University, Coimbatore, Tamil Nadu 641 003, India	1	Agricultural Sciences	1
Department of Food Science and Technology, Guru Nanak Dev University, Amritsar-143005, India	1	Agricultural Sciences	1
Department of Pharmaceutical Sciences, University Institute of Pharmaceutical Sciences, Panjab University, Chandigarh, India	1	Agricultural Sciences	1

Department of Physiology and Molecular Medicine, Medical College of Ohio, Block Health Science Building, 3035 Arlington Avenue, Toledo, OH 43614-5804, USA	1	Agricultural Sciences	1
Dept Genet & Plant Breeding, Cooch Behar, India.	1	Plant and Animal Science	1
Dhirubhai Ambani Life Sci Ctr, Reliance Life Sci Pvt Ltd, Ind Biotechnol Grp, Thane Belapur Rd, Rabale 400701, Navi Mumbai, India.	1	Pharmcology & Toxicology	1
Directorate of Oil Palm Research, Indian Council of Agricultural Research, Pedavegi, Eluru, Andhra Pradesh 534 450, India	1	Agricultural Sciences	1
Directorate of Rice Research, Rajendranagar, Hyderabad	1	Agricultural Sciences	1
Directorate Oilseeds Res, Crop Improvement Sect, Hyderabad 500030, Andhra Pradesh, India.	1	Plant and Animal Science	1
Dr A Ramachandrans Diabet Hosp, Madras 600008, Tamil Nadu, India.	1	Clinical Medicine	1
Dr Babasaheb Ambedkar Marathwada Univ, Dept Phys, Thin Film & Nanotechnol Lab, Aurangabad 431004, Maharashtra, India.	1	Engineering	1
Dr BC Roy Coll Pharm & Allied Hlth Sci, Durgapur 713206, India.	1	Pharmcology & Toxicology	1
Dr DY Patil Univ, Pune, Maharashtra, India.	1	Clinical Medicine	1
Dr Jilla Hosp, Aurangabad, Maharashtra, India.	1	Clinical Medicine	1
Dr Reddys Res Fdn, Hyderabad 500050, Andhra Pradesh, India.	1	Chemistry	1
Dr Vijay Kumar Fdn, Madras 600078, Tamil Nadu, India.	1	Materials Science	1
Dr. Shashi Prabha Jain, Obstetrics And Gynaecology at A-10 (PART-B) Ashok Nagar, Ghaziabad in Uttar Pradesh	1	Clinical Medicine	1
DRDO, INMAS, Div Radiat Biosci, Delhi 110054, India.	1	Materials Science	1
Econ Indian Inst Management Bangalore, Bangalore 560076, Karnataka, India.	1	Social Science	1
Ekjut, Ward 17,Plot 556B, Po Chakradharpur 833102, Jharkhand, India.	1	Clinical Medicine	1
Eras Lucknow Med Coll, Dept Biochem, Lucknow, Uttar Pradesh, India.	1	Molicular Biology	1

Extens Ctr Jaipur, Birla Inst Technol Mesra, Jaipur, Rajasthan, India.	1	Geosciences	1
Formulation Development, Panacea Biotec Ltd, Lalru, Punjab, India	1	Agricultural Sciences	1
Fac Pharm, Dept Pharmaceut, New Delhi 62, India.	1	Pharmcology & Toxicology	1
GB Pant Hosp, Dept Gastroenterol, New Delhi, India.	1	Clinical Medicine	1
GE Aviat India Technol Ctr, Bangalore, Karnataka, India.	1	Computer Science	1
Gen Elect John F Welch Technol Ctr, Bangalore, Karnataka, India.	1	Chemistry	1
Gen Motors R&D India Sci Lab, Bangalore 560066, Karnataka, India.	1	Engineering	1
GH Patel Coll Engn & Technol, Dept Chem Engn, Vallabh Vidyanagar 388120, Gujarat, India.	1	Engineering	1
GITAM Univ, Dept ECE, Visakhapatnam, Andhra Pradesh, India.	1	Engineering	1
Goa Univ, Dept Chem, Taleigao Plateau 403206, Goa, India.	1	Chemistry	1
Govt Autonomous Sci Coll, Dept Chem, Bose Mem Res Lab, Jabalpur 482001, MP, India.	1	Chemistry	1
Govt Coll Engn & Ceram Technol, Dept Math, Calcutta 700010, W Bengal, India.	1	Mathematics	1
Govt Coll Pharm, Bangalore	1	Agricultural Sciences	1
Govt Coll Pharm, Dept Pharmaceut, Aurangabad 431001, Maharashtra, India.	1	Chemistry	1
Govt Coll Technol, Dept Elect Engn, Coimbatore 641114, Tamil Nadu, India.	1	Engineering	1
Govt Coll, Dept Bot, Ajmer 305001, Rajasthan, India.	1	Plant and Animal Science	1
Govt Coll, Dept Zool, Palakkad, Kerala, India.	1	Biology & Biochemistry	1
Govt First Grade Coll, Dept Math, Sindhanur 584128, Karnataka, India.	1	Mathematics	1
Govt India, Minist Sci & Technol, Dept Biotechnol, New Delhi 110003, India.	1	Clinical Medicine	1
Govt Med Coll, Dept Med, Nagpur, Maharashtra, India.	1	Clinical Medicine	1

Gurrala Chavidy, Chilakaluripet 522616, Andhra Pradesh, India.	1	Clinical Medicine	1
Guru Teg Bahadur Hosp, New Delhi,	1	Clinical Medicine	1
GVK Biosci Pvt Ltd, Hyderabad 500037, Andhra Pradesh, India.	1	Chemistry	1
Harcourt Butler Technol Inst, Dept Chem, Analyt Res Lab, Kanpur 208002, Uttar Pradesh, India.	1	Engineering	1
Harvard Smithsonian Ctr Astrophys, Cambridge.	1	Physcis	1
Healis Sekhsaria Inst Publ Hlth, Bombay, Maharashtra, India.	1	Clinical Medicine	1
Heart Care Clin, Ahmadabad, Gujarat, India.	1	Clinical Medicine	1
Heat Transfer & Thermal Power Lab, Dept Mech Engn, Madras 600036, Tamil Nadu, India.	1	Engineering	1
Hellosoft Inc, Banjara Hills, Hyderabad 500034, Andhra Pradesh, India.	1	Computer Science	1
Hematooncol Clin, Ahmadabad, Gujarat, India.	1	Clinical Medicine	1
Hindu Rao Hospital, Department of Pathology, New Delhi,	1	Biology & Biochemistry	1
Hindu University, 6 SK Gupta Nagar, Lanka, Varanasi 221005, India.	1	Clinical Medicine	1
Hinduja Natl Hosp, Bombay, Maharashtra, India.	1	Clinical Medicine	1
Hindustan Coll Engn, Dept Civil Engn, Madras 21, Tamil Nadu, India.	1	Engineering	1
Homi Bhabha Ctr Sci Educ, Bombay 400088, Maharashtra, India.	1	Space Science	1
Hop Cochin, AP HP, Dept Endocrinol,Ctr Rare Adrenal Dis,Oncogenet Un, Dept Pathol,Unit Digest & Endocrine Surg, Cochin, Kerala, India	1	Clinical Medicine	1
HTS Res, 151-1 Doraisanipalya,Bannerghatta Rd, Bangalore 560076, Karnataka, India	1	Engineering	1
Indian Assoc Cultivat Sci, Raman Ctr Atom Mol & Opt Sci, Jadavpur Kolkata 700032, India.	1	Chemistry	1
Indian Inst Chem Technol, CSIC, Hyderabad 500007, Andhra Pradesh, India.	1	Chemistry	1
Indian Inst Sci, Solid State & Struct Chem Unit, Bangalore 12, Karnataka, India.	1	Chemistry	1

Indian Inst Technol, Dept Chem, Bombay 400076, Maharashtra, India.	1	Chemistry	1
IACS, Calcutta 700032, India.	1	Chemistry	1
ICAR, Div Crop Sci, New Delhi 110114, India.	1	Plant and Animal Science	1
Ideal Inst Technol, Dept Chem, Ghaziabad, UP, India.	1	Engineering	1
IISc, Ctr Condensed Matter Theory, Bangalore, Karnataka, India	1	Physics	1
IISER, Pune, Maharashtra, India.	1	Materials Science	1
IIT Delhi, Phys Dept, New Delhi 110016, India.	1	Materials Science	1
IIT, Ctr Fuel Cell Technol, Madras 600113, Tamil Nadu, India	1	Materials Science	1
INCASR, Engn Mech Unit, Bangalore 560064, Karnataka, India.	1	Engineering	1
Ind Technol Inst, Alternate Hydro Energy Ctr, Uttaranchal 247667, India.	1	Environment/ Ecology	1
Ind Technol Inst, Analyt & Environm Chem Div, Hyderabad, Andhra Pradesh, India.	1	Geosciences	1
Ind Technol Inst, Ctr Rural Dev & Technol, New Delhi 110016, India.	1	Environment/ Ecology	1
Ind Technol Inst, Dept Chem Engn, Roorkee 247667, Uttar Pradesh, India	1	Engineering	1
Indian Council Med Res, Rajendra Mem Res Inst Med Sci, Patna, Bihar, India.	1	Clinical Medicine	1
Indian Council of Agricultural Research (ICAR), KAB-II, Pusa, New Delhi-110 012, India	1	Agricultural Sciences	1
Indian Initiat Rice Genome Sequencing, New Delhi 110021, India.	1	Plant and Animal Science	1
Indian Inst Informat Technol, Bioinformat Div, Allahabad, Uttar Pradesh, India.	1	Pharmcology & Toxicology	1
Indian Inst Management, Ahmadabad 380015, Gujarat, India.	1	Clinical Medicine	1
Indian Inst Petr, Dehra Dun 248005, Uttar Pradesh, India.	1	Environment/ Ecology	1
Indian Inst Sci Educ & Res Kolkata, Mohanpur	1	Physcis	1
Indian Inst Sci Educ & Res, Dept Chem Sci, Nadia 741252, West Bengal, India.	1	Chemistry	1

Institution		Field	
Indian Inst Technol Gandhinagar, Dept Chem Engn, Ahmadabad 382424, Gujarat, India.	1	Engineering	1
Indian Inst Technol, Patna 800013, Bihar, India	1	Chemistry	1
Indian Inst Trop Meteorol Branch IITM, New Delhi, India.	1	Geosciences	1
Indian Meteorol Dept, New Delhi 110003, India.	1	Geosciences	1
Indian Natl Ctr Ocean Informat Serv, Hyderabad, Andhra Pradesh, India.	1	Geosciences	1
Indian Natl Sci Acad, New Delhi, India.	1	Plant and Animal Science	1
Indian Sch Mines, Dept Appl Geol, Dhanbad 826004, Bihar, India.	1	Geosciences	1
Indian Space Res Org, Bangalore 562140, Karnataka, India.	1	Space Science	1
Indian Vet Res Inst, Biochem Lab, Reg Stn, Palampur 176061, Himachal Prades, India.	1	Agricultural Sciences	1
Industrial Toxicology Research Centre, P.O. Box 80, M.G Marg, Lucknow	1	Agricultural Sciences	1
Innovat Ctr Appl Nanotechnol, Calcutta, India.	1	Microbiology	1
Inst Computat Biol, -2 karnataka	1	Computer Science	1
Inst Dev & Res Banking Technol, Castle Hills Rd 1,Masab Tank, Hyderabad 500057, Andhra Pradesh, India	1	Engineering	1
Inst Hlth Management & Res, Jaipur, Rajasthan, India.	1	Clinical Medicine	1
Inst Met Sci & Technol, Shenyang Natl Lab Mat Sci, New Delhi 110016, India.	1	Engineering	1
Inst Minerals & Mat Technol, Bhubaneswar 751013, Orissa, India	1	Engineering	1
Inst Mol Med, New Delhi 110020, India	1	Clinical Medicine	1
Inst Nucl Med & Allied Sci, Dept Radiol, New Delhi	1	Clinical Medicine	1
Inst Rd & Transport Technol, Erode, Tamil Nadu, India.	1	Environment/ Ecology	1
Int Adv Res Ctr Powder Met & New Mat, Balapur PO, Hyderabad 500005, Andhra Pradesh, India.	1	Materials Science	1
Int Food Policy Res Inst, New Delhi, India.	1	Social Science	1

Intel Corp, Microprocessor Technol Lab, Bangalore 560037, Karnataka, India.	1	Engineering	1
International Center for Genetic Engineering and Biotechnology, Delhi, India	1	Clinical Medicine	1
International Journal of Food Science and Technology	1	Agricultural Sciences	1
International Rice Research Institute (IRRI), IRRI-India Office, National Agriculture Science Center (NASC) Complex, New Delhi	1	Agricultural Sciences	1
Interuniv Ctr Astron & Astrophys, Pune 411007, Maharashtra, India.	1	Engineering	1
Invent Pharma Pvt Ltd, Navi Mumbai 400701, Maharashtra, India;	1	Chemistry	1
ISB, Hyderabad 500019, Andhra Pradesh, India.	1	Economics & Business	1
Islamic Univ Sci & Technol, Dept Food Technol, Awantipora, Jammu & Kashmir, India.	1	Agricultural Sciences	1
ISRO Head Quarters, Bangalore 560094, Karnataka, India.	1	Geosciences	1
IWMI, CGIAR Challenge Program Climate Change Agr & Food, New Delhi 110012, India.	1	Plant and Animal Science	1
Jamia Hamdard, Dept Med Elementol & Toxicol, New Delhi 110062, India.	1	Pharmcology & Toxicology	1
Jawaharlal Nehru Canc Hosp & Res Ctr, Bhopal, India.	1	Clinical Medicine	1
Jawaharlal Nehru Med Coll, Belgaum, India.	1	Clinical Medicine	1
Jawaharlal Nehru Technological University, Kakinada, Andhra Pradesh, India	1	Agricultural Sciences	1
JGovt Engn Coll Women, Dept Phys, Ajmer, Rajasthan, India.	1	Materials Science	1
JNU, delhi	1	Engineering	1
JUET, Dept Chem Engn, Guna 473226, Madhya Pradesh, India.	1	Social Science	1
Justice KS Hegde Acad, Dept Biochem, Mangalore, India	1	Chemistry	1
Kala Azar Res Ctr, Brahmpura, Muzaffarpur, India.	1	Clinical Medicine	1
Kanchi Mamunivar Ctr Postgrad Studies, Dept Chem, Puducherry 605008, India.	1	Engineering	1

Kandaswami Kandars Coll, P Velur 638182, Tamil Nadu, India.	1	Engineering	1
Kaveri Engine Program, Project Off Mat, Hyderabad 500058, Andhra Pradesh, India.	1	Materials Science	1
KELs CET, Dept Mech Engn, Belgaum, India.	1	Engineering	1
Kerala Agricultural University, Thrissur 680656, Kerala	1	Agricultural Sciences	1
Kerala Inst Med Sci, Trivandrum, Kerala, India.	1	Clinical Medicine	1
KM Coll Pharm, Dept Pharmaceut Chem, Madurai 625107, Tamil Nadu, India.	1	Engineering	1
Kothari Med & Res Inst, Bikaner, Rajasthan, India.	1	Clinical Medicine	1
LRS Inst TB & Allied Dis, New Delhi, India.	1	Clinical Medicine	1
Madras Christian Coll, Dept Chem, Madras 600059, Tamil Nadu, India.	1	Agricultural Sciences	1
Madras Diabetes Research Foundation and Dr Mohan's Diabetes Specialities Centre, WHO Collaborating Centre for Non-Communicable Diseases, Gopalapuram, Chennai	1	Agricultural Sciences	1
Maharaja Sayajirao Univ Baroda, Dept Pharm, Fac Technol & Engn, Vadodara 390001, Gujarat, India.	1	Materials Science	1
Maharana Pratap Univ Agr & Technol, Coll Technol & Engn, Dept Renewable Energy Sources, Udaipur 313001, India.	1	Environment/ Ecology	1
Maharani Lakshmi Ammani Coll Women,karanataka	1	Computer Science	1
Maharshi Dayanand Univ, Dept Math, Rohtak 124001, Haryana, India	1	Mathematics	1
Mahatma Gandhi Univ, Sch Chem Sci, Priyadarshini Hills PO, Kottayam 686560, Kerala, India	1	Chemistry	1
Malaviya Natl Inst Technol, Jaipur 302017, Rajasthan, India.	1	Physics	1
Mamata Hosp, Bombay, Maharashtra, India.	1	Clinical	1
Management Dev Inst, Post Box 60, Sukhrali 122001, Gurgaon, India	1	Economics & Business	1
Mangalore Univ, Dept Chem, Mangalagangothri 574199, India.	1	Chemistry	1

Manipal Hosp, Manipal Inst Neurol Disorders, Bangalore, Karnataka, India.	1	Clinical Medicine	1
Maulana Azad Coll, Calcutta 700013, India.	1	Physics	1
Max Heart & Vasc Inst, New Delhi, India.	1	Clinical Medicine	1
Med Sci Univ, Nizams Inst, Hyderabad, Andhra Pradesh, India.	1	Clinical Medicine	1
Medanta Medicity, Gurgaon, India.	1	Clinical Medicine	1
Mediciti Hosp, Div Nephrol, Hyderabad, Andhra Pradesh, India.	1	Clinical Medicine	1
Metro Hosp & Heart Inst, Noida, Uttar Pradesh, India.	1	Clinical Medicine	1
Microsoft Res India, Bangalore 560080, Karnataka, India.	1	Social Science	1
MITS, Dept Elect Engn, Gwalior, India.	1	Engineering	1
MJ Coll, Dept Biotechnol, MJ Rd, Jalgaon 425001, India.	1	Pharmcology & Toxicology	1
MLN Med Coll, Dept Microbiol, Allahabad 211002, Uttar Pradesh, India.	1	Molicular Biology	1
Mol Connect Private Ltd, Bangalore 560004, Karnataka, India.	1	Biology & Biochemistry	1
Monsanto Research Centre, Bangalore	1	Agricultural Sciences	1
Motilal Nehru Natl Inst Technol, Dept Phys, Allahabad 211004, Uttar Pradesh, India.	1	Engineering	1
MS Swaminathan Res Fdn, Madras 600113, Tamil Nadu, India.	1	Environment/ Ecology	1
MUC Womens Coll, Dept Math, Burdwan 713104, W Bengal, India.	1	Computer Science	1
MUICT, Chem Engn Sect, Bombay 400019, Maharashtra, India.	1	Environment/ Ecology	1
NIIST, Photosci & Photon Grp, Chem Sci & Technol Div, CSIR, Trivandrum, Kerala, India.	1	Chemistry	1
N Eastern Reg Inst Sci & Technol, Dept Agr Engn, Itanagar 791109, Arunachal Prade	1	Environment/ Ecology	1
N Orissa Univ, Mayurbhanj 757003, Orissa, India.	1	Environment/ Ecology	1
Nagarjuna Univ, Nagarjuna Nagar 522510, Andhra Pradesh, India.	1	Plant and Animal Science	1

Nagpur Univ, Laxminarayan Inst Technol, Nagpur 440010, Maharashtra, India	1	Engineering	1
Nano Cutting Edge Technol Pvt Ltd, Bombay 400033, Maharashtra, India.	1	Pharmcology & Toxicology	1
Nanosci Grp, Mat Chem Div, Pune 411008, Maharashtra, India.	1	Materials Science	1
Nargis Dutt Mem Canc Hosp, Tata Mem Ctr, Rural Canc Project, Barshi, India.	1	Clinical Medicine	1
Nat Remedies, Bangalore 560100, Karnataka, India.	1	Agricultural Sciences	1
National Centre for Aquatic Animal Health, Cochin University of Science and Technology, Kochi, 682 016, India;	1	Biology & Biochemistry	1
National Drug Dependence Treatment Centre and Department of Psychiatry, All India Institute of Medical Sciences, New Delhi, India	1	Clinical Medicine	1
National Institute of Nutrition, Hyderabad	1	Agricultural Sciences	1
Natl Aerosp Labs, Surface Engn Div, Post Bag 1779, Bangalore 560017, Karnataka, India.	1	Materials Science	1
Natl AIDS Control Org, New Delhi, India.;	1	Clinical Medicine	1
Natl Atmospher Res Lab, Gadanki, India.	1	Geosciences	1
Natl Bur Fish Genet Resources, Lucknow 226002, Uttar Pradesh, India.	1	Environment/ Ecology	1
Natl Coll Autonomous, Dept Phys, Tiruchchirappalli 620001, Tamil Nadu, India.	1	Physics	1
Natl Council Appl Econ Res, New Delhi 110002, India.	1	Social Science	1
Natl Ctr Cell Sci, Pune 411007, Maharashtra, India.	1	Clinical Medicine	1
Natl Ctr Dis Control, Delhi 110054, India.	1	Molicular Biology	1
Natl Ctr Plant Genome Res, New Delhi 110067, India.	1	Microbiology	1
Natl Disaster Management Author, New Delhi, India.	1	Geosciences	1
Natl Hlth Syst Resources Ctr, New Delhi, India.	1	Clinical Medicine	1
Natl Hydrol Res Inst, Roorkee, Uttar Pradesh, India.	1	Engineering	1

Natl Inst Biomed Genom, Kalyani 741251, W Bengal, India.	1	Clinical Medicine	1
Natl Inst Forged & Foundry Technol, Dept Mfg Engn, Ranchi 834003, Bihar, India.	1	Engineering	1
Natl Inst Foundry & Forge Technol, Dept Mfg Engn, Ranchi 834003, Bihar, India	1	Computer Science	1
Natl Inst Ind Engn, Bombay 400087, Maharashtra, India.	1	Engineering	1
Natl Inst Nutr, Hyderabad 500007, Andhra Pradesh, India.	1	Clinical Medicine	1
Natl Inst Oceanog, Reg Ctr, Visakhapatnam, Andhra Pradesh, India.	1	Pharmcology & Toxicology	1
Natl Inst Plant Biodivers Conservat & Res, Bhubaneswar 751015, Orissa, India.	1	Environment/ Ecology	1
Natl Inst Plant Genome Res, New Delhi 110067, India.	1	Plant and Animal Science	1
Natl Inst Res Reprod Hlth, Dept Immunol, Bombay 400012, Maharashtra, India.	1	Microbiology	1
Natl Inst Tech Teachers Training & Res, Calcutta 700106, India.	1	Engineering	1
Natl Inst Technol Hamirpur, Dept Civil Engn, Hamirpur 177005, HP, India.	1	Engineering	1
Natl Inst Technol, Dept Chem Engn, Surathkal, India	1	Engineering	1
Natl Inst Technol, Dept Chem, Rourkela 769008, India.	1	Engineering	1
Natl Inst Technol, Dept Engn Mech, Calicut 673601, Kerala, India.	1	Chemistry	1
Natl Inst Technol, Sch Elect Sci, Rourkela, Orissa, India.	1	Computer Science	1
Natl Remote Sensing Agcy, Hyderabad, Andhra Pradesh, India..	1	Space Science	1
Natl Res Ctr Agroforestry, Jhansi, Uttar Pradesh, India.	1	Environment/ Ecology	1
Natl Res Ctr Plant Biotechnol, Indian Agr Res Inst, Indian Initiat Rice Genome Sequencing, New Delhi 110112, India.	1	Plant and Animal Science	1
Naval Mat Res Lab, Ambernath 421506, Thane, India	1	Materials Science	1
NE Reg Inst Sci & Technol, Dept Agr Engn, Itanagar 791109, Arunachal Prade, India.	1	Engineering	1

NGM Coll, Dept Chem, Pollachi 642001, India.	1	Engineering	1
NIPGR, New Delhi 110067, India.	1	Molicular Biology	1
NMSS Vellaichamy Nadar Coll, Dept Microbiol, Madurai, Tamil Nadu, India.	1	Clinical Medicine	1
Novartis Healthcare Private Ltd, Hyderabad, Andhra Pradesh, India.	1	Biology & Biochemistry	1
NRC Plant Biotechnol, New Delhi 110012, India.	1	Plant and Animal Science	1
NRCPB, New Delhi 110012, India.	2	Plant and Animal Science	1
Nucl Sci Ctr, New Delhi 110067, India.	1	Engineering	1
Nutr Fdn India, New Delhi, India.	1	Clinical Medicine	1
Orissa State Pollut Control Board, Bubaneswar, Orissa, India.	1	Engineering	1
Panjabrao Deshmukh Krishi Vidyapeeth, Akola 444104, Maharasthra, India.	1	Plant and Animal Science	1
PD Hinduja Hosp, Bombay,	1	Clinical Medicine	1
Penn State Univ, Dept Phys, University Pk, PA USA.	1	Physics	1
Photosciences and Photonics Group, Chemical Sciences and Technology Division, National Institute for Interdisciplinary Science and Technology (NIIST), CSIR, Trivandrum 695019, India.	1	Chemistry	1
PIET, Dept Chem, Rourkela 770034, Orissa, India. Natl Inst Technol, Dept Chem, Rourkela 769008, India	1	Engineering	1
Polytech Coll, Dept Sci & Humanities, Karaikal 609609, Puducherry, India.	1	Engineering	1
Presidency Coll, Dept Chem, Calcutta 700073, W Bengal, India.	1	Engineering	1
Pune MIT, Pune, Maharashtra, India.	1	Clinical Medicine	1
Punjab Tech Univ, Jalandhar, India.	1	Engineering	1
Reg Res Lab, Jammu 180001, India.	1	Chemistry	1
Raidigi Coll, Dept Chem, Raidigi 743383, India.	1	Chemistry	1

Rainbow Childrens Hosp, Hyderabad 500034, Andhra Pradesh, India.	1	Molicular Biology	1
Raja Rammohun Roy Mahavidyalaya, Dept Chem, Hooghly 712406, WB, India	1	Materials Science	1
Rajendra Mem Res Inst Med Sci, Patna, Bihar, India.	1	Plant and Animal Science	1
Rajiv Gandhi Center for Biotechnology (RGCB), Trivandrum, 695014, India	1	Biology & Biochemistry	1
Ravenshaw Coll, Dept Chem, Polymer Composites Lab, Post Grad Dept Chem, Cuttack 753003, Orissa, India.	1	Materials Science	1
Reg Engn Coll, Dept Chem Engn, Warangal 506004, Andhra Pradesh, India.	1	Engineering	1
Reg Res Lab, Synthet Organ Chem Div, Jorhat 785006, Assam, India.	1	Chemistry	1
Regenerative Medicine, Reliance Life Sciences Pvt. Ltd., R-282, TTC Area of MIDC, Thane Belapur Road, Rabale, Navi Mumbai-400701, India	1	Agricultural Sciences	1
Regional Research Laboratory, CSIR, Trivandrum	1	Agricultural Sciences	1
Res Inst, Bangalore, Karnataka, India.	1	Clinical Medicine	1
RGTU, Sch Energy & Environm Management, Bhopal, Madhya Pradesh, India.	1	Engineering	1
RITS, Inst Pharm, Cardiovasc Pharmacol Div, Dept Pharmacol, Sirsa 125055, Haryana, India	1	Pharmcology & Toxicology	1
ROFEL Shri GM Bilakhia Coll Appl Sci, Vapi Namcha Rd,POB 61, Vapi 396191, India.	1	Engineering	1
RRR Mahavidyalaya, Hooghly 712406, WB, India.	1	Materials Science	1
S Gujarat Univ, Dept Chem, Surat 395007, India.	1	Engineering	1
SAL Hospital and Medical Institute, Ahmedabad, India	1	Clinical Medicine	1
Salt Lake Elect Complex, Inst Engn & Management, Calcutta 700091, W Bengal, India.	1	Engineering	1
Samata Hlth Study, Bangalore, Karnataka, India.	1	Social Science	1
Sarat Centenary Coll, Hooghly, India.	1	Computer Science	1
Sarvajanik Coll Engn & Technol, Dept Chem Engn, Surat 395001, India.	1	Engineering	1

SASTRA Univ, Dept Chem Engn, Sch Chem & Biotechnol, Thirumalaisamudram 613402, Thanjavur, India.	1	Engineering	1
Seth GS Med Coll, Bombay, Maharashtra, India.	1	Clinical Medicine	1
Sharma Ctr Heritage Educ, Madras - 2	1	Social Science	1
Shri Mata Vaishno Devi Univ, Sch Math, Katra 182320, J&K, India.	1	Engineering	1
Shri Sharanabasaveswar Coll Sci, Dept Chem, Gulbarga 585102, India.	1	Chemistry	1
Singhania Univ, Dept Chem, Jhunjhunu 333515, Rajasthan, India.	1	Materials Science	1
Sir CR Reddy Coll Engn, Dept Math, Eluru 534007, India.	1	Engineering	1
Sir Padampat Singhania Univ, Dept Basic Sci, Udaipur 313601, Rajasthan, India.	1	Mathematics	1
SMM Town Postgrad Coll, Dept Zool, Ballia 277001, India.	1	Plant and Animal Science	1
Snow & Avalance Study Estab, Manali, India.	1	Engineering	1
South Asia Network for Chronic Disease, Public Health Foundation of India, C-1/52, SDA, New Delhi 100016, India.	1	Clinical Medicine	1
SPIC Sci Fdn, Energy Res Ctr, Madras 600032, Tamil Nadu, India.	1	Engineering	1
SR Fatepuria Coll, Murshidabad, India.	1	Economics & Business	1
Sri Krishnadevaraya Univ, Dept Phys, Anantapur 515005, Andhra Pradesh, India.	1	Geosciences	1
Sri Ramachandra University, Porur, Chennai	1	Agricultural Sciences	1
Sri Sathya Sai Med Educ & Res Fdn, Coimbatore 641014, Tamil Nadu, India.	1	Nueroscience & Behaviour	1
Srinivasan Ctr Clin Neurosci, Inst Neurol Sci, Voluntary Hlth Serv, Madras, Tamil Nadu, India.	1	Social Science	1
Sriramchandra Bhanja Med Coll, Cuttack, Orissa, India.	1	Clinical Medicine	1
SSMRV Degree Coll, Chem Res Ctr, Bangalore 560041, Karnataka, India.	1	Engineering	1
St Stephens Hosp, New Delhi, India.	1	Clinical Medicine	1

STD HIV Intervent Programme, Calcutta, W Bengal, India.	1	Social Science	1
Sterling Hosp, Ahmadabad, Gujarat, India.	1	Clinical Medicine	1
Sunder Lal Jain Hosp, Dept Pediat, Delhi, India.	1	Clinical Medicine	1
Swami Ramanand Teerth Marathwada Univ, Sch Chem Sci, Organ Chem Res Lab, Vishnupuri 431606, Nanded, India	1	Chemistry	1
Tagore Arts Coll, Dept Phys, Pondicherry, India.	1	Engineering	1
Tamil Nadu Water Supply & Drainage Board, Dist Water Testing Lab, Theni 625531, India.	1	Engineering	1
Tata Consultancy Serv, Innovat Labs, Biosci Div, 1 Software Units Layout, Hyderabad 500081, Andhra Pradesh, India	1	Computer Science	1
Tata Inst Fundamental Res, Natl Ctr Radio Astrophys, Pune 411007, Maharashtra, India.	1	Space Science	1
TB Research Centre, Indian Council of Medical Research, Chennai, India	1	Clinical Medicine	1
Tezpur Univ, Dept Chem Sci, Napaam 784028, Assam, India.	1	Plant and Animal Science	1
Thagore Arts Coll, Dept Phys, Pondicherry, India. Avvaiyar Govt Coll Women, Dept Phys,	1	Engineering	1
The Heart Care Clinic, Ahmedabad, India	1	Clinical Medicine	1
Thiagarajar Coll Engn, Dept Phys, Madurai 625015, Tamil Nadu, India.	1	Social Science	1
Tissue and Cell Culture Unit, Central Drug Research Institute, Lucknow, India.	1	Clinical Medicine	1
TKM Coll Engn, Dept Elect & Commun, Kollam 691005, India.	1	Engineering	1
Torrent Pharmaceut Ltd, Gandhinagar 382428, Gujarat, India.	1	Chemistry	1
TOTALL Diabet Hormone Res Inst, Indore, Madhya Pradesh, India.	1	Clinical Medicine	1
TPL, Torrent Res Ctr, Formulat & Dev Dept, Lab Nanoparticle Res, Village Bhat 382428, Gujarat, India.	1	Clinical Medicine	1
Tuberculosis Res Ctr, Dept Immunol, Madras 600031, Tamil Nadu, India.	1	Clinical Medicine	1

Univ Hyderabad, Sch Chem, Hyderabad 500046, Andhra Pradesh, India.	1	Chemistry	1
UAS, Natl Ctr Biol Sci, Bangalore 560065, Karnataka, India.	1	Molicular Biology	1
UGC DAE Consortium Sci Res, Indore 452017, Madhya Pradesh, India.	1	Physcis	1
UICT, Div Chem Engn, Bombay 400019, Maharashtra, India.	1	Environment/ Ecology	1
UNICEF India, New Delhi, India.	1	Clinical Medicine	1
Univ Bombay, Inst Chem Technol, Bombay 400019, Maharashtra, India.	1	Engineering	1
Univ Sci Malaysia, Sch Chem Engn, Engn Campus, Penang 14300, India	1	Chemistry	1
Vasantdada Sugar Inst, Mol Biol & Genet Engn Div, Pune 412307, Maharashtra, India.	1	Plant and Animal Science	1
VBS Purvanchal Univ, Dept Biotechnol, Jaunpur 222001, UP, India	1	Molicular Biology	1
Velammal Engn Coll, Dept Chem, Madras 600066, Tamil Nadu, India.	1	Engineering	1
Vellore Inst Technol, Vellore 632014, Tamil Nadu, India.	1	Materials Science	1
Vikram Univ, Ujjain, Madhya Pradesh, India.	1	Engineering	1
Visvesvaraya Natl Inst Technol, Nagpur 440011, Maharashtra, India.	1	Engineering	1
VL Coll Pharm, Dept Pharmaceut Chem, Raichur, India.	1	Chemistry	1
VNIT, Dept Civil Engn, Nagpur 440010, Maharashtra, India.	1	Social Science	1
WHO, Collaborating Ctr Res Educ & Training Diabet, MV Hosp Diabet, Diabet Res Ctr, Madras, Tamil Nadu, India.	1	Clinical Medicine	1
Women India Assoc, Inst Canc, Madras, Tamil Nadu, India.	1	Clinical Medicine	1
World Agroforestry Ctr ICRAF Reg Off S Asia, Natl Agr Sci Ctr, New Delhi 110012, India.	1	Environment/ Ecology	1

*No. of times occurring in the given subject field corresponding to the no. of highly cited papers contributed
**Subject Category with highest number of highly cited papers as reflected in Column 2 (Occurrence Frequency)